Interdisciplinary Statistics

GENERALIZED LATENT VARIABLE MODELING

VARIABLE MODELING

Multilevel, Longitudinal, and Structural Equation Models

CHAPMAN & HALL/CRC
Interdisciplinary Statistics Series

Series editors: N. Keiding, B. Morgan, T. Speed, P. van der Heijden

Interdisciplinary Statistics

GENERALIZED LATENT VARIABLE MODELING

Multilevel, Longitudinal, and Structural Equation Models

Anders Skrondal
Sophia Rabe-Hesketh

CHAPMAN & HALL/CRC

A CRC Press Company
Boca Raton   London   New York   Washington, D.C.

Library of Congress Cataloging-in-Publication Data

Skrondal, Anders.
  Generalized latent variable modeling : multilevel, longitudinal, and structural equation models / Anders Skrondal, Sophia Rabe-Hesketh.
    p. cm. — (Chapman & Hall/CRC interdisciplinary statistics series)
    Includes bibliographical references and indexes.
    ISBN 1-58488-000-7 (alk. paper)
    1. Latent variables. 2. Latent structure analysis. I. Rabe-Hesketh, S. II. Title. III. Interdisciplinary statistics.

QA278.6.S57 2004
519.5'35—dc22   2004042808

Visit the CRC Press Web site at www.crcpress.com

© 2004 by Chapman & Hall/CRC

No claim to original U.S. Government works
International Standard Book Number 1-58488-000-7
Library of Congress Card Number 2004042808
Printed in the United States of America   4 5 6 7 8 9 0
Printed on acid-free paper

Contents

Preface

A major aim of this book is to unify and extend latent variable modeling in the widest sense. The models covered include multilevel, longitudinal and structural equation models as well as relatives and friends such as generalized linear mixed models, random coefficient models, item response models, factor models, panel models, repeated measurement models, latent class models and frailty models.

Latent variable models are used in most empirical disciplines, although often referred to by other names. In the spirit of the title of the 'Interdisciplinary Statistics Series', we attempt to synthesize approaches from different disciplines and to translate between the languages of statistics, biometrics, psychometrics and econometrics (although we do not claim to have full command of all these languages).

We strongly believe that progress is hampered by use of 'local' jargon leading to compartmentalization. For instance, econometricians and biostatisticians are rarely seen browsing each other's journals. Even more surprising is tribalism within disciplines, as reflected by lack of cross-referencing between item response theory and factor modeling in psychometrics (even within the same journal!). A detrimental effect of such lack of communication is a lack of awareness of useful developments in other areas until they are 'translated' and published in the 'correct' literature. For instance, models for drop-out (attrition in social science) have been prominent in econometrics for decades but have only quite recently been 'discovered' in the statistical literature.

The book consists of two parts; methodology and applications. In Chapter 1 we discuss the concept, uses and interpretations of latent variables. In Chapter 2 we bring together models for different response types used in different disciplines. After reviewing classical latent variable models in Chapter 3, we unify and extend these models in Chapter 4 for all response types surveyed in Chapter 2. Established and novel methods of model identification, estimation, latent variable prediction and model diagnostics are extensively covered in Chapters 5 to 8.

In the application Chapters 9 to 14 we use the methodology developed in the first part to address problems from biology, medicine, psychology, education, sociology, political science, economics, marketing and other areas. All applications are based on real data, but our analysis is often simplified for didactic reasons. We have used our Stata program gllamm, developed jointly with Andrew Pickles, for all applications.

It is our hope that ample cross-referencing between the two parts of the book will allow readers to find illustrations of methodology in the application chapters (by skipping forward) and statistical background for applications in the methodology chapters (by skipping back).

The first three and a half chapters are intended as a relatively gentle introduction to the modeling approach pervading this book. The remaining methodological chapters are somewhat more technical, partly due to the generality of the framework. However, we also consider simple special cases where notation becomes less complex, ideas more transparent and results more intuitive. Readers who are primarily interested in the interpretation and application of latent variable models might want to skip most of Chapters 4 to 8 and concentrate on the application chapters.

The book is one of the outcomes of our collaboration over the last four years on developing 'Generalized Linear Latent And Mixed Models' (GLLAMMs) and the accompanying `gllamm` software. We acknowledge the input from our collaborator Andrew Pickles. Anders would like to thank his 'boss' Per Magnus and Sophia her former 'boss' Brian Everitt for encouragement and accepting that this book took priority over other projects. Brian Everitt, Leonardo Grilli, Carla Rampichini and two anonymous reviewers have read drafts of the book and provided us with many helpful suggestions. We would also like to acknowledge constructive comments from Irit Aitkin, Bill Browne, Stephen Jenkins, Andrew Pickles, Sven Ove Samuelsen and Jeroen Vermunt.

David Clayton, Per Kragh Andersen, Anthony Heath, Andrew Pickles and Bente Træen have kindly provided data for our applications. We appreciate that Patrick Heagerty, the BUGS project, Muthén & Muthén, Journal of Applied Econometrics and the Royal Statistical Society have made data accessible via the internet. We have also used data from the UK Data Archive and the Norwegian Social Science Data Service with efficient help from Helene Roshauw. Thanks are due to Jasmin Naim and Kirsty Stroud at Chapman & Hall/CRC who have ensured steady progress through frequent but gentle reminders. We also thank the developers of LaTeX for providing this invaluable free tool for preparing manuscripts.

We have written each chapter together and contributed about equally to the book. Although writing this book has been hard work, we have had a lot of fun in the process!

The `gllamm` software, documentation, etc., can be downloaded from:

`http://www.gllamm.org`

Datasets and scripts for some of the applications in this book are available at:

`http://www.gllamm.org/books`

Oslo and Berkeley   Anders Skrondal
January 2004   Sophia Rabe-Hesketh

We are thrilled that our book is doing well and is already being reprinted. This has also given us the opportunity to make some corrections and updates. Thanks are due to Graham Dunn, Tor Haldorsen, Scott Long, Jay Magidson, William Saylor, Jeroen Vermunt and others for pointing out errors we did not discover ourselves. We would appreciate being made aware of any remaining errors. Remarks and corrections are continuously updated at:

http://www.gllamm.org/books/remarks.html

Oslo and Berkeley   Anders Skrondal
December 2004   Sophia Rabe-Hesketh

Dedication

To my children Astrid and Inge
Anders Skrondal

To Simon
Sophia Rabe-Hesketh

CHAPTER 1

The omni-presence of latent variables

1.1 Introduction

Since this book is about latent variable models it is natural to begin with a discussion of the meaning of the concept 'latent variable'. Depending on the context, latent variables have been defined in different ways, some of which will be briefly described in this chapter, although we generally find the definitions too narrow (see also Bollen, 2002). In this book we simply define a *latent variable* as a random variable whose realizations are hidden from us. This is in contrast to *manifest variables* where the realizations are observed.

Scepticism and prejudice regarding latent variable modeling are not uncommon among statisticians. Latent variable modeling is often viewed as a dubious exercise fraught with unverifiable assumptions and naive inferences regarding causality. Such a position can be rebutted on at least three counts: First, any reasonable statistical method can be abused by naive model specifications and over-enthusiastic interpretation. Second, ignoring latent variables often implies stronger assumptions than including them. Latent variable modeling can then be viewed as a sensitivity analysis of a simpler analysis excluding latent variables. Third, many of the assumptions in latent variable modeling can be empirically assessed and some can be relaxed, as we will see in later chapters.

Latent variable modeling is furthermore often viewed as a rather obscure area of statistics, primarily confined to psychometrics. However, latent variables pervade modern mainstream statistics and are widely used in different disciplines such as medicine, economics, engineering, psychology, geography, marketing and biology. This 'omni-presence' of latent variables is commonly not recognized, perhaps because latent variables are given different names in different literatures, such as random effects, common factors and latent classes.

In this chapter we will demonstrate that latent variables are used to represent phenomena such as

- 'True' variables measured with error
- Hypothetical constructs
- Unobserved heterogeneity
- Missing data
- Counterfactuals or 'potential outcomes'
- Latent responses underlying categorical variables

Latent variables are also used to

- Generate flexible multivariate distributions

- Combine information about individual units from different sources

In this chapter we give a taste of each of these uses of latent variables, referring to relevant sections in the methodology and application parts of the book for fuller explanations and examples.

1.2 'True' variable measured with error

Latent variables can represent the 'true scores' of a continuous variable measured with error. For continuous items or measures y_j for unit j, the measurement model from classical test theory (e.g. Lord and Novick, 1968) can be written as

$$y_j = \eta_j + \epsilon_j,$$

where η_j is the true score and ϵ_j is the measurement error. The measurement errors have zero means and are uncorrelated with each other and the true scores. The true score is defined as the expected value of the measured variable or item for a subject,

$$\eta_j \equiv \mathrm{E}(y_j)$$

over imagined replications (Lord and Novick, 1968). This is the *expected value definition* of latent variables.

Typically, several measurements y_{ij}, $i = 1, \ldots, n_j$ are used for each unit and the classical measurement model becomes

$$y_{ij} = \eta_j + \epsilon_{ij};$$

see also Section 3.3.1.

A path diagram for $n_j = 5$ measurements is given in Figure 1.1. Following

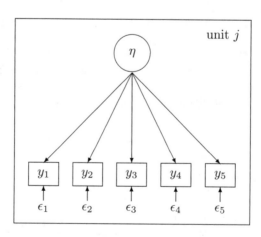

Figure 1.1 *Path diagram of the classical measurement model*

the conventions of path diagrams, the circle represents the latent variable η_j, the rectangles represent the observed measurements y_{ij} and the arrows represent linear relations (here with regression coefficients set to 1). The label 'unit j' implies that all variables inside the box vary between units and the subscript j is hence omitted in the diagram.

Measurement models are usually specified with continuous latent variables η_j. Such models are called factor models (see equation (1.1) on page 4) when the observed measures are continuous and item response models when the measures are categorical. Factor and item response models are discussed in more detail in Section 3.3.

Sometimes the true variable is instead construed as categorical, a typical example being medical diagnosis (ill versus not ill). The measurement is in this case usually also categorical with the same number of categories as the true variable. Measurement error then takes the form of misclassification. Measurement models with categorical latent and measured variables are known as latent class models and will be treated in Section 3.4. An application to the diagnosis of myocardial infarction (heart attack) is presented in Section 9.3.

A basic assumption of measurement models, both for continuous and categorical variables, is that the measurements are conditionally independent given the latent variable, i.e. the dependence among the measurements is solely due to their common association with the latent variable. This is reflected in the path diagram in Figure 1.1 where there are no arrows directly connecting the observed variables. This conditional or 'local' independence property is the basis of the *local independence definition* of latent variables (e.g. Lord, 1953; Lazarsfeld, 1959).

Measurement modeling can be used to assess measurement quality. If the true variable is continuous, measurement quality is typically assessed in terms of the reliability of individual measures; see Section 3.3. An application for fibre intake measurements is presented in Section 14.2, and other applications are given in Sections 10.3 and 10.4. If the true variable is categorical, measurement quality is typically formulated in terms of the misclassification rate, sensitivity and specificity. This is investigated in the context of diagnosis of myocardial infarction in Section 9.3 (see also Section 13.5).

Measurement models can also be combined with regression models to avoid regression dilution when a covariate has been measured with error (e.g. Carroll *et al.*, 1995a; see Section 3.5). In Section 14.2 we discuss models for the effect of dietary fibre intake on coronary heart disease when fibre intake is measured with error, with replicate measurements available for a subgroup. Sometimes a 'gold standard' measurement is available for a 'validation sample', whereas the fallible measurement is available for the whole sample. In the validation sample, the true value of the covariate is therefore observed, whereas it is represented by a latent variable in the validation sample, giving a model of the same form as missing covariate models; see also Section 1.5. We will discuss such a model for a case-control study of cervical cancer in Section 14.3.

Covariate measurement error models often make relatively strong assumptions such as conditional independence of the measures given the true value, nondifferential measurement error (that the measured covariate is conditionally independent of the response variable given the true covariate), normally distributed measurement errors and normally distributed true covariates. Many of these assumptions can be assessed and/or relaxed. For example, we will relax the nondifferential measurement error assumption in Section 14.3. In Section 14.2, we use nonparametric maximum likelihood estimation to leave the distribution of the true covariate unspecified when replicate measurements are available. Furthermore, the 'naive' analysis ignoring measurement error is likely to (but need not!) produce greater biases than a misspecified model taking measurement error into account. The latter can be seen as a sensitivity analysis for the former.

1.3 Hypothetical constructs

In contrast to true variables measured with error which are presumed to exist (be ontological), hypothetical constructs have an exclusively epistemological status (e.g. Messick, 1981). Treating hypothetical constructs as real would thus entail a reification error. The scepticism regarding 'latent variables' among many statisticians can probably be attributed to the metaphysical status of hypothetical constructs. On the other hand, communication seems impossible without relying on hypothetical constructs. For instance, the concept of a 'good statistician' is not real, but nevertheless useful and widely understood among statisticians (although not easily defined).

According to Cronbach (1971) a construct is an intellectual device by means of which one *construes* events. Thus, constructs are simply concepts. Relationships between constructs provide inductive summaries of observed relationships as a basis for elaborating networks of theoretical laws (e.g. Cronbach and Meehl, 1955). Nunnally and Durham (1975, p.305) put it the following way:

> "... the words that scientists use to denote constructs, for example, 'anxiety' and 'intelligence', have no real counterpart in the world of observables; they are only heuristic devices for exploring observables."

Since hypothetical constructs do not correspond to real phenomena, it follows that they cannot be measured directly even in principle (e.g. Torgerson, 1958; Goldberger, 1972). Instead, the construct is operationally defined in terms of a number of items or indirect 'indicators' such as answers in an intelligence test. The relationship between the latent construct and the observed indicators is usually modeled using a common factor model (Spearman, 1904),

$$y_{ij} = \lambda_i \eta_j + \epsilon_{ij}, \tag{1.1}$$

where η_j is the latent variable or 'common factor' representing the hypothetical construct, λ_i is a factor loading for item i and ϵ_{ij} is a unique factor, representing specific aspects of item i and measurement error; see also Section 3.3.2. The factor model can be represented by the same path diagram as the classical measurement model in Figure 1.1 where the paths from the factor to the indicators could now be labeled with the factor loadings.

Hypothetical constructs are prominent in psychological research. In fact, it seems fair to say that most research in psychology and similar disciplines is concerned with hypothetical constructs such as 'self-esteem', 'personality' and 'life-satisfaction' (see Section 10.4). Sociologists are often concerned with constructs such as 'aspiration' and 'alienation', whereas political scientists are interested in 'political efficacy' (see Section 10.3). In education, researchers are interested in constructs such as 'arithmetic ability' (see Section 9.4). It should be noted that hypothetical constructs are also used in 'harder' sciences such as economics to represent for instance 'permanent income' (e.g. Goldberger, 1971) and 'expectations' (e.g. Griliches, 1974). Hypothetical constructs are also important in medicine. Examples include 'depression' (e.g. Dunn *et al.*, 1993) and 'quality of life' (e.g. Fayers and Hand, 2002).

So far we have discussed continuous hypothetical constructs which appears to be the most common situation. However, it is sometimes more natural to consider categorical constructs or typologies. In sociology a prominent example is 'social class' (e.g. Marx, 1970). In psychology, 'stages of change' (precontemplation, contemplation, preparation, action, maintenance and relapse) are thought to be useful for assessing where patients are in their 'journey' to change health behaviors such as trying to quit smoking (e.g. Prochaska and DiClemente, 1983). In business, it is common practice to classify customers into 'market segments', either for targeted marketing or for tailoring products. For instance, Magidson and Vermunt (2002) used latent class models to classify bank customers and found four segments: 'value seekers', 'conservative savers', 'mainstreamers' and 'investors'. We consider an application of market segmentation for coffee makers in Section 13.6. In medicine, functional syndromes such as irritable bowel syndrome, which are characterized by a set of symptoms (whose cause is unknown), can be viewed as categorical hypothetical constructs. Here the fact that certain symptoms have high probabilities of occurring together is taken as an indication that they may be caused by the same disorder.

Instead of defining hypothetical constructs on theoretical grounds, they are sometimes 'derived' from an exploratory analysis, the classical example being the use of exploratory factor analysis when the latent variables are construed as continuous. The analogue for categorical latent variables is to use exploratory latent class analysis to derive categorical constructs or typologies. The danger of developing theory in this way has been vividly demonstrated by Armstrong (1967). He used exploratory factor analysis in an example where the underlying factors were known, the underlying model simple and providing a perfect fit to the data. While the exploratory factor analysis 'explained' a large proportion of the total variance, it failed spectacularly to recover the known factors. It is well worth citing Armstrong's summary:

"The cost of doing factor analytic studies has dropped substantially in recent years. In contrast with earlier time, it is now much easier to perform the factor analysis than to decide what to factor analyze. It is not clear that the resulting proliferation of the literature will lead us to the development of better theories.

Factor analysis may provide a means of evaluating theory or of suggesting revisions in theory. This requires, however, that the theory be explicitly specified prior to the analysis of data. Otherwise, there will be insufficient criteria for the evaluation of the results. If principal components is used for generating hypotheses without an explicit *a priori* analysis, the world will soon be overrun by hypotheses."

Indeed, a perusal of contemporary psychology journals definitely suggests that his prophecy has been fulfilled; this part of the world has already been overrun by hypotheses!

As a concrete example Armstrong mentioned a study by Cattell (1949) who attempted to discover primary dimensions of culture. The 12 basic factors obtained were rather mysterious, including gems such as 'enlightened affluence', 'thoughtful industriousness' and 'bourgeois philistinism'. Unfortunately, questionable applications of this kind still abound in psychology. A prominent recent example is the ruling 'big-five theory' in personality psychology (e.g. Costa and McRae, 1985), which ardently advocates that personality is characterized by the five dimensions 'extraversion', 'agreeableness', 'conscientiousness', 'neurotisism' and 'openness to experience'. This 'theory' has to a large extent been derived via exploratory factor analysis, but the factors have nevertheless been given ontological status (interpreted as real). Vassend and Skrondal (1995, 1997, 2004) are critical of the big-five theory and argue that the conventional analysis of personality instruments is fraught with conceptual and statistical problems.

Although continuous hypothetical constructs are usually modeled by common factors, this is not always the case. Several other multivariate statistical methods have been used to explore the 'dimensions' underlying data. Examples include principal component analysis (e.g. Joliffe, 1986), partial least squares (PLS) (e.g. Lohmöller, 1989), canonical correlations (e.g. Thompson, 1984), discriminant analysis (e.g. Klecka, 1980) and multidimensional scaling (e.g. Kruskal and Wish, 1978). Categorical 'constructs' can be derived using cluster analysis, finite mixture modeling or multidimensional scaling (e.g. Shepard, 1974). However, we do not consider the 'dimensions' or 'groupings' produced by these methods as latent variables since they merely represent transformations or geometric features of the data and not elements in a statistical model. In fact, Bentler (1982) defines a latent variable as a variable that cannot be expressed as a function of manifest variables only. Another limitation is that the methods are usually strictly exploratory, not permitting any hypothesized structure, based on research design, previous research or substantive theory, to be imposed and tested. This problem is also shared by exploratory factor and latent class analysis. Different types of statistical models are contrasted in Section 8.2.1 and different modeling strategies discussed in Section 8.2.2.

Acknowledging that hypothetical constructs are useful in many disciplines, methods for investigating *construct validity*, the measurement and interrelationships among constructs, are essential. Bentler (1978) and Bagozzi (1980, 1981), among others, argue that construct validity is best investigated by means of structural equation models with latent variables (see Section 3.5). This confirmatory approach, explicitly specifying hypothesized structures, can be viewed as a formalization of the classical notion of construct validity (e.g. Cronbach and Meehl, 1955). Cronbach and Meehl understand constructs as involved in a network of both theoretical and empirical relationships; Cronbach (1971, p.476) puts it the following way:

> "The statements connecting constructs with each other, and observable indicators with constructs, constitute a nomological network (*nomological* meaning law-like)."

'Valid' (hypothesized structure)

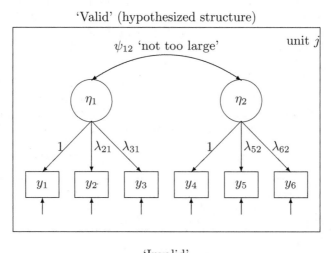

'Invalid'

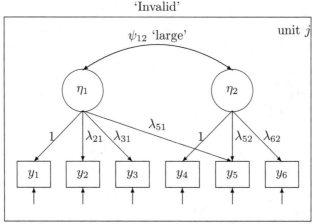

Figure 1.2 *Convergent and discriminant validity*

Several types of validity have been outlined in psychology, including 'content', 'convergent', 'concurrent', 'discriminant', 'predictive' and 'nomological' (American Psychological Association *et al.*, 1974). Construct validity has become the core of a hierarchical and unifying view of validity, integrating these types (e.g. Silva, 1993).

An advantage of latent variable modeling is that we can investigate the tenability of hypothesized structures, either by assessing model fit (see Section 8.5) or by elaborating the model in various ways. For instance, *convergent validity* can be assessed by specifying models where indicators designed to reflect a given construct only reflect that construct and not others. This is illustrated in the upper panel of Figure 1.2 where the first factor is measured by items 1 to 3 whereas the second factor is measured by items 4 to 6. If this model is rejected in favor of the model in the lower panel, where item 5 loads on both factors, then convergent validity does not hold. Two alternative courses of action could be taken in this case, either accommodating the item as in the bottom panel or discarding it. The latter approach, common in item response modeling, could be criticized for being 'self-fitting' (e.g. Goldstein, 1994) but may be justified if the theory is well founded.

Discriminant validity may be investigated by inspecting the uniqueness of the constructs, in the sense that the estimated correlations among constructs should not be too large. In the lower panel of the figure, the estimated covariance $\widehat{\psi}_{12}$ is large and the discriminant validity is questionable.

'Valid' (hypothesized)   'Invalid'

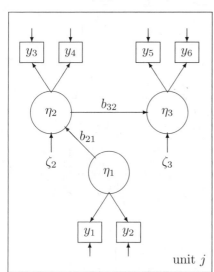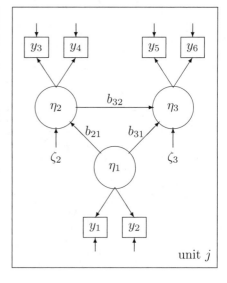

Figure 1.3 *Nomological validity*

Nomological validity is typically assessed by investigating the tenability of the structural model that incorporates the theory induced relationships among the constructs. For example, consider the model of 'complete mediation' (e.g. Baron and Kenny, 1986) represented in the left panel of Figure 1.3, positing that there is no direct effect of η_1 (e.g. 'anomalous parental bonding') on η_3 (e.g. 'depression'), but only an indirect effect via the mediator η_2 (e.g. 'personality'). If this model is rejected in favor of a model with both direct and indirect effects of η_1 on η_3, as shown in the right panel, the complete mediation model does not have 'nomological' validity.

We refer to Bagozzi (1981) for an application examining the construct validity of the expectancy-value and semantic-differential models of attitude.

1.4 Unobserved heterogeneity

A major aim of statistical modeling is to 'explain' the variability in the response variable in terms of variability in observed covariates, sometimes called 'observed heterogeneity'. However, in practice, not all relevant covariates are observed, leading to *unobserved heterogeneity*. Including latent variables, in this context typically referred to as *random effects*, in statistical models is a common way of taking unobserved heterogeneity into account.

Random effects models are widely used for a variety of problems. Examples include longitudinal analysis (e.g. Laird and Ware, 1982), meta-analysis (e.g. DerSimonian and Laird, 1986), capture-recapture studies (e.g. Coull and Agresti, 1999), conjoint analysis (e.g. Green and Srinivasan, 1990), biometrical genetics (e.g. Neale and Cardon, 1992) and disease mapping (e.g. Clayton and Kaldor, 1987). Applications of models for unobserved heterogeneity are given in Sections 9.2 and 11.3 for longitudinal studies, Section 9.5 for meta-analysis, Section 9.7 for capture-recapture studies, Section 11.4 for small area estimation and disease mapping, and Section 13.6 for conjoint analysis in marketing.

Note that unobserved heterogeneity is not a hypothetical construct since it merely represents the combined effect of all unobserved covariates, and is not given any meaning beyond this. The random effects in genetic studies perhaps occupy an intermediate position, since they are interpreted as shared and unshared genetic and environmental influences.

When the units are clustered, shared unobserved heterogeneity may induce 'intra-cluster' dependence among the responses, even after conditioning on observed covariates. This is illustrated in Figure 1.4 for ten clusters with two units each and no covariates. Here, heterogeneity is reflected in the scatter of the cluster means (shown as horizontal bars) around the overall mean (horizontal line), leading to within-cluster correlations because both responses for a cluster tend to lie on the same side of the overall mean. This phenomenon is common for longitudinal or panel data, where observations for the same unit are influenced by the same (shared) unit-specific unobserved heterogeneity. An example involving repeated measurements of respiratory infection in Indonesian children is given in Section 9.2. Other examples of clustered data include individuals in households, or children in schools.

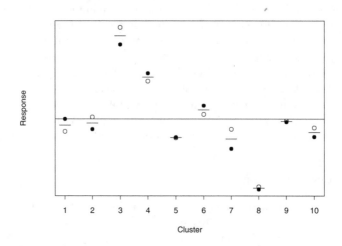

Figure 1.4 *Between-cluster heterogeneity and within-cluster correlation*

A different kind of clustered data are counts, such as number of epileptic fits for a person over a week. Although there is a single count for each unit, each count comprises several events whose occurrence is likely to be influenced by shared unit-specific covariates. The resulting variability in 'proneness' to experiencing the event can lead to 'overdispersion', in a Poisson model meaning that the variance is larger than the mean (see Sections 2.3.1 and 11.2). These two consequences of unobserved heterogeneity, within-cluster dependence and overdispersion, lead to incorrect inferences if not properly accounted for. One way of accounting for unobserved heterogeneity is to include a random intercept in a regression model. In the case of clustered data, units in the same cluster must share the same value or realization of the random effects. In Figure 1.4, the random intercept would represent the deviations of the cluster means (horizontal bars) from the overall mean.

In multilevel or hierarchical data there are often several levels of clustering, an example being panel data on individuals in households. We can then use latent variables at each of the higher levels to represent unobserved heterogeneity at that level. A simple three-level random intercept regression model for panel wave i, individual j and household k can be written as

$$y_{ijk} = \eta_{0jk} + \beta_1 x_{ijk} + \epsilon_{ijk},$$

$$\eta_{0jk} = \gamma_{00} + \zeta_{jk}^{(2)} + \zeta_k^{(3)}.$$

In the level-1 model for y_{ijk}, x_{ijk} is a covariate with 'fixed' regression coefficient β_1 and η_{0jk} is a random intercept with mean γ_{00} and residuals $\zeta_{jk}^{(2)}$ and

$\zeta_k^{(3)}$ at levels 2 and 3, respectively. The random part of this model (not showing $\beta_1 x_{ijk}$ and γ_{00}) is presented in path diagram form in Figure 1.5 for a household with three individuals participating at 3, 1 and 2 occasions, respectively. Multilevel models are discussed in greater detail in Section 3.2.

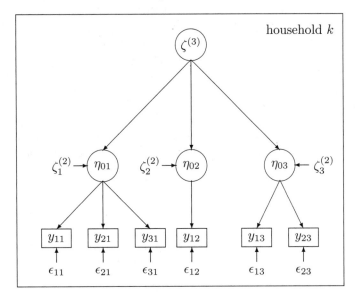

Figure 1.5 *Path diagram of three-level random intercept model*

The effect of a covariate on the response can also differ between clusters which can be modeled by including cluster-specific 'random coefficients'. For example, the change in epileptic seizure rate over time may vary across subjects as investigated in Section 11.3. Analogously, the effect of political distance (between party and voter) on party preference may vary across constituencies as discussed in Section 13.4. Sometimes the unobserved heterogeneity is discrete with units falling into distinct clusters (see Section 12.4 for an application).

An important consequence of unobserved heterogeneity is that relationships between the response and the observed covariates are usually different at the unit (or cluster) and population levels. A prominent example is frailty in survival analysis (e.g. Aalen, 1988), where the population-level hazard can differ drastically from the unit-level hazards due to unexplained variability in the latter. The reason for this is that some individuals are more 'frail' than others, being more susceptible to the event than can by explained by their observed covariates. These individuals will tend to experience the event early on, leaving behind the less frail. Consequently, even if individual hazards are constant, the population average hazard will decline over time. We consider multivariate frailty models for the treatment of angina in Section 12.4.

The distinction between effects at the unit (or cluster) and population levels is also important for dichotomous responses. If the unit-specific models are probit regressions with different intercepts (due to unobserved heterogeneity) but sharing a common coefficient for a single covariate, then the population averaged model is a probit regression with an attenuated coefficient. This is illustrated in Figure 1.6 where the population averaged curve, shown in bold, has a smaller 'slope' than the unit-specific curves (see Section 4.8.1 for a derivation). Whether unit-specific or population-averaged effects are of interest will depend on the context. For example, population averaged effects are often of concern in public health where the focus is on the population level. In a clinical setting, on the other hand, patient-specific effects are obviously more important for the individual patient and her physician. Importantly, since causal processes necessarily operate at the unit and not the population level, it follows that investigation of causality requires unit-specific effects. If there are repeated observations on the units, unit-specific effects can be

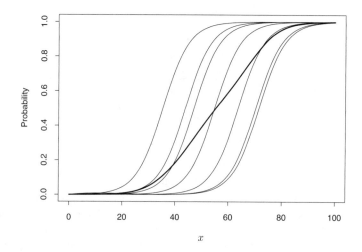

Figure 1.6 *Unit-specific versus population-average probit regression*

estimated by including random effects in the models.

In the preceding examples, we have considered units observed on multiple occasions and we will now return to the more general case of units (possibly occasions) nested in clusters (possibly units). It is important to note that relationships between covariates and response variables may be different at the unit and cluster level. Inferences regarding effects at the unit level based on aggregated data at the cluster level may therefore lead to the so called 'ecological fallacy' (e.g. Robinson, 1950). Robinson's classical example concerned the correlation between the percentage of black people and illiteracy at the region level, estimated as 0.95, which was very different from the estimated individual-level correlation between being black and individual illiteracy, estimated as 0.20.

Figure 1.7 illustrates that within-cluster effects can be very different from between-cluster effects, possibly having opposite directions. The clusters in the

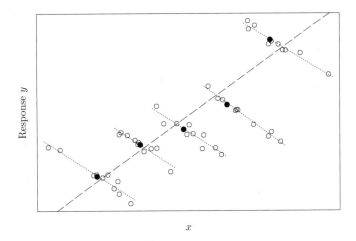

Figure 1.7 *Within-cluster and between-cluster effects*

figure could represent countries, the response y could be a health outcome such as length of life and the explanatory variable x could be exposure to unhealthy luxuries such as red meat. Within a country, increasing exposure is associated with decreasing health (as reflected by the downward slopes of the dotted lines). Between countries, on the other hand, increasing average exposure is associated with increasing average health (as reflected by the upward slope of the dashed line) since it is also associated with increasing average living standards.

Another important example would be longitudinal data where the within-unit decline could represent an age affect, whereas the between-unit increase could be a cohort effect. If only cross-sectional data are available, we cannot distinguish between age and cohort effects. For example, greater conservatism in older people as compared to younger people could be due to being at a later stage in their lives or due to being born into a different epoch. More formally, consider the longitudinal model

$$y_{ij} = \beta_0 + \beta_C x_{1j} + \beta_L(x_{ij} - x_{1j}) + \zeta_j + \epsilon_{ij},$$

where y_{ij} is a response for unit j at occasion i, x_{ij} could be age, and ζ_j is a subject-specific random intercept. The longitudinal design allows separate estimation of the cross-sectional (or cohort) effect β_C and the longitudinal effect β_L. This distinction is closely related to the problem of correlation between random effects and covariates discussed on page 52.

In random effects modeling it is typically assumed that the random effects have multivariate normal distributions. Model diagnostics (see Section 8.6) can be used to assess this assumption, although they may not be sufficiently sensitive. Fortunately, inferences are in many cases quite robust to misspecifications of the random effects distribution (e.g. Bartholomew, 1988, 1994). We can moreover relax the distributional assumption by using nonparametric maximum likelihood estimation (see Section 6.5). This approach is used in modeling faulty teeth in children in Section 11.2, epileptic seizures in Section 11.3 and diet and heart disease in Section 14.2.

1.5 Missing values and counterfactuals

Latent variables can represent missing values of partially observed variables, that is variables that are observed on a subset of the units. Usually, the missing values are presumed to have been 'realized' but for some reason not recorded. However, missing values are sometimes values that would have been realized under 'counterfactual' circumstances, for instance if a covariate had had a different value.

If a *covariate* is missing for some units, these units typically cannot contribute to parameter estimation. This loss of units leads to reduced efficiency which can be overcome by filling in missing covariate values using (multiple) imputation (e.g. Rubin, 1987; Schafer, 1997). The units can then contribute information on the relationship between the responses and the other covariates. Instead of imputing covariate values, we could jointly estimate the imputation model with the model of interest, integrating the likelihood over the 'imputation distribution' for the missing values. Here the missing values can be represented by a latent variable assumed to have the same distribution as the observed values and the same relationship to the responses (e.g. regression parameter) in the model of interest. An example of a missing covariate problem is covariate measurement error when there is a validation sample in which the true covariate is observed (see Section 14.3). Another example is estimation of 'complier average causal effects' (Imbens and Rubin, 1997b) in randomized interventions with noncompliance where compliance is not observed in the control group (see Section 14.4). Here compliance status in the control group can also be viewed as 'counterfactual'.

If *responses* are missing for some units, the units can contribute to parameter estimation as long as there is at least one observed response, leading to consistent parameter estimates if the data are missing at random (MAR) (e.g. Rubin, 1976; Little and Rubin, 2002). However, if the responses are not missing at random (NMAR), ignoring the missing data mechanism can lead to biased parameter estimates. This can be addressed by joint modeling of the substantive and missingness processes. In the 'selection model' approach (e.g. Little, 1995), the dependence of the missingness process on the unobserved response is explicitly modeled.

For example, in a longitudinal setting where responses are missing due to dropout or attrition, Hausman and Wise (1979) introduced a model in econometrics that was later rediscovered in the statistical literature by Diggle and Kenward (1994). Here the dropout at each time-point (given that it has not yet occurred) is modeled using a logistic (or probit) regression model with previous responses and the contemporaneous response as covariates. If dropout occurs, the contemporaneous response is not observed but represented by a latent variable. The likelihood is the joint likelihood of the response and dropout status after integrating out the latent variable. A simple version of the model for the response of interest y_{ij} at time i for unit j can be written as

$$y_{ij} \;=\; \mathbf{x}'_{ij}\boldsymbol{\beta} + \eta_j + \epsilon_{ij},$$

where η_j is a unit-specific random intercept. The dropout variable, $d_{ij} = 1$ if unit j drops out at time i and 0 otherwise, can be modeled as

$$\mathrm{logit}[\Pr(d_{ij} = 1 | \mathbf{y}_j)] \;=\; \alpha_1 y_{ij} + \alpha_2 y_{i-1,j},$$

where y_{ij} is replaced by a latent variable y^*_{ij} when it is unobserved. Figure 1.8 shows path diagrams for a unit with complete data across three time-points and a unit dropping out at time 2. In the latter case, the response at time 2 is represented by a latent variable. Note that we have used arrows for logistic regressions although they are not linear. This analysis can be viewed literally

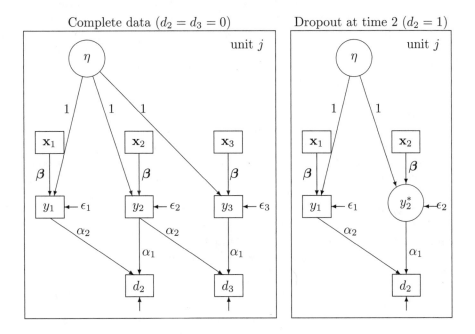

Figure 1.8 *Path diagram of Hausman-Wise-Diggle-Kenward dropout model*

as correcting for missing data that are NMAR. However, since estimation of α_1 relies heavily on the distributional assumption for y_{ij}, it might be advisable to instead interpret the results as a sensitivity analysis of the MAR assumption.

1.6 Latent responses

Latent variables can represent continuous variables underlying observed 'coars-ened' responses such as dichotomous or ordinal responses (see Section 2.4). The latent response interpretation of dichotomous responses was introduced by Pearson (1901) for normally distributed latent responses. Although most commonly used for such probit models, the latent response formulation is just as applicable for logit and complementary log-log models. In the dichotomous case, the observed response y_i is modeled as resulting from combining a re-gression model for an underlying continuous response y_i^*

$$y_i^* = \mathbf{x}_i'\boldsymbol{\beta} + \epsilon_i,$$

with a threshold model

$$y_i = \begin{cases} 1 & \text{if } y_i^* > 0 \\ 0 & \text{otherwise.} \end{cases}$$

This model corresponds to a probit model if ϵ_i is standard normal, a logit model if ϵ_i has a logistic distribution and a complementary log-log model if ϵ_i has a Gumbel distribution. A general interpretation of a latent response is the 'propensity' to have a positive observed response $y_i = 1$. In genetics, the latent response is interpreted as the 'liability' to develop a qualitative trait or phe-notype such as diabetes type I (e.g. Falconer, 1981). Heckman (1978) modeled whether or not American states had introduced fair-employment legislation and described the corresponding latent response as the 'sentiment' favoring fair-employment legislation. In toxicology, a unit's 'tolerance' to a drug is the maximum dose the unit can tolerate, so that exceeding the dose results in death (e.g. Finney, 1971).

In the decision context where individuals choose the most preferred alter-native or rank alternatives in order of preference, the latent responses can be interpreted as utility differences. For example, consider the scenario that commuters must choose between car or bus. The utilities u_i^{car} and u_i^{bus} for car and bus may depend on the respective travelling times for the commuter and other covariates. The commuter decides to travel by car if $u_i^{\text{car}} > u_i^{\text{bus}}$, or alternatively, if $y_i^* = u_i^{\text{car}} - u_i^{\text{bus}} > 0$. Models for such comparative responses and several applications are presented in Chapter 13.

Not surprisingly, the introduction of latent responses attracted criticism. For instance, Yule (1912, p.611-612) remarked:

> "...all those who have died of smallpox are equally dead: no one is more dead or less dead than another, and the dead are quite distinct from the survivors."

with the response by Pearson and Heron (1913, p.159):

> "...if Mr Yule's views are accepted, irreparable damage will be done to the growth of modern statistical theory."

Leaving this philosophical (and personal?) debate aside, the latent response formulation of probit models is undoubtedly useful regardless of whether the latent response can be given a real meaning. For example, in the latent re-sponse formulation, we can specify models for dependence between dichotomous variables by simply allowing the latent responses to be correlated ('tetra-choric correlations'). Similarly, we can model the dependence between dichoto-mous and continuous variables by allowing latent responses to be correlated with observed continuous responses ('biserial correlations'), as in selection models (e.g. Heckman, 1979). We refer to Section 4.8.2 for details.

Several estimation methods are furthermore based on the latent response formulation (see Chapter 6). This includes the limited information methods suggested by Muthén (e.g. Muthén, 1984; Muthén and Satorra, 1995) and discussed in Section 6.7 and the EM algorithms discussed by Schoenberg (1985). Some estimation methods simulate latent responses. Examples include the Monte Carlo EM algorithms discussed in Section 6.4.1 (e.g. Meng and Schilling, 1996), Markov Chain Monte Carlo methods of the kind exemplified in Section 6.11.5 (e.g. Albert and Chib, 1993), the method of simulated mo-ments (e.g. McFadden, 1989), and the GHK method for simulated maximum likelihood (e.g. Train, 2003) discussed in Section 6.3.3.

Moreover, it is much easier to investigate model identification using the latent response formulation as we will see in Chapter 5. Finally, procedures for model diagnostics based on latent responses have been developed by for instance Chesher and Irish (1987) and Gourieroux *et al.* (1987a) in the fre-quentist setting and by Albert and Chib (1995) in Bayesian modeling (see Section 8.6).

1.7 Generating flexible distributions

Latent variables are useful for generating distributions with the desired vari-ance function and shape, or multivariate distributions with a particular de-pendence structure.

As discussed in Section 1.4, overdispersion of counts can be addressed by including a random intercept in the model (e.g. Breslow and Clayton, 1993). If there are an excess number of zero counts, zero-inflated Poisson (ZIP) mod-els can be used which are a mixture of a Poisson model and a mass at zero (e.g. Lambert, 1992). This kind of model may for instance be useful if the re-sponse is the number of alcoholic drinks consumed in a week. In this setting a zero response could be due to the person being a 'teetotaller' (nondrinker) or simply a random fluctuation for a drinker. In the ZIP model, the component 'membership' label (or latent class) can be viewed as a realization of a dis-crete latent variable. An application of this model for the number of decayed, missing or filled teeth is given in Section 11.2. Many other types of mixture models have been used to generate flexible distributions, including mixtures of normals and mixtures of Poisson distributions, see for example Everitt and Hand (1981), Böhning (2000) and McLachlan and Peel (2000).

Another use of latent variables is as a parsimonious way of inducing depen-dence between responses. A typical example is the analysis of longitudinal data as discussed in Section 3.6. Latent variables are also used to induce depen-dence between nonnormal responses where flexible multivariate distributions do not exist. For example, Coull and Agresti (2000) use correlated random intercepts in the linear predictors of logistic regression models to model the dependence between dichotomous responses. They refer to these models as binomial logit-normal models (BLN). Latent variables are also often used to induce dependence between different processes. A common example is joint modeling of a response of interest and a missing data process, in so-called 'shared random effects' models (e.g. Wu and Carroll, 1988). This idea is used for endogenous treatment modeling in Section 14.5 and joint modeling of repeated measurements and survival in Section 14.6.

1.8 Combining information about individual units from different sources

It is often of interest to assign values to latent variables, taking the form of scores for continuous latent variables and categories and classes in the categorical case. Such prediction, scoring or classification is important for all types of latent variables discussed. For instance, true classification of categorical variables measured with error (or misclassified) is crucial in medical diagnosis, and is performed in Section 9.3 for myocardial infarction. Scoring of hypothetical constructs such as ability is central in education as illustrated in Section 9.4, whereas prediction of unit-specific effects is the purpose of small area estimation and disease mapping as discussed in Section 11.4.

It is beneficial to base scoring and classification on explicit latent variable models. This approach allows empirical assessment of the reliability and validity of the measurements and provides optimal means of combining information. As explained in Chapter 7, predictions for a unit are not solely based on the measurement for that unit, but are also influenced by the estimated distribution of the latent variables for the population of units. In the empirical Bayesian approach, the latent variable distribution represents the (empirical) prior, whereas the conditional distribution of the measurements given the latent variables represents the 'likelihood'. Since the latent variable distribution is estimated using information from all units, prediction for a given unit 'borrows strength' from the measurements of other units (e.g. Rubin, 1983; Morris, 1983). For instance, in disease mapping adjacent geographical units provide useful information improving estimates of disease rates that are based on small numbers of events.

1.9 Summary

We have demonstrated that latent variables pervade modern statistics and described how they are used to represent widely different phenomena such as true variables measured with error, hypothetical constructs, unobserved heterogeneity, missing data, counterfactuals and latent responses underlying categorical variables. Latent variables can also be used to generate flexible multivariate distributions and to combine information about individual units from different sources.

In the next chapter we describe the class of generalized linear models which, although not including latent variables, provide an essential stepping stone to the latent variable models that are the core of the rest of the book.

CHAPTER 2

Modeling different response processes

2.1 Introduction

In this chapter we describe a wide range of response processes, producing the following types of observed responses:

- Continuous or metric
- Dichotomous
- Grouped
- Censored
- Ordinal
- Unordered polytomous or nominal
- Pairwise comparisons
- Rankings or permutations
- Counts
- Durations or survival

The aim of statistical modeling is to capture the main features of the empirical process under investigation (see Section 8.2 for further discussion). Typically, a first simplifying step is to focus on a restricted set of response variables and to consider the data generating process of these variables given a set of explanatory variables. Univariate models have one response variable whereas multivariate models have several, possibly including intervening or intermediate variables serving as both response and explanatory variables. The response variables are sometimes called 'dependent', 'endogenous' or 'outcome' variables whereas the explanatory variables are called 'independent', 'exogenous' or 'predictor' variables. The explanatory variables of primary interest are sometimes called 'exposures' or '(risk) factors' and the others 'confounders' or 'covariates'. However, the term 'covariate' is often used as a generic term for explanatory variable.

The variables can be further classified according to their 'measurement levels'. For explanatory variables it is usually sufficient to distinguish between continuous or categorical variables since we do not model these variables but merely condition on them. If the values of a variable are ordered and differences between values are meaningful, the variable is typically treated as continuous, otherwise as categorical.

For response variables, we must also consider the process that may have generated the response since this is crucial for formulating an appropriate statistical model. For example, a variable taking on ordered discrete values 1,2,3, etc. could represent an ordinal response such as the level of pain (none, mild, moderate, etc.), a count such as the number of headache-free days in a week, or a discrete-time duration such as the number of months from diagnosis of some condition to death. The *response processes* generating these same values are obviously of a widely different nature and require different statistical models.

There are two general approaches to modeling response processes. In statistics and biostatistics, the most common approach is generalized linear modeling, whereas a latent response formulation is popular in econometrics and psychometrics. Although very different in appearance, the approaches can generate equivalent models for many response types. However, as we will see in later chapters, the choice of formulation can have implications for estimation and identification. The latent response formulation is useful even for applications where interpretation in terms of a latent response appears contrived.

We start by describing generalized linear models and their extensions. We then introduce the latent response formulation and point out correspondences between approaches. Finally, durations or survival data are discussed separately because they do not fit entirely into either of the frameworks. Both continuous and discrete time models are considered.

In this chapter we do not yet introduce random coefficients or common factors. However, the models discussed represent an essential building block for the general model framework to be presented in Chapter 4.

2.2 Generalized linear models

2.2.1 Introduction

In generalized linear models (e.g. Nelder and Wedderburn, 1972) the explanatory variables affect the response only through the *linear predictor* ν_i for unit i,

$$\nu_i = \mathbf{x}'_i\boldsymbol{\beta},$$

where $\mathbf{x}_i$ is a vector of explanatory variables and $\boldsymbol{\beta}$ contains the corresponding regression parameters.

Both continuous and categorical explanatory variables can be accommodated. For continuous variables such as age, a single term is often used to represent the linear effect of age on the linear predictor. More flexible ways of modeling effects include polynomials, splines or other smooth functions. For categorical variables such as nationality (e.g. Norwegian, German, other), a *dummy variable* would typically be specified for each category except for a reference category, for instance 'other'. For example, a dummy variable for Norwegian equals 1 if the person is Norwegian and 0 otherwise so that the corresponding coefficient represents the effect of being Norwegian compared with 'other'.

The response process is fully described by specifying the conditional probability (density) of y_i given the linear predictor. The simplest response process is the continuous. A *linear regression* model

$$y_i = \nu_i + \epsilon_i \qquad (2.1)$$

is usually specified in this case, where the residuals ϵ_i are independently normally distributed with zero mean and constant variance σ^2,

$$\epsilon_i \sim N(0, \sigma^2). \qquad (2.2)$$

Special cases of linear regression models include analysis of variance (ANOVA) and analysis of covariance (ANCOVA) models.

The linear regression model can alternatively be defined by setting the conditional expectation of the response, given the linear predictor ν_i, equal to ν_i,

$$\mu_i \equiv E(y_i|\nu_i) = \nu_i,$$

and specifying that the y_i are independently normally distributed with mean μ_i and variance σ^2.

For dichotomous or binary responses taking on values 0 or 1, the conditional probability of response 1, $\Pr(y_i=1|\nu_i)$, is just the conditional expectation μ_i of y_i. This can be modeled as a *logistic regression*

$$\mu_i = \frac{\exp(\nu_i)}{1 + \exp(\nu_i)} \quad \text{or} \quad \ln\left(\frac{\mu_i}{1 - \mu_i}\right) = \nu_i,$$

or a *probit regression*

$$\mu_i = \Phi(\nu_i) \quad \text{or} \quad \Phi^{-1}(\mu_i) = \nu_i,$$

where $\Phi(\cdot)$ is the standard normal cumulative distribution function. Conditional on ν_i, the y_i are independently Bernoulli distributed.

Counts are discrete non-negative integer valued responses $(0,1,..)$. The standard model for counts is *Poisson regression* with expectation

$$\mu_i = \exp(\nu_i) \quad \text{or} \quad \ln(\mu_i) = \nu_i$$

and Poisson distribution

$$\Pr(y_i|\mu_i) = \frac{\exp(-\mu_i)\mu_i^{y_i}}{y_i!}. \qquad (2.3)$$

Counts have a Poisson distribution if the events being counted for a unit occur at a constant rate in continuous time and are mutually independent.

If a count corresponds to the number of events in a given number n of 'trials' (or opportunities for an event), the count has a binomial distribution if the events for a unit are independent and equally probable. The probability of a proportion y_i out of n then is

$$\Pr(y_i|\mu_i) = \binom{n}{y_i n}\mu^{y_i n}(1 - \mu_i)^{(1-y_i)n},$$

where the binomial coefficient $\binom{n}{y_i n} = n!/[(y_i n)!(n - y_i n)!]$ is the number of ways of choosing $y_i n$ out of n objects regardless of their ordering.

2.2.2 Model structure

All the models described above have a common structure and represent special cases of generalized linear models defined by two components:

1. The functional relationship between the expectation of the response and the linear predictor is

$$\mu_i = g^{-1}(\nu_i) \quad \text{or} \quad g(\mu_i) = \nu_i,$$

where $g(\cdot)$ is a *link function*. We have already encountered the identity, logit, probit and log link for linear, logistic, probit and Poisson regression, respectively. These and other common links are given in Table 2.1.

Table 2.1 Common links

Link	$g(\mu)$	$g^{-1}(\nu)$	range of $g^{-1}(\nu)$
Identity	μ	ν	$-\infty, \infty$
Reciprocal	$1/\mu$	$1/\nu$	$-\infty, \infty$
Logarithm	$\ln(\mu)$	$\exp(\nu)$	$0, \infty$
Logit	$\ln\left(\frac{\mu}{1-\mu}\right)$	$\frac{\exp(\nu)}{1+\exp(\nu)}$	$0, 1$
Probit	$\Phi^{-1}(\mu)$	$\Phi(\nu)$	$0, 1$
Scaled probit	$\sigma\Phi^{-1}(\mu)$	$\Phi(\nu/\sigma)$	$0, 1$
Complementary log-log	$\ln(-\ln(1-\mu))$	$1 - \exp(-\exp(\nu))$	$0, 1$

2. The conditional probability distribution of the responses is a member of the exponential family with expectation μ_i and, possibly, a common scale parameter ϕ,

$$f(y_i|\theta_i, \phi) = \exp\left\{\frac{y_i\theta_i - b(\theta_i)}{\phi} + c(y_i, \phi)\right\}. \tag{2.4}$$

Here, θ_i is the canonical or natural parameter, ϕ is the scale or dispersion parameter and $b(\cdot)$ and $c(\cdot)$ are functions depending on the member of the exponential family. We have already encountered the normal or Gaussian, the Bernoulli, Poisson and binomial distributions. Table 2.2 gives details on these and other important members of the exponential family.

Table 2.2 *Members of the exponential family*

Distribution	Canonical link $\theta(\mu)$	Cumulant function $b(\theta)$	Dispersion parameter ϕ	Expectation $b'(\theta)$	Variance $\phi b''(\theta)$	Probability or density $f(y\|\theta,\phi)$
Bernoulli	$\ln(\mu/(1-\mu))$	$\ln(1+\exp(\theta))$	1	$\frac{\exp(\theta)}{1+\exp(\theta)}$	$\mu(1-\mu)$	μ
Binomial[†]	$\ln(\mu/(1-\mu))$	$\ln(1+\exp(\theta))$	$1/n$	$\frac{\exp(\theta)}{1+\exp(\theta)}$	$\mu(1-\mu)/n$	$\binom{n}{yn}\mu^{yn}(1-\mu)^{(1-y)n}$
Poisson	$\ln(\mu)$	$\exp(\theta)$	1	$\exp(\theta)$	μ	$\frac{\mu^y\exp(-\mu)}{y!}$
Normal	μ	$\theta^2/2$	σ^2	θ	σ^2	$\frac{1}{\sqrt{2\pi\sigma^2}}\exp\frac{-(y-\mu)^2}{2\sigma^2}$
Gamma	$-1/\mu$	$-\ln(-\theta)$	α^{-1}	$-1/\theta$	$\mu^2\alpha^{-1}$	$\frac{1}{\Gamma(\alpha)}\left(\frac{\alpha}{\mu}\right)^\alpha y^{\alpha-1}\exp\left(-\frac{\alpha y}{\mu}\right)$
Inverse Gaussian	$1/\mu^2$	$-(-2\theta)^{1/2}$	σ^2	$(-2\theta)^{-1/2}$	$\mu^3\sigma^2$	$\frac{1}{\sqrt{2\pi y^3\sigma^2}}\exp\left\{\frac{-(y-\mu)^2}{2(\mu\sigma)^2 y}\right\}$

[†] y is the proportion of 'successes' out of n 'trials'

Assuming that the responses of units are independent, the likelihood for generalized linear models is

$$l = \prod_{i=1}^{N} l_i,$$

where $l_i \equiv f(y_i|\theta_i, \phi)$ is the likelihood contribution from unit i, $i=1, 2, \ldots, N$, and the log-likelihood becomes

$$\ell = \sum_{i=1}^{N} \ell_i,$$

where $\ell_i \equiv \ln l_i$. The first and second derivatives of the log-likelihood contributions with respect to θ_i are

$$\frac{\partial \ell_i}{\partial \theta_i} = [y_i - b'(\theta_i)]/\phi, \qquad (2.5)$$

and

$$\frac{\partial^2 \ell_i}{\partial \theta_i^2} = -b''(\theta_i)/\phi, \qquad (2.6)$$

where $b'(\theta_i)$ and $b''(\theta_i)$ are the first and second derivatives of $b(\cdot)$ evaluated at θ_i. Maximum likelihood estimation of generalized linear models using iteratively reweighted least squares is described in Section 6.8.1.

2.2.3 Mean function and choice of link function

From standard likelihood theory the expected scores are zero,

$$E\left(\frac{\partial \ell}{\partial \theta_i}\right) = 0,$$

so it follows from (2.5) that

$$\mu_i = b'(\theta_i).$$

Writing θ_i as a function of μ_i gives the *canonical link* function

$$g(\mu_i) = \theta_i(\mu_i).$$

The canonical link has convenient statistical properties. However, the choice of link should be guided by theoretical considerations and model fit. One consideration in choosing a link function is the range of values it generates for the mean $\mu_i = g^{-1}(\nu_i)$ when $-\infty \leq \nu_i \leq \infty$ (see Table 2.1). For example, the logit and probit links are popular for dichotomous responses because they restrict the probability μ_i to lie in the permissible interval $[0,1]$. In contrast, use of the identity link may in this case lead to predicted probabilities that are negative or larger than one.

    Another important consideration relates to the interpretation of the regression parameters. Since $\nu_i = \mathbf{x}_i'\boldsymbol{\beta}$, using an identity link corresponds to additive effects of the covariates on the mean, $\mu_i = \mathbf{x}_i'\boldsymbol{\beta}$, and a log link corresponds to multiplicative effects, $\mu_i = \exp(\mathbf{x}'_i \boldsymbol{\beta})$. Using a logit link for dichotomous responses gives a multiplicative model for the odds, $\mu_i/(1-\mu_i) = \exp(\mathbf{x}'_i \boldsymbol{\beta})$. This link is particularly useful in case-control studies since odds-ratios are invariant with respect to retrospective or 'choice-based' sampling (e.g. Farewell, 1979).

Use of the identity link for dichotomous responses has been advocated in epidemiology, motivated by a particular notion of causality (see Skrondal (2003) and the references therein). This illustrates that there are sometimes reasons for departing from the canonical links of the exponential family.

2.2.4 Variance function and choice of distribution

The choice of distribution depends on the type of response variable, the process that may have generated the response variable, and the shape of the empirical distribution. For binary responses, the obvious choice is the Bernoulli distribution. Counts can be shown to have a Poisson distribution if the process generating the events has certain characteristics (constant incidence rate and independence).

The choice of distribution also determines the conditional variance of the responses as a function of the mean. It follows from standard likelihood theory that

$$-\mathrm{E}\left(\frac{\partial^2 \ell}{\partial \theta_i^2}\right) = \mathrm{E}\left(\frac{\partial \ell}{\partial \theta_i}\frac{\partial \ell}{\partial \theta_i}\right),$$

and substitution of terms from (2.5) and (2.6) gives

$$\mathrm{Var}(y_i|\nu_i) = \phi b''(\theta_i) = \phi V(\mu_i),$$

where $V(\mu_i)$ is known as the *variance function* and ϕ is the dispersion parameter. For example, the variance equals the mean for the Poisson distribution whereas the variance is given by a constant parameter $\phi = \sigma^2$ for the normal distribution (see Table 2.2).

2.3 Extensions of generalized linear models

2.3.1 Modeling underdispersion and overdispersion

Counts are typically modeled by the binomial or Poisson distribution (the binomial if the event can occur only at a predetermined number of 'trials', n). For both distributions the conditional variance, given the explanatory variables, is determined by the mean. However, the conditional variance observed in practice is often larger or smaller than that implied by the model, phenomena known as *overdispersion* or *underdispersion*, respectively. Overdispersion could be due to variability in the binomial probabilities or Poisson rates not fully accounted for by the included covariates and is more common than underdispersion.

An ad hoc solution to the problems of overdispersion and underdispersion is to introduce an extra proportionality parameter ϕ^* for the variance, giving

$$\mathrm{Var}(y_i|\nu_i) \;=\; \phi^*\mu_i(1-\mu_i)/n$$

for the binomial and

$$\mathrm{Var}(y_i|\nu_i) \;=\; \phi^*\mu_i$$

for the Poisson distribution. Note that these specifications need not correspond to probability models but can nevertheless be estimated using so-called *quasi-likelihood* methods (see Section 6.8).

An alternative 'proper' modeling approach to overdispersion is to allow the mean to vary randomly between units for fixed covariate values. Combining the binomial response distribution with a beta distribution for the probabilities gives the beta-binomial distribution. The negative binomial distribution results from combining the Poisson response distribution with a gamma distribution for the rate. Another possibility is to include a normally distributed random intercept in the linear predictor, a special case of the general model framework to be presented in Chapter 4. Lindsey (1999, p.197-220) discusses these and further methods for modeling overdispersion and underdispersion.

2.3.2 Modeling heteroscedasticity

A classical assumption in linear regression models is homoscedasticity, i.e. the residual standard deviation σ is assumed to be constant over units. However, σ may depend on categorical or continuous covariates. For example, when comparing the heights of boys and girls aged 11, we would expect the girls' heights to be more variable because many of the girls would have entered puberty while (nearly) all the boys would be prepubertal. Since the standard deviation must be positive, it is convenient to model heteroscedasticity using a log link,

$$\ln \sigma_i \;=\; \mathbf{x}_i' \boldsymbol{\iota} \quad \text{or} \quad \sigma_i \;=\; \exp(\mathbf{x}_i' \boldsymbol{\iota}), \tag{2.7}$$

where $\mathbf{x}_i$ are covariates and $\boldsymbol{\iota}$ parameters. Such 'multiplicative heteroscedasticity' was suggested by Harvey (1976). This specification can also be used for other models with scale or dispersion parameters; see also page 31.

2.3.3 Models for polytomous responses

Polytomous responses are unordered categorical responses such as political party voted for in an election. In econometrics such responses are often referred to as *discrete choices*. Terms used in statistics include qualitative or nominal responses since the categories are not quantitative and have no inherent ordering. Other terms include quantal, polychotomous or multinomial responses.

A separate linear predictor is specified for each category a_s, $s = 1, \ldots, S$. In this respect, the response can be viewed as multivariate and is sometimes represented as a vector having a one for the realized category and zeros for the other categories (e.g. Fahrmeir and Tutz, 2001). The probability of the sth category or alternative a_s is typically modeled as a *multinomial logit*

$$\Pr(y_i = a_s) = \frac{\exp(\nu_i^s)}{\sum_{t=1}^{S} \exp(\nu_i^t)}, \qquad (2.8)$$

where ν_i^s is the linear predictor for unit i and category a_s and the sum is over all S categories. Thissen and Steinberg (1986) refer to models of this type as *divide-by-total models*. Note that $\Pr(y_i = a_s)$ is a conditional probability given the linear predictors, although we have suppressed the conditioning here and in the remainder of the chapter for simplicity.

We can include unit and category-specific covariates or attributes in the linear predictor. For instance, consider the case where the response categories are supermarkets and a customer's response represents his choice of super-market a_s. The linear predictors could include customer specific variables $\mathbf{x}_i$ such as income as well as customer and supermarket specific variables $\mathbf{x}_i^s$ such as travelling time to the supermarket. The linear predictor becomes

$$\nu_i^s = m^s + \mathbf{x}_i' \boldsymbol{\beta}^s + \mathbf{x}_i^{s'} \boldsymbol{\beta}, \qquad (2.9)$$

where m^s is a category-specific constant, $\boldsymbol{\beta}^s$ are category-specific effects of unit-specific covariates $\mathbf{x}_i$ and $\boldsymbol{\beta}$ are constant effects of unit and category-specific covariates $\mathbf{x}_i^s$. The coefficients of $\mathbf{x}_i^s$ could also differ between categories if for example the effect of travelling time is greater for small supermarkets than for large ones.

Note that adding a term B_i to the linear predictors for unit i, ν_i^s, $s = 1, \ldots, S$ does not change the probability in (2.8) since it amounts to multiply-ing both numerator and denominator by $\exp(B_i)$. For this reason, we could add constants to m^s and $\boldsymbol{\beta}^s$ without changing the model. This is an example of an *identification problem* (see also Chapter 5). The problem can be over-come by taking one category as 'base category' (typically the first, a_1) and imposing $m^1 = 0$ and $\boldsymbol{\beta}^1 = \mathbf{0}$.

The *conditional logit model*, standard in econometrics, arises as the special case where there are no unit-specific covariates $\mathbf{x}_i$ and no constants m^s (e.g. McFadden, 1973). The *polytomous logistic regression model*, a standard model in for instance biostatistics (e.g. Hosmer and Lemeshow, 2000), results as the special case where there are no category-specific covariates $\mathbf{x}_i^s$.

2.3.4 Models for ordinal responses

Cumulative models

Models for ordered categorical or ordinal responses can be defined by linking the cumulative probability $\Pr(y_i \leq a_s)$ to the linear predictor (e.g. McCullagh, 1980),

$$g[\Pr(y_i \leq a_s)] = \kappa_s - \nu_i, \qquad s = 1, \ldots, S-1 \qquad (2.10)$$

where $a_1 < a_2 < \ldots < a_S$ are ordered response categories, $\Pr(y_i \leq a_S) = 1$ and κ_s are *threshold parameters*, $\kappa_1 < \kappa_2 < \ldots < \kappa_{S-1}$. Typical choices of link function include the probit, logit and complementary log-log. Such models are often called *cumulative models* or *graded response models* (Samejima, 1969).

Consider the right-hand side of (2.10), $\kappa_s - \nu_i$. Since adding an arbitrary constant β_0 to the linear predictor can be counteracted by adding the same constant to each κ_s, it is clear that we cannot simultaneously estimate the constant and all thresholds. This identification problem can be overcome by setting $\kappa_1 = 0$, making the parametrization identical to that used for dichotomous response models in the previous section (if $S = 2$, $a_1 = 0$ and $a_2 = 1$). Alternatively, κ_1 could be a model parameter if we instead omit the constant from the linear predictor.

It follows from (2.10) that the probability of a particular response y_s becomes

$$\Pr(y_i = a_s) = \Pr(y_i \leq a_s) - \Pr(y_i \leq a_{s-1}), \qquad (2.11)$$

prompting Thissen and Steinberg (1986) to refer to these models as *difference models*.

The effects of the covariates on the cumulative response probabilities in (2.10) are constant across categories s, a feature called the parallel regression assumption. With a logit link, the odds of y exceeding a_s become

$$\frac{\Pr(y_i > a_s)}{\Pr(y_i \leq a_s)} = \frac{1 - \Pr(y_i \leq a_s)}{\Pr(y_i \leq a_s)} = \exp(\mathbf{x}'_i\boldsymbol{\beta} - \kappa_s).$$

The ratio of these odds for two units i and i', $\exp[(\mathbf{x}_i - \mathbf{x}_{i'})'\boldsymbol{\beta}]$, is the same for all s, a property known as *proportional odds*. A useful feature of cumulative models is that the estimated regression parameters are approximately invariant to merging of the categories.

The assumption of constant effects of the covariates across response categories can be relaxed by allowing the thresholds to depend on covariates $\mathbf{x}_{2i}$ (e.g. Terza, 1985)

$$\kappa_{is} = \mathbf{x}'_{2i}\boldsymbol{\varsigma}_s,$$

where $\boldsymbol{\varsigma}_s$ is a parameter vector. Considering the probit version, the model then becomes

$$P(y_i \leq a_s) = \Phi(\kappa_{si} - \nu_i) = \Phi(\mathbf{x}'_{2i}\boldsymbol{\varsigma}_s - \mathbf{x}'_i\boldsymbol{\beta}), \qquad (2.12)$$

where $\mathbf{x}_{2i}$ are covariates with category-specific effects and $\mathbf{x}_i$ are covariates with constant effects. It is clear that the coefficients of any variables included in both $\mathbf{x}_{2i}$ and $\mathbf{x}_i$ are not separately identified. A problem with this parametrization is that for some covariate values the thresholds will not necessarily satisfy the order constraint $\kappa_{i1} < \kappa_{i2} < \ldots < \kappa_{iS-1}$, so the probabilities in (2.11) are not constrained to be nonnegative. The order constraint can be imposed by using the parametrization

$$\kappa_{i1} = 0, \quad \kappa_{is} = \kappa_{is-1} + \exp(\mathbf{x}'_{2i}\boldsymbol{\varsigma}_s), \quad s = 2, \ldots, S,$$

see for example Fahrmeir and Tutz (2001).

An alternative device for relaxing the parallel regression assumption is to use a *scaled ordinal probit link* in which the scale parameter is modeled as in (2.7)

$$\ln \sigma_i = \mathbf{x}'_{2i} \boldsymbol{\iota},$$

where $\boldsymbol{\iota}$ is a parameter vector. The model then becomes

$$P(y_i \leq a_s) = \Phi[(\kappa_s - \nu_i)/\sigma_i] = \Phi(\kappa^*_{si} - \nu^*_i), \qquad (2.13)$$

where $\kappa^*_{si} = \kappa_s/\sigma_i$ and $\nu^*_i = \nu_i/\sigma_i$. Such a heteroscedastic probit model is discussed for dichotomous responses in Greene (2003, p.680) and was used for ordinal responses by Skrondal (1996). The scaled thresholds have the form

$$\kappa^*_{si} = \kappa_s \exp(-\mathbf{x}'_{2i} \boldsymbol{\iota})$$

and the scaled linear predictor has the form

$$\nu^*_i = \mathbf{x}'_i \boldsymbol{\beta} \exp(-\mathbf{x}'_{2i} \boldsymbol{\iota}).$$

For identification, the intercept in the model for σ is omitted since adding a constant ι_0 to $\mathbf{x}'_{2i} \boldsymbol{\iota}$ can be counteracted by dividing κ_s and $\boldsymbol{\beta}$ by $\exp(-\iota_0)$. If $\mathbf{x}_i$ and $\mathbf{x}_{2i}$ overlap, the model may be identified due to different functional forms specified for $\mathbf{x}_i$ and $\mathbf{x}_{2i}$, but the identification is likely to be fragile.

Stereotype and adjacent category logit models

Instead of modeling the cumulative probabilities, we can specify models for the probabilities $\Pr(y_i = a_s)$ using the multinomial logit model in (2.8) and (2.9) with additional parameter constraints reflecting the ordering of the categories.

The *stereotype model* (Anderson, 1984) is obtained by imposing the constraints $\boldsymbol{\beta}^s = \alpha^s \boldsymbol{\beta}$ and $\alpha^1 < \ldots < \alpha^S$ so that the linear predictor becomes

$$\nu^s_i = m^s + \alpha^s (\mathbf{x}'_i \boldsymbol{\beta}).$$

In addition to the usual constraints $m^1 = 0$ and $\alpha^1 = 0$, a further constraint such as $\alpha^2 = 1$ is necessary to set the scale of α^s since only the scale of the product $\alpha^s \boldsymbol{\beta}$ is identified. Note that the stereotype model does not include category-specific covariates.

Setting $\alpha^s = s$ in the stereotype model gives the *adjacent category logit model*. In this model the odds for adjacent categories become

$$\frac{\Pr(y_i = a_s)}{\Pr(y_i = a_{s-1})} = \exp(m^s - m^{s-1} + \mathbf{x}'_i \boldsymbol{\beta}),$$

so that the odds ratio of adjacent categories for two individuals i and i' is

$$\exp[(\mathbf{x}_i - \mathbf{x}_{i'})' \boldsymbol{\beta}],$$

which is the same for all s. Note that the adjacent category logit model assumes proportionality of the adjacent category odds whereas the proportional odds model assumes proportionality of the cumulative odds. The linear predictor of the adjacent category logit,

$$\nu^s_i = m^s + s\mathbf{x}'_i \boldsymbol{\beta},$$

has the same form as that of the standard multinomial logit model with categegory specific covariates $\mathbf{x}_i^s = s\mathbf{x}_i$. Goodman (1983) referred to this model as the parallel odds model, whereas the model is known as the *partial credit model* (Masters, 1982) in item response theory (see Section 3.3.4).

Instead of assigning equally spaced scores to α^s, we may specify scores reflecting the 'distances' between the ordered categories. The stereotype model can be thought of as a generalization of this model where the scores are estimated instead of fixed. Models in which scores are assigned to both the α^s and to an ordered explanatory variable correspond to the linear by linear association model (e.g. Goodman, 1979; Agresti, 2002).

Continuation ratio logit model

Another possibility is to assume proportionality of the odds of exceeding category a_s given that y_i is at least equal to a_s ($y_i \geq a_s$),

$$\frac{\Pr(y_i > a_s)}{\Pr(y_i = a_s)} = \exp(m^s + \mathbf{x}_i'\boldsymbol{\beta}),$$

interpretable as the odds of continuing beyond 'stage' a_s versus stopping at that stage. An equivalent model is not obtained from reversing the ordering of the categories, suggesting that the model should only be used for sequential stages. Since this *continuation ratio logit model* is often used for discrete-time durations (sequential stages), we return to it in Section 2.5.2.

2.3.5 Composite links

Composite links are of the form

$$\mu_i = \sum_k c_{ik}\, g_k^{-1}(\nu_{ik}), \tag{2.14}$$

where c_{ik} are known constants. Such link functions are useful for modeling count data where some observed counts represent the sums of counts for different covariate values, typically due to missing covariate information.

A famous example is the blood-type problem where the offspring inherits blood-type (phenotype) A if the mother and father contribute genes A and O in any of the combinations AA, AO or OA. If the gene frequencies for A and O are p and r, respectively, the expected frequency of blood-type A is

$$N(p^2 + 2pr) = \exp(\ln N + 2\ln p) + \exp(\ln N + \ln p + \ln r) + \exp(\ln N + \ln p + \ln r),$$

where the three terms on the right-hand side correspond to the expected frequencies for genotypes AA, AO and OA. In this example, i in (2.14) indexes the observed phenotypes whereas k indexes the unobserved genotypes. The coefficients c_{ik} are 1 for all genotypes k consistent with phenotype i and 0 otherwise, g^{-1} is the exponential function, the log of the total number of units $\ln N$ is used as an offset (a covariate with coefficient set to 1) and $\ln p$ and $\ln r$ are model parameters.

Another example of a composite link is the probability of a response category in cumulative models for ordinal responses, which is equal to a difference between cumulative probabilities (see equation (2.11)). In Section 9.4 we will use a composite link to specify a three-parameter item response model. We refer to Thompson and Baker (1981), Rindskopf (1992) and Skrondal and Rabe-Hesketh (2004) for many applications of composite links.

2.4 Latent response formulation

A response variable y_i can often be viewed as a partial observation or coarsening of a continuous *latent response* y_i^* (e.g. Pearson, 1901). Let the latent response be modeled as

$$y_i^* = \nu_i + \epsilon_i,$$

where ν_i is a linear predictor and ϵ_i an error term or disturbance. For continuous responses the latent response simply equals the observed response; $y_i = y_i^*$. Other response types arise when the latent response is coarsened by applying different kinds of threshold functions described in the following subsections.

2.4.1 Grouped, interval-censored, ordinal and dichotomous responses

The observed response y_i takes on one of S response categories a_s, $s = 1, \ldots, S$, and the relationship between observed and latent response can be written as

$$y_i = \begin{cases} a_1 & \text{if} & \kappa_{i0} < y_i^* \leq \kappa_{i1} \\ a_2 & \text{if} & \kappa_{i1} < y_i^* \leq \kappa_{i2} \\ \vdots & \vdots & \vdots \\ a_S & \text{if} & \kappa_{iS-1} < y_i^* \leq \kappa_{iS}, \end{cases} \tag{2.15}$$

where $\kappa_{i0} = -\infty$ and $\kappa_{iS} = \infty$. For $S = 3$ this is illustrated in Figure 2.1 for normally distributed ϵ_i.

For *grouped responses* the thresholds do not vary between units, $\kappa_{is} = \kappa_s$, and are known a priori. An example of grouped data are salaries grouped into prespecified income brackets with boundaries κ_s, a situation considered by Stewart (1983).

For *interval-censored responses* the κ_{is} vary between units and are known a priori. For example, time of onset of an illness may not be known exactly but only to lie within a censoring interval between two clinic visits, with the timing of visits varying between individuals.

For *ordinal responses* the thresholds κ_s are unknown parameters and usually do not vary between units. For example, severity of pain may be described as 'none', 'moderate' or 'severe'. These outcomes may literally be considered as resulting from pain severity, an unobserved continuous latent response, exceeding certain thresholds. Sometimes we can relax the assumption of constant thresholds to model individual differences in pain tolerance.

Dichotomous responses can often be viewed as ordinal responses with 2

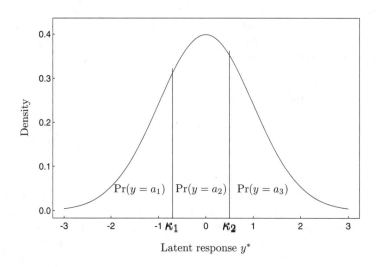

Figure 2.1 *Latent response formulation for grouped, interval-censored or ordinal responses with $S = 3$*

categories, but are sometimes better interpreted as comparative responses (see Section 2.4.3).

If a normal distribution is assumed for the error term, the model is the dichotomous probit model in the case of two categories (Bliss, 1934) and the ordinal probit model for more than two categories (see Section 2.3.4). The latent response formulation of the ordinal probit model was suggested by Aitchison and Silvey (1957) and rediscovered by McKelvey and Zavoina (1975). If the error term is assumed to have a logistic cumulative distribution function,

$$F(\tau) \equiv \Pr(\epsilon_i \leq \tau) = \frac{\exp(\tau)}{1 + \exp(\tau)},$$

the dichotomous logit model is obtained in the case of two categories and the ordinal logit model, proportional odds model or graded response model in the case of more than two categories (see Section 2.3.4).

To show the correspondence between the latent response and generalized linear model formulations, we consider the dichotomous case ($S = 2$, $a_1 = 0$, $a_2 = 1$, $\kappa_1 = 0$). The latent response model is

$$y_i^* = \nu_i + \epsilon_i, \qquad y_i = \begin{cases} 1 & \text{if } y_i^* > 0 \\ 0 & \text{otherwise.} \end{cases}$$

It follows that

$$
\begin{aligned}
\mathrm{E}(y_i|\nu_i) &= \Pr(y_i = 1|\nu_i) = \Pr(y_i^* > 0|\nu_i) = \Pr(\nu_i + \epsilon_i > 0) \\
&= \Pr(-\epsilon_i \le \nu_i) = F(\nu_i).
\end{aligned}
$$

Here $F = g^{-1}$ is the cumulative distribution function; the standard normal for the probit, the logistic for the logit, and the (asymmetric) Gumbel for the complementary log-log link.

2.4.2 Censored responses

The threshold model for doubly censored responses can be written as

$$
y_i = \begin{cases}
\kappa_{i1} & \text{if} & y_i^* \le \kappa_{i1} \\
y_i^* & \text{if} & \kappa_{i1} < y_i^* \le \kappa_{i2} \\
\kappa_{i2} & \text{if} & \kappa_{i2} < y_i^*.
\end{cases}
\tag{2.16}
$$

For the special cases of right-censored responses, $\kappa_{i1} = -\infty$, and for left-censored responses, $\kappa_{i2} = \infty$.

Different types of censored responses are prominent in duration or survival analysis. Right-censoring is typically due to the event not having occurred by the end of the observation period. Left-censoring occurs if all we know is that the event had already happened before observation began. If both types of censoring can occur, the responses are doubly censored. Other examples of censoring are ceiling and floor effects. For example, when measuring ability using the percentage of correctly solved problems we cannot differentiate between candidates achieving 100%. All we know is that their ability is greater than or equal to that required to achieve the maximum score (ceiling effect). Analogously, a floor effect occurs if some candidates cannot solve any problems. For normal latent responses (conditional on covariates), one-sided censoring was introduced by Tobin (1958) and is hence denoted the *Tobit*. The analogous model for double censoring is due to Rosett and Nelson (1975) and is often denoted the *two-limit probit*.

Censoring should not be confused with *truncation*. Left-truncation occurs if units with a response below a certain threshold are excluded from the sample. An example is a clinical trial for treatment of hypertension where baseline blood-pressure must exceed a threshold for inclusion into the study. In the duration literature, left-truncation due to a delay between becoming at risk and being included in a study (so that units with short durations are excluded) is often called late entry or delayed entry. Right-truncation occurs if only those units are included in the sample whose response falls below some threshold. A classical example from econometrics are negative income tax experiments where families with income levels above a certain limit, for instance 1.5 times the poverty line, are excluded from the study of earnings. Unlike censoring, truncation means that we have no information on the explanatory variables of those individuals whose response is beyond the threshold.

The different coarsening processes leading to different variable types are summarized in Table 2.3. The columns specify the type of coarsening and

Table 2.3 *Response types and types of coarsening*

Type of variable	Coarsening				Threshold(s) constant/ varying	known/ unknown
	y_i^*	$y_i^* \leq \kappa_{is-1}$	$y_i^* > \kappa_{is}$	$\kappa_{is-1} < y_i^* \leq \kappa_{is}$		
Continuous	$\checkmark$				-	-
Grouped				$\checkmark$	constant	known
Interval-censored				$\checkmark$	varying	known
Ordinal/dichotom.				$\checkmark$	constant	unknown
Right-censored	$\checkmark$		$\checkmark$		varying	known
Left-censored	$\checkmark$	$\checkmark$			varying	known
Doubly censored	$\checkmark$	$\checkmark$	$\checkmark$		varying	known

the type of threshold. The thresholds can either vary or be constant across units and either be known or unknown model parameters. The resulting types of variable are given in the rows. Apart from continuous and ordinal variables, the variable types are often referred to as *limited dependent variables* in econometrics.

2.4.3 Comparative responses

In Section 2.3.3 we considered models for polytomous responses. Polytomous responses can be construed as *comparative* in the sense that the realized category 'dominates' the others.

This interpretation is particularly apt for *first choice* or *discrete choice* data where decision makers choose from a set of alternatives (categories). For instance, in election studies a central outcome variable is the first choice of a voter, say Conservatives, among a set of alternatives (say Labour, Conservatives and Liberals).

Another type of comparative response is *pairwise comparisons* where the responses are the dominant categories in each pair of categories for a unit. For instance, Labour could be preferred to Liberal in the first pair, Liberal preferred to Conservatives in the second pair, etc.

A *permutation* of categories is also a comparative response. In the decision context permutations can be interpreted as rankings, where alternatives are ordered according to preference. Political parties may for instance be ranked, say Liberals preferred to Labour preferred to Conservatives. In contrast to pairwise comparison data, the pairwise comparisons implied by ranking data are necessarily transitive (Liberals preferred to Conservatives follows from Liberals preferred to Labour and Labour preferred to Conservatives). Comparative responses are nominal in the sense that the categories do not possess an inherent ordering shared by all units as is assumed for ordinal variables. We find it useful to use the decision terminology for comparative responses even when decisions are not actually involved.

Comparative responses can be modeled by assuming that each unit assigns a utility u_i^s to each alternative a_s. The term utility should be broadly construed as popularity or attractiveness of alternatives. For polytomous responses, it is assumed that the alternative with the greatest utility is chosen, i.e.

$$y_i = a_s \quad \text{if } u_i^s - u_i^t > 0 \qquad \forall t, \ t \neq s.$$

If we model the utilities as

$$u_i^s = \nu_i^s + \epsilon_i^s,$$

where the linear predictors ν_i^s can take on different values for different alternatives and the ϵ_i^s are independently Gumbel (extreme value Type I) distributed

$$\Pr(\epsilon_i^s \leq \tau) = \exp[-\exp(-\tau)], \tag{2.17}$$

it can be shown (McFadden, 1973; Yellott, 1977) that the probability of a particular choice a_s is

$$\Pr(y_i = a_s) = \frac{\exp(\nu_i^s)}{\sum_{t=1}^{S} \exp(\nu_i^t)}. \tag{2.18}$$

This is the multinomial logit model introduced in Section 2.3.3.

The *multinomial probit model* (e.g. Daganzo, 1979) instead assumes that the vector containing the ϵ_i^s has a multivariate normal distribution with variances ω_s^2 and covariances $\omega_{ss'}$. Consider for simplicity the case of 3 alternatives, $S = 3$, and define $\nu_i^{1k} \equiv \nu_i^1 - \nu_i^k$ and $\epsilon_i^{1k} \equiv \epsilon_i^1 - \epsilon_i^k$. The probability of choosing the first alternative a_1 then becomes

$$\Pr(y_i = a_1) = \int_{-\infty}^{\nu_i^{12}} \int_{-\infty}^{\nu_i^{13}} \varphi(\epsilon_i^{12}, \epsilon_i^{13}) \, d\epsilon_i^{12} \, d\epsilon_i^{13},$$

where $\varphi(\epsilon_i^{12}, \epsilon_i^{13})$ is bivariate normal with expectation vector zero and covariance matrix

$$\Omega_1 = \begin{bmatrix} \omega_1^2 + \omega_2^2 - 2\omega_{12} \\ \omega_1^2 - \omega_{13} - \omega_{12} + \omega_{23} & \omega_1^2 + \omega_3^2 - 2\omega_{13} \end{bmatrix}.$$

Note that the choice probabilities cannot be expressed in closed form and require integration over $S-1$ dimensions. On the other hand, the multinomial probit allows the utilities to be dependent in contrast to the multinomial logit model. It thus relaxes the so-called independence from irrelevant alternatives (IIA) property of the latter model (see Section 13.2).

For pairwise comparisons, let $y_{ist} = 1$ if unit i prefers alternative s to alternative t and 0 otherwise. Under the Gumbel specification the corresponding probability becomes

$$\Pr(y_{ist} = 1) = \frac{\exp(\nu_i^s - \nu_i^t)}{1 + \exp(\nu_i^s - \nu_i^t)},$$

where ν_i^1 is typically set to zero to ensure identification. This model is often called the Bradley-Terry-Luce model after Bradley and Terry (1952) and Luce (1959). The joint probability of a set of pairwise comparisons for unit i is often assumed to be the product of these probabilities. This unrealistic independence assumption may be relaxed by introducing latent variables.

Turning to rankings, let r_i^ℓ be the alternative given rank ℓ among S alternatives and let $\mathbf{R}_i \equiv (r_i^1, r_i^2, \ldots, r_i^S)$ be the ranking of unit i. The probability of a ranking can be construed in terms of a *utility ordering* and expressed in terms of $S-1$ binary utility comparisons, where the utility of the alternative ranked first is larger than the utility of that ranked second, which is larger than that ranked third and so on.

For Gumbel distributed random utilities this leads to the logistic model for rankings (e.g. Luce, 1959; Plackett, 1975)

$$\Pr(\mathbf{R}_i) \; = \; \prod_{\ell=1}^{S-1} \frac{\exp(\nu_i^{r_i^\ell})}{\sum_{m=\ell}^{S} \exp(\nu_i^{r_i^m})}. \tag{2.19}$$

The model is often denoted the *exploded logit* (Chapman and Staelin, 1982) since the ranking probability is written as a product of first choice probabilities for successively remaining alternatives. That such an explosion results was proven by Luce and Suppes (1965) and Beggs *et al.* (1981). The latent response perspective reveals that the exploded logit can be derived without making the behavioral assumption that the choice process is sequential. Importantly, an analogous explosion is not obtained under normally distributed utilities. The Gumbel model is not reversible, that is, successive choices starting with the worst alternative would lead to a different ranking probability. Another essential feature of the model is independence from irrelevant alternatives.

Different alternative sets for different units, for instance different eligible parties in different constituencies, can simply be handled by substituting S_i for S in the above formulae. *Partial rankings* result when unit i only ranks a subset of the full set of alternatives, for example when experimental designs are used in presenting specific subsets of alternatives to different units (e.g. Durbin, 1951; Böckenholt, 1992). Such designs are easily handled by letting the alternative sets vary over units. Another kind of partial ranking is *top-rankings* where not all alternatives are ranked but only the subset of the $P_i < S_i$ most preferred alternatives. The probability of a top-ranking is simply the product of the first P_i terms in equation (2.19). Note that the first choice probability is obtained as the special case of the ranking probability when $P_i = 1$ for all i.

Two or more alternatives are said to be *tied* if they are given the same rank. Although the probability of tied rankings is theoretically zero since the utilities are continuous, ties are often observed in practice. As we will see in Section 2.5, equation (2.19) has the same form as the partial likelihood of a stratum in Cox regression for survival analysis. Exploiting this duality, we can utilize methods for handling ties previously suggested in the survival literature (see Section 2.5).

Assembling comparative response data is the natural approach when studying choice behavior. Comparative responses can also fruitfully replace rating or thermometer scales in studying the popularity of objects. Use of scales would invoke the unrealistic assumption that individuals use the scale in the same way (e.g. Brady, 1989, 1990). However, some subjects tend to use the high end of the scale whereas others use the low end. In addition there could be differences in the range of ratings used. Alwin and Krosnick (1985) discuss pros and cons of ranking and rating designs for the measurement of values.

2.5 Modeling durations or survival

The response of interest may be time to some event. In medicine, the archetypal example is *survival* time from the onset of a condition or treatment to death. In studies of the reliability of products or components, for instance light bulbs, lifetime to *failure* is often investigated. Instead of using such application specific terms, economists usually refer to *durations* between events. Generally, we will adhere to this terminology but occasionally we lapse by referring to survival or failure times.

There are some important distinguishing features of duration data. Durations are always nonnegative and some durations are typically not known because the event has not occurred before the end of the observation period (right-censoring). Furthermore, the values of covariates may change as time elapses and the effect of covariates may change over time. These features imply that one cannot simply apply standard models for continuous responses.

Another common phenomenon in survival analysis is left-truncation or delayed entry where units that have already experienced the event of interest (e.g. death) when observation begins are excluded from the study (see e.g. Keiding, 1992). Note that this phenomenon is called 'stock sampling with follow up' in econometrics since only those in the 'alive state' at a given time are sampled (e.g. Lancaster, 1990). Left-censoring and right-truncation are less common. Left-censoring occurs if the event is only known to have occurred before a given time-point, for instance when observation begins. Right-truncation can occur in retrospective studies, for example when studying the incubation period for AIDS in patients who have already developed the disease.

In this section we assume that censoring and truncation processes are ignorable in the sense that the probability of being censored or truncated is independent of the risk of experiencing the event given the covariates. We will also confine the discussion to so called *absorbing events* or states where the unit can only experience the event once. Treatment of multiple events, such as recurring headaches, is deferred to Chapter 12 since the dependence among events for the same unit must be accommodated. For simplicity, we only consider single absorbing events and will not explicitly treat competing risk models where there are several absorbing states.

Duration models are usually not defined as generalized linear models. However, it turns out that generalized linear models can often be adapted to yield likelihoods that are proportional to those of duration models.

Durations are either considered in continuous or discrete time, and these cases will be discussed in the two subsequent sections.

2.5.1 Continuous time durations

Let T_i be a random variable representing the duration for a unit i from becoming at risk until it experiences the event. The realized duration is denoted t_i, with a corresponding indicator variable δ_i taking the value 1 if the event is experienced by the unit and 0 if the duration is censored.

The density function for unit i is denoted $f_i(t)$ and the cumulative distribution function becomes

$$F_i(t) \equiv \int_0^t f_i(u)\mathrm{d}u.$$

The survival function, the probability that duration exceeds t, is then defined as

$$S_i(t) \equiv 1 - F_i(t).$$

The *hazard*, sometimes also called the *incidence rate* or *instantaneous risk*, is defined as

$$h_i(t) \equiv \lim_{\Delta \to 0} \left\{ \frac{\Pr(t \le T_i < t + \Delta | T_i \ge t)}{\Delta} \right\}.$$

Somewhat loosely, this is the 'risk' of an event at time t for unit i given that the event has not yet occurred and that the unit is therefore still 'at risk'.

It follows from these definitions that

$$h_i(t) = \frac{f_i(t)}{S_i(t)} = -\frac{\partial \ln S_i(t)}{\partial t}. \tag{2.20}$$

Since $S_i(0) = 1$, it also follows that

$$S_i(t) = \exp[-\int_0^t h_i(u)\mathrm{d}u].$$

Defining the *integrated hazard* or *cumulative hazard* as

$$H_i(t) \equiv \int_0^t h_i(u)\mathrm{d}u,$$

unit i's contribution to the likelihood is

$$l_i = S_i(t_i)h_i(t_i)^{\delta_i} = \exp[-H_i(t_i)]h_i(t_i)^{\delta_i}. \tag{2.21}$$

Accelerated failure time models

One approach to specifying a duration model is as

$$\ln T_i = \beta_0 + \nu_i + \epsilon_i,$$

where ν_i does not include a constant, or alternatively as

$$T_i = \exp(\beta_0)\exp(\nu_i)\exp(\epsilon_i).$$

We see that the covariates act multiplicatively directly on the time scale, thus accelerating or decelerating durations. The survival function of such accelerated failure time models can therefore be written as

$$S_i(t) = S_0\left(\exp(\nu_i)t\right),$$

where $S_0(t)$ is the baseline survival function for $\mathbf{x}_i = \mathbf{0}$. Examples of accelerated failure time models include the *log-normal duration model* where ϵ_i is normally distributed, the *log-logistic duration model* with a logistic ϵ_i, and the *Weibull duration model* if ϵ_i has a Gumbel distribution, see (2.17) in the previous section.

Proportional hazards models

The hazards can be modeled as

$$h_i(t) = h^0(t)\exp(\nu_i), \qquad (2.22)$$

where $h^0(t)$ is the 'baseline' hazard, the hazard when all covariates are zero (the linear predictor does not include a constant). Considering two units i and i', we obtain

$$\frac{h_i(t)}{h_{i'}(t)} = \exp(\nu_i - \nu_{i'}), \qquad (2.23)$$

the hazard functions of any two units are proportional over time.

The Weibull duration model introduced above is unique in that it possesses both the proportional hazards and accelerated failure time properties (e.g. Cox and Oakes, 1984). The *exponential duration model* is the special case of the Weibull model where the baseline hazard is constant $h^0(t) = h^0$. This property is relaxed in the *piecewise exponential duration model* where the baseline hazard function is assumed to be piecewise constant over intervals s, with $h^0(t) = h_s$ for $\tau_{s-1} \leq t < \tau_s$, $s = 1, 2, \ldots, S$.

Interestingly, it turns out that the likelihood of the piecewise exponential duration model is proportional to that of a Poisson model when the data are expanded appropriately (Holford, 1980). Let $\theta_i = \exp(\nu_i)$ so that unit i has hazard $h_s\theta_i$. For a unit that was censored or failed in the sth interval its contribution to the likelihood becomes (see (2.21))

$$l_i = \underbrace{(h_s\theta_i)^{\delta_i}}_{h_i(t_i)}\exp(-\underbrace{\sum_{r=1}^{s} h_r\theta_i d_{ir}}_{H_i(t_i)}), \qquad (2.24)$$

where d_{ir} is the time unit i spent in interval r, $d_{ir} = \min(t_i, \tau_r) - \tau_{r-1}$. This can be rewritten as

$$l_i = \prod_{r=1}^{s}(h_r\theta_i)^{y_{ir}}\exp(-h_r\theta_i d_{ir}), \qquad (2.25)$$

where $y_{ir} = 0$ for $r < s$ and $y_{is} = \delta_i$. This is proportional to the contribution to the likelihood of s independent Poisson variates y_{ir} with means $h_r \theta_i d_{ir}$, see (2.3). See also Clayton (1988).

We represent each unit by a number of observations (or 'risk sets') equal to the number of time intervals preceding or including that unit's failure (or censoring) time as shown in the upper panel of Display 2.1 on page 43. The model may then be fitted by Poisson regression with a log link, using y_{ir} as response variable, $\ln(d_{ir})$ as an offset (a covariate with regression coefficient set to 1) and dummies for the time intervals as explanatory variables to allow for different piecewise constant hazards h_r. Explicitly,

$$y_{ir} \sim \text{Poisson}(\mu_{ir}),$$

where

$$\ln(\mu_{ir}) = \nu_{ir}$$

and

$$\nu_{ir} = \ln(d_{ir}) + \ln(h_r) + \mathbf{x}_i' \boldsymbol{\beta}.$$

Therefore, one approach to survival modeling is to divide the follow-up period into intervals over which the hazard can be assumed to be piecewise constant and use Poisson regression. Breslow and Day (1987, p.137) show how Poisson regression can be implemented for identity and power links. Assuming a piecewise linear log hazard corresponds to a piecewise Gompertz distribution (e.g. Lillard, 1993).

Another approach is to define as many intervals as there are unique failure times with each interval starting at (just after) a failure time and ending at (just after) the next failure time (Holford, 1976; Whitehead, 1980; Laird and Olivier, 1981). This corresponds to the famous *Cox proportional hazards model* since a 'saturated' or nonparametric model with a separate constant for each risk set is used for the baseline hazard.

The data expansion necessary to estimate the Cox model using Poisson regression with log link is shown for an artificial dataset in the lower panel of Display 2.1. It can be shown that the corresponding likelihood is proportional to the *partial likelihood* of Cox regression

$$l_P = \prod_r \frac{\exp(\nu_{(r)})}{\sum_{i \in R(t_{(r)})} \exp(\nu_i)},$$

where $\nu_{(r)}$ is the linear predictor for the unit that failed at the rth ordered failure time $t_{(r)}$ and $R(t_{(r)})$ is the risk set for the rth failure time (see Display 2.1 where the unit that fails and contributes to the numerator is enclosed in a box). This partial likelihood can be derived by eliminating the baseline hazard using a profile likelihood approach (Johansen, 1983). Note that the partial likelihood is equivalent to the conditional likelihood of logistic regression with pair-specific intercepts that is widely used for matched case-control studies (e.g. Breslow and Day, 1980).

Instead of allowing the baseline hazard to take on a different value for each failure as in Cox regression, the baseline hazard could be modeled as a

Display 2.1 Continuous durations: original and expanded data

Piecewise exponential model:

Poisson regression

Original data				i	r	τ_r	d_{ir}	y_{ir}
i	t_i	δ_i		1	1	1.5	1.5	0
			$\tau_0=0,\ \tau_1=1.5,\ \tau_2=5,\ \tau_3=8$	1	2	5	3.5	0
1	5.2	0		1	3	8	0.2	0
2	4.5	1		2	1	1.5	1.5	0
				2	2	5	3	1

Cox proportional hazards model:

		Poisson regression					Risk sets	
		i	r	$t_{(r)}$	d_{ir}	y_{ir}	r	$t_{(r)}$	$i \in R(t_{(r)})$

Original data			i	r	$t_{(r)}$	d_{ir}	y_{ir}	r	$t_{(r)}$	$i \in R(t_{(r)})$
i	t_i	δ_i	1	1	2.8	2.8	0	1	2.8	1
			1	2	4.2	1.4	0	1	2.8	2
1	5.6	0	1	3	5.1	0.9	0	1	2.8	3
2	4.2	1	2	1	2.8	2.8	0	1	2.8	4
3	2.8	1	2	2	4.2	1.4	1	2	4.2	1
4	5.1	1	3	1	2.8	2.8	1	2	4.2	2
			4	1	2.8	2.8	0	2	4.2	4
			4	2	4.2	1.4	0	3	5.1	1
			4	3	5.1	0.9	1	3	5.1	4

smooth function of time using the Poisson regression approach. A polynomial in time could be used for the log hazard, or alternatively splines or fractional polynomials. The estimated function can be interpreted as the baseline hazard function only if $\ln(d_{ir})$ is included as an offset, a point ignored in Goldstein (2003, Chapter 10). Note that although the hazard is modeled as a smooth function of the failure times $t_{(r)}$, it is constant between adjacent failure times.

Tied or identical durations often arise in practice although inconsistent with continuous time duration models. The appropriate way of handling ties is to sum the likelihood contributions for all possible permutations of the tied durations (e.g. Kalbfleisch and Prentice, 2002), but this can become very involved. Thus, approximate methods have been suggested, the most commonly used being the Peto-Breslow method (Peto, 1972; Breslow, 1974). This method amounts to assuming that all tied units are still at risk when any of the units fail. The Peto-Breslow method appears to work well as long as the number of ties is moderate (Farewell and Prentice, 1980). A better approximation has been proposed by Efron (1977), where the contribution of the tied units in the denominator is successively downweighted to reflect the decreasing risk sets. When there are many ties, it may be more appropriate to treat durations as discrete rather than continuous.

2.5.2 Discrete time durations

Discrete time duration data most commonly arise from interval-censoring of processes in continuous time. Another source are discrete time processes where events can only occur at discrete time points, for instance durations of party loyalty in terms of number of elections. In either case, it can be useful to model discrete durations as interval-censored. Let τ_s be the censoring limits so that all we know is that

$$\tau_{s-1} \leq T_i < \tau_s.$$

In addition, the discrete survival time may be right-censored,

$$\tau_{s-1} < T_i.$$

Proportional odds model

The proportional odds model introduced in Section 2.3.4 for ordinal responses can also be used for modeling discrete time durations. The probability that T_i is less than τ_s becomes

$$\Pr(T_i < \tau_s) = \frac{\exp(\mathbf{x}_i'\boldsymbol{\beta} + \kappa_s)}{1 + \exp(\mathbf{x}_i'\boldsymbol{\beta} + \kappa_s)},$$

and the probability that the survival time lies in the kth interval $\tau_{s-1} \leq T_i < \tau_s$ is

$$\Pr(\tau_{s-1} \leq T_i < \tau_s) = \Pr(T_i < \tau_s) - \Pr(T_i < \tau_{s-1}),$$

with $\Pr(T_i < \tau_0) = 0$. It should be noted that the present $\boldsymbol{\beta}$ have opposite signs to the $\boldsymbol{\beta}$ in Section 2.3.4 so that large coefficients imply increased risk. In the absence of right-censoring, this is the likelihood contribution of all observations whose survival times lie in the sth interval. For observations that are censored after the sth interval, the likelihood contribution is $\Pr(T_i \geq \tau_s) = 1 - \Pr(T_i < \tau_s)$. The proportional odds model has been used for discrete survival time data by Bennett (1983), Han and Hausman (1990), Ezzet and Whitehead (1991) and Hedeker *et al.* (2000).

The proportional odds model can also be given a latent response interpretation (see Section 2.4),

$$y_i^* = \mathbf{x}_i'\boldsymbol{\beta} + \epsilon_i \tag{2.26}$$

where ϵ_i has a logistic distribution. If a standard normal distribution is assumed for ϵ_i, the ordinal probit model is obtained. The event occurs in the

sth interval if $\kappa_{s-1} \leq y_i^* < \kappa_s$, i.e.,

$$\Pr(\tau_{s-1} \leq T_i < \tau_s) = \Pr(\kappa_{s-1} \leq y_i^* < \kappa_s).$$

The latent response y_i^* can therefore be thought of as a monotonic transformation of T_i so that $y_i^* = \kappa_s$ corresponds to $T_i = \tau_s$. By constraining the threshold parameters κ_s to be equally spaced so that the transformation from T_i to y_i^* is linear, the appropriateness of a linear regression model for T_i can be assessed.

Models based on the discrete time hazard

The discrete time hazard for the sth interval is defined as the probability that the event occurs in the sth interval given that it has not already occurred,

$$h_i(s) \equiv \Pr(\tau_{s-1} \leq T_i < \tau_s | T_i \geq \tau_{s-1}) = \frac{\Pr(\tau_{s-1} \leq T_i < \tau_s)}{\Pr(T_i \geq \tau_{s-1})}.$$

The likelihood contribution of a unit whose survival time lies in the sth interval is

$$l_i = h_i(s) \prod_{r=1}^{s-1} [1 - h_i(r)] = \prod_{r=1}^{s} h_i(r)^{y_{ir}} [1 - h_i(r)]^{(1-y_{ir})} \quad \text{with } y_{is} = 1. \quad (2.27)$$

Here, y_{ir} is an indicator variable that is equal to 1 if the event occurred in the rth interval and equal to 0 otherwise, i.e. $y_{ir} = 0$ when $r < s$ and $y_{ir} = 1$ when $r = s$. The likelihood contribution of a unit who was censored after the sth interval has the same form,

$$l_i = \prod_{r=1}^{s} [1 - h_i(r)] = \prod_{r=1}^{s} h_i(r)^{y_{ir}} [1 - h_i(r)]^{(1-y_{ir})} \quad \text{with } y_{is} = 0. \quad (2.28)$$

The likelihood contributions of both censored and noncensored observations are just the likelihood contributions of s independent binary responses y_{ir}, $r = 1, \ldots, s$ with Bernoulli probabilities $h_i(r)$. Therefore, for a unit that either fails or is censored in the sth interval we expand the data to s records and construct the indicator variable y_{ir} as shown in Display 2.2 on page 46. Discrete time survival models can then be written as generalized linear models for binary responses.

One possibility is to use logistic regression with a separate constant for each interval,

$$\ln \frac{h_i(r)}{1 - h_i(r)} = \nu_i + \kappa_r. \quad (2.29)$$

This model, proposed by Cox (1972), is often referred to as a proportional odds model. However, whereas proportionality here applies to the conditional odds of the event happening in an interval given that it has not already happened, proportionality in the proportional odds model presented in the previous section applied to the odds of the event happening in a given interval or earlier. The above logistic model for discrete time survival data is equivalent to the continuation ratio logit model introduced in Section 2.3.4 except for the sign

---

Display 2.2 Discrete time durations: original and expanded data

	Original data			Logistic or compl. log-log regression		
i	s	δ_i		i	r	y_{ir}
1	3	0	$\longrightarrow$	1	1	0
2	2	1		1	2	0
3	1	1		1	3	0
4	3	1		2	1	0
				2	2	1
				3	1	1
				4	1	0
				4	2	0
				4	3	1

---

of the linear predictor. Continuation ratio models are useful for sequential processes in which stages (such as educational attainment levels) cannot be skipped and interest focuses on the odds of (not) continuing beyond a stage given that the stage has been reached. See Jenkins (1995) and Singer and Willett (1993) for introductions to this model.

If a Cox proportional hazards model is assumed for the unobserved continuous survival times and the observed discrete survival times are treated as interval-censored, it can be shown that the likelihood contributions are equal to those in (2.27) and (2.28) if a complementary log-log link is used for the discrete time hazard (e.g. Thompson, 1977), i.e.

$$\ln\{-\ln[1 - h_i(r)]\} = \nu_i + \kappa_r. \tag{2.30}$$

2.6 Summary and further reading

We have introduced a wide range of response processes, including continuous, dichotomous, grouped, censored, ordinal and comparative responses, as well as counts and durations in discrete and continuous time. Most of the processes can more or less directly be expressed as generalized linear models, and many as latent response models. The models for the response processes will serve as building blocks for the more general models introduced later. Furthermore, the application chapters 9 to 14 are structured according to type of response process.

Useful books on generalized linear models include McCullagh and Nelder (1989), Aitkin *et al.* (1989, 2004) and Fahrmeir and Tutz (2001).

For categorical responses we recommend Long (1997), Agresti (1996, 2002), Collett (2002) and Andersen (1980). Ordinal responses are discussed in Clogg and Shihadeh (1994) and Johnson and Albert (1999), counts in Cameron and Trivedi (1998) and Winkelmann (2003). Maddala (1983) considers categorical responses as well as limited-dependent responses such as censored and truncated responses.

The following books treat comparative responses: Marden (1995) consider rankings, David (1988) pairwise comparisons and Train (1986, 2003) polytomous responses or discrete choices.

Useful books for modeling durations or 'survival' include (approximately in increasing order of difficulty) Singer and Willett (2003), Allison (1984), Hosmer and Lemeshow (1999), Collett (2003), Breslow and Day (1987), Klein and Moeschberger (2003), Therneau and Grambsch (2000), Vermunt (1997), Cox and Oakes (1984), Kalbfleisch and Prentice (2002), and Andersen *et al.* (1993).

CHAPTER 3

Classical latent variable models

3.1 Introduction

In this chapter we survey classical latent variable models. The point of describing these models is threefold: first, to give an overview of how latent variables are traditionally used in various branches of statistics; second, to familiarize the reader with basic ideas, notation and terminology used later in the book; third, to start unifying different approaches in preparation for the general model framework discussed in Chapter 4.

The latent variable models considered here include

- Multilevel regression models
- Factor models
- Item response models
- Structural equation models
- Latent class models
- Models for longitudinal data

It should be evident that latent variable models are being used for various purposes in different academic disciplines, a point that will be repeatedly illustrated in the application chapters. Moreover, some of the models are used across subject areas whereas others are little known outside specific disciplines.

3.2 Multilevel regression models

Multilevel data arise when units are nested in clusters. Examples include students in classes, patients in hospitals and left and right eyes of individuals. We refer to the elementary units (e.g. students, patients or eyes) as level-1 units and the clusters (e.g. classes, hospitals or heads) as level-2 units. If the clusters are themselves clustered into 'higher level' (super)clusters, for example if students are nested in classes and classes nested in schools, the data have a three-level structure. Important developments in multilevel modeling were initiated in the school setting; see for instance Aitkin *et al.* (1981) and Aitkin and Longford (1986).

The units belonging to the same cluster share the same cluster-specific influences. For example, students in the same class are taught by the same teacher and students in the same school have parents who send them to that school (by choice or due to place of residence). However, we cannot expect to include all cluster-specific influences as covariates in an analysis. This is because we often have limited knowledge regarding relevant covariates and our dataset may furthermore lack information on these covariates. As a result there is cluster-level unobserved heterogeneity leading to dependence between responses for units in the same cluster after conditioning on covariates. This was illustrated in Figure 1.4 on page 10.

In multilevel regression, unobserved heterogeneity is modeled by including random effects in a multiple regression model. There are two types of random effect, random intercepts and random coefficients. Whereas random intercepts represent unobserved heterogeneity in the overall response, random coefficients represent unobserved heterogeneity in the effects of explanatory variables on the response variable.

3.2.1 Two-level random intercept models

Let level-2 units, say schools, be indexed $j = 1, \ldots, J$, and level-1 units, say students, be indexed $i = 1, \ldots, n_j$. Consider a two-level random intercept model with a single student-specific covariate x_{ij}

$$y_{ij} = \eta_{0j} + \beta_1 x_{ij} + \epsilon_{ij}, \qquad (3.1)$$

where η_{0j} are school-specific intercepts, β_1 is a regression coefficient and ϵ_{ij} are level-1 residual terms. The η_{0j} are modeled as

$$\eta_{0j} = \gamma_{00} + \zeta_{0j}, \qquad (3.2)$$

where γ_{00} is the mean intercept and ζ_{0j} is the deviation of the school-specific intercept η_{0j} from the mean. Defining $\theta \equiv \mathrm{Var}(\epsilon_{ij})$ and $\psi \equiv \mathrm{Var}(\zeta_{0j})$, it is typically assumed that the clusters j are independent and

$$\epsilon_{ij} | x_{ij} \sim \mathrm{N}(0, \theta),$$

$$\mathrm{Cov}(\epsilon_{ij}, \epsilon_{i'j}) = 0, \quad i \neq i'$$

$$\zeta_{0j} | x_{ij} \sim \mathrm{N}(0, \psi),$$

$$\mathrm{Cov}(\zeta_{0j}, \epsilon_{ij}) = 0.$$

Note that the first assumption implies that the ϵ_{ij} are uncorrelated with the covariate x_{ij} and the third assumption that the ζ_{0j} are uncorrelated with x_{ij}. Somewhat carelessly, the conditioning on x_{ij} is typically omitted when these assumptions are stated. We will adhere to this convention from now on, keeping in mind that expectations of random terms should be interpreted as conditional on covariates.

The *reduced form* of the model is obtained by substituting the level-2 model (3.2) for η_{0j} into the level-1 model (3.1) for y_{ij}, giving

$$y_{ij} = \gamma_{00} + \zeta_{0j} + \beta_1 x_{ij} + \epsilon_{ij}. \qquad (3.3)$$

This model resembles a conventional analysis of covariance (ANCOVA) model but the ζ_{0j} are random effects of the 'factor' school instead of fixed effects of school. In the ANOVA terminology, school would be referred to as a *random factor* (not to be confused with 'common factors' to be discussed in Section 3.3.2). By assuming a distribution for the intercepts, the school effect is captured by a single parameter, the variance ψ, instead of a separate parameter for each school (except one). Treating school as a random factor is appropriate if we wish to make inferences regarding the population of schools rather than the specific schools in the dataset. We return to the distinction between fixed and random effects on pages 81 to 84 of this chapter and in Sections 6.1 and 6.10 of the estimation chapter. Interactions between the random factor school and fixed factors produce random coefficient models, discussed in Section 3.2.2.

The random intercept model is an example of a *mixed effects* model or *linear mixed model* since it includes both fixed effects γ_{00} and β_1 and a random effect ζ_{0j} in addition to the residual ϵ_{ij}. We can partition the reduced form into a fixed and random part as follows

$$y_{ij} = \underbrace{\gamma_{00} + \beta_1 x_{ij}}_{\text{fixed part}} + \underbrace{\zeta_{0j} + \epsilon_{ij}}_{\text{random part}},$$

where the sum of the terms in the random part can be thought of as a total residual $\xi_{ij} = \zeta_{0j} + \epsilon_{ij}$. Due to this composition of the error the model is sometimes called an error-component model. The variance of this total residual, or equivalently the conditional variance of the responses given x_{ij}, is

$$\text{Var}(y_{ij}|x_{ij}) = \text{Var}(\zeta_{0j} + \epsilon_{ij}) = \psi + \theta.$$

This variance is composed of two *variance components*, the between-school variance ψ and the within-school variance θ. If x_{ij} is omitted in (3.1), the model is therefore called a *variance components model*.

Any two responses y_{ij} and $y_{i'j}$ in the same level-2 unit are conditionally independent given the random intercept ζ_{0j} and covariate values x_{ij} and $x_{i'j}$,

$$\text{Cov}(y_{ij}, y_{i'j}|\zeta_{0j}, x_{ij}, x_{i'j}) = \text{Cov}(\epsilon_{ij}, \epsilon_{i'j}) = 0, \quad i' \neq i.$$

However, because the random intercepts are shared among students in the same school, they induce dependence between responses from students within the same school after conditioning on the covariate. This dependence is often expressed in terms of the correlation within a cluster, the so called *intraclass correlation*. The intraclass correlation ρ becomes

$$\rho \equiv \text{Cor}(y_{ij}, y_{i'j}|x_{ij}, x_{i'j}) = \text{Cor}(\zeta_{0j} + \epsilon_{ij}, \zeta_{0j} + \epsilon_{i'j}) = \frac{\psi}{\psi + \theta}, \quad i' \neq i.$$

The intraclass correlation thus represents the proportion of the total residual variance $\psi + \theta$ that is due to the between-school residual variance ψ.

We can attempt to 'explain' the between-school variability by including a school-level covariate w_j, such as teacher to student ratio, in the school-level model (3.2),

$$\eta_{0j} = \gamma_{00} + \gamma_{01}w_j + \zeta_{0j}, \tag{3.4}$$

where γ_{00} and γ_{01} are fixed coefficients and ζ_{0j} now becomes a school-level residual or disturbance. Substituting this model into (3.1), we obtain the reduced form

$$y_{ij} = \gamma_{00} + \gamma_{01}w_j + \zeta_{0j} + \beta_1 x_{ij} + \epsilon_{ij}. \tag{3.5}$$

Endogeneity

One of the assumptions of the random intercept model is that $E(\zeta_{0j}|x_{ij})=0$, which implies that $\text{Cov}(\zeta_{0j}, x_{ij}) = 0$. If the random intercept represents the effect of missing covariates, this assumption will often be violated since these missing covariates may well be correlated with the observed covariate. The observed covariate is in this case called *endogenous*; see Engle *et al.* (1983) for a discussion of endogeneity and exogeneity.

We will discuss the situation where the specified model is

$$y_{ij} = \gamma_{00} + \zeta'_{0j} + \beta_1 x_{ij} + \epsilon_{ij}, \tag{3.6}$$

whereas the correct model is that in equation (3.5). In other words, we have omitted a cluster-level covariate w_j. If the random intercept represents the effects of omitted covariates, it should therefore have expectation $\gamma_{01}w_j$ and can be written as

$$\zeta'_{0j} = \gamma_{01}w_j + \zeta_{0j}. \tag{3.7}$$

Omission of w_j can be problematic if w_j and x_{ij} are dependent. The dependence can be expressed as the regression

$$w_j = \alpha_0 + \alpha_1 \bar{x}_{.j} + u_j,$$

where $\bar{x}_{.j}$ is the cluster mean of x_{ij}. Note that we have regressed w_j on the cluster mean $\bar{x}_{.j}$ instead of x_{ij} because $x_{ij} = (x_{ij}-\bar{x}_{.j})+\bar{x}_{.j}$ and the regression coefficient of w_j on $(x_{ij} - \bar{x}_{.j})$ is zero.

Substituting for w_j in (3.7), we see that the random intercept depends on $\bar{x}_{.j}$,

$$\zeta'_{0j} = \underbrace{\gamma_{01}\alpha_0}_{\delta_0} + \underbrace{\gamma_{01}\alpha_1}_{\delta_1}\bar{x}_{.j} + \underbrace{\gamma_{01}u_j + \zeta_{0j}}_{\varpi_j}$$

$$= \delta_0 + \delta_1\bar{x}_{.j} + \varpi_j.$$

Substituting for ζ'_{0j} in (3.6), the reduced form model becomes

$$y_{ij} = \gamma_{00} + \delta_0 + \delta_1\bar{x}_{.j} + \beta_1 x_{ij} + \varpi_j + \epsilon_{ij}. \tag{3.8}$$

We see that by including the cluster mean $\bar{x}_{.j}$ as a separate covariate, the coefficient of x_{ij} becomes the required parameter β_1 in the correctly specified model although we have omitted the cluster-level covariate w_j. However, omitting $\bar{x}_{.j}$ from the model will yield a biased estimate of β_1 if $\delta_1=\gamma_{01}\alpha_1 \neq 0$.

It may be useful to revisit the hypothetical example briefly discussed in Section 1.4. Here the units i are people and the clusters j countries. The response y_{ij} is length of life, x_{ij} exposure to red meat and w_j some index of the country's standard of living. The country-level random intercept, representing unexplained variability between different countries' life expectancies, could well include the effect of the omitted variable standard of living w_j which in turn is correlated with the average red meat consumption $\bar{x}_{.j}$. We would expect $\gamma_{01} > 0$, $\alpha_1 > 0$ so that $\delta_1 > 0$. This positive effect of the country-mean $\bar{x}_{.j}$ on the country-specific intercept can be seen in Figure 1.7 on page 13. Using the misspecified model (not including $\bar{x}_{.j}$ as a separate covariate), the negative true effect of red meat on life expectancy would therefore be underestimated due to the country-level positive relationship between red meat and life expectancy induced by the omitted covariate standard of living.

The model in (3.8) can alternatively be written as

$$y_{ij} = \gamma_{00} + \delta_0 + (\delta_1 + \beta_1)\bar{x}_{.j} + \beta_1(x_{ij} - \bar{x}_{.j}) + \varpi_j + \epsilon_{ij},$$

where $\delta_1 + \beta_1$ is the between-cluster effect and β_1 the within-cluster effect (this parameterization may be preferable because the covariates $\bar{x}_{.j}$ and $x_{ij} - \bar{x}_{.j}$ are uncorrelated). A Wald test of the equality of the between-cluster and within-cluster regression effects (i.e., a test of the null hypothesis that $\delta_1 = 0$) is identical to the Hausman specification test for the random intercept model (e.g. Hausman, 1978); see also Section 8.5.1. Some economists believe that a significant Hausman test implies that the random intercept model must be abandoned in favor of a fixed effects model. However, this is misguided since β_1 can be estimated without bias as long as the cluster mean $\bar{x}_{.j}$ is included as covariate in addition to x_{ij} as shown above. See Snijders and Berkhof (2004) for further discussion where it is also shown that analogous results hold for models with several random effects described in the next section.

Note that correlations between the level-1 residual ϵ_{ij} and covariates remain a potential problem as in any linear regression model. Unfortunately, this problem is not as easily discovered and overcome as for the random intercept.

3.2.2 Two-level random coefficient models

The two-level random intercept model can be extended to a *random coefficient model* by allowing the slope of a covariate x_{ij} to vary between clusters, for instance schools. The extended model can be expressed as

$$y_{ij} = \eta_{0j} + \eta_{1j}x_{ij} + \epsilon_{ij}, \tag{3.9}$$

where x_{ij} is a student-level covariate such as gender and η_{0j} and η_{1j} are the intercept and slope for the jth school, respectively. The between-school variability of the intercept is modeled as before and we add a similar model for the slope,

$$\begin{aligned}
\eta_{0j} &= \gamma_{00} + \gamma_{01}w_j + \zeta_{0j}, \\
\eta_{1j} &= \gamma_{10} + \gamma_{11}w_j + \zeta_{1j}.
\end{aligned} \tag{3.10}$$

The school-level residuals or disturbances ζ_{0j}, ζ_{1j} are specified as bivariate normal with zero means, variances ψ_1 and ψ_2 and covariance ψ_{21}.

Substituting the school-level models for the coefficients η_{0j} and η_{1j} in (3.10) into the student-level model in (3.9), we obtain the reduced form

$$y_{ij} \;=\; \underbrace{\gamma_{00} + \gamma_{01}w_j + \zeta_{0j}}_{\eta_{0j}} + \underbrace{(\gamma_{10} + \gamma_{11}w_j + \zeta_{1j})}_{\eta_{1j}} x_{ij} + \epsilon_{ij}$$

$$\;=\; \gamma_{00} + \gamma_{01}w_j + \gamma_{10}x_{ij} + \gamma_{11}(w_j x_{ij}) + \zeta_{0j} + \zeta_{1j}x_{ij} + \epsilon_{ij}.$$

In contrast to the random intercept model, the random coefficient model induces heteroscedastic responses since the conditional variance,

$$\mathrm{Var}(y_{ij}|x_{ij}, w_j) \;=\; \psi_1 + 2\psi_{21}x_{ij} + \psi_2 x_{ij}^2 + \theta, \tag{3.11}$$

depends on x_{ij}. It follows that the intraclass correlation in this case also depends on covariates.

It is important to note that the random intercept variance and the correlation between intercept and slope are not invariant to translation of x_{ij}. This can be seen in Figure 3.1 where identical cluster-specific regression lines are shown in two panels, but with the explanatory variable $x_{ij}' = x_{ij} - 3.5$ in the right panel translated relative to the explanatory variable x_{ij} in the left panel. The intercepts are the intersections of the regression lines with the vertical line at zero. It is clear that these intercepts vary more in the left panel than the right panel, whereas the correlation between intercepts and slopes is negative in the left panel and positive in the right panel.

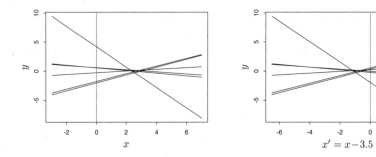

Figure 3.1 *Cluster-specific regression lines for random coefficient model, illustrating lack of invariance under translation of explanatory variable*

It is convenient to formulate the reduced form of the random coefficient model in vector notation

$$y_{ij} \;=\; \underbrace{\gamma_{00}}_{\beta_0} + \underbrace{\gamma_{01}w_j}_{\beta_1 x_{1ij}} + \underbrace{\gamma_{10}x_{ij}}_{\beta_2 x_{2ij}} + \underbrace{\gamma_{11}(w_j x_{ij})}_{\beta_3 x_{3ij}} + \zeta_{0j} + \underbrace{\zeta_{1j}x_{ij}}_{\zeta_{1j}z_{1ij}} + \epsilon_{ij}$$

$$\;=\; (\beta_0 + \beta_1 x_{1ij} + \beta_2 x_{2ij} + \beta_3 x_{3ij}) + (\zeta_{0j} + \zeta_{1j}z_{1ij}) + \epsilon_{ij}$$

$$\;=\; \mathbf{x}_{ij}'\boldsymbol{\beta} + \mathbf{z}_{ij}'\boldsymbol{\zeta}_j + \epsilon_{ij}, \tag{3.12}$$

where the x_{qij} denote covariates with fixed coefficients and z_{qij} covariates with random coefficients. The covariate vector $\mathbf{x}'_{ij} = (1, x_{1ij}, x_{2ij}, x_{3ij})$ for the fixed effects $\boldsymbol{\beta}' = (\beta_0, \beta_1, \beta_2, \beta_3)$ includes both student and school-level covariates as well as products of student and school-level variables representing *cross-level interactions*, $\mathbf{z}'_{ij} = (1, z_{1ij})$ and $\boldsymbol{\zeta}'_j = (\zeta_{0j}, \zeta_{1j})$.

Two-level models can either be specified through separate models at levels 1 and 2 as in (3.9) and (3.10) or directly through their reduced form as in (3.12). The former approach is used for instance by Raudenbush and Bryk (2002) whereas the latter is used for instance by Goldstein (2003) and Rabe-Hesketh *et al.* (2004a).

See also page 85 for a discussion of random coefficient models or 'growth curve models' for longitudinal data.

Two-level model in matrix notation

A two-level model can be written in matrix notation by stacking all responses into a single vector $\mathbf{y}$. Correspondingly, we stack the row vectors $\mathbf{x}'_{ij}$ into the matrix $\mathbf{X}$. We will in the sequel adhere to standard terminology and denote this matrix as a 'design' matrix, although we acknowledge that its values are not necessarily determined from an experimental design (see Kempthorne, 1980). Letting $\boldsymbol{\zeta}_{(D)}$ denote the vector of all random effects, the matrix equation for the entire sample becomes

$$\mathbf{y} = \mathbf{X}\boldsymbol{\beta} + \mathbf{Z}_{(D)}\boldsymbol{\zeta}_{(D)} + \boldsymbol{\epsilon}, \tag{3.13}$$

where the subscript (D) denotes that the matrix or vector contains *elements for the entire dataset* (here $\mathbf{y} \equiv \mathbf{y}_{(D)}$, $\mathbf{X} \equiv \mathbf{X}_{(D)}$ and $\boldsymbol{\epsilon} \equiv \boldsymbol{\epsilon}_{(D)}$). Note that $\mathbf{Z}_{(D)}$ is a block-diagonal matrix with blocks corresponding to level-2 units. To see this, consider a model with two covariates having both fixed and random coefficients, where $x_{1ij} = z_{1ij}$ and $x_{2ij} = z_{2ij}$. For two level-2 units, the first containing 3 level-1 units and the second 4 level-1 units, the matrix equation is written out in full for the individual level-1 units in Display 3.1 B.i on page 56.

For a single level-2 unit j, the model becomes

$$\mathbf{y}_{j(2)} = \mathbf{X}_{j(2)}\boldsymbol{\beta} + \mathbf{Z}_{j(2)}\boldsymbol{\zeta}_j + \boldsymbol{\epsilon}_{j(2)}, \tag{3.14}$$

where $\mathbf{y}_{j(2)}$, $\mathbf{X}_{j(2)}$, $\boldsymbol{\zeta}_j$ and $\boldsymbol{\epsilon}_{j(2)}$ are the rows in the unit-level representation in Display 3.1 B.i corresponding to the jth level-2 unit whereas $\mathbf{Z}_{j(2)}$ is the pertinent block from the block diagonal design matrix for the random effects, see Display 3.1 C.i. Here the subscript $j(2)$ indicates that the vectors and matrices contain *all elements for the jth level-2 unit*. Note that $\boldsymbol{\zeta}_j \equiv \boldsymbol{\zeta}_{j(2)}$.

Letting $\boldsymbol{\Psi}$ be the covariance matrix of $\boldsymbol{\zeta}_j$, the conditional covariance structure for the responses of a level-2 unit can be written as

$$\boldsymbol{\Omega}_{j(2)} \equiv \mathrm{Cov}(\mathbf{y}_{j(2)}|\mathbf{X}_{j(2)}, \mathbf{Z}_{j(2)}) = \mathbf{Z}_{j(2)}\boldsymbol{\Psi}\mathbf{Z}'_{j(2)} + \sigma_\epsilon^2\mathbf{I}. \tag{3.15}$$

Display 3.1 Matrix notation for a two-level random coefficient model.

A. Representation for a single unit:

$$y_{ij} = \beta_0 + \beta_1 x_{1ij} + \beta_2 x_{2ij} + \zeta_{0j} + \zeta_{1j} z_{1ij} + \zeta_{2j} z_{2ij} + \epsilon_{ij}$$

B. Representation for the entire sample:

$$\mathbf{y} = \mathbf{X}\boldsymbol{\beta} + \mathbf{Z}_{(D)}\boldsymbol{\zeta}_{(D)} + \boldsymbol{\epsilon}$$

i. Unit-level representation (level 1), for responses y_{ij}:

$$
\underbrace{\begin{bmatrix} y_{11} \\ y_{21} \\ y_{31} \\ \hline y_{12} \\ y_{22} \\ y_{32} \\ y_{42} \end{bmatrix}}_{\mathbf{y}}
=
\underbrace{\begin{bmatrix} 1 & x_{111} & x_{211} \\ 1 & x_{121} & x_{221} \\ 1 & x_{131} & x_{231} \\ \hline 1 & x_{112} & x_{212} \\ 1 & x_{122} & x_{222} \\ 1 & x_{132} & x_{232} \\ 1 & x_{142} & x_{242} \end{bmatrix}}_{\mathbf{X}}
\underbrace{\begin{bmatrix} \beta_0 \\ \beta_1 \\ \beta_2 \end{bmatrix}}_{\boldsymbol{\beta}}
$$

$$
+ \;
\underbrace{\begin{bmatrix} 1 & z_{111} & z_{211} & 0 & 0 & 0 \\ 1 & z_{121} & z_{221} & 0 & 0 & 0 \\ 1 & z_{131} & z_{231} & 0 & 0 & 0 \\ \hline 0 & 0 & 0 & 1 & z_{112} & z_{212} \\ 0 & 0 & 0 & 1 & z_{122} & z_{222} \\ 0 & 0 & 0 & 1 & z_{132} & z_{232} \\ 0 & 0 & 0 & 1 & z_{142} & z_{242} \end{bmatrix}}_{\mathbf{Z}_{(D)}}
\underbrace{\begin{bmatrix} \zeta_{01} \\ \zeta_{11} \\ \zeta_{21} \\ \hline \zeta_{02} \\ \zeta_{12} \\ \zeta_{22} \end{bmatrix}}_{\boldsymbol{\zeta}_{(D)}}
+ \;
\underbrace{\begin{bmatrix} \epsilon_{11} \\ \epsilon_{21} \\ \epsilon_{31} \\ \hline \epsilon_{12} \\ \epsilon_{22} \\ \epsilon_{32} \\ \epsilon_{42} \end{bmatrix}}_{\boldsymbol{\epsilon}}
$$

ii. Cluster-level representation (level 2), for responses $\mathbf{y}_{j(2)}$:

$$
\begin{bmatrix} \mathbf{y}_{1(2)} \\ \hline \mathbf{y}_{2(2)} \end{bmatrix}
=
\begin{bmatrix} \mathbf{X}_{1(2)} \\ \hline \mathbf{X}_{2(2)} \end{bmatrix} \boldsymbol{\beta}
+
\begin{bmatrix} \mathbf{Z}_{1(2)} & \mathbf{0} \\ \hline \mathbf{0} & \mathbf{Z}_{2(2)} \end{bmatrix}
\begin{bmatrix} \boldsymbol{\zeta}_1 \\ \boldsymbol{\zeta}_2 \end{bmatrix}
+
\begin{bmatrix} \boldsymbol{\epsilon}_{1(2)} \\ \boldsymbol{\epsilon}_{2(2)} \end{bmatrix}
$$

C. Representation for a single cluster:

$$\mathbf{y}_{j(2)} = \mathbf{X}_{j(2)}\boldsymbol{\beta} + \mathbf{Z}_{j(2)}\boldsymbol{\zeta}_j + \boldsymbol{\epsilon}_{j(2)}$$

i. Unit-level representation (level 2), for responses y_{ij} (here for cluster $j=1$):

$$
\underbrace{\begin{bmatrix} y_{1j} \\ y_{2j} \\ y_{3j} \end{bmatrix}}_{\mathbf{y}_{j(2)}}
=
\underbrace{\begin{bmatrix} 1 & x_{11j} & x_{21j} \\ 1 & x_{12j} & x_{22j} \\ 1 & x_{13j} & x_{23j} \end{bmatrix}}_{\mathbf{X}_{j(2)}}
\underbrace{\begin{bmatrix} \beta_0 \\ \beta_1 \\ \beta_2 \end{bmatrix}}_{\boldsymbol{\beta}}
+
\underbrace{\begin{bmatrix} 1 & z_{11j} & z_{21j} \\ 1 & z_{12j} & z_{22j} \\ 1 & z_{13j} & z_{23j} \end{bmatrix}}_{\mathbf{Z}_{j(2)}}
\underbrace{\begin{bmatrix} \zeta_{0j} \\ \zeta_{1j} \\ \zeta_{2j} \end{bmatrix}}_{\boldsymbol{\zeta}_j}
+
\underbrace{\begin{bmatrix} \epsilon_{1j} \\ \epsilon_{2j} \\ \epsilon_{3j} \end{bmatrix}}_{\boldsymbol{\epsilon}_{j(2)}}
$$

3.2.3 Three-level models

A two-level model for students nested in schools may be unrealistic since students are also nested within classes (which are themselves nested within schools). We would hence expect the correlation between responses of two students from the same school to be higher if the students also belong to the same class. This can be modeled using a three-level model.

Extending (3.12), a general three-level model for level-1 units i (e.g. students), level-2 units j (e.g. classes) and level-3 units k (e.g. schools) can be written in reduced form as

$$y_{ijk} = \mathbf{x}'_{ijk}\boldsymbol{\beta} + \mathbf{z}_{ijk}^{(2)\prime}\boldsymbol{\zeta}_{jk}^{(2)} + \mathbf{z}_{ijk}^{(3)\prime}\boldsymbol{\zeta}_{k}^{(3)} + \epsilon_{ijk}. \qquad (3.16)$$

The terms represent, respectively, the fixed part of the model, the level-2 random part, the level-3 random part and the level-1 residual. $\mathbf{x}_{ijk}$ is a vector of explanatory variables (including the constant) with fixed regression coefficients $\boldsymbol{\beta}$, $\mathbf{z}_{ijk}^{(2)}$ is an M_2-dimensional vector of explanatory variables with random coefficients $\boldsymbol{\zeta}_{jk}^{(2)}$ at level 2 and $\mathbf{z}_{ijk}^{(3)}$ is an M_3-dimensional vector of explanatory variables with random coefficients $\boldsymbol{\zeta}_{k}^{(3)}$ at level 3. The superscripts attached to the random effects indicate the level at which they vary whereas the superscripts attached to the covariates indicate the level at which the corresponding random coefficients vary. The random effects at each level have a multivariate normal distribution and random effects at different levels are mutually independent and independent of the level-1 residual.

Extending (3.13), the three-level model can be written in matrix notation for the entire sample as

$$\mathbf{y} = \mathbf{X}\boldsymbol{\beta} + \mathbf{Z}_{(D)}^{(2)}\boldsymbol{\zeta}_{(D)}^{(2)} + \mathbf{Z}_{(D)}^{(3)}\boldsymbol{\zeta}_{(D)}^{(3)} + \boldsymbol{\epsilon}. \qquad (3.17)$$

It is sometimes convenient to use a single design matrix

$$\mathbf{Z}_{(D)} \equiv \left[\mathbf{Z}_{(D)}^{(2)}, \mathbf{Z}_{(D)}^{(3)}\right],$$

with a corresponding vector of random effects

$$\boldsymbol{\zeta}_{(D)} \equiv [\boldsymbol{\zeta}_{(D)}^{(2)\prime}, \boldsymbol{\zeta}_{(D)}^{(3)\prime}]',$$

where $\boldsymbol{\zeta}_{(D)}$ stands for all random effects for the entire dataset, including both level-2 and level-3 random effects. The model can then be expressed as

$$\mathbf{y} = \mathbf{X}\boldsymbol{\beta} + \mathbf{Z}_{(D)}\boldsymbol{\zeta}_{(D)} + \boldsymbol{\epsilon}. \qquad (3.18)$$

This formulation is shown in detail in Display 3.2 B.i on page 59 for a random intercept model. Here two level-3 units each contain two level-2 units containing two level-1 units each. We have permuted the columns of $\mathbf{Z}_{(D)}$ to obtain a block-diagonal form with blocks $\mathbf{Z}_{1(3)}$ for the units in the first level-3 cluster and $\mathbf{Z}_{2(3)}$ for the units in the second level-3 cluster. The vector of random effects $\boldsymbol{\zeta}_{k(3)}$ is correspondingly permuted. This allows the model for the kth level-3 unit to be written as

$$\mathbf{y}_{k(3)} = \mathbf{X}_{k(3)}\boldsymbol{\beta} + \mathbf{Z}_{k(3)}\boldsymbol{\zeta}_{k(3)} + \boldsymbol{\epsilon}_{k(3)};$$

see Display 3.2 C.i.

The model now looks algebraically equivalent to a two-level model. This has the advantage that we can apply any results specifically developed for two-level models to higher-level models. For example, we can directly apply equation (3.15) to obtain the conditional covariance matrix for the responses of the kth level 3 unit,

$$\boldsymbol{\Omega}_{k(3)} \equiv \mathrm{Cov}(\mathbf{y}_{k(3)}|\mathbf{X}_{k(3)}, \mathbf{Z}_{k(3)}) = \mathbf{Z}_{k(3)}\boldsymbol{\Psi}_{k(3)}\mathbf{Z}'_{k(3)} + \sigma_\epsilon^2\mathbf{I}, \qquad (3.19)$$

where $\boldsymbol{\Psi}_{k(3)}$ is the covariance matrix of $\boldsymbol{\zeta}_{k(3)}$, all random effects for the kth level-3 unit.

3.2.4 Higher-level models

A general L-level model can be written as

$$y = \mathbf{x}'\boldsymbol{\beta} + \sum_{l=2}^{L} \mathbf{z}^{(l)\prime}\boldsymbol{\zeta}^{(l)} + \epsilon, \qquad (3.20)$$

where the fixed part is as before, $\mathbf{z}^{(l)}$ is an M_l-dimensional vector of explanatory variables with random coefficients $\boldsymbol{\zeta}^{(l)}$ at level l, and we have dropped the unit and cluster indices to simplify notation. The random effects at a given level l are usually assumed to have a multivariate normal distribution with zero mean and covariance matrix $\boldsymbol{\Psi}^{(l)}$. The random effects at different levels are assumed to be mutually independent and independent of the residual error term.

3.2.5 Generalized linear mixed models

The multilevel models discussed so far have been for continuous responses. However, all the response types discussed in Chapter 2 can be accommodated by specifying the conditional distribution of the responses given the random effects as a generalized linear model with a linear predictor ν of the same form as the conditional mean in (3.20),

$$\nu = \mathbf{x}'\boldsymbol{\beta} + \sum_{l=2}^{L} \mathbf{z}^{(l)\prime}\boldsymbol{\zeta}^{(l)}.$$

The resulting model is called a *generalized linear mixed model*. The linear mixed model is the special case with an identity link and conditionally normally distributed responses.

In generalized linear mixed models the regression coefficients represent *conditional effects* of covariates, given the values of the random effects. These effects can be interpreted as cluster-specific effects. In contrast, marginal or

Display 3.2 Matrix notation for a three-level random intercept model.

A. Representation for a single unit:

$$y_{ijk} = \mathbf{x}'_{ijk}\boldsymbol{\beta} + \zeta_{jk}^{(2)} + \zeta_k^{(3)} + \epsilon_{ijk}$$

B. Representation for the entire sample:

$$\mathbf{y} = \mathbf{X}\boldsymbol{\beta} + \mathbf{Z}_{(D)}^{(2)}\boldsymbol{\zeta}_{(D)}^{(2)} + \mathbf{Z}_{(D)}^{(3)}\boldsymbol{\zeta}_{(D)}^{(3)} + \boldsymbol{\epsilon} = \mathbf{X}\boldsymbol{\beta} + \mathbf{Z}_{(D)}\boldsymbol{\zeta}_{(D)} + \boldsymbol{\epsilon}$$

i. Unit-level representation (level 1), for responses y_{ijk}:

$$
\underbrace{\begin{bmatrix} y_{111} \\ y_{211} \\ y_{121} \\ y_{221} \\ \hline y_{112} \\ y_{212} \\ y_{122} \\ y_{222} \end{bmatrix}}_{\mathbf{y}} = \mathbf{X}\boldsymbol{\beta} + \underbrace{\begin{bmatrix} 1 & 0 & 0 & 0 \\ 1 & 0 & 0 & 0 \\ 0 & 1 & 0 & 0 \\ 0 & 1 & 0 & 0 \\ \hline 0 & 0 & 1 & 0 \\ 0 & 0 & 1 & 0 \\ 0 & 0 & 0 & 1 \\ 0 & 0 & 0 & 1 \end{bmatrix}}_{\mathbf{Z}_{(D)}^{(2)}} \underbrace{\begin{bmatrix} \zeta_{11}^{(2)} \\ \zeta_{21}^{(2)} \\ \zeta_{12}^{(2)} \\ \zeta_{22}^{(2)} \end{bmatrix}}_{\boldsymbol{\zeta}_{(D)}^{(2)}} + \underbrace{\begin{bmatrix} 1 & 0 \\ 1 & 0 \\ 1 & 0 \\ 1 & 0 \\ \hline 0 & 1 \\ 0 & 1 \\ 0 & 1 \\ 0 & 1 \end{bmatrix}}_{\mathbf{Z}_{(D)}^{(3)}} \underbrace{\begin{bmatrix} \zeta_1^{(3)} \\ \zeta_2^{(3)} \end{bmatrix}}_{\boldsymbol{\zeta}_{(D)}^{(3)}} + \boldsymbol{\epsilon}
$$

$$
= \mathbf{X}\boldsymbol{\beta} + \underbrace{\begin{bmatrix} 1 & 0 & 1 & 0 & 0 & 0 \\ 1 & 0 & 1 & 0 & 0 & 0 \\ 0 & 1 & 1 & 0 & 0 & 0 \\ 0 & 1 & 1 & 0 & 0 & 0 \\ \hline 0 & 0 & 0 & 1 & 0 & 1 \\ 0 & 0 & 0 & 1 & 0 & 1 \\ 0 & 0 & 0 & 0 & 1 & 1 \\ 0 & 0 & 0 & 0 & 1 & 1 \end{bmatrix}}_{\mathbf{Z}_{(D)}} \underbrace{\begin{bmatrix} \zeta_{11}^{(2)} \\ \zeta_{21}^{(2)} \\ \zeta_1^{(3)} \\ \hline \zeta_{12}^{(2)} \\ \zeta_{22}^{(2)} \\ \zeta_2^{(3)} \end{bmatrix}}_{\boldsymbol{\zeta}_{(D)}} + \boldsymbol{\epsilon}
$$

ii. Cluster-level representation (level 3), for responses $\mathbf{y}_{k(3)}$:

$$
\begin{bmatrix} \mathbf{y}_{1(3)} \\ \mathbf{y}_{2(3)} \end{bmatrix} = \begin{bmatrix} \mathbf{X}_{1(3)} \\ \mathbf{X}_{2(3)} \end{bmatrix}\boldsymbol{\beta} + \begin{bmatrix} \mathbf{Z}_{1(3)} & \mathbf{0} \\ \mathbf{0} & \mathbf{Z}_{2(3)} \end{bmatrix}\begin{bmatrix} \boldsymbol{\zeta}_{1(3)} \\ \boldsymbol{\zeta}_{2(3)} \end{bmatrix} + \begin{bmatrix} \boldsymbol{\epsilon}_{1(2)} \\ \boldsymbol{\epsilon}_{2(2)} \end{bmatrix}
$$

C. Representation for a top-level cluster k:

$$\mathbf{y}_{k(3)} = \mathbf{X}_{k(3)}\boldsymbol{\beta} + \mathbf{Z}_{k(3)}\boldsymbol{\zeta}_{k(3)} + \boldsymbol{\epsilon}_{k(3)}$$

i. Unit-level representation (level 1), for responses y_{ijk}:

$$
\underbrace{\begin{bmatrix} y_{11k} \\ y_{21k} \\ y_{12k} \\ y_{22k} \end{bmatrix}}_{\mathbf{y}_{k(3)}} = \mathbf{X}_{k(3)}\boldsymbol{\beta} + \underbrace{\begin{bmatrix} 1 & 0 & 1 \\ 1 & 0 & 1 \\ 0 & 1 & 1 \\ 0 & 1 & 1 \end{bmatrix}}_{\mathbf{Z}_{k(3)}} \underbrace{\begin{bmatrix} \zeta_{1k}^{(2)} \\ \zeta_{2k}^{(2)} \\ \zeta_k^{(3)} \end{bmatrix}}_{\boldsymbol{\zeta}_{k(3)}} + \underbrace{\begin{bmatrix} \epsilon_{11k} \\ \epsilon_{21k} \\ \epsilon_{12k} \\ \epsilon_{22k} \end{bmatrix}}_{\boldsymbol{\epsilon}_{k(3)}}
$$

population averaged effects are effects of covariates after integrating over the random effects. The difference between conditional and marginal effects for a random intercept probit model was shown graphically in Figure 1.6 on page 12. Note that the present notion of marginal effects is different from that sometimes used in econometrics where it typically refers to the effect of a small change in a covariate (given the others) on the expected response in models without latent variables (e.g. Greene, 2003). With an identity link, the conditional effects are equal to the marginal effects, but this is not generally the case. We refer to Section 3.6.5 for further discussion, Section 4.8.1 for derivations and Section 9.2 for an example.

In generalized linear models the variance of the responses given the random effects and covariates, the 'level-1 variance', is determined by the variance function of the specified conditional distribution. If the responses are counts, modeled as conditionally Poisson or binomial, overdispersion at level 1 may be modeled by including a random intercept at level 1 (e.g. Breslow and Clayton, 1993). Goldstein (1987), Schall (1991) and others adopt a quasi-likelihood approach by including an extra dispersion parameter in the level-1 variance function (see Section 2.3.1).

Since the level-1 variance is generally not constant, the correlation between observed responses in the same cluster is also not constant even in a simple random intercept model (e.g. Goldstein *et al.*, 2002). For dichotomous and ordinal responses, the intraclass correlation is therefore often expressed in terms of the correlation between the latent responses y^* – which is constant. For a random intercept probit model, this type of intraclass correlation becomes

$$\rho \equiv \mathrm{Cor}(y_{ij}^*, y_{i'j}^* | \mathbf{x}_{ij}, \mathbf{x}_{i'j}) = \frac{\psi}{\psi + 1},$$

known as the *tetrachoric correlation* in the dichotomous case without covariates. For a logit model, the '1' in the denominator is replaced by $\pi^2/3$, the variance of the logistic level-1 error.

3.2.6 Models with nonhierarchical random effects

Models with crossed random effects

So far, we have discussed hierarchical models where units are classified by some factor (for instance school) into top-level clusters at level L. The units in each top-level cluster are then (sub)classified by a further factor (for instance class) into clusters at level $L-1$, etc. The factors defining the classifications are nested in the sense that a lower-level cluster can only belong to one higher-level cluster (for instance a class can only belong to one school).

We now discuss non-hierarchical models where units are *cross-classified* by two or more factors, with each unit potentially belonging to any combination of 'levels' of the different factors. A prominent example is panel data where the factor 'individual' (or country, firm, etc.) is crossed with another factor 'time' or occasion. While unit-specific unobserved heterogeneity is often accommodated using random effects (see Section 3.6.1), random effects modeling of occasion-specific unobserved heterogeneity, due to shared experience of events at each occasion such as strikes, new legislation or weather conditions, appears to be confined to econometrics. If both factors are treated as random, econometricians call the model a two-way error component model (e.g. Baltagi, 2001, Chapter 3). Models with cross-classified random effects also arise in 'generalizability theory'. Here a simple design is a two-way cross-classification of subjects to be rated and raters; see also Section 3.3.1.

Consider students who are cross-classified by elementary school and high school. Ignoring elementary school, a two-level random intercept model for students i nested in high schools j would be specified as

$$\nu_{ij} = \mathbf{x}'_{ij}\boldsymbol{\beta} + \zeta_j^{(2)}. \tag{3.21}$$

The corresponding hierarchical structure of students nested in high schools is shown in Figure 3.2 where each student is represented by a short vertical line to the right of the long vertical line representing his or her high school.

Including an additional random intercept ζ_p for the students' elementary school p destroys the nesting since students from the same high school do not necessarily come from the same elementary school and vice versa. This can be seen in the figure where the lines connecting students with their elementary schools are crossed. Note that students cannot be reshuffled to untangle the crossings. The model with crossed random effects can be written as

$$\nu_{ijp} = \mathbf{x}'_{ijp}\boldsymbol{\beta} + \zeta_j^{(2)} + \zeta_p,$$

where the p subscript has been added in the fixed part to accommodate elementary school-specific covariates.

Goldstein (1987) described a trick for expressing a model with crossed random effects as a hierarchical model with a larger number of random effects. This is important because many estimation methods are confined to models with nested random effects. In the present example, we must first introduce a 'virtual' level within which both elementary schools and high schools are nested, such as towns, and call this level 3. Note that this level does not need to be 'natural' in any sense; for instance, if one child moved to another town to attend high school, these two towns could be merged into a single level-3 unit. If it is not possible to find a virtual third level, this level could be defined as a single unit encompassing the whole dataset. In Figure 3.2 the virtual level is shown by vertical lines spanning groups of high schools and elementary schools. Students within a virtual level-3 unit can only belong to high schools and elementary schools within that unit (no crossing between level-3 units).

Now label the elementary schools arbitrarily within each level-3 unit (e.g. town) as $p' = 1, \ldots, n_{\max}$, where $n_{\max}$ is the maximum number of elementary schools within a level-3 unit. Using the k subscript for level 3, we can write the model above equivalently as

$$\nu_{ijp} = \mathbf{x}'_{ijp}\boldsymbol{\beta} + \zeta_j^{(2)} + \sum_{p'=1}^{n_{\max}} \zeta_{kp'}^{(3)} d_{ip'},$$

where $\zeta_{kp'}^{(3)}$ is the p'th random 'slope' at level 3 and $d_{ip'}$ equals 1 if student i went to any of the elementary schools labeled p' and zero otherwise. Here the random intercepts for all elementary schools numbered p' are represented by (different realizations of) a random slope (varying at the third level) of a dummy variable for elementary schools p'. The variances of the random slopes are constrained to be equal (to achieve a constant random intercept variance in the original model) and their covariances set to zero (since elementary schools are mutually independent). The covariances between the random slopes and the random intercept for high school are also zero.

In the figure $n_{\max} = 3$ and the elementary schools within a virtual level-3 unit are labeled from 1 to at most 3. The three long vertical lines for virtual level 3 represent values of the random effects for elementary schools labeled 1, 2 and 3. The dots indicate for which students the corresponding dummy variables d_{i1} to d_{i3} are equal to 1. For instance, the first three dots from the top signify that the first three students from the top belong to elementary schools 1, 2 and 3, respectively. The reason a single random effect can be used for several primary schools (all primary schools with the same label) is that the random effect takes on a different value for each primary school.

Unfortunately, formulating models with crossed effects as multilevel models becomes unfeasible when $n_{\max}$ becomes too large. In this case Markov chain Monte Carlo methods such as the AIP algorithm discussed in Section 6.11.5 may be used.

See Snijders and Bosker (1999, Chapter 11), Raudenbush and Bryk (2002, Chapter 12) and Goldstein (2003, Appendix 11.1) for further discussion of models with crossed random effects.

Multiple-membership models

Ignoring elementary schools, we now return to the model in (3.21) with a random intercept for high schools. If a student i has attended several high schools $\{j\}$, spending time t_{ih} in school $h \in \{j\}$, a reasonable 'multiple-membership' model might be

$$\nu_{i\{j\}} = \mathbf{x}'_i\boldsymbol{\beta} + \sum_{h \in \{j\}} \zeta_h^{(2)} t_{ih},$$

where $\zeta_h^{(2)}$ represents the effect, per time unit 'exposed', of attending school h. (Sometimes t_{ih} is replaced by the proportion of time spent in school h, by dividing by the total time $\sum_h t_{ih}$.)

For simplicity we have assumed that there are no school-specific fixed effects. There are two problems making this model nonnested. First, for students attending (at least) two schools, the first and second schools attended are

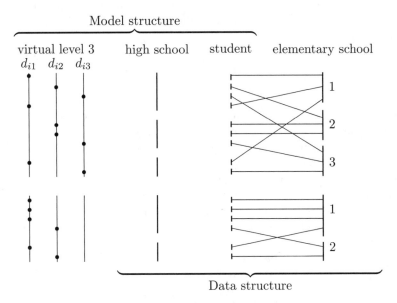

Figure 3.2 *Model structure and data structure for students in high schools crossed with elementary schools*

crossed, i.e. students attending the same first school could end up in different second schools and vice versa. Second, the random effect of a particular school must take on the same value, whether it is a given student's first, second or third school. This aspect is different from crossed-effects models.

Hill and Goldstein (1998) proposed a trick for representing these models as hierarchical models; see also Rasbash and Browne (2001). For the present example, we first need to find a third level so that students only cross between schools within a level but not between levels (e.g. towns). Labeling the schools within each level-3 unit arbitrarily as $h = 1, \ldots, n_{\max}$, where $n_{\max}$ is the maximum number of schools per level-3 unit, the model can be written as

$$\nu_{i\{j\}} = \mathbf{x}_i'\boldsymbol{\beta} + \sum_{h=1}^{n_{\max}} \zeta_{kh}^{(3)} t_{ih} d_{ih},$$

where $\zeta_{kh}^{(3)}$ is the hth random slope at level 3 and d_{ih} is a dummy variable equal to 1 if student i ever attended any of the schools labeled h and 0 otherwise.

3.3 Factor models and item response models

3.3.1 Platonic true scores: measurement models

Platonic true scores exist, can in principle be measured and would in this case represent a *gold standard*. However, error prone measures are often used instead of the gold standard for practical reasons. For instance, Roeder *et*

al. (1996) considered an application where 'low-density lipoprotein cholesterol' was measured using the less costly and time-consuming measure 'total
cholesterol'.

A simple measurement model, standard in 'classical test theory' (see e.g.
Lord and Novick, 1968), can be written as

$$y_{ij} = \eta_j + e_{ij},\qquad(3.22)$$

where y_{ij} is the ith measurement on the jth unit, η_j is the *true score* for unit
j with variance ψ and e_{ij} are measurement errors with variance θ.

It is usually assumed that the e_{ij} are mutually independent so that the y_{ij}
are conditionally independent given η_j. It is furthermore assumed that the e_{ij}
have zero expectation and are independent of the true score. However, if we
have a validation sample for which the gold-standard is available in addition
to the fallible measures, we can assess these assumptions. In particular, we
say that a measurement is *biased* if the expectation of the measurement error
is not zero.

The *reliability* ρ can be defined as the proportion of the total variance of
the measurements that is due to the true score variance

$$\rho = \frac{\text{Var}(\eta_j)}{\text{Var}(\eta_j) + \text{Var}(e_{ij})} = \frac{\psi}{\psi + \theta}.$$

Note that this is just the intraclass correlation for a random intercept or *one-
way random effects* model discussed in Section 3.2.1.

The simple one-way random effects model is appropriate if the measurements on each person can be considered exchangeable replicates, but this may
be unrealistic. For instance, if exams are marked by a panel of examiners or
raters i, a *two-way* model

$$y_{ij} = \beta_i + \eta_j + e_{ij}$$

may be more appropriate, where β_i represents the *response bias* for rater i.

If the raters are considered a random sample of possible raters, β_i can
be treated as random, giving a *two-way random effects* model. Note that the
random effects for subjects and raters are not nested but *crossed* if each person
is assessed by each rater (see Section 3.2.6).

If the raters are considered fixed, β_i are fixed effects and the model is a *two-
way mixed effects model*. Subject-rater interactions can be included in both
types of model if each rater provides replicate measurements for each subject.
These models and the different types of reliability coefficients that can be
derived from them are discussed in Shrout and Fleiss (1979) and McGraw
and Wong (1996).

Treating β_i as fixed, we can allow the measurement scales and reliabilities
to differ between methods (e.g. raters). The *congeneric measurement model*
(Jöreskog, 1971b) is specified as

$$y_{ij} = \beta_i + \lambda_i \eta_j + e_{ij},\qquad(3.23)$$

where β_i are fixed parameters, $\text{E}(\eta_j) = 0$, $\psi \equiv \text{Var}(\eta_j)$ and $\theta_{ii} \equiv \text{Var}(e_{ij})$. In this model β_i represents rater bias as before whereas λ_i is a rater-specific scale parameter. We can interpret $\lambda_i \eta_j$ as the true score measured in the units of rater i.

The scale of the true score η_j is typically fixed to that of y_{1j} by setting $\lambda_1 = 1$, an identification restriction of the kind to be discussed in Section 5.2.3. The measurement error variances θ_{ii} may differ between raters. The reliability for rater i becomes

$$\rho_i = \frac{\text{Var}(\lambda_i \eta_j)}{\text{Var}(\lambda_i \eta_j) + \text{Var}(e_{ij})} = \frac{\lambda_i^2 \psi}{\lambda_i^2 \psi + \theta_{ii}}.$$

The congeneric measurement model implies the following covariance structure for the vector of measurements $\mathbf{y}_j$ for unit j,

$$\Omega \equiv \text{Cov}(\mathbf{y}_j) = \begin{bmatrix} \lambda_1^2 \psi + \theta_{11} & & & \\ \lambda_2 \psi \lambda_1 & \lambda_2^2 \psi + \theta_{22} & & \\ \vdots & \vdots & \ddots & \\ \lambda_I \psi \lambda_1 & \lambda_I \psi \lambda_2 & \cdots & \lambda_I^2 \psi + \theta_{II} \end{bmatrix}.$$

Three common special cases of the congeneric measurement model are

- the *essentially tau-equivalent measurement model* where $\lambda_i = \lambda$

- the *tau-equivalent measurement model* where $\lambda_i = \lambda$ and $\beta_i = \beta$

- the *parallel measurement model* where $\lambda_i = \lambda$, $\beta_i = \beta$ and $\theta_{ii} = \theta$, giving the simple measurement model in (3.22).

The interpretation of e_{ij} as measurement error presupposes that the raters are measuring the same thing. However, if the raters are influenced by different idiosyncrasies, such as handwriting or spelling, the expectation for rater i could be expressed as $\beta_i + \lambda_i \eta_j + s_{ij}$, where s_{ij} is the subject-specific bias (e.g. due to handwriting or spelling) or, in other words, a rater by subject interaction (e.g. Dunn, 1992). As mentioned above, the variances of the specific factors cannot be separated from the measurement error variances unless replicate measurements are available for each rater and subject.

It is crucial to distinguish the *Berkson measurement model* (Berkson, 1950) from the classical measurement models discussed above. In the Berkson model it is assumed that the true scores η_{ij} are normally distributed around the measured score y_j

$$\eta_{ij} = y_j + \epsilon_{Bij},$$

where ϵ_{Bij} is the *Berkson error*. A situation where such a model may be appropriate is when y_j is a controlled variable. For example, an experimenter may aim to administer a given dose of a drug y_j, but the actual true dose given on occasion i, η_{ij}, differs from y_j due to measurement error. The Berkson measurement model therefore assumes the measured response to be independent of the measurement error, whereas the classical measurement model does not. This has important implications for regression with covariate measurement errors as discussed in Section 3.5.

3.3.2 Classical true scores: common factor models

Unlike platonic true scores, classical true scores or hypothetical constructs cannot be measured directly even in principle, intelligence being a typical example. The construct is instead measured by different indicators or items such as problems in an intelligence test. In contrast to a measurement model, the individual items cannot be said to measure intelligence per se in the sense that the expectation could be interpreted as 'true' intelligence. Instead, different aspects of intelligence are measured by different items; for instance items will often require different 'blends' of verbal, quantitative, abstract and visual reasoning. The answer to a particular item is therefore a reflection of both general intelligence and an item-specific aspect, referred to as the common and specific factor, respectively. We refer to Section 1.3 for an extensive discussion of hypothetical constructs.

A unidimensional common factor model for items $i = 1, \ldots, I$ can be written as

$$y_{ij} = \beta_i + \lambda_i \eta_j + \epsilon_{ij}. \tag{3.24}$$

Here, η_j is the common factor or *latent trait* for subject j, λ_i is a factor loading for the ith item and ϵ_{ij} are unique factors. We define $\psi \equiv \mathrm{Var}(\eta_j)$ and $\theta_{ii} \equiv \mathrm{Var}(\epsilon_{ij})$ and let η_j be independent of ϵ_{ij}.

The scale of the common factor η_j is either fixed by 'anchoring' (typically fixing the first factor loading, $\lambda_1 = 1$) or 'factor standardization' (fixing the factor variance to a positive constant, $\psi = 1$). Although the models resulting from either identification restriction are equivalent (see Section 5.3), anchoring is beneficial from the point of view of 'factorial invariance' (see e.g. Meredith, 1964, 1993; Skrondal and Laake, 1999). For instance, assume that model (3.24) holds for a population but we consider the subpopulation of units with negative factor values. In this case the original factor loadings are recovered in the subpopulation under anchoring (with a reduced variance ψ) but not under factor standardization.

Note that the intercept β_i cannot be interpreted as measurement bias in the present context. The intercept can be omitted if the item-specific mean $\bar{y}_{i\cdot}$ has been subtracted from y_{ij}. Also note that the unidimensional common factor model and the congeneric measurement model presented in (3.23) are mathematically identical.

The unique factor can be further decomposed as

$$\epsilon_{ij} = s_{ij} + e_{ij},$$

the sum of a specific factor and measurement error taken to be mutually independent and independent of η_j. Note that the specific factor has a similar interpretation to the rater by subject interaction in a measurement model. The specific factor is generally considered to be part of the true score for an item in which case the reliability becomes

$$\rho_i = \frac{\lambda_i^2 \psi + \mathrm{Var}(s_{ij})}{\lambda_i^2 \psi + \mathrm{Var}(s_{ij}) + \mathrm{Var}(e_{ij})}.$$

Unfortunately, in most designs the variances of the specific factors and measurement errors are not separately identified because there are no replicates for the individual items. Replication in terms of longitudinal or multimethod-multitrait designs is sometimes used in an attempt to decompose the unique factor into specific factors and measurement errors (e.g. Alwin, 1989 and the references therein). In the absence of replicates, the reliabilities for factor models are often somewhat carelessly expressed as

$$\rho_i = \frac{\lambda_i^2 \psi}{\lambda_i^2 \psi + \text{Var}(s_{ij}) + \text{Var}(e_{ij})},$$

which then represent lower bounds of the true reliabilities.

Closely related to this reliability is Cronbach's α which can be interpreted as the maximum likelihood estimator of the reliability of an unweighted sumscore, estimated without replicates, by assuming that the items are parallel measurements (e.g. Novick and Lewis, 1967). See Greene and Carmines (1980) for a discussion of α and other reliability measures for sumscores.

The factor model can also be written in matrix notation as

$$\mathbf{y}_j = \boldsymbol{\beta} + \boldsymbol{\Lambda}\eta_j + \boldsymbol{\epsilon}_j,$$

where $\boldsymbol{\beta}$ is a $I \times 1$ vector of intercepts, $\boldsymbol{\Lambda}$ is a $I \times 1$ vector of factor loadings, $\boldsymbol{\epsilon}_j$ a $I \times 1$ vector of unique factors and I is the total number of items.

The covariance structure, in this case called a *factor structure*, becomes

$$\boldsymbol{\Omega} \equiv \text{Cov}(\mathbf{y}_j) = \boldsymbol{\Lambda}\psi\boldsymbol{\Lambda}' + \boldsymbol{\Theta} = \begin{bmatrix} \lambda_1^2\psi + \theta_{11} & & & \\ \lambda_2\psi\lambda_1 & \lambda_2^2\psi + \theta_{22} & & \\ \vdots & \vdots & \ddots & \\ \lambda_I\psi\lambda_1 & \lambda_I\psi\lambda_2 & \cdots & \lambda_I^2\psi + \theta_{II} \end{bmatrix},$$

where $\boldsymbol{\Theta}$ is a diagonal matrix with the θ_{ii} placed on the diagonal. Note that the covariance structures for the unidimensional common factor model and the congeneric measurement model are identical. To fix the scale of the factor we typically either fix a factor loading to one or the factor variance to one (see Section 5.2.3).

The models discussed in this section have all been *reflective* with the items interpreted as reflecting or being 'caused' by a latent variable. However, it sometimes makes more sense to construe latent variables as *formative*, being 'caused' by the items. A standard example is measurement of socio-economic status (SES) for a family, based on the education and income of adult family members. In this case a reflective model is dubious; we would expect education and income to affect SES and not the other way around. Using factor models in the formative case would entail a misspecification. We refer to Edwards and Bagozzi (2000) for an overview of different types of relations between items and constructs.

3.3.3 Multidimensional factor models

The unidimensional factor model imposes a rather restrictive structure on the covariances. In structuring $I(I + 1)/2$ variances and covariances only $2I$ parameters are used. Hence, less restrictive multidimensional factor models are often useful. An M-dimensional factor model can be formulated as

$$
\begin{aligned}
y_{1j} &= \beta_1 + \lambda_{11}\eta_{1j} + \ldots + \lambda_{1M}\eta_{Mj} + \epsilon_{1j} \\
&\ \vdots \qquad \vdots \qquad \vdots \qquad \ddots \qquad \vdots \qquad \vdots \\
y_{Ij} &= \beta_I + \lambda_{I1}\eta_{1j} + \ldots + \lambda_{IM}\eta_{Mj} + \epsilon_{Ij}.
\end{aligned}
\tag{3.25}
$$

Such a model can alternatively be expressed in matrix form as

$$
\mathbf{y}_j = \boldsymbol{\beta} + \boldsymbol{\Lambda}_y\boldsymbol{\eta}_j + \boldsymbol{\epsilon}_j,
$$

where $\boldsymbol{\beta}$ is a vector of constants (usually omitted if $\mathbf{y}_j$ is mean-centered), $\boldsymbol{\Lambda}_y$ is a factor loading matrix, $\boldsymbol{\eta}_j$ a vector of M common factors with covariance matrix $\boldsymbol{\Psi}$ and $\boldsymbol{\epsilon}_j$ a vector of unique factors with diagonal covariance matrix $\boldsymbol{\Theta}$. The covariance matrix of the responses becomes

$$
\boldsymbol{\Omega} = \boldsymbol{\Lambda}_y\boldsymbol{\Psi}\boldsymbol{\Lambda}_y' + \boldsymbol{\Theta}.
\tag{3.26}
$$

Confirmatory factor analysis

If prior information is available, in terms of substantive theory, previous results or employed research design, *confirmatory* factor analysis (CFA) should be used where particular parameters are set to prescribed values, typically zero. For instance, $\boldsymbol{\Lambda}$ is often specified as an independent clusters structure (e.g. Jöreskog, 1969; McDonald, 1985) where each item loads on one and only one common factor.

For example, Mulaik (1988b) considered 9 subjective rating-scale variables designed to measure two 'dimensions' or factors in connection with a soldier's conception of firing a rifle in combat. The first factor, supposed to be *fear*, had as indicators the four scales 'frightening', 'nerve-shaking', 'terrifying', and 'upsetting'. The second factor, *optimism about outcome*, had as indicators the five scales 'useful', 'hopeful', 'controllable', 'successful', and 'bearable'. The loadings of variables on irrelevant factors were hypothesized to be zero, whereas the factors were expected to be (negatively) correlated.

An independent clusters two-factor model where each factor is measured by three nonoverlapping items can be written as

$$
\begin{bmatrix} y_{1j} \\ y_{2j} \\ y_{3j} \\ y_{4j} \\ y_{5j} \\ y_{6j} \end{bmatrix}
=
\begin{bmatrix} \beta_1 \\ \beta_2 \\ \beta_3 \\ \beta_4 \\ \beta_5 \\ \beta_6 \end{bmatrix}
+
\begin{bmatrix} 1 & 0 \\ \lambda_{21} & 0 \\ \lambda_{31} & 0 \\ 0 & 1 \\ 0 & \lambda_{52} \\ 0 & \lambda_{62} \end{bmatrix}
\begin{bmatrix} \eta_{1j} \\ \eta_{2j} \end{bmatrix}
+
\begin{bmatrix} \epsilon_{1j} \\ \epsilon_{2j} \\ \epsilon_{3j} \\ \epsilon_{4j} \\ \epsilon_{5j} \\ \epsilon_{6j} \end{bmatrix},
\tag{3.27}
$$

where we have fixed the scale of each factor by setting one factor loading to 1. A path diagram of this model is given in Figure 3.3. Here circles repre-

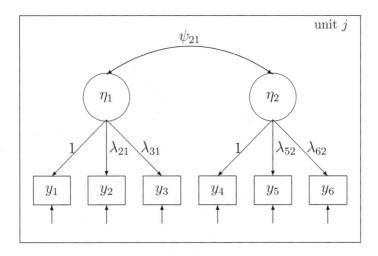

Figure 3.3 *Path diagram of independent clusters two-factor model*

sent latent variables, rectangles observed variables, arrows connecting circles and/or rectangles represent linear relations and short arrows pointing at circles or rectangles represent residual variability. Curved double-headed arrows connecting two variables (here the factors) indicate that the variables are correlated.

Factors are sometimes also specified as uncorrelated by setting pertinent off-diagonal elements of $\mathbf{\Psi}$ to zero. Confirmatory factor analysis is thus a hypotheticist procedure designed to test hypotheses about the relationship between items and factors, whose number and interpretation are determined in advance.

Exploratory factor analysis

A fundamentally different approach is *exploratory* factor analysis (EFA). Following Mulaik (1988a), exploratory factor analysis can be construed as an inductivist method designed to *discover* an optimal set of factors, their number to be determined in the analysis, that accounts for the covariation among the items (see also Holzinger and Harman, 1941). Each factor is then interpreted and 'named' according to the subset of items having high loadings on the factor.

The factor model in (3.25) is not identified without some restrictions on the parameters since we can multiply the factors by an arbitrary nonsingular transformation matrix $\mathbf{R}$

$$\boldsymbol{\eta}_j^* = \mathbf{R}\boldsymbol{\eta}_j,$$

and obtain the same covariance matrix as in (3.26) by multiplying the factor loading matrix $\mathbf{\Lambda}_y$ by $\mathbf{R}^{-1}$,

$$\mathbf{\Omega} = (\mathbf{\Lambda}_y \mathbf{R}^{-1}) \mathbf{R} \mathbf{\Psi} \mathbf{R} (\mathbf{R}^{-1} \mathbf{\Lambda}'_y) + \mathbf{\Theta}.$$

If $\mathbf{R}$ is orthogonal, the transformation is a rotation or reflection.

In confirmatory factor analysis, restrictions on the factor loadings serve to fix the factor rotation, and combined with constraints for the factor scales (either by fixing factor variances or by fixing one factor loading for each factor) will often suffice to identify the model. In exploratory factor analysis, a standard but arbitrary way of identifying the model is to set the factor covariance matrix equal to the identity matrix, $\mathbf{\Psi} = \mathbf{I}$, and fix the rotation for instance by requiring that $\mathbf{\Lambda}'_y \mathbf{\Theta} \mathbf{\Lambda}_y$ is diagonal (e.g. Lawley and Maxwell, 1971).

Exploratory factor analysis is often confused with *principal component analysis* and we therefore give a brief description of the latter. Principal components are linear combinations $\mathbf{a}'\mathbf{y}$ of the responses where $\mathbf{a}'\mathbf{a} = 1$. The coefficients or 'component weights' for the first principal component are determined to maximize the variance of the principal component. The coefficients of each subsequent principal component are determined to maximize the variance of the principal component subject to the constraint that it is uncorrelated with the previous components. The covariance matrix of the responses is therefore decomposed as

$$\mathrm{Cov}(\mathbf{y}) = \mathbf{A}\mathbf{\Psi}^*\mathbf{A}',$$

where the rows of $\mathbf{A}$ are the coefficient vectors $\mathbf{a}$ and $\mathbf{\Psi}^*$ is the diagonal covariance matrix of the principal components. The rows of $\mathbf{A}$ are the eigenvectors of $\mathrm{Cov}(\mathbf{y})$ and the diagonal elements of $\mathbf{\Psi}^*$ are the corresponding eigenvalues. Important differences from the factor structure in (3.26) are that there is no unique factor covariance matrix $\mathbf{\Theta}$ and that the components cannot be correlated. Principal component analysis is a data reduction technique since the first few principal components may capture the main features of the original data - in terms of the 'percentage of variance explained'. Unlike factor analysis, there is no statistical model underlying principal component analysis – it is merely a transformation of the data. An advantage of factor analysis as compared to principal component analysis is that there is a simple relationship between estimates based on different scalings of the responses (e.g. Krane and McDonald, 1978).

An exploratory factor analysis usually proceeds through the following rather ad hoc steps. First the number of factors is determined based on a principal component analysis of the correlation matrix. The number of factors is typically chosen to be equal to the number of eigenvalues that are larger than one, the so-called Kaiser-Guttmann criterion. Sometimes, however, a so-called *scree-plot* is used where the eigenvalues are plotted against their rank and the number of factors is indicated by the 'elbow' of the curve (Cattell, 1966). Second, a factor model with the chosen number of factors is estimated. There are a number of methods for this including maximum likelihood (e.g. Bartholomew and Knott, 1999). However, some software packages actually use the principal components as factors and the component weights as factor loadings.

It is typically difficult to ascribe meaning to the factors at this stage since most items will have nonnegligible loadings on most factors. Third, a transformation matrix $\mathbf{R}$ is therefore used to produce more interpretable loadings according to some criteria such as loadings being either 'small' or 'large' (e.g. Harman, 1976; McDonald, 1985). If uncorrelated factors are required, $\mathbf{R}$ must be orthogonal and the transformation is a rotation; otherwise it is referred to as the misnomer 'oblique rotation'. The fourth and final step is to retain only the 'salient' loadings, interpreting as zero any loadings falling below an arbitrary threshold, typically 0.3 or 0.4.

The tenability of this final model is never assessed and it may not even fit the data used for exploration. For this reason, and since modeling is purely exploratory, it is hardly surprising that such models are usually falsified by confirmatory factor analyses in other samples (e.g. Vassend and Skrondal, 1995, 1997, 1999, 2004). We return to the philosophical differences between the exploratory and confirmatory approaches in Section 8.2.2.

A confirmatory factor model equivalent to the traditional exploratory factor model with M factors (in the sense to be defined in Section 5.3) can be specified by judiciously imposing M^2 restrictions as in the exploratory model. The 'reference solution' of Jöreskog (1969, 1971a) is obtained in the following way:

- Fix the factor variances by imposing the M restrictions $\psi_{11} = \psi_{22} = \ldots = \psi_{MM} = 1$ (there are no restrictions on the correlations)
- Pick an 'anchor' item i_m for each factor m, preferably one with a large loading for the factor and small loadings for the other factors. Impose $\lambda_{i_m,k} = 0$ for all other factors η_k, $k \neq m$

This is useful since exploratory factor analysis can then be performed via confirmatory factor analysis, taking full advantage of the facilities for statistical inference within the latter approach. We use this approach for investigating the dimensionality of political efficacy in Section 10.3.

3.3.4 Item response models

The unidimensional factor model can be extended to dichotomous and ordinal responses using two different approaches (e.g. Takane and de Leeuw, 1987). Factor analysts typically use a latent response formulation as described in Section 2.4. In this case *latent* responses y^*_{ij} simply take the place of the observed responses y_{ij} in the conventional factor model. In *item response theory* (IRT), on the other hand, the generalized linear model formulation is typically used. Here the conditional probability of a particular response given the latent trait (or factor), the so-called *item characteristic curve*, is specified by a link function, typically a logit or probit.

In this formulation, the single factor model is known as a *two-parameter item response model* since there are two parameters associated with each item, an intercept and a factor loading. The classical application of these models is in ability testing, where the items i represent questions or problems in a test and

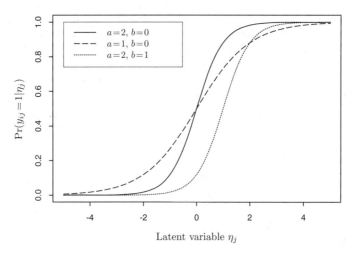

Figure 3.4 *Item characteristic curves for two-parameter logistic model with different a and b*

the answers are scored as right (1) or wrong (0). In this setting η_j represents the ability of person j and the model is typically parameterized as

$$\nu_{ij} = a_i(\eta_j - b_i).$$

Here, b_i can be interpreted as the item difficulty, giving a 50% chance of a correct answer when ability equals difficulty, whereas a_i is an item discrimination parameter (or factor loading) determining how well the item discriminates between subjects with different abilities. Figure 3.4 shows examples of item characteristic curves for a two-parameter logistic (2-PL) item response model (Birnbaum, 1968). The solid and dashed curves are for items with the same difficulty b but different discrimination parameters (slopes) a, whereas the solid and dotted curves are for items with the same discrimination parameter a but different difficulties (horizontal shifts) b. We will estimate a two-parameter item response model for items testing arithmetic reasoning in Section 9.4.

In the two-parameter model the probability of answering correctly tends to zero as ability goes to minus infinity. However, this is unrealistic if multiple choice formats are used, since guessing would produce a nonzero probability of answering correctly even for people with dismal abilities. An extra parameter is therefore sometimes introduced into the two-parameter model leading to the three-parameter logistic item response model (Birnbaum, 1968), see Section 9.4 for an example. Unfortunately, huge samples appear to be required to obtain reliable estimates of this model (e.g. Wainer and Thissen, 1982). We expect this problem to be exacerbated for the four-parameter model suggested by McDonald (1967). This model introduces an extra parameter to capture that even extremely able examinees commit errors.

The factor loadings a_i in the two-parameter item response model are often constrained equal, and without loss of generality set to 1, giving

$$\nu_{ij} = \eta_j - b_i,$$

a one-parameter model. In the logistic case the one-parameter model is often abbreviated as 1-PL. Note that a one-parameter item response model is just a random intercept model for dichotomous items without covariates. It is evi-

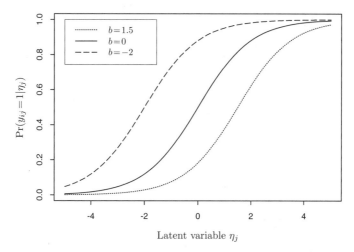

Figure 3.5 *Item characteristic curves for one-parameter logistic model with different* b

dent from Figure 3.5 that the one-parameter model has a property sometimes called 'double-monotonicity': for each ability, performance decreases with difficulty and for each difficulty, performance increases with ability, i.e. items and subjects are strictly ordered.

If the η_j in the one-parameter model are taken as fixed parameters and a logit link is used, the famous *Rasch model* (Rasch, 1960) is obtained. This model has a number of attractive theoretical properties (e.g. Fischer, 1995). For instance, the Rasch model is equivalent to the requirement that the unweighted sum-score of the responses is a sufficient statistic for η_j given the item-parameters b_i. This implies that conditional maximum likelihood estimation can be used for the item-parameters (see Section 6.10.2). Furthermore, the Rasch model is equivalent to a particular notion of generalizability of scientific statements, dubbed 'specific objectivity' by Rasch (1967). Broadly speaking, specific objectivity means that comparison of the ability of two subjects should only depend on the ability of these subjects (and not the ability of others) and that the comparison should yield the same result whatever item the comparison is based on.

As a development of the Rasch model, Fischer (1977, 1995) suggested the *linear logistic test model* where the item parameters in the Rasch model are structured in terms of item-specific covariates $\mathbf{x}_i$,

$$b_i = \mathbf{x}_i'\boldsymbol{\beta}.$$

The *nominal response model* (e.g. Rasch, 1961; Bock, 1972; Samejima, 1972; Andersen, 1973) handles polytomous responses such as answers $a_s = 1, \ldots, S$ to multiple-choice questions. It is a multinomial logit model

$$\Pr(y_{ij} = a_s) = \frac{\exp(\nu_{ij}^s)}{\sum_{t=1}^{S_i} \exp(\nu_{ij}^s)},$$

where

$$\nu_{ij}^s = \beta_i^s + \lambda_i^s \eta_j.$$

Identification can be achieved by setting $\beta_i^1 = 0$, $\lambda_i^1 = 0$ and $\mathrm{Var}(\eta_j) = 1$.

Famous models for ordinal responses in item response theory, including the *partial credit model* (Masters, 1982) and *rating scale model* (Andrich, 1978), can be obtained by imposing restrictions in the nominal response model (e.g. Thissen and Steinberg, 1986). In the partial credit model the linear predictor can be written as

$$\nu_{ij}^s = \beta_i^s + (s-1)\eta_j = -\sum_{t=0}^{s-1} \Delta_i^t + (s-1)\eta_j,$$

where equidistant category scores $s-1$ are substituted for the unknown factor loadings of the nominal response model. The Δ_i^t are sometimes called the 'item step difficulties', with $\Delta_i^0 = 0$.

The rating scale model is a special case of the partial credit model where the item step difficulty Δ_i^t is replaced by

$$\delta_i + \tau^t, \qquad t = 0, \ldots, S-1,$$

a sum of an item-specific component δ_i and a category-specific component τ^t, with $\tau^0 = 0$.

3.4 Latent class models

In latent class models the units are assumed to belong to one of C discrete classes $c = 1, \ldots, C$ where class membership is unknown. Thus, the classes can be viewed as the categories of a categorical latent variable. The (prior) probability that a unit j is in class c, π_{jc}, is a model parameter.

Latent class models are traditionally used when dichotomous or polytomous responses i are observed on each unit j. If unit j is in class c, the conditional response probability that item i takes on the value a_s, $s = 1, \ldots, S$, is modeled as a multinomial logit

$$\Pr(y_{ij} = a_s | c) = \frac{\exp(\nu_{ijc}^s)}{\sum_{t=1}^{S} \exp(\nu_{ijc}^t)}.$$

The responses to the items are assumed to be conditionally independent given membership in a given latent class. This is analogous to item response and factor models where the responses are conditionally independent given a continuous latent trait.

The unconditional response probabilities become finite mixtures

$$\Pr(y_{ij} = a_s) = \sum_{c=1}^{C} \pi_{jc} \Pr(y_{ij} = a_s | c),$$

and the probability of a response pattern $\mathbf{y}_j = (y_{1j}, \ldots, y_{Ij})$ is

$$\Pr(\mathbf{y}_j) = \sum_{c=1}^{C} \pi_{jc} \prod_{i=1}^{I} \Pr(y_{ij} | c).$$

In the conventional exploratory latent class model (e.g. McCutcheon, 1987), the linear predictor for item i and category s is a free parameter,

$$\nu_{ijc}^s = \beta_{ic}^s,$$

with the constraint $\beta_{ic}^1 = 0$. In Section 9.3 we estimate an exploratory latent class model for four dichotomous diagnostic tests for myocardial infarction. The two classes represent patients with and without myocardial infarction and the parameters in the linear predictor relate to the sensitivities and specificities of the tests.

In the case of ordinal responses, β_{ic}^s is often structured as $\beta_{ic}^s = b_s \beta_c$ for some scores b_s, giving the adjacent category logit if $b_s = s$. Other parameterizations are also possible, see Section 2.3.4. Confirmatory latent class models impose restrictions on the parameters, typically setting some conditional response probabilities equal across latent classes. Latent class models can be formulated as log-linear models for contingency tables where one of the categorical variables, the latent class variable, is unobserved (e.g. Goodman, 1974; Haberman, 1979).

Finite mixture models have the same structure as latent class models. For continuous responses and counts, these models are often used when there is only one response per unit to obtain a flexible model for the probability distribution. In the case of multivariate continuous responses, the conditional response distribution is often specified as multivariate normal, thus relaxing the usual conditional independence assumption. Such model-based cluster analysis is discussed in Banfield and Raftery (1993) and Bensmail *et al.* (1997).

3.5 Structural equation models with latent variables

Measurement and factor models are important in their own right but also as building blocks in structural equation models where the relations among latent variables are modeled. These relationships are often of main scientific interest whereas the relationships between the observed items and the latent variables are of secondary interest. An important advantage of modeling the relationships among latent variables directly is that detrimental effects of measurement error, such as regression dilution (e.g. Rosner *et al.*, 1990), may potentially be corrected (see for example Fuller, 1987 and Carroll *et al.*, 1995a).

Consider the simple 'errors in variables' problem where a single covariate ξ_j is measured with error according to a conventional measurement model

$$x_j = \xi_j + \delta_j, \tag{3.28}$$

with reliability $\rho < 1$. We wish to study the regression of y_j on the true covariate (the platonic score) ξ_j,

$$y_j = \gamma_0 + \gamma_1 \xi_j + \zeta_j, \tag{3.29}$$

and are particularly interested in the regression parameter γ_1. However, using conventional regression, we have to rely on the regression on the observed but error prone covariate x_j instead,

$$y_j = \gamma_0^* + \gamma_1^* x_j + \zeta_j^*.$$

In this simple case it can be shown that the consequence of ignoring measurement error is that the estimated regression parameter is *attenuated* relative to the true regression parameter

$$\mathrm{E}(\widehat{\gamma}_1^*) = \gamma_1 \rho.$$

When there are several covariates measured with error the consequences are less clear cut.

Importantly, γ_1 can be consistently estimated by jointly modeling (3.28) and (3.29), a simple example of a structural equation model. It should also be noted that the attenuation problem does not arise under the Berkson measurement model or when only the response variable is measured with error.

We now describe traditional structural equation modeling with latent variables, also often referred to as covariance structure analysis. As the latter term suggests, interest focuses on the covariance structure whereas the mean structure is typically eliminated by subtracting the mean from each variable. Having defined common factor models, a structural model specifying relations among the latent variables can be constructed. In this structural model, there could be both latent dependent variables and latent explanatory variables. As an example, consider a structural equation model for two latent dependent variables η_{1j}, η_{2j} and two latent explanatory variables ξ_{1j}, ξ_{2j}. The measurement model for the dependent variables is specified as an independent clusters model where the latent dependent variables are each measured by three non-overlapping items as in (3.27), written in vector notation as

$$\mathbf{y}_j = \mathbf{\Lambda}_y \boldsymbol{\eta}_j + \boldsymbol{\epsilon}_j.$$

Similarly, the measurement model for the explanatory variables can be written as

$$\mathbf{x}_j = \mathbf{\Lambda}_x \boldsymbol{\xi}_j + \boldsymbol{\delta}_j.$$

Note that we have omitted the constants assuming that both $\mathbf{y}_j$ and $\mathbf{x}_j$ are mean-centered.

We now specify a structural model letting both latent dependent variables be regressed on both latent explanatory variables. In addition, one latent dependent variable is regressed on the other,

$$\left[\begin{array}{c} \eta_{1j} \\ \eta_{2j} \end{array}\right] = \left[\begin{array}{cc} 0 & 0 \\ \beta_{21} & 0 \end{array}\right] \left[\begin{array}{c} \eta_{1j} \\ \eta_{2j} \end{array}\right] + \left[\begin{array}{cc} \gamma_{11} & \gamma_{12} \\ \gamma_{21} & \gamma_{22} \end{array}\right] \left[\begin{array}{c} \xi_{1j} \\ \xi_{2j} \end{array}\right] + \left[\begin{array}{c} \zeta_{1j} \\ \zeta_{2j} \end{array}\right].$$

This is an example of a *recursive model* where reciprocal or feedback relations are absent among the latent dependent variables. In general form the model becomes

$$\boldsymbol{\eta}_j = \mathbf{B}\boldsymbol{\eta}_j + \boldsymbol{\Gamma}\boldsymbol{\xi}_j + \boldsymbol{\zeta}_j. \qquad (3.30)$$

Structural equation models are often presented as path diagrams, an example of which is given in Figure 3.6 for the model just described, assuming that the ξ_{1j} and ξ_{2j} are correlated whereas the ζ_{1j} and ζ_{2j} are uncorrelated.

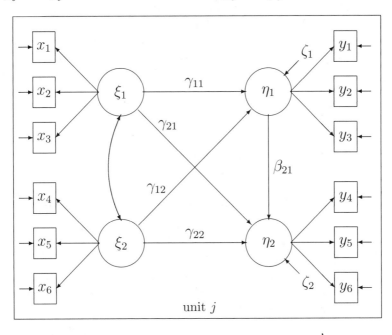

Figure 3.6 *Path diagram for structural equation model*

The above LISREL model (Jöreskog, 1973) is the dominant parameterization for structural equation models with latent variables, but several alternative formulations have been proposed (see e.g. Bollen, 1989, pp.395). Sometimes all measurements are stacked into a long vector $\mathbf{y}_j$, with the latent variables correspondingly stacked into a single vector $\boldsymbol{\eta}_j$ and the error terms stacked in $\boldsymbol{\epsilon}_j$. The combined measurement model can then be written as

$$\mathbf{y}_j = \boldsymbol{\Lambda}\boldsymbol{\eta}_j + \boldsymbol{\epsilon}_j, \qquad (3.31)$$

where

$$\boldsymbol{\Lambda} \equiv \left[\begin{array}{cc} \boldsymbol{\Lambda}_y & \mathbf{0} \\ \mathbf{0} & \boldsymbol{\Lambda}_x \end{array}\right].$$

The structural model becomes

$$
\begin{bmatrix} \eta_{1j} \\ \eta_{2j} \\ \xi_{1j} \\ \xi_{2j} \end{bmatrix} = \begin{bmatrix} 0 & 0 & \gamma_{11} & \gamma_{12} \\ \beta_{21} & 0 & \gamma_{21} & \gamma_{22} \\ 0 & 0 & 0 & 0 \\ 0 & 0 & 0 & 0 \end{bmatrix} \begin{bmatrix} \eta_{1j} \\ \eta_{2j} \\ \xi_{1j} \\ \xi_{2j} \end{bmatrix} + \begin{bmatrix} \zeta_{1j} \\ \zeta_{2j} \\ \zeta_{3j} \\ \zeta_{4j} \end{bmatrix},
$$

which can be written compactly as

$$
\boldsymbol{\eta}_j = \mathbf{B}\boldsymbol{\eta}_j + \boldsymbol{\zeta}_j. \tag{3.32}
$$

If there is an observed covariate, it can be included by introducing an artificial 'latent' explanatory variable, with factor loading equal to one for the covariate and zero for all other observed variables and setting the unique factor variance to zero. A disadvantage of this approach is that the explanatory variables are in effect treated as response variables so that the assumption of multivariate normality is often invoked. This is obviously unreasonable for many continuous covariates and even more so for dichotomous covariates such as gender. Some frameworks, for instance that of Muthén (1984), include an extra term $\boldsymbol{\Gamma}\mathbf{x}_{1j}$ for regressions of latent variables on observed covariates:

$$
\boldsymbol{\eta}_j = \boldsymbol{\alpha} + \mathbf{B}\boldsymbol{\eta}_j + \boldsymbol{\Gamma}\mathbf{x}_{1j} + \boldsymbol{\zeta}_j, \tag{3.33}
$$

where $\boldsymbol{\alpha}$ is an intercept vector. Muthén specifies the model conditional on the covariates so that distributional assumptions are not required for the covariates. In the measurement model, the additional term $\mathbf{K}\mathbf{x}_{2j}$ is included by Muthén and Muthén (1998) to represent regressions of observed responses on observed covariates

$$
\mathbf{y}_j = \boldsymbol{\nu} + \boldsymbol{\Lambda}\boldsymbol{\eta}_j + \mathbf{K}\mathbf{x}_{2j} + \boldsymbol{\epsilon}_j, \tag{3.34}
$$

where $\boldsymbol{\nu}$ is a vector of intercepts (often $\mathbf{x}_{1j} = \mathbf{x}_{2j}$).

A popular structural equation model with observed covariates is the Multiple-Indicator Multiple-Cause (MIMIC) model, a one-factor model where the factor is measured by multiple indicators and regressed on several observed covariates or 'causes' (e.g. Zellner, 1970; Hauser and Goldberger, 1971; Goldberger, 1972). Here the structural model is simply

$$
\eta_j = \alpha + \boldsymbol{\gamma}'\mathbf{x}_{1j} + \zeta_j.
$$

A path diagram of a MIMIC model with three indicators and three covariates is shown in Figure 3.7.

Robins and West (1977) considered a MIMIC model for handling measurement error in the estimation of home value. Three measures of home value were used in the measurement part of the model; 'appraised value by a private firm', 'estimated value by owner' and 'assessed value by county for tax purposes'. In the structural part home value was regressed on twelve property characteristics, including 'construction grade', 'type of garage' and 'finished area'. For MIMIC models with several factors we refer to Robinson (1974).

Returning to the general model in equation (3.33), the structural model can

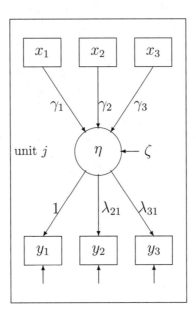

Figure 3.7 *Path diagram of MIMIC model*

be solved for the latent variables giving

$$\boldsymbol{\eta}_j \;=\; (\mathbf{I} - \mathbf{B})^{-1}[\boldsymbol{\alpha} + \boldsymbol{\Gamma}\mathbf{x}_{1j} + \boldsymbol{\zeta}_j]. \qquad (3.35)$$

Substituting $\boldsymbol{\eta}_j$ into equation (3.34) gives the reduced form

$$\mathbf{y}_j \;=\; \boldsymbol{\nu} + \boldsymbol{\Lambda}(\mathbf{I} - \mathbf{B})^{-1}[\boldsymbol{\alpha} + \boldsymbol{\Gamma}\mathbf{x}_{1j} + \boldsymbol{\zeta}_j] + \mathbf{K}\mathbf{x}_{2j} + \boldsymbol{\epsilon}_j.$$

The conditional expectation structure given $\mathbf{x}_{1j}$ and $\mathbf{x}_{2j}$ becomes

$$\mathrm{E}(\mathbf{y}_j|\mathbf{x}_{1j}, \mathbf{x}_{2j}) \;=\; \boldsymbol{\nu} + \boldsymbol{\Lambda}(\mathbf{I} - \mathbf{B})^{-1}[\boldsymbol{\alpha} + \boldsymbol{\Gamma}\mathbf{x}_{1j}] + \mathbf{K}\mathbf{x}_{2j},$$

and the conditional covariance structure becomes

$$\boldsymbol{\Omega} \equiv \mathrm{Cov}(\mathbf{y}_j|\mathbf{x}_{1j}, \mathbf{x}_{2j}) \;=\; \boldsymbol{\Lambda}(\mathbf{I} - \mathbf{B})^{-1}\boldsymbol{\Psi}(\mathbf{I} - \mathbf{B})^{-1\prime}\boldsymbol{\Lambda}' + \boldsymbol{\Theta},$$

where $\boldsymbol{\Psi}$ is the covariance matrix of $\boldsymbol{\zeta}_j$ and $\boldsymbol{\Theta}$ the covariance matrix of $\boldsymbol{\epsilon}_j$. See Chapter 5 for further examples of traditional structural equation models with latent variables.

Structural equation models are often given *causal* interpretations. For instance, Goldberger (1972) talks about 'causal links' in this context, and the methodology was previously often called 'causal modeling' (e.g. Bentler, 1978, 1980; James *et al.*, 1982). The causal parlance attached to simultaneous equations is undoubtedly a major reason both for the attractiveness of these kinds of models among social scientists and the scepticism from many statisticians. Causal interpretations should of course be conducted with extreme caution in the context of observational designs as has been stressed by Guttmann (1977), Cliff (1983), Freedman (1985, 1986, 1992), Holland (1988) and Sobel (1995), among others.

The specification of structural equation models and drawing of corresponding path diagrams is nevertheless indispensable for reasoning about how causal processes operate. For instance, in epidemiology a simple path diagram will often reveal which variables are best treated as confounders (and 'controlled' for) and which variables should be treated as intermediate variables in the causal pathway (and not 'controlled' for). It is important to note that the structural equation models discussed here are closely related to graphical models and models for potential outcomes (see Greenland and Brumback, 2002). Pearl (2000) provides a lucid discussion of modern 'causal modeling'.

3.6 Longitudinal models

Longitudinal data, often called repeated measurements in medicine, panel data in the social sciences and cross-sectional time-series data in economics, arise when units provide responses on multiple occasions. Two important features of longitudinal data are the clustering of responses within units and the chronological ordering of responses. A typical problem is to investigate predictors of the overall levels of the responses as well as predictors of changes in the responses over time. Longitudinal designs allow the separation of cross-sectional and longitudinal effects, as demonstrated in Section 1.4.

In addition to accommodating the mean structure, longitudinal models must also allow for dependence among responses on the same unit. The reasons for this dependence include unobserved heterogeneity between units inducing within-unit dependence (as in two-level models) as well as unobserved time-varying influences inducing greater dependence between responses occurring closer together in time. Both types of unobserved heterogeneity can be modeled explicitly in an attempt to explain the conditional covariance structure (given the covariates).

In the following subsections we distinguish between two types of longitudinal data; data with *balanced* or *unbalanced* occasions. The occasions are balanced if all units are measured at the same sets of time points t_i, $i = 1, \ldots, I$ and unbalanced if different sets of time points, $t_{ij}, i = 1, \ldots, n_j$ are used for different units. Missing data are possible in either case. If different units are measured at different sets of time points, but at each time-point there are measurements from a considerable number of units, the occasions can be treated as balanced with missing data.

In the case of either balanced or unbalanced occasions, longitudinal data can be thought of as two-level data with occasions i at level 1 and units j at level 2. In the case of balanced occasions, the data can also be viewed as single-level multivariate data where responses at different occasions are treated as different variables. In this case models for the mean and covariance structure can include occasion-specific parameters, for instance occasion-specific residual variances. However, in the case of unbalanced occasions the mean and covariance structures are typically modeled as a function of the time associated with the occasions or as a function of time-varying covariates.

3.6.1 Models with unit-specific effects

The primary reason for collecting information at multiple occasions for each unit is that it allows investigation of change within individual units; unit-specific (constant over time) effects can be controlled for, and we can investigate between-unit variability in the nature and degree of change or growth over time. Perhaps the most natural approach to longitudinal modeling is therefore to model individual growth trajectories using a combination of common (across units) fixed effects to summarize the average features of the trajectories and unit-specific fixed or random effects to represent variability between units. The random effects then induce and hence explain the conditional covariance structure.

Fixed effects models

Consider the response y_{ij} of unit j on occasion i. A simple linear fixed effects model for longitudinal data is of the form

$$y_{ij} = \mathbf{x}'_{ij}\boldsymbol{\beta} + \alpha_j + \epsilon_{ij}, \qquad (3.36)$$

where $\mathbf{x}_{ij}$ are time-varying covariates, sometimes including a time variable t_{ij}, with regression parameters $\boldsymbol{\beta}$, α_j are unit-specific intercepts or 'fixed effects' and ϵ_{ij} are identically and independently normally distributed residuals with $\mathrm{E}(\epsilon_{ij}|\mathbf{x}_{ij}) = 0$. The fixed effects α_j represent unit-specific effects that, if ignored, could lead to confounding and induce dependence among the residuals producing bias.

As a consequence of including α_j in the model, the effects $\boldsymbol{\beta}$ are interpretable as within-unit effects. This can be seen by considering the cluster means of model (3.36),

$$\bar{y}_j = \bar{\mathbf{x}}'_j\boldsymbol{\beta} + \alpha_j + \bar{\epsilon}_j, \qquad (3.37)$$

where the responses are the means for the units (over occasions). With a separate parameter α_j for each response, giving a saturated model, the responses do not provide any information on $\boldsymbol{\beta}$. Subtracting (3.37) from (3.36), we obtain the within-unit regression model

$$y_{ij} - \bar{y}_j = (\mathbf{x}_{ij} - \bar{\mathbf{x}}_j)'\boldsymbol{\beta} + \epsilon_{ij} - \bar{\epsilon}_j, \qquad (3.38)$$

which eliminates α_j.

Estimates of $\boldsymbol{\beta}$ and α_j can be obtained from ordinary least squares (OLS) estimation of the fixed effects model (3.36), which simultaneously produces the estimates $\widehat{\boldsymbol{\beta}}_{\mathrm{FE}}$ and $\widehat{\alpha}_j$. Alternatively, and equivalently, OLS estimation of $\boldsymbol{\beta}$ may be based on the within-unit model (3.38) producing

$$\widehat{\boldsymbol{\beta}}_{\mathrm{W}} = \mathbf{W}_{xx}^{-1}\mathbf{W}_{xy},$$

Display 3.3 Common dependence structures for longitudinal data.

A. Random intercept structure:

$$\Omega = \psi \mathbf{1}\mathbf{1}' + \theta\mathbf{I} = \begin{bmatrix} \psi+\theta & & & \\ \psi & \psi+\theta & & \\ \vdots & \vdots & \ddots & \\ \psi & \psi & \cdots & \psi+\theta \end{bmatrix}$$

B. Random coefficient structure:

$$\Omega_j = \mathbf{Z}_j \mathbf{\Psi} \mathbf{Z}_j' + \theta\mathbf{I}_{N_j}$$

C. Autoregressive residual structure AR(1):

$$\Omega = \frac{\sigma_\delta^2}{1-\alpha^2} \begin{bmatrix} 1 & & & \\ \alpha & 1 & & \\ \vdots & \vdots & \ddots & \\ \alpha^{I-1} & \alpha^{I-2} & \cdots & 1 \end{bmatrix}$$

D. Moving average residual structure MA(1):

$$\Omega = \sigma_\delta^2 \begin{bmatrix} 1+a^2 & & & & \\ a & 1+a^2 & & & \\ 0 & a & 1+a^2 & & \\ \vdots & \vdots & \vdots & \ddots & \\ 0 & 0 & 0 & \cdots & 1+a^2 \end{bmatrix}$$

E. Autoregressive response structure AR(1):

$$\Omega = \frac{\sigma_\epsilon^2}{1-\gamma^2} \begin{bmatrix} 1 & & & \\ \gamma & 1 & & \\ \vdots & \vdots & \ddots & \\ \gamma^{I-1} & \gamma^{I-2} & \cdots & 1 \end{bmatrix}$$

F. Factor structure:

i. One factor:

$$\Omega = \Lambda\psi\Lambda' + \Theta = \begin{bmatrix} \lambda_1^2\psi+\theta_{11} & & & \\ \lambda_2\psi\lambda_1 & \lambda_2^2\psi+\theta_{22} & & \\ \vdots & \vdots & \ddots & \\ \lambda_I\psi\lambda_1 & \lambda_I\psi\lambda_2 & \cdots & \lambda_I^2\psi+\theta_{II} \end{bmatrix}$$

ii. Multidimensional factor:

$$\Omega = \Lambda\mathbf{\Psi}\Lambda' + \Theta.$$

where $\mathbf{W}_{xx} \equiv \sum_{i,j} (\mathbf{x}_{ij} - \bar{\mathbf{x}}_j)(\mathbf{x}_{ij} - \bar{\mathbf{x}}_j)'$ and $\mathbf{W}_{xy} \equiv \sum_{i,j} (\mathbf{x}_{ij} - \bar{\mathbf{x}}_j)(y_{ij} - \bar{y}_j)$. The unit-specific intercepts are subsequently estimated as

$$\widehat{\alpha}_j = \bar{y}_j - \mathbf{x}_{ij}' \widehat{\boldsymbol{\beta}}_{\mathrm{W}}.$$

If it is assumed that ϵ_{ij} is normally distributed, it can also be shown that $\widehat{\boldsymbol{\beta}}_{\mathrm{W}}$ is obtained from conditional maximum likelihood estimation given $\sum_{i=1}^{n} y_{ij}$. In practice, where the number of occasions n is fixed and of a moderate magnitude, $\widehat{\boldsymbol{\beta}}_{\mathrm{W}}$ is a best linear unbiased estimator and consistent, whereas $\widehat{\alpha}_j$ is inconsistent as $N \to \infty$.

In univariate *repeated measures* ANOVA, or ANOVA for a balanced split plot design (e.g. Hand and Crowder, 1996), within-unit effects can be estimated using the above model whereas between-unit effects can be estimated by specifying an ordinary linear model for the mean responses $\bar{y}_j$. The first analysis corresponds to a partitioning of the within-unit sums of squares whereas the second uses the between-unit sums of squares.

A problem with the fixed effects model is that regression parameters for time-constant covariates such gender or treatment group where $\mathbf{x}_{ij} = \bar{\mathbf{x}}_j$ are not identified; see (3.38). Conditional maximum likelihood estimation of fixed effects models is also discussed in Section 6.10.2.

Random intercept models

Instead of treating the unit-specific effects as fixed, we can assume that the effects are realizations of a random variable ζ_j,

$$y_{ij} = \mathbf{x}_{ij}' \boldsymbol{\beta} + \zeta_j + \epsilon_{ij},$$

where ζ_j and ϵ_{ij} are independently distributed $\zeta_j \sim \mathrm{N}(0, \psi)$ and $\epsilon_{ij} \sim \mathrm{N}(0, \theta)$. This random intercept model is often called a 'one-way error component model' in econometrics.

The random intercept or 'permanent component' ζ_j allows the level of the response to vary across units. An advantage of this model compared with the fixed effects model is that the between-unit model is no longer saturated and we can include time-constant covariates. However, these advantages are purchased at the cost of relying on several assumptions, such as zero correlations between the random intercept and the covariates (see Section 3.2.1).

The maximum likelihood estimator of $\boldsymbol{\beta}$ under normality of ζ_j and ϵ_{ij} cannot be expressed in closed form. We will instead discuss the generalized least squares (GLS) estimator $\widehat{\boldsymbol{\beta}}_{\mathrm{GLS}}$, since it can be written in closed form and is asymptotically equivalent to the maximum likelihood estimator. This estimator also has the advantage that normality of ζ_j and ϵ_{ij} need not be assumed.

Although the within-estimator $\widehat{\boldsymbol{\beta}}_{\mathrm{W}}$ in (3.6.1) is an unbiased and consistent estimator of $\boldsymbol{\beta}$ in the random intercept model, the GLS estimator $\widehat{\boldsymbol{\beta}}_{\mathrm{GLS}}$ is a best linear unbiased estimator (BLUE). The GLS estimator is a matrix weighted average of the within-estimator $\widehat{\boldsymbol{\beta}}_{\mathrm{W}}$ and the between-estimator $\widehat{\boldsymbol{\beta}}_{\mathrm{B}}$, where the weights are the inverses of the covariance matrices of the respective estimators. The between-estimator is the OLS estimator

$$\widehat{\beta}_{\mathrm{B}} = \mathbf{B}_{xx}^{-1}\mathbf{B}_{xy},$$

for β in the between-unit model

$$\bar{y}_j - \bar{y} = (\bar{\mathbf{x}}_j - \bar{\mathbf{x}})'\beta + (\bar{\epsilon}_j - \bar{\epsilon}), \tag{3.39}$$

where $\mathbf{B}_{xx} \equiv \sum_j (\bar{\mathbf{x}}_j - \bar{\mathbf{x}})(\bar{\mathbf{x}}_j - \bar{\mathbf{x}})'$ and $\mathbf{B}_{xy} \equiv \sum_j (\bar{\mathbf{x}}_j - \bar{\mathbf{x}})(\bar{y}_j - \bar{y})$. Note that the between-estimator only uses variation between units and ignores the additional information from the longitudinal design as compared to a cross-sectional design.

The GLS estimator (e.g. Maddala, 1971) can then be expressed as

$$\widehat{\beta}_{\mathrm{GLS}} = \mathbf{V}_W\,\widehat{\beta}_{\mathrm{W}} + \mathbf{V}_B\,\widehat{\beta}_{\mathrm{B}},$$

where the weight matrices are

$$\mathbf{V}_W = (\mathbf{W}_{xx} + \omega^2\mathbf{B}_{xx})^{-1}\mathbf{W}_{xx},$$
$$\mathbf{V}_B = \mathbf{I} - \mathbf{V}_W,$$

and

$$\omega = \frac{\theta}{\theta + n\psi}.$$

The GLS estimator can alternatively be written as

$$\widehat{\beta}_{\mathrm{GLS}} = (\mathbf{W}_{xx} + \omega\mathbf{B}_{xx})^{-1}(\mathbf{W}_{xy} + \omega\mathbf{B}_{xy}),$$

from which we see that ω essentially represents the weight given to the between-unit variation. In fixed effects OLS, $\omega = 0$ and this source of variation is ignored. OLS for a 'naive' model without unit-specific effects corresponds to $\omega = 1$ so that all between-unit variation is added to the within-unit variation. Treating the unit-specific effects as random thus provides an intermediate approach between these extreme treatments of the between-unit variation. Also note that $1 - \omega$ corresponds to the shrinkage factor to be discussed in Section 7.3.1; see (7.5). In the special case of balanced occasions, no missing data and balanced covariates $\mathbf{x}_{ij} = \mathbf{x}_i$, we obtain $\widehat{\beta}_{\mathrm{B}} = \mathbf{0}$. It follows that the GLS estimator in this case is identical to the between-estimator $\widehat{\beta}_{\mathrm{W}}$.

The conditional variances of the responses $\mathbf{y}_j$, or the variances of the total residuals $\xi_{ij} = \zeta_j + \epsilon_{ij}$, are equal to $\psi + \theta$ and constant across occasions. The conditional covariances for any two occasions are just ψ and the corresponding correlation is the intraclass correlation previously encountered. This *random intercept* covariance structure is shown in Display 3.3A on page 82. Note that this covariance structure is the special case of a one-factor structure where $\lambda_i = 1$ and $\theta_{ii} = \theta$ for all i. The covariance structure is sometimes referred to as *exchangeable* since the joint distribution of the residuals for a given person remains unchanged if the residuals are exchanged across occasions. The covariance structure is also consistent with the sphericity assumption that the conditional variances $\mathrm{Var}(y_{ij} - y_{i'j}|\mathbf{x}_{ij})$ of all pairwise differences are equal. Note that the covariances ψ are restricted to be nonnegative in the random intercept model. If this restriction is relaxed, the above configuration of the covariance structure is often called *compound symmetric*. In the case of balanced occasions, we can allow the variance of ϵ_{ij} to take on a different value for each occasion, θ_{ii}.

Random coefficient models - Growth curve models

The random coefficient model (e.g. Laird and Ware, 1982) allows both the level of the response and the effects of covariates to vary randomly across units. The model was previously specified as a two-level model in (3.12),

$$y_{ij} = \mathbf{x}'_{ij}\boldsymbol{\beta} + \mathbf{z}'_{ij}\boldsymbol{\zeta}_j + \epsilon_{ij},$$

where $i = 1, 2, \ldots, n_j$. Here, $\mathbf{x}_{ij}$ denotes both time-varying and time-constant covariates with fixed coefficients $\boldsymbol{\beta}$ and $\mathbf{z}_{ij}$ time-varying covariates with random coefficients $\boldsymbol{\zeta}_j$. Since the random coefficients have zero means, $\mathbf{x}_{ij}$ will typically contain all elements in $\mathbf{z}_{ij}$, with the corresponding fixed effects interpretable as the mean effects. The first element of the vectors is typically equal to one corresponding to a fixed and random intercept. Letting $\boldsymbol{\Psi} \equiv \mathrm{Cov}(\boldsymbol{\zeta}_j)$, the covariance structure of the vector $\mathbf{y}_j$ is presented in Display 3.3B on page 82. The special case where the residual variances are set equal across occasions, $\theta_{ii} = \theta$, is common.

A useful version of the random coefficient model for longitudinal data is a *growth curve model* where individuals are assumed to differ not only in their intercepts but also in other aspects of their trajectory over time, for example in the linear growth (or decline) of the response over time. These models include random coefficients for (functions of) time. For example, a linear growth curve model can be written as

$$y_{ij} = \mathbf{x}'_{ij}\boldsymbol{\beta} + \zeta_{0j} + \zeta_{1j}t_{ij} + \epsilon_{ij}, \tag{3.40}$$

where t_{ij}, the time at the ith occasion for individual j, is one of the covariates in $\mathbf{x}_{ij}$. The random intercept and slope should not be specified as uncorrelated, because translation of the time scale t_{ij} changes the magnitude of the correlation as illustrated in Figure 3.1 on page 54 (see also Elston (1964) and Longford (1993)).

The covariance structure is the same as for the two-level random coefficient model in (3.15), shown explicitly for the variance in (3.11). A path diagram of this model is shown in the first panel of Figure 3.8 on page 86, where there are three occasions with times $t_1 = 0$, $t_2 = 1$ and $t_3 = 2$. The second diagram represents the unbalanced case, where all variables inside the box labelled 'unit j' have a j subscript and vary between units. Variables that are also in the box labelled 'occasion i' vary between occasions and units and have both an i and j subscript. The arrow from t to y therefore represents a regression of y_{ij} on t_{ij}. The latent variable ζ_{1j} modifies this regression or interacts with t_{ij} and therefore represents the random slope.

In the case of balanced occasions, the linear growth curve model can also be formulated as a two-factor model,

$$y_{ij} = \lambda_{0i}\eta_{0j} + \lambda_{1i}\eta_{1j} + \epsilon_{ij}.$$

Here

$$\eta_{0j} = \beta_0 + \zeta_{0j}, \quad \eta_{1j} = \beta_1 + \zeta_{1j},$$

the loadings for the intercept factor η_{0j} are fixed to $\lambda_{0i} = 1$ and the loadings for the slope factor η_{1j} are set equal to t_i. Note that the means of the factors cannot be set to zero here as is usually done in factor models.

Meredith and Tisak (1990) suggest using a two-factor model similar to that in the first diagram of Figure 3.8 but with *free* factor loadings for η_{1j} (subject to identification restrictions, such as $\lambda_{11} = 0$ and $\lambda_{12} = 1$) to model nonlinear growth. Traditionally, estimation of this factor model would require balanced occasions without missing data, but this is no longer a limitation. However, if the occasions are very unbalanced, with few responses at a given occasion, factor models can no longer be used since reliable estimation of occasion-specific factor loadings would be precluded.

Balanced occasions   Unbalanced occasions

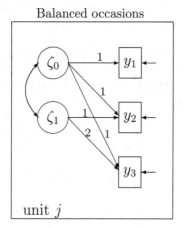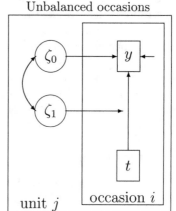

Figure 3.8 *Path diagrams for growth curve models with balanced and unbalanced occasions*

The models can easily be extended to noncontinuous responses by using generalized linear mixed models. We will estimate a random coefficient model for longitudinal count data in Section 11.3.

Longitudinal models with discrete latent variables - latent trajectory models

It is sometimes believed that the population consists of different types or classes of units characterized by different patterns of development or development trajectories over time. The models are latent class models having the same form as the latent growth models discussed above except that the random effects are now discrete.

For instance, in a linear latent trajectory model analogous to (3.40), the linear predictor for a unit in class c is given by

$$\nu_{ijc} = e_{0c} + e_{1c}t_{ij}.$$

Each latent class is therefore characterized by a pair of coefficients e_{0c} and e_{1c}, representing the intercept and slope of the latent trajectory. For balanced occasions, we do not have to assume that the latent trajectories are linear or have another particular shape but can instead specify an unstructured model with latent trajectory

$$\nu_{ijc} = e_{ic}, \quad i = 1, \ldots, I$$

for class c, $c = 1, \ldots, C$. In the case of categorical responses, latent trajectory models are typically referred to as latent class growth models (e.g. Nagin and Land, 1993). They are an application of mixture regression models (e.g. Quandt, 1972) to longitudinal data.

If the responses are continuous, the models are known as latent profile models (e.g. Gibson, 1959),

$$y_{ijc} = \nu_{ijc} + \epsilon_{ijc}.$$

Here the variance of the residuals ϵ_{ijc} could be allowed to differ between classes. Both latent class and latent profile models assume that the responses on a unit are conditionally independent given latent class membership. Muthén and Shedden (1999) relax this assumption for continuous responses in their *growth mixture models* by allowing the residuals ϵ_{ijc} to be correlated conditional on latent class membership with covariance matrices differing between classes.

3.6.2 Models with correlated residuals

Random intercept models include two random terms, the random intercept and the occasion-specific residual. While the random intercept represents effects of random influences or omitted covariates that remain constant over time, the residuals represent effects of random influences that are immediate and do not persist over more than a single occasion. The resulting compound symmetric correlation structure often does not reflect what is observed in practice, namely that (conditional) correlations between pairs of responses tend to be greater if the responses occurred closer together in time.

Such correlation structures can be induced by allowing the effects of omitted variables to be distributed over time, leading to autocorrelated errors. It should be noted that this omitted variable interpretation requires that the total effect of the influences represented by ϵ_{ij} averages out to zero over units and also that it is uncorrelated with $\mathbf{x}_{ij}$ (e.g. Maddala, 1977). In the following subsections, we discuss the case of continuous responses, sometimes indicating how the models are modified for other response types.

Autoregressive residuals

When occasions are equally spaced in time, a first order autoregressive model AR(1) can be expressed as

$$\epsilon_{ij} = \alpha\epsilon_{i-1,j} + \delta_{ij}, \tag{3.41}$$

where $\epsilon_{i-1,j}$ is independently distributed from the 'innovation errors' δ_{ij}, $\delta_{ij} \sim N(0, \sigma_\delta^2)$. This is illustrated in path diagram form in the first panel of Figure 3.9. Note that a 'random walk' is obtained if $\alpha = 1$ in the AR(1) model.

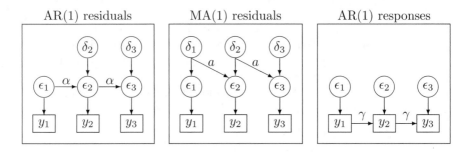

Figure 3.9 *Path diagrams for autoregressive responses and autoregressive and moving average residuals*

Assuming that the process is *weakly stationary*, $|\alpha| < 1$, the covariance structure is as shown in Display 3.3C on page 82. It follows that the correlations between responses at different occasions are structured as

$$\text{Cor}(\epsilon_{ij}, \epsilon_{i+k,j}) = \alpha^k.$$

For non-equally spaced occasions, the correlation structure is often specified as

$$\text{Cor}(y_{ij}, y_{i+k,j}) = \alpha^{|t_{i+k} - t_i|},$$

where the correlation structure for unbalanced occasions is simply obtained by replacing t_i by t_{ij} (e.g. Diggle, 1988).

These first order autoregressive covariance structures are often as unrealistic as compound symmetry since the correlations fall off too rapidly with increasing time-lags. One possibility is to specify a higher order autoregressive process of order k, AR(k),

$$\epsilon_{ij} = \alpha_1\epsilon_{i-1,j} + \alpha_2\epsilon_{i-2,j} + \ldots + \alpha_k\epsilon_{i-k,j} + \delta_{ij}.$$

Another is to add a random intercept to the AR(1) model (see 'Hybrid specifications' on page 91). In the case of balanced occasions, we can also specify a different parameter α_i for each occasion, giving an *antedependence* structure (e.g. Gabriel, 1962) for the residuals.

Moving average residuals

Random shocks disturb the response variable for some fixed number of periods before disappearing. Such a process can be modeled by means of moving averages (see Box *et al.*, 1994). A first order moving average process MA(1) for the residuals can be specified as

$$\epsilon_{ij} = \delta_{ij} + a\delta_{i-1,j}.$$

A path diagram for this model is given in the second panel of Figure 3.9 and the covariance structure is presented in Display 3.3D on page 82. We see that the process 'forgets' what happened more than one period in the past, in contrast to the autoregressive processes.

The moving average model of order k, MA(k), is given as

$$\epsilon_{ij} = \delta_{ij} + a_1\delta_{i-1,j} + a_2\delta_{i-2,j} + \ldots + a_k\delta_{i-k,j},$$

with 'memory' extending k periods in the past.

3.6.3 Models with lagged responses

In these models lags of the response y_{ij} are included as covariates in addition to $\mathbf{x}_{ij}$. The dependence on previous responses is called 'state dependence'; see Section 9.6 for elaboration and an application. The models are also referred to as transition models (e.g. Diggle *et al.*, 2002, Chapter 10). When occasions are equally spaced in time, a first order autoregressive model for the responses y_{ij} can be written as

$$y_{ij} = \mathbf{x}'_{ij}\boldsymbol{\beta} + \gamma y_{i-1,j} + \epsilon_{ij}.$$

Assuming that the process is *weakly stationary*, $|\gamma| < 1$, the covariance structure is shown in Display 3.3E on page 82. A path diagram for this model is shown in the third panel of Figure 3.9.

As for the residual autoregressive structure, the first order autoregressive structure for responses is often deemed unrealistic, since the correlations fall off too rapidly with increasing time-lags. Once again, this may be rectified by specifying a higher order autoregressive process AR(k)

$$y_{ij} = \mathbf{x}'_{ij}\boldsymbol{\beta} + \gamma_1 y_{i-1,j} + \gamma_2 y_{i-2,j} + \ldots + \gamma_k y_{i-k,j} + \epsilon_{ij}.$$

An extension of the autoregressive model is the *antedependence* model for responses y_{ij} which specifies a different parameter γ_i for each occasion.

Apart from being of interest in its own right, the lagged response model is useful in distinguishing between different longitudinal models. Consider two simple models; a state dependence model with a lagged response and lagged covariate but independent residuals ϵ_{ij}

$$y_{ij} = \gamma y_{i-1,j} + \beta_1 x_{ij} + \beta_2 x_{i-1,j} + \epsilon_{ij}, \tag{3.42}$$

and an autocorrelation model without lagged response or lagged covariate

$$y_{ij} = \beta x_{ij} + \epsilon_{ij},$$

but residuals ϵ_{ij} having a AR(1) structure. Substituting first for $\epsilon_{ij} = \alpha\epsilon_{i-1,j} + \delta_{ij}$ from (3.41), then for $\epsilon_{i-1,j} = y_{i-1,j} - \beta x_{i-1,j}$ and reexpressing, the auto-correlation model can alternatively be written as

$$y_{ij} = \alpha y_{i-1,j} + \beta x_{ij} + \alpha\beta x_{i-1,j} + \delta_{ij}.$$

Note that this model is equivalent to the state dependence model (3.42) with the restriction $\gamma\beta_1 + \beta_2 = 0$. Importantly, this means that we can use the state dependence model to discriminate between autocorrelated residuals and state dependence in longitudinal models. The distinction between true and 'spurious' state dependence (apparent state dependence that disappears when appropriately modeling residual dependence) is crucial in many applications, see Section 9.6 for an example. It also follows that the ritual of performing a Durbin-Watson test (Durbin and Watson, 1950) for autocorrelation in longitudinal modeling should be preceded by ruling out the state dependence model. Otherwise, a large Durbin-Watson statistic is ambiguous, indicating state dependence and/or autocorrelation.

Use of lagged response models should be conducted with caution. First, lags should be avoided if the lagged effects do not have a 'causal' interpretation since the interpretation of β changes when $y_{i-1,j}$ is included as an additional covariate. Second, the models require balanced data in the sense that all units are measured on the same occasions. If the response for a unit is missing at an occasion, the entire unit must be discarded. Third, lagged response models reduce the sample size. This is because the y_{ij} on the first occasions can only serve as covariates and cannot be regressed on lagged responses (which are missing). Fourth, an initial condition problem arises for the common situation where the process is ongoing when we start observing it (e.g. Heckman, 1981b).

An advantage of lagged response models as compared to models with autoregressive residuals is that they can easily be used for response types other than the continuous.

3.6.4 Other covariance structures

Unrestricted

Instead of attempting to model the covariance structure, we can simply specify

$$y_{ij} = \mathbf{x}'_{ij}\boldsymbol{\beta} + \epsilon_{ij},$$

where the vector of residuals $\boldsymbol{\epsilon}_j$ is multivariate normal with an unrestricted covariance matrix. This unrestricted model requires balanced data with $n_j = I$ and this specification corresponds to a repeated measures MANOVA (e.g. Hand and Crowder, 1996, Chapter 2).

This approach provides a safeguard against false specifications of the dependence among the responses within a unit, the only assumption being that the responses of all units are multinormal with the same residual covariance matrix. However, the specification requires large sample sizes when there are many occasions I since $I \times (I+1)/2$ covariance parameters need to be estimated along with the regression coefficients. Estimation of the unrestricted model is also inefficient if structured versions are valid.

Factor models

We can induce dependence between responses by including factor structures in the linear predictor. This approach can also be useful for generalized linear mixed models where we often cannot freely specify conditional correlations (e.g. Rabe-Hesketh and Skrondal, 2001).

For a continuous response, a one-factor model for the residual is specified as

$$y_{ij} = \mathbf{x}'_{ij}\boldsymbol{\beta} + \lambda_i\eta_j + \epsilon_{ij},$$

where $\mathbf{x}_{ij}$ denotes covariates with fixed coefficients $\boldsymbol{\beta}$, λ_i a factor loading for occasion i and η_j a factor. $\xi_{ij} \equiv \lambda_i\eta_j + \epsilon_{ij}$ can be viewed as the total residual. Note that use of the model requires a certain degree of balance, since a factor loading is estimated for each occasion. The covariance structure of $\mathbf{y}_j$, called a *factor structure*, is given in 3.3F.i. on page 82. Note that the random intercept model arises as the special case where $\lambda_i = 1$ and the restricted random intercept model (producing compound symmetry) results when the additional restrictions $\theta_{ii} = \theta$ are imposed. For three occasions $I = 3$, the one factor model is equivalent to the unrestricted model. Special cases for $I = 3$ include a stationary first-order autoregressive residual process and a first order moving average residual process with a random intercept (e.g. Heckman, 1981c).

A multidimensional factor model for the residuals can be specified as

$$y_{ij} = \mathbf{x}'_{ij}\boldsymbol{\beta} + \boldsymbol{\lambda}_i\boldsymbol{\eta}_j + \epsilon_{ij},$$

where $\mathbf{x}_{ij}$ denote covariates with fixed coefficients $\boldsymbol{\beta}$, and $\boldsymbol{\lambda}_i$ a vector of factor loadings for occasion i and $\boldsymbol{\eta}_j$ factors. The multidimensional factor structure is shown in Display 3.3Fii on page 82. In the longitudinal setting, in contrast to measurement modeling, we believe that the choice between exploratory or confirmatory factor models should be made on a pragmatic basis, since no meaning is attributed to the factor.

For nonnormal responses, the multidimensional factor model is specified for the linear predictor as

$$\nu_{ij} = \mathbf{x}'_{ij}\boldsymbol{\beta} + \boldsymbol{\lambda}_i\boldsymbol{\eta}_j.$$

Hybrid specifications

The different dependence specifications we have surveyed can be combined. A famous example is the ARMA model which combines autoregressive and moving average models. Another possibility is to combine the random intercept model with a first order autoregressive process for the responses (e.g. Jöreskog, 1978), or with a first order autoregressive process for the residuals (e.g. Diggle, 1988), thereby relaxing the conditional independence assumption usually made in multilevel models. Other approaches include ARIMA models, where differencing is used in order to obtain stationarity (e.g. Box *et al.*, 1994) and their special cases.

3.6.5 Generalized estimating equations for nonnormal responses

Most models discussed so far are based on the notion that the dependence among responses (conditional on the covariates) can be modeled and in some sense explained by latent variables. For instance, in growth curve modeling, the random effects capture individual differences in growth trajectories and simultaneously induce residual dependence.

A radically different approach is to focus on the mean structure and relegate the dependence to a nuisance, by using generalized estimating equations (GEE) (e.g. Liang and Zeger, 1986; Zeger and Liang, 1986); see Section 6.9. The simplest version is to estimate the mean structure as if the responses were independent and then adjust standard errors for the dependence using the so-called sandwich estimator (see Section 8.3.3). The parameter estimates can be shown to be consistent, but if the responses are correlated, they are not efficient. To increase efficiency a 'working correlation matrix' is therefore specified within a multivariate extension of the iteratively reweighted least squares algorithm for generalized linear models (see Section 6.9 for details). Typically, one of the structures listed in Display 3.3 is used for the working correlation matrix of the residuals $y_{ij} - g^{-1}(\mathbf{x}'_{ij}\boldsymbol{\beta})$, as well as unrestricted and independence correlation structures. The working correlation matrix is combined with the variance function of an appropriate generalized linear model, typically allowing for overdispersion if the responses are counts. It is important to note that, apart from continuous responses, the specified correlation structures generally cannot be derived from a statistical model. Thus, there is no likelihood and GEE is a multivariate quasi-likelihood approach.

In general the regression coefficients estimated using GEE have a different interpretation than those of models including latent variables. The latter represent the *conditional effects* of covariates given the latent variables, *unit-specific effects* in longitudinal settings. GEE, on the other hand, provides *marginal* or *population averaged* effects, where the individual differences are averaged over instead of modeled by latent variables. In probit and logistic regression the marginal effects tend to be attenuated (closer to zero) compared with the conditional effects, as was shown for the probit case in Figure 1.6 on page 12. Differences between marginal and conditional effects also arise for other links and models with random coefficients, exceptions being models with an identity link and models with a log link and a normally distributed random intercept (see also Section 4.8.1).

Note that there are also 'proper' marginal statistical models with corresponding likelihoods. Examples include the Bahadur model (Bahadur, 1961) which parameterizes dependence via marginal correlations and the Dale model (Dale, 1986) which parameterizes dependence via marginal bivariate odds-ratios; see Fitzmaurice *et al.* (1993) and Molenberghs (2002) for introductions.

Whether conditional or marginal effects are of interest will depend on the context. For example, in public health, population averaged effects may be of interest, whereas conditional effects are important for the patient and clinician. Importantly, marginal effects can be derived from conditional models by integrating out the latent variables. Unfortunately, conditional effects cannot generally be derived from marginal effects. Conditional effects are more likely to be stable across populations. However, if the conditional effect is the same in two populations, but the random intercept variance differs, the marginal effects will be different.

Note that Heagerty and Zeger (2000) introduce latent variable models where the marginal mean is regressed on covariates as in GEE. In these models the relationship between the conditional mean (given the latent variables) and the covariates is found by solving an integral equation linking the conditional and marginal means (see equation (4.28) on page 123). Interestingly, the integral involved can be written as a unidimensional integral over the distribution of the sum of the terms in the random part of the model.

3.7 Summary and further reading

We have described classical latent variable models such as multilevel regression models, measurement models, exploratory and confirmatory factor models, item response models, structural equation models, latent class models and several models for longitudinal data. A unifying framework for these classical latent variable models, combining them with the response processes described in Chapter 2, is presented in the next chapter. Some classical latent variable models are employed in the Application part of this book, particularly in Chapter 9, although most applications are based on extended models.

There are a large number of books on multilevel or mixed models, see for example (in approximate order of difficulty) Kreft and de Leeuw (1998), Hox (2002), Raudenbush and Bryk (2002), Snijders and Bosker (1999), Aitkin *et al.* (2004), Goldstein (2003), Longford (1993), Cox and Solomon (2002), McCulloch and Searle (2001) and Demidenko (2004).

We have only presented a simplified version of measurement theory, not going into for instance generalizability theory (see e.g. Cronbach *et al.*, 1972; Shavelson and Webb, 1991; Brennan, 2001). For introductory reading we recommend Streiner and Norman (1995). Intermediate treatments include Crocker and Algina (1986) and Dunn (2004). An advanced and authoritative treatment is provided by Lord and Novick (1968). Lawley and Maxwell (1971) and Mulaik (1972) are useful books on factor models for continuous responses whereas Bartholomew and Knott (1999) also consider dichotomous, polytomous and mixed responses.

Books on item response theory include Lord and Novick (1968), Lord (1980), Hambleton and Swaminathan (1985), Hambleton *et al.* (1991), van der Linden and Hambleton (1997), Embretson and Reise (2000) and De Boeck and Wilson (2004). We have not discussed nonparametric item response theory (e.g. Sijtsma and Molenaar, 2002) or unfolding (ideal point) models where the item characteristic curves are nonmonotonic (e.g. Coombs, 1964; Roberts *et al.*, 2000).

An introduction to latent class modeling is given by McCutcheon (1987) and a survey is given by Clogg (1995). Books on mixture models include Everitt and Hand (1981), McLachlan and Peel (2000) and Böhning (2000).

Books on structural equation models include Dunn *et al.* (1993), Bollen (1989) and, in econometrics, Wansbeek and Meijer (2002).

We have not discussed state-space models for longitudinal data (e.g. Jones, 1993) or hidden Markov (latent transition) models (e.g. MacDonald and Zucchini, 1997; van de Pol and Langeheine, 1990). Useful books on modelling longitudinal data include Hand and Crowder (1996), Crowder and Hand (1990), Vonesh and Chinchilli (1997), Hsiao (2002), Baltagi (2001), Diggle *et al.* (2002), Lindsey (1999), Everitt and Pickles (1999) and Fitzmaurice *et al.* (2004).

CHAPTER 4

General model framework

4.1 Introduction

The general model framework unifies and generalizes the multilevel, factor, item response, latent class, structural equation and longitudinal models discussed in Chapter 3.

In that chapter we were mainly concerned with models having continuous responses. Here we describe latent variable models accommodating all the response processes discussed in Chapter 2. As we shall see, and in contrast to the models in Chapter 3, random coefficients and factors can now be included in the same model. Latent variables are also allowed to vary at several levels, yielding for instance multilevel factor models. Multilevel structural equations can be specified to regress latent variables on same and higher level latent and observed variables. We will also relax the assumption of multivariate normality of the latent variables by using other continuous or discrete distributions or nonparametric maximum likelihood. Different kinds of latent class models are also accommodated. The model framework mostly corresponds to the class of Generalized Linear Latent And Mixed Models (GLLAMM) described in Rabe-Hesketh et al. (2004a); see also Rabe-Hesketh et al. (2001a). However, we also discuss model types not accommodated within that class such as multilevel latent class models.

The essence of the general model formulation is the specification of hierarchical conditional relationships: The response model specifies the distribution of the observed responses conditional on the latent variables and covariates (via a linear predictor and link function) and in the structural model the latent variables themselves may be regressed on other latent and observed covariates. Finally, the distribution of the disturbances in the structural model is specified. Sections 4.2 to 4.4 of this chapter are therefore on:

- the response model

- the structural model for the latent variables

- the distribution of the disturbances in the structural model

An essential part of model specification concerns imposing appropriate restrictions on model parameters. Hence, different types of parameter restrictions are presented and the related notion of fundamental parameters introduced in Section 4.5.

In order to fully understand a latent variable model, it is important to consider the moment structure of the observed responses, marginal with respect to latent variables but conditional on observed covariates. To derive this, we start by deriving the reduced forms of the latent variables and the linear predictor in Section 4.6. We then derive the moment structure of the latent variables in Section 4.7 by integrating out the disturbances of the structural model. (This section and the previous are somewhat technical and might be skipped if desired.) Having obtained this moment structure, we derive the moment structure of the responses in Section 4.8 by integrating out the latent variables. This helps clarify the crucial distinction between conditional and marginal covariate effects in models with latent variables. Finally, we derive the reduced form distribution in Section 4.9, the conditional distribution of the observed responses given the explanatory variables, which represents the basis for the likelihood. The concept of reduced form parameters, which is important for the discussion of identification and equivalence in Chapter 5, is introduced in Section 4.10.

4.2 Response model

Conditional on the latent variables, the response model is a generalized linear model specified via a linear predictor ν_i, a link function $g(\cdot)$ and a distribution from the exponential family

$$f(y_i|\theta_i, \phi) = \exp\left\{\frac{y_i\theta_i - b(\theta_i)}{\phi} + c(y_i, \phi)\right\},$$

where θ_i is a function of the mean $\mu_i = g^{-1}(\nu_i)$ and ν_i depends on latent variables. Any of the conditional densities for a generalized linear model can be specified for the responses, including the extensions introduced in Chapter 2. Models for scale parameters and thresholds may also be specified. Table 4.1 lists the response types that can be handled and the application chapters discussing each type.

In Section 4.2.1 we unify conventional random coefficient and factor models, leading to a 'generalized factor' (GF) formulation of the general model described in Section 4.2.2. The GF formulation is *multivariate*; a matrix expression specifies a vector of linear predictors for a multivariate response. This multivariate formulation is useful for deriving covariance structures of the observed responses (continuous case) or of the latent responses (dichotomous, ordinal or comparative case).

The linear predictor can also be defined using the 'generalized random coefficient' (GRC) formulation described in Section 4.2.3. This formulation is *univariate* and resembles the univariate formulation of multilevel random coefficient models described in Chapter 3. An important advantage of the GRC formulation is that it includes separate terms for the parameters and covariates, making the structure of the model more explicit than the GF formulation. In Section 4.2.4, both the GF and GRC formulations are used to specify a two-level factor model. Exploratory latent class models are specified using both formulations in Section 4.2.5.

Table 4.1 *Response types handled and corresponding application chapters*

Response Type	Chapter
Continuous	
Dichotomous	Chapter 9
Ordinal	Chapter 10
Counts	Chapter 11
Durations	Chapter 12
Discrete time durations	
Continuous time durations	
Comparative	Chapter 13
Nominal	
Rankings	
Pairwise comparisons	
Mixed responses	Chapter 14

4.2.1 Unifying conventional random coefficient and factor models

Conventional random coefficient and factor models, discussed in Sections 3.2 and 3.3, are more similar than generally realized. Recall the random coefficient model from equation (3.14)

$$\mathbf{y}_j = \mathbf{X}_j \boldsymbol{\beta} + \mathbf{Z}_j \boldsymbol{\eta}_j + \boldsymbol{\epsilon}_j, \tag{4.1}$$

and the measurement part of the structural equation model in equation (3.34)

$$\mathbf{y}_j = (\boldsymbol{\nu} + \mathbf{K}\mathbf{x}_j) + \boldsymbol{\Lambda} \boldsymbol{\eta}_j + \boldsymbol{\epsilon}_j, \tag{4.2}$$

where some subscripts and superscripts have been omitted to simplify notation.

Although different in interpretation, these models have a similar structure. In the random coefficient model $\mathbf{y}_j$ represents the vector of responses for the level-1 units within the jth level-2 unit whereas $\mathbf{y}_j$ represents the items in the common factor model. To facilitate the subsequent development, we will refer to the elementary units i as *level-1 units* whether they are the lowest level units in a multilevel setting or items in a factor model. The clusters j in random coefficient models or units in common factor models are then *level-2 units*. The disturbances $\boldsymbol{\epsilon}_j$ in random coefficient models correspond to unique factors in common factor models, henceforth referred to as 'errors'. The random effects $\boldsymbol{\eta}_j$ in random coefficient models correspond to the common factors in common factor models. We will use the term *latent variables* for either random effects or common factors $\boldsymbol{\eta}_j$.

The design matrix $\mathbf{Z}_j$ for the random effects corresponds to the factor loading matrix $\boldsymbol{\Lambda}$. There are two differences between $\mathbf{Z}_j$ and $\boldsymbol{\Lambda}$. First, $\mathbf{Z}_j$ is a known matrix of covariates and constants whereas $\boldsymbol{\Lambda}$ is an unknown parameter matrix. Second, while $\mathbf{Z}_j$ can differ between level-2 units j, whereas $\boldsymbol{\Lambda}$

is generally constant. Nevertheless, we will refer to both $\mathbf{Z}_j$ and $\mathbf{\Lambda}$ as the *structure matrix*, denoted $\mathbf{\Lambda}_j$.

The fixed parts $\mathbf{X}_j\boldsymbol{\beta}$ and $(\boldsymbol{\nu} + \mathbf{K}\mathbf{x}_j)$ in the two models can be used to specify the same mean structure. In the case of a single covariate, the terms for the ith row or ith level-1 unit are $\beta_0 + x_{ij}\beta$ and $\nu_i + k_i x_j$, respectively. Whereas the former assumes a constant effect β of a level-1 specific covariate x_{ij}, the latter assumes level-1 specific effects k_i of a level-2 specific covariate x_j. However, this difference is superficial: In the random coefficient model, interactions with dummy variables for the level-1 units i can be used to allow coefficients to depend on i. In the factor model, different covariates can be used for different i to represent a level-1 unit-specific covariate.

A *response model* unifying and generalizing both random coefficient and factor models can now be written as

$$\mathbf{y}_j = \mathbf{X}_j\boldsymbol{\beta} + \mathbf{\Lambda}_j\boldsymbol{\eta}_j + \boldsymbol{\epsilon}_j, \tag{4.3}$$

where the structure matrix $\mathbf{\Lambda}_j$ can contain both variables and parameters. The unifying notation and terminology is summarized in Display 4.1.

Display 4.1 Unifying notation and terminology.

Unified model		Random coefficient model		Factor model	
symbol	term	symbol	interpretation	symbol	interpretation
i	level-1 units	i	level-1 units	i	items
j	level-2 units	j	level-2 units	j	units
$\mathbf{y}_j$	responses	$\mathbf{y}_j$	responses	$\mathbf{y}_j$	responses
$\boldsymbol{\epsilon}_j$	errors	$\boldsymbol{\epsilon}_j$	disturbances	$\boldsymbol{\epsilon}_j$	unique factors
$\boldsymbol{\eta}_j$	latent variables	$\boldsymbol{\eta}_j$	random coefficients	$\boldsymbol{\eta}_j$	common factors
$\mathbf{\Lambda}_j$	structure matrix	$\mathbf{Z}_j$	design matrix	$\mathbf{\Lambda}$	factor loading matrix
$\mathbf{X}_j\boldsymbol{\beta}$	fixed part	$\mathbf{X}_j\boldsymbol{\beta}$	fixed part	$(\boldsymbol{\nu} + \mathbf{K}\mathbf{x}_j)$	fixed part

This model framework can also be used to formulate different kinds of latent class models if the latent variables are discrete. A mixture regression model, for instance a latent class growth model (see page 86), is obtained simply by using discrete latent variables in a random coefficient model; see Section 9.5, page 304, for an example in meta-analysis. Section 4.2.5 shows how exploratory latent class models are formulated using this framework.

Note that treating the items of a factor model or the variables comprising any multivariate response as level-1 units and the original units as level-2 clusters is a common approach in multivariate multilevel regression modeling (e.g. Goldstein, 2003, Chapter 6). An advantage of this approach is that missing responses then merely result in varying cluster sizes which can be handled by multilevel modeling software if responses are missing at random (MAR) (see Section 8.3.1 for types of missingness). The same approach is adopted

by Raudenbush and Sampson (1999ab) and Raudenbush and Bryk (2002) for
one-parameter item response models and Rijmen *et al.* (2003) and De Boeck
and Wilson (2004) for more general item response models.

4.2.2 Linear predictor in generalized factor (GF) formulation

The main advantage of considering the linear predictor is that all response
processes considered in Chapter 2 are accommodated.

The unified model in (4.3) can be written in generalized factor (GF) formulation by writing the vector of linear predictors for the responses on unit
j as

$$\boldsymbol{\nu}_j = \mathbf{X}_j\boldsymbol{\beta} + \boldsymbol{\Lambda}_j\boldsymbol{\eta}_j$$

and specifying an identity link and a normal density for $\mathbf{y}_j$ given $\boldsymbol{\nu}_j$.

Before introducing the multilevel extension, we will reintroduce the subscripts and superscripts for the levels of the model used in Section 3.2.1 to
write the model as

$$\boldsymbol{\nu}_{j(2)} = \mathbf{X}_{j(2)}\boldsymbol{\beta} + \boldsymbol{\Lambda}_{j(2)}\boldsymbol{\eta}_j^{(2)}. \tag{4.4}$$

Vectors with the $j(2)$ subscript contain *all elements for the jth level-2 unit*
whereas latent variables with the (2) superscript *vary at level 2*. Note that
$\boldsymbol{\eta}_{j(2)} = \boldsymbol{\eta}_j^{(2)}$ for any two-level model.

Display 4.2A on page 100 uses the GF formulation to represent the random
part of a single-level two-factor model for five items, where the first three
items load on factor 1 whereas the last two load on factor 2. Display 4.3A on
page 101 uses the same notation to represent the random part of a two-level
random coefficient model. Here there are three level-1 units in the jth level-
2 unit and the model includes a random intercept and a random slope of a
covariate t_{ij}.

We can now generalize the model to L levels as

$$\boldsymbol{\nu}_{z(L)} = \mathbf{X}_{z(L)}\boldsymbol{\beta} + \sum_{l=2}^{L} \boldsymbol{\Lambda}_{z(L)}^{(l)}\boldsymbol{\eta}_{z(L)}^{(l)}, \tag{4.5}$$

where $\boldsymbol{\nu}_{z(L)}$ is the *vector of linear predictors for all units in a particular level-L
(top level) unit z* and $\boldsymbol{\eta}_{z(L)}^{(l)}$ is the *vector of all (realizations of) the level-l latent
variables for that level-L unit* (see bottom right-hand panel of Display 4.4 on
page 102). The (l) superscript of $\boldsymbol{\Lambda}_{z(L)}^{(l)}$ denotes that the matrix is specific to
the level-l latent variables. As in Chapter 3, Display 3.2 on page 59, we can
alternatively write this model as a two-level model

$$\boldsymbol{\nu}_{z(L)} = \mathbf{X}_{z(L)}\boldsymbol{\beta} + \boldsymbol{\Lambda}_{z(L)}\boldsymbol{\eta}_{z(L)}, \tag{4.6}$$

where $\boldsymbol{\Lambda}_{z(L)} = [\boldsymbol{\Lambda}_{z(L)}^{(2)}, \dots, \boldsymbol{\Lambda}_{z(L)}^{(L)}]$ and $\boldsymbol{\eta}_{z(L)} = (\boldsymbol{\eta}_{z(L)}^{(2)\prime}, \dots, \boldsymbol{\eta}_{z(L)}^{(L)\prime})'$ is the vector
of all (realizations of all) latent variables for the zth level-L unit (see top
right-hand panel of Display 4.4).

Display 4.2 Random part of a single-level two-factor model.

A. GF formulation:

$$\mathbf{\Lambda}_{j(2)}\boldsymbol{\eta}_j^{(2)} = \begin{bmatrix} \lambda_{11} & 0 \\ \lambda_{21} & 0 \\ \lambda_{31} & 0 \\ 0 & \lambda_{42} \\ 0 & \lambda_{52} \end{bmatrix} \begin{bmatrix} \eta_{1j}^{(2)} \\ \eta_{2j}^{(2)} \end{bmatrix} = \begin{bmatrix} \lambda_{11}\eta_{1j}^{(2)} \\ \lambda_{21}\eta_{1j}^{(2)} \\ \lambda_{31}\eta_{1j}^{(2)} \\ \lambda_{42}\eta_{2j}^{(2)} \\ \lambda_{52}\eta_{2j}^{(2)} \end{bmatrix}$$

B. GRC formulation in matrix form:

$$\sum_{m=1}^{M} \eta_{mj}^{(2)}\mathbf{Z}_{mj}^{(2)}\boldsymbol{\lambda}_m^{(2)} = \eta_{1j}^{(2)} \underbrace{\begin{bmatrix} 1 & 0 & 0 \\ 0 & 1 & 0 \\ 0 & 0 & 1 \\ 0 & 0 & 0 \\ 0 & 0 & 0 \end{bmatrix}}_{\mathbf{Z}_{1j}^{(2)}} \underbrace{\begin{bmatrix} \lambda_{11} \\ \lambda_{21} \\ \lambda_{31} \end{bmatrix}}_{\boldsymbol{\lambda}_1^{(2)}} + \eta_{2j}^{(2)} \underbrace{\begin{bmatrix} 0 & 0 \\ 0 & 0 \\ 0 & 0 \\ 1 & 0 \\ 0 & 1 \end{bmatrix}}_{\mathbf{Z}_{2j}^{(2)}} \underbrace{\begin{bmatrix} \lambda_{42} \\ \lambda_{52} \end{bmatrix}}_{\boldsymbol{\lambda}_2^{(2)}}$$

$$= \begin{bmatrix} \lambda_{11}\eta_{1j}^{(2)} \\ \lambda_{21}\eta_{1j}^{(2)} \\ \lambda_{31}\eta_{1j}^{(2)} \\ \lambda_{42}\eta_{2j}^{(2)} \\ \lambda_{52}\eta_{2j}^{(2)} \end{bmatrix}$$

C. GRC formulation:

$$\sum_{m=1}^{M} \eta_{mj}^{(2)}\mathbf{z}_{mij}^{(2)\prime}\boldsymbol{\lambda}_m^{(2)} = \eta_{1j}^{(2)}\boldsymbol{\delta}_{1i}'\boldsymbol{\lambda}_1^{(2)} + \eta_{2j}^{(2)}\boldsymbol{\delta}_{2i}'\boldsymbol{\lambda}_2^{(2)},$$

where $\boldsymbol{\delta}_{1i}'$ is the ith row of $\mathbf{Z}_{1j}^{(2)}$ and $\boldsymbol{\delta}_{2i}'$ is the ith row of $\mathbf{Z}_{2j}^{(2)}$.

For $i=1$:

$$\eta_{1j}^{(2)}\boldsymbol{\delta}_{1i}'\boldsymbol{\lambda}_1^{(2)} + \eta_{2j}^{(2)}\boldsymbol{\delta}_{2i}'\boldsymbol{\lambda}_2^{(2)} = \eta_{1j}^{(2)} \begin{bmatrix} 1 & 0 & 0 \end{bmatrix} \begin{bmatrix} \lambda_{11} \\ \lambda_{21} \\ \lambda_{31} \end{bmatrix} + \eta_{2j}^{(2)} \begin{bmatrix} 0 & 0 \end{bmatrix} \begin{bmatrix} \lambda_{42} \\ \lambda_{52} \end{bmatrix} = \lambda_{11}\eta_{1j}^{(2)}$$

Display 4.3 Random part of a two-level random coefficient model.

A. GF formulation:

$$\mathbf{\Lambda}_{j(2)}\boldsymbol{\eta}_j^{(2)} = \begin{bmatrix} 1 & t_{1j} \\ 1 & t_{2j} \\ 1 & t_{3j} \end{bmatrix} \begin{bmatrix} \eta_{1j}^{(2)} \\ \eta_{2j}^{(2)} \end{bmatrix} = \begin{bmatrix} \eta_{1j}^{(2)} + t_{1j}\eta_{2j}^{(2)} \\ \eta_{1j}^{(2)} + t_{2j}\eta_{2j}^{(2)} \\ \eta_{1j}^{(2)} + t_{3j}\eta_{2j}^{(2)} \end{bmatrix}$$

B. GRC formulation in matrix form:

$$\sum_{m=1}^{M}\eta_{mj}^{(2)}\mathbf{Z}_{mj}^{(2)}\boldsymbol{\lambda}_m^{(2)} = \eta_{1j}^{(2)}\underbrace{\begin{bmatrix} 1 \\ 1 \\ 1 \end{bmatrix}}_{\mathbf{Z}_{1j}^{(2)}}\underbrace{\begin{bmatrix} 1 \end{bmatrix}}_{\lambda_1^{(2)}} + \eta_{2j}^{(2)}\underbrace{\begin{bmatrix} t_{1j} \\ t_{2j} \\ t_{3j} \end{bmatrix}}_{\mathbf{Z}_{2j}^{(2)}}\underbrace{\begin{bmatrix} 1 \end{bmatrix}}_{\lambda_2^{(2)}} = \begin{bmatrix} \eta_{1j}^{(2)} + t_{1j}\eta_{2j}^{(2)} \\ \eta_{1j}^{(2)} + t_{2j}\eta_{2j}^{(2)} \\ \eta_{1j}^{(2)} + t_{3j}\eta_{2j}^{(2)} \end{bmatrix}$$

C. GRC formulation:

$$\sum_{m=1}^{M}\eta_{mj}^{(2)}\mathbf{z}_{mij}^{(2)\prime}\boldsymbol{\lambda}_m^{(2)} = \eta_{1j}^{(2)} \times 1 \times 1 + \eta_{2j}^{(2)}t_{ij} \times 1 = \eta_{1j}^{(2)} + \eta_{2j}^{(2)}t_{ij}$$

4.2.3 Linear predictor in generalized random coefficient (GRC) formulation

For simplicity, we begin by considering a two-level model. In the GF formulation (4.4), the structure matrix $\mathbf{\Lambda}_{j(2)}$ is neither a pure design matrix nor a pure parameter matrix. Instead it contains both variables and parameters. We can spell out the form of $\mathbf{\Lambda}_{j(2)}\boldsymbol{\eta}_j^{(2)}$ by expanding it in terms of pure design matrices $\mathbf{Z}_{mj}^{(2)}$ and pure parameter vectors $\boldsymbol{\lambda}_m^{(2)}$ so that (4.4) becomes

$$\boldsymbol{\nu}_{j(2)} = \mathbf{X}_{j(2)}\boldsymbol{\beta} + \underbrace{\sum_{m=1}^{M}\eta_{mj}^{(2)}\mathbf{Z}_{mj}^{(2)}\boldsymbol{\lambda}_m^{(2)}}_{\mathbf{\Lambda}_{j(2)}\boldsymbol{\eta}_j^{(2)}} \qquad (4.7)$$

where $\eta_{mj}^{(2)}$ is the mth latent variable, $\mathbf{Z}_{mj}^{(2)}$ is an $n_j^{(2)} \times p_m^{(2)}$ (design) matrix of covariates and fixed known constants and $\boldsymbol{\lambda}_m^{(2)}$ is a vector of $p_m^{(2)}$ parameters associated with the mth latent variable. The product $\mathbf{Z}_{mj}^{(2)}\boldsymbol{\lambda}_m^{(2)}$ represents the mth column of $\mathbf{\Lambda}_{j(2)}$; $\boldsymbol{\lambda}_m^{(2)}$ is therefore not a vector in the matrix $\mathbf{\Lambda}_{j(2)}$. We will refer to this formulation as the 'GRC formulation in matrix form'.

A latent variable $\eta_{mj}^{(2)}$ can typically be interpreted as a *factor* if all elements in the corresponding matrix $\mathbf{Z}_{mj}^{(2)}$ are zero or one. In this case $\boldsymbol{\lambda}_m^{(2)}$ contains the $p_m^{(2)}$ nonzero factor loadings for that factor and the role of $\mathbf{Z}_{mj}^{(2)}$ is to assign the correct factor loadings to the different items. This can be seen in Display 4.2B,

Display 4.4 Different types of latent variable vectors.

Consider a level-3 unit k containing three level-2 units. There is one latent variable at level-3 ($M_3 = 1$) and two latent variables at level-2 ($M_2 = 2$). The elements are $\eta^{(2)}_{mjk}$, $m = 1, 2$, $j = 1, 2, 3$ and $\eta^{(3)}_{mk}$, $m = 1$.

$\boldsymbol{\eta}_{2k}$: Level-3 element k and level-2 element $j = 2$ in k:

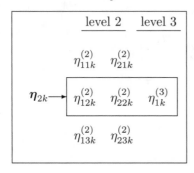

$\boldsymbol{\eta}_{k(3)}$: Level-3 element k and all level-2 elements in k:

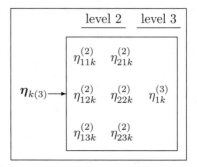

$\boldsymbol{\eta}^{(2)}_{2k}$: Level-2 element $j = 2$ in k

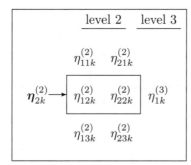

$\boldsymbol{\eta}^{(2)}_{k(3)}$: All level-2 elements in k

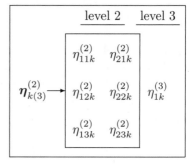

where the columns of $\mathbf{Z}^{(2)}_{mj}$ are simply dummy variables for the items loading on the mth factor, i.e. vectors equal to zero except in the position corresponding to the relevant item. This formulation of factor models is natural if the items are treated as level-1 units and all responses are stacked in a single response vector $\mathbf{y}_{j(2)}$ since the obvious way of 'referring' to a particular item in this case is by dummy variables. For the example in Display 4.2B, the 'GRC formulation in matrix form' can simply be read as 'common factor 1 is measured by items 1 to 3 with factor loadings in $\boldsymbol{\lambda}^{(2)}_1$' and 'common factor 2 is measured by items 4 to 5 with factor loadings in $\boldsymbol{\lambda}^{(2)}_2$'. Note that in Display 4.2 the elements of $\boldsymbol{\lambda}^{(2)}_m$ have been indexed as λ_{rm} for consistency with the conventional notation for factor models, but will be indexed $\lambda_{mr}^{(2)}$ for the rth element of $\boldsymbol{\lambda}_m^{(2)}$ in the remainder of the book.

A latent variable $\eta_{mj}^{(2)}$ is a *random coefficient* of some variable z_{ij} if the corresponding matrix $\mathbf{Z}_{mj}^{(2)}$ is a column vector with corresponding scalar 'factor loading' $\lambda_m^{(2)}$ set to 1. This is illustrated in Display 4.3B.

The ith row of (4.7) becomes

$$\nu_{ij} = \mathbf{x}_{ij}'\boldsymbol{\beta} + \sum_{m=1}^{M} \eta_{mj}^{(2)} \mathbf{z}_{mij}^{(2)\prime} \boldsymbol{\lambda}_m^{(2)}, \tag{4.8}$$

where $\mathbf{x}_{ij}'$ is the ith row of $\mathbf{X}_{j(2)}$ and $\mathbf{z}_{mij}^{(2)\prime}$ is the ith row of $\mathbf{Z}_{mj}^{(2)}$. This is the GRC formulation for a two-level model.

As shown in Display 4.2C on page 100, expressing factor models in this notation requires dummy vectors $\boldsymbol{\delta}_{mi}$ with $p_m^{(2)}$ elements (where $p_m^{(2)}$ is the number of items measuring or 'loading on' the mth factor), equal to 1 for the element of $\boldsymbol{\lambda}_m^{(2)}$ that represents the factor loading for item i on factor m and 0 otherwise. Display 4.3C shows how a random coefficient model can be expressed using this notation. Here $\mathbf{z}_{mij}^{(2)}$ are scalars with corresponding parameters $\lambda_m = 1$.

The model can be extended to L levels as

$$\nu_i = \mathbf{x}_i'\boldsymbol{\beta} + \sum_{l=2}^{L} \sum_{m=1}^{M_l} \eta_m^{(l)} \mathbf{z}_{mi}^{(l)\prime} \boldsymbol{\lambda}_m^{(l)}, \tag{4.9}$$

where M_l is the number of latent variables at level l and we have omitted higher-level observation indices to simplify notation.

See Rabe-Hesketh and Pickles (1999) and Rabe-Hesketh *et al.* (2000) for further examples of the GRC formulation.

4.2.4 Example: Two-level factor model in GF and GRC formulation

We can now define a two-level factor model. Such a model is useful if the units providing responses to the items are nested in clusters, for instance pupils in schools. A single-level factor model would not be appropriate in this case since responses from different units in the same cluster are likely to be correlated. For continuous or latent responses, it is typically assumed that

$$\mathbf{y}_{j(2)}^* \sim \mathrm{N}(\boldsymbol{\mu}_j, \boldsymbol{\Sigma}_1)$$

$$\boldsymbol{\mu}_j \sim \mathrm{N}(\boldsymbol{\mu}, \boldsymbol{\Sigma}_2),$$

where $\boldsymbol{\mu}$ and $\boldsymbol{\mu}_j$ are vectors of intercepts. Separate common factor models are then specified to structure the covariance matrices $\boldsymbol{\Sigma}_1$ and $\boldsymbol{\Sigma}_2$ at the unit and cluster levels (e.g., Longford and Muthén, 1992; Poon and Lee, 1992; Longford, 1993, Linda *et al.*, 1993; Muthén, 1994; Lee and Shi, 2001). The common factors at the cluster-level can then be interpreted as cluster-level constructs which may have a different factor structure than unit-level constructs.

Note that the models are conventionally called two-level factor models since the units are treated as level-1 and the clusters as level-2 with the vector of responses for each unit treated as a multivariate response. If the items are treated as level-1 units, as in our GF and GRC formulations, the model becomes a three-level model with units at level 2 and clusters at level 3. For simplicity, we will assume that there are single common factors $\eta_{1jk}^{(2)}$ and $\eta_{1k}^{(3)}$ at levels 2 and 3. The unique factors at level 3 are denoted $\eta_{mk}^{(3)}$, $m = 2, \ldots, I+1$. A graphical representation of the model for $I = 3$ is given in Figure 4.1. The items and level-2 factor vary at the unit level as indicated by the inner

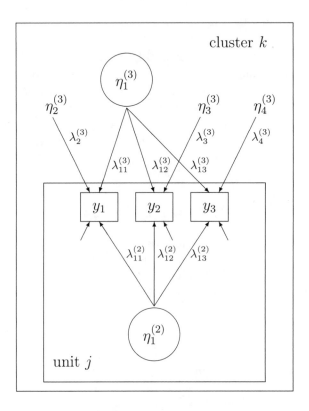

Figure 4.1 *Path diagram of 'two-level' factor model*

box labelled 'unit j'. The model in this box is an ordinary unidimensional factor model with unit-level specific factors indicated by the unlabelled arrows representing level-1 variability. The latent variables outside the inner box vary at the cluster-level as indicated by the 'cluster k' label in the outer box. These latent variables represent a common factor measured by all items and unique factors measured by one item each. Unlike the common factor, the unique factors are not shown as circles since they merely represent residual terms (without factor loadings). In contrast to conventional path diagrams, an arrow pointing at the observed response y represents a possibly nonlinear relation, for instance a logit link function, in the diagrams presented in this book. The short unlabelled arrows pointing at the observed responses do not necessarily represent additive error terms. They could for instance represent Poisson variability for counts.

The GF formulation of this model is shown in Display 4.5A on page 106 for a level-3 unit k with two level-2 units $j = 1, 2$. Note that the factor loadings in the figure and display are labelled according to the GRC formulation shown in Display 4.5B. For the ith item, each common factor is multiplied by the dummy vector $\mathbf{d}_i'$, with ith element equal to one and all other elements equal to zero, to pick the ith factor loadings from $\boldsymbol{\lambda}_1^{(2)}$ and $\boldsymbol{\lambda}_1^{(3)}$. The mth unique factor is multiplied by $d_{im}\lambda_{m+1}^{(3)}$, equal to $\lambda_{i+1}^{(3)}$ if $m = i$ and 0 otherwise. Note that parameter restrictions are necessary to identify the model, for example, setting the factor loadings $\lambda_{11}^{(2)}$, $\lambda_{11}^{(3)}$, $\lambda_2^{(3)}$, $\lambda_3^{(3)}$ and $\lambda_4^{(3)}$ to 1. See Section 5.2 for a detailed discussion of identification. A simpler version of this model is discussed on page 110; see also Figure 4.3(a).

4.2.5 Example: Exploratory latent class model in GF and GRC formulation

We have so far implicitly assumed that the latent variables are continuous. However, we can combine the same response model with discrete latent variables to define latent class models. Here, two-level models are usually sufficient. Let $\boldsymbol{\eta}_j^{(2)}$ take on discrete values $\mathbf{e}_c$ with probabilities π_c, where we constrain the mean to zero,

$$\sum_c \pi_c \mathbf{e}_c = \mathbf{0}.$$

For an exploratory latent class model for dichotomous responses, the linear predictor for item i, unit j and class c can be written as

$$\nu_{ijc} = \beta_i + e_{ic}, \; i = 1, \ldots, I.$$

Using the multivariate GF formulation, this becomes

$$\boldsymbol{\nu}_{j(2)} = \boldsymbol{\beta} + \boldsymbol{\eta}_j^{(2)}, \tag{4.10}$$

where $\boldsymbol{\eta}_j^{(2)} = \mathbf{e}_c$, $c = 1, \ldots, C$, is an I-dimensional latent variable and $\mathbf{X}_{j(2)} = \mathbf{I}$ and $\boldsymbol{\Lambda}_{j(2)} = \mathbf{I}$ in this case.

$\mathbf{I}$ is an I-dimensional identity matrix. Using the GRC formulation, the linear predictor is

$$\nu_{ij} = \mathbf{d}_i' \boldsymbol{\beta} + \sum_{m=1}^{I} \eta_{jm}^{(2)} d_{mi}, \tag{4.11}$$

where $\mathbf{d}_i'$ is the ith row of the I-dimensional unit matrix and d_{mi} is the mth element of $\mathbf{d}_i$, a dummy variable for $m = i$.

In the case of polytomous or other comparative responses with S_i categories

Display 4.5 Random part of two-level factor model.

For a level-3 unit k with two level-2 units $j = 1, 2$:

A. GF Formulation

$$
\mathbf{\Lambda}^{(2)}_{k(3)}\boldsymbol{\eta}^{(2)}_{k(3)} + \mathbf{\Lambda}^{(3)}_{k(3)}\boldsymbol{\eta}^{(3)}_{k(3)} =
\underbrace{\begin{bmatrix}
\lambda^{(2)}_{11} & 0 \\
\vdots & \vdots \\
\lambda^{(2)}_{1I} & 0 \\
\hline
0 & \lambda^{(2)}_{11} \\
\vdots & \vdots \\
0 & \lambda^{(2)}_{1I}
\end{bmatrix}}_{\mathbf{\Lambda}^{(2)}_{k(3)}}
\underbrace{\begin{bmatrix}
\eta^{(2)}_{11k} \\
\eta^{(2)}_{12k}
\end{bmatrix}}_{\boldsymbol{\eta}^{(2)}_{k(3)}}
$$

$$
+
\underbrace{\begin{bmatrix}
\lambda^{(3)}_{11} & \lambda^{(3)}_{2} & \cdots & 0 \\
\vdots & \vdots & \ddots & \vdots \\
\lambda^{(3)}_{1I} & 0 & \cdots & \lambda^{(3)}_{I+1} \\
\hline
\lambda^{(3)}_{11} & \lambda^{(3)}_{2} & \cdots & 0 \\
\vdots & \vdots & \ddots & \vdots \\
\lambda^{(3)}_{1I} & 0 & \cdots & \lambda^{(3)}_{I+1}
\end{bmatrix}}_{\mathbf{\Lambda}^{(3)}_{k(3)}}
\underbrace{\begin{bmatrix}
\eta^{(3)}_{1k} \\
\eta^{(3)}_{2k} \\
\vdots \\
\eta^{(3)}_{I+1,k}
\end{bmatrix}}_{\boldsymbol{\eta}^{(3)}_{k(3)}}
$$

$$
\mathbf{\Lambda}_{k(3)}\boldsymbol{\eta}_{k(3)} =
\underbrace{\begin{bmatrix}
\lambda^{(2)}_{11} & 0 & \lambda^{(3)}_{11} & \lambda^{(3)}_{2} & \cdots & 0 \\
\vdots & \vdots & \vdots & \vdots & \ddots & \vdots \\
\lambda^{(2)}_{1I} & 0 & \lambda^{(3)}_{1I} & 0 & \cdots & \lambda^{(3)}_{I+1} \\
\hline
0 & \lambda^{(2)}_{11} & \lambda^{(3)}_{11} & \lambda^{(3)}_{2} & \cdots & 0 \\
\vdots & \vdots & \vdots & \vdots & \ddots & \vdots \\
0 & \lambda^{(2)}_{1I} & \lambda^{(3)}_{1I} & 0 & \cdots & \lambda^{(3)}_{I+1}
\end{bmatrix}}_{\mathbf{\Lambda}_{k(3)}}
\underbrace{\begin{bmatrix}
\eta^{(2)}_{11k} \\
\eta^{(2)}_{12k} \\
\eta^{(3)}_{1k} \\
\eta^{(3)}_{2k} \\
\vdots \\
\eta^{(3)}_{I+1,k}
\end{bmatrix}}_{\boldsymbol{\eta}_{k(3)}}
$$

B. GRC Formulation

$$
\eta^{(2)}_{1jk}\mathbf{z}^{(2)\prime}_{1i}\boldsymbol{\lambda}^{(2)}_{1} + \sum_{m=1}^{I+1}\eta^{(3)}_{mk}\mathbf{z}^{(3)\prime}_{mi}\boldsymbol{\lambda}^{(3)}_{m} = \eta^{(2)}_{1jk}\,\underbrace{\mathbf{d}_i'}_{\mathbf{z}^{(2)\prime}_{1i}}\underbrace{\begin{bmatrix}\lambda^{(2)}_{11} \\ \vdots \\ \lambda^{(2)}_{1I}\end{bmatrix}}_{\boldsymbol{\lambda}^{(2)}_{1}} + \eta^{(3)}_{1k}\,\underbrace{\mathbf{d}_i'}_{\mathbf{z}^{(3)\prime}_{1i}}\underbrace{\begin{bmatrix}\lambda^{(3)}_{11} \\ \vdots \\ \lambda^{(3)}_{1I}\end{bmatrix}}_{\boldsymbol{\lambda}^{(3)}_{1}}
$$

$$
+ \quad \eta^{(3)}_{2k}\,\underbrace{d_{i1}}_{z^{(3)}_{2i}}\lambda^{(3)}_{2} + \cdots + \eta^{(3)}_{I+1,k}\,\underbrace{d_{i,I}}_{z^{(3)}_{I+1,i}}\lambda^{(3)}_{I+1}
$$

$$
= \quad \eta^{(2)}_{1jk}\lambda^{(2)}_{1i} + \eta^{(3)}_{2k}\lambda^{(3)}_{1i} + \eta^{(3)}_{i+1,k}\lambda^{(3)}_{i+1}
$$

where $d_{ir} = \begin{cases} 1 & \text{if } r = i \\ 0 & \text{otherwise} \end{cases}$ and $\mathbf{d}_i' = (d_{i1}, d_{i2}, \ldots, d_{iI})$, for item i, the linear predictor for unit j, item i, response category s and class c is

$$\nu_{sijc} = \beta_i^s + e_{ic}^s,$$

where $\beta_i^1 = e_{ic}^1 = 0$ for $i = 1, \ldots, I$, $c = 1, \ldots, C$. We stack the linear predictors for the different response categories and items into a single vector $\boldsymbol{\nu}_{j(2)}$. The model can then be written as in (4.10) except that $\boldsymbol{\eta}_j^{(2)}$ now has dimension $R = \sum_i S_i - I$ and the identity matrices are replaced by $(R+I) \times R$ dimensional structure matrices, equal to $(R+I) \times (R+I)$ dimensional identity matrices with those I columns removed that correspond to the first response categories for each item. In the GRC formulation in (4.11), $\mathbf{d}_i'$ is then replaced by the ith row of this structure matrix. We will use an exploratory latent class model to analyze dichotomous diagnostic tests in Section 9.3 and rankings of political values in Section 13.5.

4.2.6 Relaxing conditional independence

A basic assumption of the model framework is that the responses are conditionally independent given the latent variables and covariates, an assumption also known as 'local independence'. While this may appear restrictive, we can always generate more complex dependence structures by including further latent variables. For example, in a common factor model we can induce a correlation among two responses, conditional on the common factor, by making the responses load on a further latent variable with factor loadings set equal to 1. Similarly, Qu *et al.* (1996) relax the conditional independence assumption in a latent class model by including a common factor for all items with class-specific factor loadings.

To induce dependence between the residuals ϵ_{1j} of a latent response in a probit regression and ϵ_{2j} of an observed response in a linear regression (as in the famous Heckman selection model, e.g. Heckman, 1979), we can specify

$$\epsilon_{ij} = \eta_j^{(2)} \lambda_i + e_{ij}, \quad \lambda_1 = 1$$

where the level-1 residuals e_{1j} and e_{2j} are independently normally distributed with zero means and variances θ_{11} and θ_{22}. Two further restrictions need to be imposed on the four parameters $(\psi, \lambda_2, \theta_{11}$ and $\theta_{22})$ since only the residual variance of the linear regression model,

$$\mathrm{Var}(\epsilon_{2j}) = \lambda_2^2 \psi + \theta_{22}, \tag{4.12}$$

and the correlation between the total residuals of the two models

$$\mathrm{Cor}(\epsilon_{1j}, \epsilon_{2j}) = \frac{\lambda_2 \psi}{\sqrt{(\psi + \theta_{11})(\lambda_2^2 \psi + \theta_{22})}}$$

are identified (see Section 5.2 for a general treatment of identification). We cannot set λ_2 to a constant because this would determine the sign of the correlation. An obvious choice would therefore be to set $\theta_{11} = 1$ (as usual in probit regression) and $\psi = 1$. However, the correlation between the total residuals of the two models then becomes

$$\mathrm{Cor}(\epsilon_{1j}, \epsilon_{2j}) = \frac{\lambda_2}{\sqrt{(2)(\lambda_2^2 + \theta_{22})}} \leq \frac{\lambda_2}{\sqrt{2\lambda_2^2}} = \frac{1}{\sqrt{2}},$$

where the upper bound results if θ_{22} in the linear regression model is zero. To avoid an upper bound (less than 1) for the correlation, we therefore suggest the restrictions

$$\psi = 1, \ \theta_{11} = \theta_{22}.$$

For categorical responses, we can also relax conditional independence without including further latent variables in the model. Consider two dichotomous responses y_1 and y_2. We can treat the four possible response patterns $(0,0)$, $(1,0)$, $(0,1)$ and $(1,1)$ as a single multinomial response and model the probabilities as

$$\mathrm{Pr}(y_1, y_2) = \frac{\exp(\beta_1 y_1 + \beta_2 y_2 + \beta_{12} y_1 y_2)}{\sum_{z_1=0,1} \sum_{z_2=0,1} \exp(\beta_1 z_1 + \beta_2 z_2 + \beta_{12} z_1 z_2)}.$$

Extra dependence, in addition to that induced by latent variables, results if $\beta_{12} \neq 0$. This way of introducing 'local' dependence has been suggested for latent class models by Harper (1972) and Hagenaars (1988) among others, but is generally applicable to latent variable models with categorical responses. For instance, in item response modeling, local dependence among a group of items is sometimes accommodated by combining the items into a single response called 'testlet' (Wainer and Kiely, 1987) or 'item bundle' (Wilson and Adams, 1995).

4.3 Structural model for the latent variables

4.3.1 Continuous latent variables

In order to define the structural model, we first define the latent variable vector $\boldsymbol{\eta}_{j...z} = (\boldsymbol{\eta}_{jk...z}^{(2)\prime}, \boldsymbol{\eta}_{k...z}^{(3)\prime}, \ldots, \boldsymbol{\eta}_z^{(L)\prime})'$ containing *all latent variables for the jth level-2 unit*, where $k \ldots z$ are the indices for units at levels 3 to L. The vector could also be denoted $\boldsymbol{\eta}_{jk...z(2)}$ or simply $\boldsymbol{\eta}_j$. For a three-level model, $\boldsymbol{\eta}_{jk}$ is shown for $j = 2$ in the top left panel of Display 4.4 on page 102. The structural model for the latent variables has the form (omitting higher-level subscripts)

$$\boldsymbol{\eta}_j = \mathbf{B}\boldsymbol{\eta}_j + \boldsymbol{\Gamma}\mathbf{w}_j + \boldsymbol{\zeta}_j, \qquad (4.13)$$

where $\mathbf{B}$ is an $M \times M$ matrix of regression parameters, $M = \sum_l M_l$, $\mathbf{w}_j$ is a vector of R covariates, $\boldsymbol{\Gamma}$ is an $M \times R$ matrix of regression parameters and $\boldsymbol{\zeta}_j$ is a vector of M errors or disturbances. This model is essentially a generalization of the conventional single-level structural model (e.g. Muthén, 1984) to a multilevel setting. The crucial difference is that latent and observed variables may vary at different levels in our framework. Each element of $\boldsymbol{\zeta}_j$ varies at the same level as the corresponding element of $\boldsymbol{\eta}_j$.

It would not make sense to regress a higher level latent variable on a lower level latent or observed variable since this would force the higher level variable to vary at a lower level. In terms of the blocks of $\mathbf{B}$ corresponding to the vectors of latent variables at each level, $\boldsymbol{\eta}^{(l)}$, the matrix $\mathbf{B}$ is therefore upper block-diagonal. Similarly, if the covariate vector $\mathbf{w}_j$ is written as $\mathbf{w}_j = (\mathbf{w}_{jk\ldots z}^{(2)\prime}, \mathbf{w}_{k\ldots z}^{(3)\prime}, \ldots, \mathbf{w}_z^{(L)\prime})'$, the matrix $\boldsymbol{\Gamma}$ is upper block-diagonal:

$$
\begin{bmatrix} \boldsymbol{\eta}_{jk\ldots z}^{(2)} \\ \boldsymbol{\eta}_{k\ldots z}^{(3)} \\ \vdots \\ \boldsymbol{\eta}_z^{(L)} \end{bmatrix} = \begin{bmatrix} \mathbf{B}^{(22)} & \mathbf{B}^{(23)} & \cdots & \mathbf{B}^{(2L)} \\ \mathbf{0} & \mathbf{B}^{(33)} & \cdots & \mathbf{B}^{(3L)} \\ \vdots & \vdots & \ddots & \vdots \\ \mathbf{0} & \mathbf{0} & \cdots & \mathbf{B}^{(LL)} \end{bmatrix} \begin{bmatrix} \boldsymbol{\eta}_{jk\ldots z}^{(2)} \\ \boldsymbol{\eta}_{k\ldots z}^{(3)} \\ \vdots \\ \boldsymbol{\eta}_z^{(L)} \end{bmatrix}
$$

$$
+ \begin{bmatrix} \boldsymbol{\Gamma}^{(22)} & \boldsymbol{\Gamma}^{(23)} & \cdots & \boldsymbol{\Gamma}^{(2L)} \\ \mathbf{0} & \boldsymbol{\Gamma}^{(33)} & \cdots & \boldsymbol{\Gamma}^{(3L)} \\ \vdots & \vdots & \ddots & \vdots \\ \mathbf{0} & \mathbf{0} & \cdots & \boldsymbol{\Gamma}^{(LL)} \end{bmatrix} \begin{bmatrix} \mathbf{w}_{jk\ldots z}^{(2)} \\ \mathbf{w}_{k\ldots z}^{(3)} \\ \vdots \\ \mathbf{w}_z^{(L)} \end{bmatrix} + \begin{bmatrix} \boldsymbol{\zeta}_{jk\ldots z}^{(2)} \\ \boldsymbol{\zeta}_{k\ldots z}^{(3)} \\ \vdots \\ \boldsymbol{\zeta}_z^{(L)} \end{bmatrix} . \qquad (4.14)
$$

Block $\mathbf{B}^{(ab)}$ contains regression parameters for the regressions of $\boldsymbol{\eta}^{(a)}$ on $\boldsymbol{\eta}^{(b)}$ and similarly for $\boldsymbol{\Gamma}^{(ab)}$. Note, however, that unlike $\boldsymbol{\eta}_j$, $\mathbf{w}_j$ need not contain subvectors for each level. There may for example be a single covariate at level L ($R=1$, $\mathbf{w}_j = w_z^{(L)}$). We will henceforth omit the superscript from the covariates w.

The model becomes easier to estimate, and easier to understand, if the regressions among latent variables at a particular level are *recursive*. In this case the elements of $\boldsymbol{\eta}^{(l)}$ can be permuted in such a way that the blocks $\mathbf{B}^{(aa)}$ on the diagonal are strictly upper diagonal. The expression for $\eta_{M_l}^{(l)}$ can then be substituted into the expression for $\eta_1^{(l)}$ to $\eta_{M_l-1}^{(l)}$, the expression for $\eta_{M_l-1}^{(l)}$ into the regression for $\eta_1^{(l)}$ to $\eta_{M_l-2}^{(l)}$, etc., until all $\eta_m^{(l)}$ are eliminated from the right-hand side of the equation. Substituting the final expressions into the linear predictor then yields what we will call the reduced form for the latent variables (see Section 4.6), where the only latent variables remaining on the right-hand side are the disturbances $\boldsymbol{\zeta}$. An implication of restricting the relations to be recursive is that we cannot have simultaneous effects with a particular latent variable regressed on another and vice versa. However, such models are rarely used in practice, possibly due to a combination of conceptual complexity and identification restrictions that are often deemed unpalatable.

Examples

An example of a structural model involving two latent variables at level 2 and one latent variable at level 3 is given by

$$
\begin{bmatrix} \eta_{1jk}^{(2)} \\ \eta_{2jk}^{(2)} \\ \eta_{1k}^{(3)} \end{bmatrix} = \begin{bmatrix} 0 & b_{12} & b_{13} \\ 0 & 0 & 0 \\ 0 & 0 & 0 \end{bmatrix} \begin{bmatrix} \eta_{1jk}^{(2)} \\ \eta_{2jk}^{(2)} \\ \eta_{1k}^{(3)} \end{bmatrix} + \begin{bmatrix} \gamma_{11} & 0 \\ 0 & \gamma_{22} \\ 0 & 0 \end{bmatrix} \begin{bmatrix} w_{jk} \\ w_k \end{bmatrix} + \begin{bmatrix} \zeta_{1jk}^{(2)} \\ \zeta_{2jk}^{(2)} \\ \zeta_{1k}^{(3)} \end{bmatrix} ,
$$

which is shown in path diagram form in Figure 4.2. Here $\eta_{1jk}^{(2)}$ is regressed both

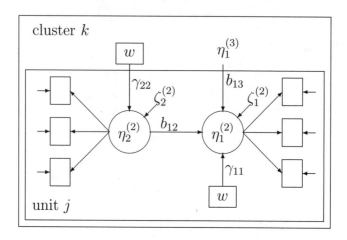

Figure 4.2 *Example of a multilevel structural equation model*

on a same-level latent variable $\eta_{2jk}^{(2)}$ and a higher-level latent variable $\eta_{1k}^{(3)}$, as well as an observed covariate w_{jk} varying at level 2. Reversing the path b_{13} to b_{31} would not make sense since $\eta_{1k}^{(3)}$ would then be forced to vary at level 2. Adding a path b_{21} from $\eta_{1jk}^{(2)}$ to $\eta_{2jk}^{(2)}$ would render the relations at level 2 nonrecursive.

We will now consider an alternative to the two-level factor model considered in Section 4.2.4 which is shown again Figure 4.3(b). As illustrated in Figure 4.3(a), we simply retain the level-2 model (inside the inner box) and allow the level-2 factor to vary at level 3 by adding a regression of the level-2 factor on the level-3 factor. Such a model is referred to as a *variance components factor model* in Rabe-Hesketh *et al.* (2004a). The model is arguably much easier to interpret than the less structured alternative in Figure 4.3(b). The common factor, defined through its relationship to the observed items at the unit level, simply has a component of variation at the cluster level. The model is analogous to a MIMIC model, with the crucial difference that the common factor is regressed on a latent variable varying at a higher level instead of an observed variable. Including unique factors at the higher level is analogous to including direct effects in a MIMIC model. An advantage of the variance components factor model is that it is easier to incorporate within a structural equation model than the general two-level factor model. In fact, such a model formed part of the previous example in Figure 4.2, where $\eta_{1jk}^{(2)}$ is the common factor at level 2 and $\eta_{1k}^{(3)}$ is the variance component at level 3. A variance components factor model is used to analyze attitudes to abortion in Section 9.8.

Structural models that are nonlinear in the latent variables have also been

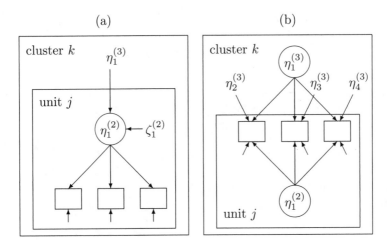

Figure 4.3 *(a) A variance components factor model and (b) a general two-level factor model*

proposed (e.g. Busemeyer and Jones, 1983; Kenny and Judd, 1984). Arminger and Muthén (1998) discuss a nonlinear version of the LISREL model for continuous responses (see Section 3.5). The structural model for latent response variables $\boldsymbol{\eta}_j$ in terms of the latent explanatory variables $\boldsymbol{\xi}_j$ is given by

$$\boldsymbol{\eta}_j = \mathbf{B}\boldsymbol{\eta}_j + \boldsymbol{\Gamma}\boldsymbol{\alpha}_j + \boldsymbol{\zeta}_j, \qquad \boldsymbol{\alpha}_j \equiv \mathbf{g}(\boldsymbol{\xi}_j).$$

Here, $\mathbf{g}(\boldsymbol{\xi}_j)$ is a known deterministic vector function and normality is assumed for $\boldsymbol{\xi}_j$ and $\boldsymbol{\zeta}_j$. Special cases include polynomial regression models where

$$\boldsymbol{\alpha}_j = [\xi_j, \xi_j^2, \dots, \xi_j^p],$$

and (first order) interaction models where

$$\boldsymbol{\alpha}_j = [\xi_{1j}, \xi_{2j}, \dots, \xi_{qj}, \xi_{1j}\xi_{2j}, \xi_{1j}\xi_{3j}, \dots, \xi_{q-1,j}\xi_{qj}].$$

Setting $\boldsymbol{\eta}_j = \boldsymbol{\alpha}_j$ produces a nonlinear factor model. We refer to Jöreskog (1998) for an overview of nonlinear structural equation modeling; see also other contributions in Schumaker and Marcoulides (1998).

4.3.2 Discrete latent variables

For discrete latent variables, the structural model is the model for the (prior) probabilities that the units belong to the corresponding latent classes. For a level-2 unit j, let the probability of belonging to class c be denoted as π_{jc},

$$\pi_{jc} \equiv \Pr(\boldsymbol{\eta}_j^{(2)} = \mathbf{e}_c).$$

This probability may depend on covariates $\mathbf{w}_j$ through a multinomial logit model

$$\pi_{jc} = \frac{\exp(\mathbf{w}_j' \varrho^c)}{\sum_d \exp(\mathbf{w}_j' \varrho^d)}, \qquad (4.15)$$

where ϱ^c are regression parameters with $\varrho^1 = \mathbf{0}$ imposed for identification. Such a 'concomitant variable' latent class model is used for instance by Dayton and MacReady (1988) and Formann (1992). The multinomial logit parametrization is useful even if the class membership does not depend on covariates since it forces latent class probabilities to sum to one.

It is sometimes useful to use the following constraint for the locations. Let π_c denote the probabilities when the covariates $\mathbf{w}_j$ are zero (except for the constant). Then the $\mathbf{e}_c$ for $c = 1, \ldots, C - 1$ are free parameters and $\mathbf{e}_C$ is determined by setting the mean location to zero

$$\sum_{c=1}^{C} \pi_c \mathbf{e}_c = \mathbf{0}.$$

An advantage of this parametrization is that the mean structure can be specified in the fixed part of the model $\mathbf{x}_{ij}' \boldsymbol{\beta}$ as in continuous latent variable models.

If the latent classes are ordered along a dimension, as in a discrete one-factor model, ordinal models can be specified for the latent class probabilities either by constraining parameters in (4.15) or using cumulative models (see Section 2.3.4).

If there are several discrete latent variables, we can either use model (4.15) where c labels the combinations of categories for the latent variables, or we can parameterize the model as a log-linear model with main effects and interactions of the latent variables. Hagenaars (1993) and Vermunt (1997) considered regressions of discrete latent variables on other discrete latent variables at the same level. Vermunt (2003) extends the structural model to include higher level continuous or discrete latent variables in the linear predictor of (4.15); see also Section 4.4.4.

4.4 Distribution of the disturbances

To complete model specification we must specify the distribution of the disturbances $\boldsymbol{\zeta}$ in the structural model. If there is no structural model the latent variables simply equal the disturbances; $\boldsymbol{\eta} = \boldsymbol{\zeta}$.

The dependence structure of the disturbances is specified by the number of levels L and the number of disturbances M_l at each level. A particular level may coincide with a level of clustering in the hierarchical dataset. However, there will often not be a direct correspondence between the levels of the model and the levels of the data hierarchy. For instance, in factor models items were treated as units at level 1 and subjects as units at level 2.

The terms 'unit at a level' and 'disturbance at a level' are defined as follows:

- a unit at level 1 is an elementary unit of observation,

- a unit k at level $l > 1$ is a cluster of level-1 units,
- the level-1 units in cluster k at level $l > 1$ fall into $n_k^{(l-1)}$ subsets representing units at level $l - 1$,
- a disturbance $\zeta^{(l)}$ at level l varies between the units at level l but not within the units,
- the units at level l are conditionally independent given the disturbances at levels $l + 1$ and above and any explanatory variables.

The basic assumption is that disturbances at the same level may be dependent, whereas disturbances at different levels are independent. In the following subsections we describe different specifications of the distribution of ζ.

4.4.1 Continuous distributions

In the case of continuous disturbances, the predominant distributional assumption is multivariate normality with mean zero and covariance matrix $\mathbf{\Psi}^{(l)}$ at level l. An advantage of this distribution is that the means, variances and covariances are explicitly parameterized and can be freely specified. Importantly, the likelihood cannot be expressed in closed form in this case unless the responses are conditionally normally distributed. However, as discussed in Section 6.2, closed form expressions exist for some combinations of latent variable and response distributions in the case of simple random intercept models with between-cluster covariates only. Wedel and Kamakura (2001) discuss factor models with independent factors having any distribution from the exponential family, although these models generally do not have closed form likelihoods.

Fortunately, in many cases inferences appear to be surprisingly robust to departures from normal disturbances (e.g. Bartholomew, 1988, 1994; Seong, 1990; Neuhaus et al., 1992; Kirisci and Hsu, 2001; Wedel and Kamakura, 2001). Several attempts have nevertheless been made to 'robustify' the disturbance distribution. Pinheiro et al. (2001) consider multivariate t-distributions and find these more robust against outliers than the multivariate normal.

In order to avoid making strong assumptions about the distribution of the disturbances, flexible parametric distributions can be used such as finite mixtures of (multivariate) normal distributions (e.g. Uebersax, 1993; Uebersax and Grove, 1993; Magder and Zeger, 1996; Verbeke and Lesaffre, 1996; Allenby et al., 1998; Carroll et al., 1999; Lenk and DeSarbo, 2000; Richardson et al., 2002). In some case the components of the finite mixture are interpreted as subpopulations, for instance those with or without a disease in Qu et al. (1996); see also Section 4.4.4. In the Bayesian setting there has recently been considerable interest in modeling disturbances via semiparametric mixtures of Dirichlet processes (e.g. Müller and Roeder, 1997; Chib and Hamilton, 2002).

Another approach is to use a truncated Hermite series expansion as suggested by Gallant and Nychka (1987) and Davidian and Gallant (1992).

4.4.2 Nonparametric distributions

Instead of making distributional assumptions regarding the distrurbances, we can use 'nonparametric maximum likelihood estimation' (NPMLE) (Laird, 1978). Principally in the context of random intercept models, Simar (1976) and Laird (1978), and more generally Lindsay (1983), have shown that the nonparametric maximum likelihood estimator (NPMLE) of the unspecified (possibly continuous) distribution is a discrete distribution with nonzero probabilities π_c at a finite set of locations e_c, $c = 1, \ldots, C$ as shown in the upper panel of Figure 4.4 (see also Lindsay *et al.*, 1991; Aitkin, 1996, 1999a; Rabe-Hesketh *et al.*, 2003a). For this reason the model is often referred to as a semiparametric mixture model. The cumulative distribution function of the disturbance is a step function as shown in the lower panel of Figure 4.4.

For a multivariate disturbances with M elements, the masses are located at points $\mathbf{e}_c$ in M dimensions (e.g. Davies and Pickles, 1987; Aitkin, 1999a). See Section 9.5, page 304, for an example with $M = 2$, where the two-dimensional mass-point distribution is displayed in two different ways in Figure 9.7. Vermunt (2004) describes NPMLE for three-level models.

For a given number of masses, the locations and probabilities can be estimated jointly with the other parameters using maximum likelihood estimation. The number of masses can then be increased until the largest maximized likelihood is achieved. Alternatively, a model with a very large number of masses can be estimated so that redundant masses will either merge with other masses (sharing the same location) or have zero probabilities. A method for determining if a given C corresponds to the NPMLE, based on the directional derivative, is discussed in Section 6.5. Maximum likelihood theory for NPMLE is reviewed by Lindsay (1995) and Böhning (2000). Heinen (1996) denotes NPMLE 'fully semiparametric' and refers to the simpler approach where masses are estimated but locations fixed a priori as 'semiparametric'.

An important advantage of NPMLE is that it is appropriate regardless of the disturbance distribution. The true distribution could be continuous (normal or nonnormal), discrete or continuous with discrete components. Relying on NPMLE, we can concentrate on the specification of other model components and need not worry about the nature of the disturbance distribution. We use NPMLE for instance in Sections 9.5, 11.2, 11.3.3 and 14.2.

4.4.3 Discrete distributions

If the latent variables are discrete, we model their distribution using multinomial logit models as discussed in Section 4.3.2.

4.4.4 Mixed continuous and discrete distributions

Models with both continuous and discrete latent variables can take different forms.

The first incudes both types of latent variable in the response model. In a

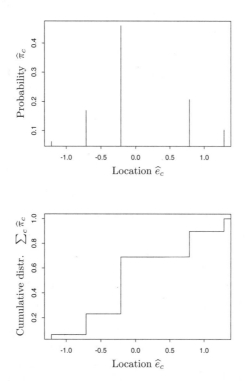

Figure 4.4 *Discrete distribution and cumulative distribution*

model for rankings Böckenholt (2001b) includes a discrete alternative-specific random intercept as well as continuous common factors and random coefficients. Similarly, McCulloch *et al.* (2002) specify a 'latent class mixed model' with both discrete and continuous random coefficients for joint modeling of continuous longitudinal responses and survival (see Section 14.6 for a similar application). The latent classes are interpreted as subpopulations of men differing both in their mean trajectories of (log) prostate specific antigen and in their time to onset of prostate cancer. Variability among men within the same latent class is accommodated by the (continuous) random effects. Both Böckenholt (2001b) and McCulloch *et al.* (2002) treat the discrete and continuous latent variables as independent of each other. Note that the sum of a discrete and continuous (zero mean, normally distributed) latent variable is just a finite mixture of normal densities with equal variances, the discrete variable representing the component means.

The second kind of model has only discrete latent variables in the response model and there are continuous latent variables in the structural model. Such a model was proposed by Vermunt (2003) in the multilevel setting. The item-level model is a conditional response model for item i, unit j, cluster k, given class membership c. For unordered categorical responses with categories $s = 1, \ldots, S$, the model can be written as

$$\Pr(y_{ijk} = s | \boldsymbol{\eta}_{jk}^{(2)} = \mathbf{e}_c) = \frac{\exp(\nu_{ijkc}^s)}{\sum_{t=1}^S \exp(\nu_{ijkc}^t)}, \tag{4.16}$$

where ν_{ijkc}^s is the linear predictor for category s. The unit-level model is a multinomial logit model for class membership,

$$\Pr(\boldsymbol{\eta}_{jk}^{(2)} = \mathbf{e}_c) = \frac{\exp(\alpha_{jk}^c)}{\sum_b \exp(\alpha_{jk}^b)},$$

where the linear predictor α_{jk}^c of the structural model includes a cluster-level random intercept,

$$\alpha_{jk}^c = \mathbf{w}_{jk}' \boldsymbol{\varrho}^c + \eta_k^{(3)}.$$

Here, $\mathbf{w}_{jk}$ are unit- and cluster-specific covariates with fixed class-specific coefficients $\boldsymbol{\varrho}^c$. A normal distribution is specified for the cluster-level random intercept. Vermunt remarks that it is often useful to assume that the conditional response probabilities do not depend on the clusters (by dropping the k subscript in (4.16)). He also points out that the cluster-level random intercept can be specified as discrete.

Third, the response model could contain only continuous latent variables whereas discrete latent variables appear in the structural model. The structural model, where continuous latent variables are regressed on discrete latent variables, is usually more complex than the conventional structural models (only including continuous latent variables). For instance, the covariance matrix of the disturbances may depend on the discrete latent variables. The most common structural model is a finite mixture of multivariate normal distributions

$$\sum_{c=1}^C \pi_c h_c(\boldsymbol{\zeta}),$$

where c indexes the components, π_c are the component weights and $h_c(\boldsymbol{\zeta})$ is a multivariate normal density with component-specific mean and covariance parameters. Such a model was used by Verbeke and Lesaffre (1996), Allenby et al. (1998) and Lenk and DeSarbo (2000) for random coefficient models, by Magder and Zeger (1996), Carroll et al. (1999) and Richardson et al. (2002) for covariate measurement error and Uebersax (1993) and Uebersax and Grove (1993) for measurement models with dichotomous and ordinal responses.

Finally, the most general model allows any of the parameters of conventional structural equation models with continuous responses to depend on discrete latent variables. Both the response model and structural model can therefore differ between latent classes, giving a multiple-group structural equation model of the kind proposed by Jöreskog (1971a), with the crucial difference that group membership is unknown.

Yung (1997) and Fokoué and Titterington (2003), among others, consider the special case of finite mixture factor models. Yung's model can be written as

$$\mathbf{y}_{jc} = \boldsymbol{\beta}_c + \boldsymbol{\Lambda}_c \boldsymbol{\eta}_{jc} + \boldsymbol{\epsilon}_{jc}, \tag{4.17}$$

where $\boldsymbol{\eta}_{jc}$ are continuous common factors with class-specific variances

$$\boldsymbol{\Psi}_c \equiv \mathrm{Var}[\boldsymbol{\eta}_{jc}],$$

and the unique factors have class-specific covariance matrices,

$$\boldsymbol{\Theta}_c \equiv \mathrm{Var}[\boldsymbol{\epsilon}_{jc}].$$

Fokoué and Titterington assume that $\boldsymbol{\Theta}_c = \boldsymbol{\Theta}$ and $\boldsymbol{\Psi}_c = \mathbf{I}$. In the context of diagnostic test agreement, Qu *et al.* (1996) specify a unidimensional probit version of this model. They interpret the model as a latent class model with a 'random effect' (the common factor) to relax conditional independence.

Blåfield (1980), Jedidi *et al.* (1997), Dolan and van der Maas (1998), Arminger *et al.* (1999), McLachlan and Peel (2000), Wedel and Kamakura (2000), Muthén (2002) and others specify 'finite mixture structural equation models' by including a structural model

$$\boldsymbol{\eta}_{jc} = \mathbf{B}_c \boldsymbol{\eta}_{jc} + \boldsymbol{\Gamma}_c \mathbf{w}_{jc} + \boldsymbol{\zeta}_{jc}, \quad \boldsymbol{\Psi}_c \equiv \mathrm{Var}[\boldsymbol{\zeta}_{jc}]$$

for the factors in (4.17).

4.5 Parameter restrictions and fundamental parameters

The parameters of the 'data generating model', presumed to have generated the observed data, are called *structural parameters*. Let $\boldsymbol{\theta}$ be the vector of all structural parameters including the regression coefficients $\boldsymbol{\beta}$, the factor loadings $\boldsymbol{\lambda}_m^l$, $m = 1, \ldots, M_l$, $l = 1, \ldots, L$, the nonduplicated elements of the covariance matrices $\boldsymbol{\Psi}^{(l)}$, $l = 1, \ldots, L$, the parameters $\boldsymbol{\iota}$ for modelling level-1 heteroscedasticity, the threshold parameters $\boldsymbol{\varsigma}$ and the class membership parameters $\boldsymbol{\varrho}$. Note that the structural parameter vector $\boldsymbol{\theta}$ should not be confused with the residual covariance matrix $\boldsymbol{\Theta}$ with elements $\theta_{ii'}$.

More or less complex restrictions, such as the sign of a parameter or equality between parameters, are often required. These restrictions can be imposed via reparameterization in terms of so called *fundamental parameters* $\boldsymbol{\vartheta}$ which are unrestricted. The resulting reparameterized model is equivalent (in the sense of Chapter 5) to the original structural model with parameter restrictions. The main idea of this approach is to solve the implicit functions among structural parameters for a subset of fundamental parameters. An important merit of this approach, apart from its generality, is that *unconstrained* optimization procedures can be used in the estimation phase. This avoids the more complex estimation approach with restrictions imposed using for instance Lagrange multipliers (e.g. Fletcher, 1987).

Each structural parameter θ_k is specified as a *known* one time differentiable function of the fundamental parameters

$$\theta_k = h_k(\boldsymbol{\vartheta}).$$

Observe that $h_k(\theta_k) = \vartheta_l$ for some l whenever θ_k is unconstrained, which is usually the case for most structural parameters. The full set of restrictions can be written in terms of a vector function

$$\boldsymbol{\theta} = \mathbf{h}(\boldsymbol{\vartheta}).$$

The following lists different types of restrictions and their implementation via reparameterization of the structural parameters $\boldsymbol{\theta}$ in terms of fundamental parameters $\boldsymbol{\vartheta}$. Let there be K structural parameters θ_k, $k = 1, \ldots, K$ and let a_k, b_k and a be specified constants, including zero.

1. **Identity** restrictions are perhaps the most common. For instance in growth curve modeling, the residual variances are often constrained equal across occasions, corresponding to the assumption of homoscedasticity. The restrictions are of the form

$$\theta_k = \theta_l, \ k \neq l$$

and are implemented as

$$\theta_k = \vartheta_k$$
$$\theta_l = \vartheta_k.$$

2. **Linear** restrictions are of the form

$$\sum_{k=1}^{K} a_k \theta_k = a.$$

Such restrictions are often useful for simplifying the model structure. For example, in a threshold model for ordinal responses, linear restrictions can be used to force the thresholds to be equally spaced. Linear restrictions are implemented as

$$\theta_k = \vartheta_k; \ \forall k \neq K \quad \text{and} \quad \theta_K = \frac{a - \sum_{k=1}^{K-1} a_k \theta_k}{a_K}.$$

3. **Inequality** restrictions of the form

$$\theta_k \geq a_k \quad \text{and} \quad \theta_k > a_k$$

are frequently required, a typical example being that a variance is positive. These restrictions are implemented as

$$\theta_k = a_k + (\vartheta_k)^2$$

and

$$\theta_k = a_k + \exp(\vartheta_k).$$

Note that in situations where the unconstrained parameter would be estimated as less than or equal to a_k, so that the constrained parameter must be a_k to maximize the likelihood, the second parametrization can cause difficulties with convergence. This is because ϑ_k will take on very large negative values and the likelihood will appear flat with respect to ϑ_k.

Another type of inequality restriction are order restrictions of the form

$$\theta_1 \leq \theta_2 \leq \cdots \leq \theta_K \quad \text{and} \quad \theta_1 < \theta_2 < \cdots < \theta_K$$

which can be implemented as

$$\theta_1 = \vartheta_1, \quad \theta_k = \vartheta_{k-1} + (\vartheta_k)^2, \quad k > 1$$

and

$$\theta_1 = \vartheta_1, \quad \theta_k = \vartheta_{k-1} + \exp(\vartheta_k), \quad k > 1.$$

Such restrictions are required for example for the stereotype model described in Section 2.3.4 where $\alpha^1 < \cdots < \alpha^{S-1}$ and for the cumulative models in the same section where $\kappa_{i1} < \cdots < \kappa_{iS-1}$. Note that estimates for these models usually obey these restrictions even when they are not explicitly enforced.

4. **Domain** restrictions of the form

$$a_k < \theta_k < b_k$$

can be implemented as

$$\theta_k = \frac{a_k + b_k \exp(\vartheta_k)}{1 + \exp(\vartheta_k)}.$$

For example, probabilities must lie in the range $[0,1]$, giving the familiar multinomial logit transformation, useful for example for latent class probabilities. Correlations can be restricted to lie in the permitted range by setting $a_k = -1$ and $b_k = 1$.

5. **Nonlinear** restrictions are often of the form

$$\sum_{k=1}^{K} a_k \prod_{l=1}^{L} \theta_l^{b_l} = a.$$

A simple example is $\frac{\theta_1}{\theta_2} = \frac{\theta_3}{\theta_4}$ which can simply be imposed as

$$\theta_1 = \frac{\vartheta_3}{\vartheta_4} \vartheta_2 \quad \text{where} \quad \theta_k = \vartheta_k; \; k = 2, 3, 4.$$

An application of this would be the restriction that two reliabilities (ratio of true score to total variance) are equal. This is illustrated in the life satisfaction example in Section 10.4; see also Table 10.13.

Nonlinear inequality restrictions are implied by the requirement that a matrix $\mathbf{M}$ is positive semi-definite, (i.e. $\mathbf{a'Ma} \geq 0$ for all $\mathbf{a}$), the classical example being a covariance matrix $\mathbf{M}$. We can use the *Cholesky decomposition* $\mathbf{L}$ of $\mathbf{M}$,

$$\mathbf{M} = \mathbf{LL'}$$

to impose the restriction where $\boldsymbol{\theta}$ are the nonduplicated elements of $\mathbf{M}$ and $\boldsymbol{\vartheta}$ are the lower diagonal elements of $\mathbf{L}$.

4.6 Reduced form of the latent variables and linear predictor

4.6.1 Reduced form for $\boldsymbol{\eta}_j$

Remember that $\boldsymbol{\eta}_j$ is shorthand notation for $\boldsymbol{\eta}_{jk\cdots z}$, comprising all latent variables for the jth level-2 unit. Assuming that $(\mathbf{I} - \mathbf{B})$ is invertible, the structural model in (4.13),

$$\boldsymbol{\eta}_j = \mathbf{B}\boldsymbol{\eta}_j + \boldsymbol{\Gamma}\mathbf{w}_j + \boldsymbol{\zeta}_j,$$

can be solved for the latent variables in terms of the explanatory variables and disturbances giving a reduced form model for the latent variables,

$$\begin{aligned}
\boldsymbol{\eta}_j &= (\mathbf{I} - \mathbf{B})^{-1}\boldsymbol{\Gamma}\mathbf{w}_j + (\mathbf{I} - \mathbf{B})^{-1}\boldsymbol{\zeta}_j \\
&= \boldsymbol{\Pi}_1\mathbf{w}_j + \boldsymbol{\Pi}_2\boldsymbol{\zeta}_j,
\end{aligned} \tag{4.18}$$

where $\boldsymbol{\Pi}_1 \equiv (\mathbf{I} - \mathbf{B})^{-1}\boldsymbol{\Gamma}$ and $\boldsymbol{\Pi}_2 \equiv (\mathbf{I} - \mathbf{B})^{-1}$ are parameters.

4.6.2 Reduced form for $\boldsymbol{\eta}^{(l)}$

We will now consider the reduced form models for the latent variables at each level of the model. Recall from (4.14) that $\mathbf{B}$ and $\boldsymbol{\Gamma}$ are upper block diagonal since latent variables varying at a given level l can only be regressed on latent or observed variables varying at the same or higher level. It follows that $\boldsymbol{\Pi}_1$ and $\boldsymbol{\Pi}_2$ are also upper block diagonal,

$$\boldsymbol{\Pi}_1 = \begin{bmatrix}
\boldsymbol{\Pi}_1^{(22)} & \boldsymbol{\Pi}_1^{(23)} & \cdots & \boldsymbol{\Pi}_1^{(2L)} \\
\mathbf{0} & \boldsymbol{\Pi}_1^{(33)} & \cdots & \boldsymbol{\Pi}_1^{(3L)} \\
\vdots & \vdots & \ddots & \vdots \\
\mathbf{0} & \mathbf{0} & \cdots & \boldsymbol{\Pi}_1^{(LL)}
\end{bmatrix} \tag{4.19}$$

and similarly for $\boldsymbol{\Pi}_2$.

Reduced form models for the latent variables at each of the levels $l = 2, \ldots, L$ can then be written as

$$\boldsymbol{\eta}^{(l)} = \boldsymbol{\Pi}_1^{(ll+)}\mathbf{w}^{(l+)} + \boldsymbol{\Pi}_2^{(ll+)}\boldsymbol{\zeta}^{(l+)}. \tag{4.20}$$

Here, $\mathbf{w}^{(l+)} = (\mathbf{w}^{(l)\prime}, \ldots, \mathbf{w}^{(L)\prime})^\prime$ is a subvector of $\mathbf{w}$ containing all variables varying at level l or above, and similarly for $\boldsymbol{\zeta}^{(l+)}$. $\boldsymbol{\Pi}_1^{(ll+)}$ is the submatrix of $\boldsymbol{\Pi}_1$ with rows corresponding to block l and columns corresponding to blocks l to L, and analogously for $\boldsymbol{\Pi}_2^{(ll+)}$.

4.6.3 Reduced form for the linear predictor $\boldsymbol{\nu}_{z(L)}$

We will now derive the reduced form of the vector of linear predictors $\boldsymbol{\nu}_{z(L)}$ for the zth L-level unit. From the 'two-level representation' in (4.6), the vector of linear predictors for a top-level unit can be written as

$$\boldsymbol{\nu}_{z(L)} = \mathbf{X}_{z(L)}\boldsymbol{\beta} + \boldsymbol{\Lambda}_{z(L)}\boldsymbol{\eta}_{z(L)}.$$

Recall that $\boldsymbol{\eta}_{z(L)} = (\boldsymbol{\eta}_{z(L)}^{(2)}, \boldsymbol{\eta}_{z(L)}^{(3)}, \ldots, \boldsymbol{\eta}_{z(L)}^{(L)})'$, the vector of all (realizations of all) latent variables for the zth top-level unit. Let $\boldsymbol{\Pi}_{2(L)}^{(ll)} = (\boldsymbol{\Pi}_2^{(ll)}, \ldots, \boldsymbol{\Pi}_2^{(ll)})$, the same matrix column-appended as many times as there are level-l units in the zth level-L unit. Letting $\boldsymbol{\Pi}_{2z(L)}$ be an upper block-diagonal matrix analogous to (4.19) but with blocks $\boldsymbol{\Pi}_{2z(L)}^{(ll)}$,

$$\boldsymbol{\eta}_{z(L)} = \boldsymbol{\Pi}_{2z(L)}\boldsymbol{\zeta}_{(L)}.$$

The linear predictors can then be expressed as (omitting the z subscript)

$$\begin{aligned}
\boldsymbol{\nu}_{(L)} &= \mathbf{X}_{(L)}\boldsymbol{\beta} + \boldsymbol{\Lambda}_{(L)}\boldsymbol{\Pi}_{1(L)}\mathbf{w}_{(L)} + \boldsymbol{\Lambda}_{(L)}\boldsymbol{\Pi}_{2(L)}\boldsymbol{\zeta}_{(L)} \\
&= \mathbf{X}_{(L)}\boldsymbol{\beta} + \mathbf{A}_{1(L)}\mathbf{w}_{(L)} + \mathbf{A}_{2(L)}\boldsymbol{\zeta}_{(L)},
\end{aligned} \tag{4.21}$$

where $\boldsymbol{\beta}$, $\mathbf{A}_{1(L)} \equiv \boldsymbol{\Lambda}_{(L)}\boldsymbol{\Pi}_{1(L)}$ and $\mathbf{A}_{2(L)} \equiv \boldsymbol{\Lambda}_{(L)}\boldsymbol{\Pi}_{2(L)}$ are parameters of the reduced form linear predictor.

4.7 Moment structure of the latent variables

4.7.1 Moment structure of $\boldsymbol{\eta}_j$

From (4.18), the mean structure of $\boldsymbol{\eta}_j$ becomes

$$\mathrm{E}(\boldsymbol{\eta}_j|\mathbf{w}_j) = \boldsymbol{\Pi}_1\mathbf{w}_j, \tag{4.22}$$

and the covariance structure becomes

$$\mathrm{Cov}(\boldsymbol{\eta}_j|\mathbf{w}_j) = \boldsymbol{\Pi}_2\mathrm{Cov}(\boldsymbol{\zeta}_j)\boldsymbol{\Pi}_2'. \tag{4.23}$$

The covariance matrix of the disturbances $\mathrm{Cov}(\boldsymbol{\zeta}_j)$ has a block diagonal form with blocks $\boldsymbol{\Psi}^{(l)}$ for level l. Since the disturbances are independent across levels, the covariance matrix in (4.23) can be written as

$$\mathrm{Cov}(\boldsymbol{\eta}_j|\mathbf{w}_j) = \sum_{l=2}^{L} \boldsymbol{\Pi}_2^{(ll)}\boldsymbol{\Psi}^{(l)}\boldsymbol{\Pi}_2^{(ll)'}, \tag{4.24}$$

where $\boldsymbol{\Pi}_2^{(ll)}$ are blocks of $\boldsymbol{\Pi}_2$ as shown for $\boldsymbol{\Pi}_1$ in (4.19).

4.7.2 Moment structure of $\boldsymbol{\eta}^{(l)}$

It follows from (4.20) that the conditional mean structure of the latent variables $\boldsymbol{\eta}^{(l)}$ at level l given $\mathbf{w}^{(l+)}$ becomes

$$\mathrm{E}(\boldsymbol{\eta}^{(l)}|\mathbf{w}^{(l+)}) = \boldsymbol{\Pi}_1^{(ll+)}\mathbf{w}^{(l+)}.$$

The covariance structure at level l is

$$\begin{aligned}
\mathrm{Cov}(\boldsymbol{\eta}^{(l)}|\mathbf{w}^{(l+)}) &= \boldsymbol{\Pi}_2^{(ll+)}\mathrm{Cov}(\boldsymbol{\zeta}^{(l+)})(\boldsymbol{\Pi}_2^{(ll+)})' \\
&= \sum_{c=l}^{L} \boldsymbol{\Pi}_2^{(lc)}\boldsymbol{\Psi}^{(c)}\boldsymbol{\Pi}_2^{(lc)'},
\end{aligned} \tag{4.25}$$

and the covariance structure between latent variables at levels a and b, $a < b$, is

$$\mathrm{Cov}(\boldsymbol{\eta}^{(a)}, \boldsymbol{\eta}^{(b)} | \mathbf{w}^{(a+)}) = \boldsymbol{\Pi}_2^{(ab+)} \mathrm{Cov}(\boldsymbol{\zeta}^{(b+)})(\boldsymbol{\Pi}_2^{(bb+)})'$$

$$= \sum_{c=b}^{L} \boldsymbol{\Pi}_2^{(ac)} \boldsymbol{\Psi}^{(c)} \boldsymbol{\Pi}_2^{(bc)'}, \qquad (4.26)$$

where the second equalities in (4.25) and (4.26) follow from the block diagonal form of $\mathrm{Cov}(\boldsymbol{\zeta}^{(l+)})$.

4.7.3 Example: 3-level model

It is instructive to consider the special case of a 3-level model. The structural model can in this case be expressed as

$$\begin{bmatrix} \boldsymbol{\eta}_{jk}^{(2)} \\ \boldsymbol{\eta}_k^{(3)} \end{bmatrix} = \begin{bmatrix} \mathbf{B}^{(22)} & \mathbf{B}^{(23)} \\ \mathbf{0} & \mathbf{B}^{(33)} \end{bmatrix} \begin{bmatrix} \boldsymbol{\eta}_{jk}^{(2)} \\ \boldsymbol{\eta}_k^{(3)} \end{bmatrix} + \begin{bmatrix} \boldsymbol{\Gamma}^{(22)} & \boldsymbol{\Gamma}^{(23)} \\ \mathbf{0} & \boldsymbol{\Gamma}^{(33)} \end{bmatrix} \begin{bmatrix} \mathbf{w}_{jk}^{(2)} \\ \mathbf{w}_k^{(3)} \end{bmatrix} + \begin{bmatrix} \boldsymbol{\zeta}_{jk}^{(2)} \\ \boldsymbol{\zeta}_k^{(3)} \end{bmatrix}.$$

From (4.20), the reduced forms for the latent variables are

$$\boldsymbol{\eta}_{jk}^{(2)} = \boldsymbol{\Pi}_1^{(22+)} \mathbf{w}_{jk}^{(2+)} + \boldsymbol{\Pi}_2^{(22+)} \boldsymbol{\zeta}_{jk}^{(2+)}$$

$$= \begin{bmatrix} \boldsymbol{\Pi}_1^{(22)} & \boldsymbol{\Pi}_1^{(23)} \end{bmatrix} \begin{bmatrix} \mathbf{w}_{jk}^{(2)} \\ \mathbf{w}_k^{(3)} \end{bmatrix} + \begin{bmatrix} \boldsymbol{\Pi}_2^{(22)} & \boldsymbol{\Pi}_2^{(23)} \end{bmatrix} \begin{bmatrix} \boldsymbol{\zeta}_{jk}^{(2)} \\ \boldsymbol{\zeta}_k^{(3)} \end{bmatrix},$$

and

$$\boldsymbol{\eta}_k^{(3)} = \boldsymbol{\Pi}_1^{(33)} \mathbf{w}_k^{(3)} + \boldsymbol{\Pi}_2^{(33)} \boldsymbol{\zeta}_k^{(3)}.$$

We can find expressions for the parameter matrices by first solving for $\boldsymbol{\eta}_k^{(3)}$ and then substituting the reduced form of $\boldsymbol{\eta}_k^{(3)}$ in the structural model for $\boldsymbol{\eta}_{jk}^{(2)}$ and solving for $\boldsymbol{\eta}_{jk}^{(2)}$, giving

$$\boldsymbol{\Pi}_1^{(22)} = (\mathbf{I} - \mathbf{B}^{(22)})^{-1} \boldsymbol{\Gamma}^{(22)},$$

$$\boldsymbol{\Pi}_1^{(23)} = (\mathbf{I} - \mathbf{B}^{(22)})^{-1} \left[\mathbf{B}^{(23)} (\mathbf{I} - \mathbf{B}^{(33)})^{-1} \boldsymbol{\Gamma}^{(33)} + \boldsymbol{\Gamma}^{(23)} \right],$$

$$\boldsymbol{\Pi}_2^{(22)} = (\mathbf{I} - \mathbf{B}^{(22)})^{-1},$$

$$\boldsymbol{\Pi}_2^{(23)} = (\mathbf{I} - \mathbf{B}^{(22)})^{-1} \mathbf{B}^{(23)} (\mathbf{I} - \mathbf{B}^{(33)})^{-1},$$

$$\boldsymbol{\Pi}_1^{(33)} = (\mathbf{I} - \mathbf{B}^{(33)})^{-1} \boldsymbol{\Gamma}^{(33)},$$

$$\boldsymbol{\Pi}_2^{(33)} = (\mathbf{I} - \mathbf{B}^{(33)})^{-1}.$$

The conditional expectations are

$$\mathrm{E}(\boldsymbol{\eta}_{jk}^{(2)} | \mathbf{w}_{jk}^{(2)}, \mathbf{w}_k^{(3)}) = \boldsymbol{\Pi}_1^{(22)} \mathbf{w}_{jk}^{(2)} + \boldsymbol{\Pi}_1^{(23)} \mathbf{w}_k^{(3)}$$

$$= (\mathbf{I} - \mathbf{B}^{(22)})^{-1} \times$$

$$\left\{ \left[\mathbf{B}^{(23)} (\mathbf{I} - \mathbf{B}^{(33)})^{-1} \boldsymbol{\Gamma}^{(33)} + \boldsymbol{\Gamma}^{(23)} \right] \mathbf{w}_k^{(3)} + \boldsymbol{\Gamma}^{(22)} \mathbf{w}_{jk}^{(2)} \right\}$$

and

$$E(\eta_k^{(3)}|\mathbf{w}_k^{(3)}) = \mathbf{\Pi}_1^{(33)}\mathbf{w}_k^{(3)}$$
$$= (\mathbf{I} - \mathbf{B}^{(33)})^{-1}\mathbf{\Gamma}^{(33)}\mathbf{w}_k^{(3)}.$$

The conditional covariance matrices are

$$\text{Cov}(\,\eta_{jk}^{(2)}|\,\mathbf{w}_{jk}^{(2)},\mathbf{w}_k^{(3)}) = \mathbf{\Pi}_2^{(22)}\mathbf{\Psi}^{(2)}\mathbf{\Pi}_2^{(22)\prime} + \mathbf{\Pi}_2^{(23)}\mathbf{\Psi}^{(3)}\mathbf{\Pi}_2^{(23)\prime}$$
$$= (\mathbf{I} - \mathbf{B}^{(22)})^{-1} \times$$
$$\left[\mathbf{B}^{(23)}(\mathbf{I} - \mathbf{B}^{(33)})^{-1}\mathbf{\Psi}^{(3)}(\mathbf{I} - \mathbf{B}^{(33)})^{-1\prime}\mathbf{B}^{(23)\prime} + \mathbf{\Psi}^{(2)}\right] \times$$
$$(\mathbf{I} - \mathbf{B}^{(22)})^{-1\prime}$$

$$\text{Cov}(\eta_k^{(3)}|\mathbf{w}_k^{(3)}) = \mathbf{\Pi}_2^{(33)}\mathbf{\Psi}^{(3)}\mathbf{\Pi}_2^{(33)\prime}$$
$$= (\mathbf{I} - \mathbf{B}^{(33)})^{-1}\mathbf{\Psi}^{(3)}(\mathbf{I} - \mathbf{B}^{(33)})^{-1\prime},$$

and

$$\text{Cov}(\,\eta_{jk}^{(2)},\eta_k^{(3)}|\mathbf{w}_{jk}^{(2)},\mathbf{w}_k^{(3)}) = \mathbf{\Pi}_2^{(23)}\mathbf{\Psi}^{(3)}\mathbf{\Pi}_2^{(33)\prime}$$
$$= (\mathbf{I} - \mathbf{B}^{(22)})^{-1}\mathbf{B}^{(23)}(\mathbf{I} - \mathbf{B}^{(33)})^{-1}\mathbf{\Psi}^{(3)}(\mathbf{I} - \mathbf{B}^{(33)})^{-1\prime}.$$

4.8 Marginal moment structure of observed and latent responses

Recall that conditional on the latent variables, the response model is a generalized linear model with link function $g(\nu)$. The regression parameters therefore represent *conditional effects* of covariates given the latent variables. In some instances the *population averaged effects* or marginal effects (averaged over the latent variable distribution) are of interest. We first consider the mean structure of the responses, the expectation of the responses as a function of the covariates but marginal with respect to the latent variables. Subsequently, we derive the marginal covariance structure.

4.8.1 Marginal mean structures and population average effects

We now consider the expectation of the response conditional on the covariates, but marginal with respect to the latent variables. Applying the double expectation rule,

$$E(\mathbf{y}_{(L)}|\mathbf{X}_{(L)},\mathbf{w}_{(L)}) = E_\zeta\{E(\mathbf{y}_{(L)}|\mathbf{X}_{(L)},\mathbf{w}_{(L)},\boldsymbol{\zeta}_{(L)})\} \qquad (4.27)$$
$$= \int\cdots\int g^{-1}(\boldsymbol{\nu}_{(L)})\left[\prod_{l=2}^{L}h^{(l)}(\boldsymbol{\zeta}_{(L)}^{(l)})\right]\,d\boldsymbol{\zeta}_{(L)}^{(2)}\cdots d\boldsymbol{\zeta}_{(L)}^{(L)},$$

where $h^{(l)}(\cdot)$ is the density of the disturbances at level l. This is simply the 'population averaged' response for given covariate values. So-called *population averaged effects* or *marginal effects* (with respect to the latent variables) are obtained by considering the relationship between this expectation and the covariates.

We will consider the general form of the linear predictor in (4.21), written for a single level-1 unit as

$$\nu = \mathbf{x}'\boldsymbol{\beta} + \mathbf{a}_1'\mathbf{w}_{(L)} + \mathbf{a}_2'\boldsymbol{\zeta}_{(L)},$$

where $\mathbf{a}_1'$ is a row of $\mathbf{A}_{1(L)}$ and similarly for $\mathbf{a}_2'$. For instance, in a two-level random coefficient model with continuous latent variables,

$$\nu_{ij} = \mathbf{x}_{ij}'\boldsymbol{\beta} + \mathbf{z}_{ij}'\boldsymbol{\zeta}_j^{(2)},$$

$\mathbf{a}_1 = \mathbf{0}$, $\mathbf{a}_2 = \mathbf{z}_{ij}$ and $\boldsymbol{\zeta}_{(L)} = \boldsymbol{\zeta}_j^{(2)}$. A factor model has the same structure except that $\mathbf{a}_2'$ is a row of the factor loading matrix $\boldsymbol{\Lambda}$.

For an <u>identity link</u>, $g^{-1}(\nu) = \nu$, the expectation simplifies to

$$E(y|\mathbf{x}, \mathbf{w}_{(L)}) = \mathbf{x}'\boldsymbol{\beta} + \mathbf{a}_1'\mathbf{w}_{(L)}, \qquad (4.28)$$

so that the link function is retained and the marginal effects are equal to the conditional effects, whatever the latent variable distribution.

For a <u>log link</u>, $g^{-1}(\nu) = \exp(\nu)$, the marginal effects are equal to the conditional effects, apart from the intercept, regardless of the latent variable distribution. For normal latent variables we obtain

$$E(y|\mathbf{x}, \mathbf{w}_{(L)}) = \exp(\mathbf{x}'\boldsymbol{\beta} + \mathbf{a}_1'\mathbf{w}_{(L)} + \mathbf{a}_2'\boldsymbol{\Psi}_{(L)}\mathbf{a}_2/2),$$

which in the random intercept case reduces to

$$E(y_{ij}|\mathbf{x}_{ij}, \mathbf{z}_{ij}) = \exp(\mathbf{x}_{ij}'\boldsymbol{\beta} + \psi/2),$$

so that the marginal regression parameters are equal to the conditional parameters except for the intercept which increases by $\psi/2$.

The subsequent results are confined to normal latent variables. For a <u>probit link</u>, it is convenient to derive the expectation using the latent response formulation (see also Section 2.4). The model can be written as

$$y^* = \mathbf{x}'\boldsymbol{\beta} + \mathbf{a}_1'\mathbf{w}_{(L)} + \underbrace{\mathbf{a}_2'\boldsymbol{\zeta}_{(L)} + \epsilon}_{\xi}, \qquad (4.29)$$

where ξ is the 'total residual' which is normally distributed with zero mean and variance $\mathbf{a}_2'\boldsymbol{\Psi}_{(L)}\mathbf{a}_2 + 1$ and $y = 1$ if $y^* > 0$ and $y = 0$ otherwise. Note that the mean structure for latent responses is identical to that presented in Section 4.28 for the identity link.

The expectation of the observed response becomes

$$
\begin{aligned}
E(y|\mathbf{x}, \mathbf{w}) &= \Pr(y = 1|\mathbf{x}, \mathbf{w}) = \Pr(y^* > 0|\mathbf{x}, \mathbf{w}) \\
&= \Pr(-\xi \le \mathbf{x}'\boldsymbol{\beta} + \mathbf{a}_1'\mathbf{w}_{(L)}) = \Pr(\xi \le \mathbf{x}'\boldsymbol{\beta} + \mathbf{a}_1'\mathbf{w}_{(L)}) \\
&= \Pr(\frac{\xi}{\sqrt{\mathbf{a}_2'\boldsymbol{\Psi}_{(L)}\mathbf{a}_2 + 1}} \le \frac{\mathbf{x}'\boldsymbol{\beta} + \mathbf{a}_1'\mathbf{w}_{(L)}}{\sqrt{\mathbf{a}_2'\boldsymbol{\Psi}_{(L)}\mathbf{a}_2 + 1}})
\end{aligned}
$$

$$= \Phi\left(\frac{\mathbf{x}'\boldsymbol{\beta} + \mathbf{a}_1'\mathbf{w}_{(L)}}{\sqrt{\mathbf{a}_2'\boldsymbol{\Psi}_{(L)}\mathbf{a}_2 + 1}}\right) = \Phi(\mathbf{x}'\boldsymbol{\beta}^* + \mathbf{a}_1^{*'}\mathbf{w}_{(L)}), \qquad (4.30)$$

where $\boldsymbol{\beta}^*$ and $\mathbf{a}_1^*$ are the original parameter vectors divided by the total residual standard deviation of the latent response. For the probit link, the link function is thus retained, but the marginal effects are attenuated relative to the conditional effects. Following the same reasoning, it is evident that an analogous result holds for the cumulative probit model for ordinal responses.

Sometimes the variance of the latent response y^* in the probit model is set to one, $\mathbf{a}_2'\boldsymbol{\Psi}_{(L)}\mathbf{a}_2 + \theta = 1$, instead of fixing the variance of the error term ϵ as above; see for instance Muthén (1984). Note that marginal effects are not attenuated in this case.

For a <u>logit link</u>, the model is as in (4.29) but with ϵ specified as logistic, yielding

$$\text{Var}(\xi) = \mathbf{a}_2'\boldsymbol{\Psi}_{(L)}\mathbf{a}_2 + \pi^2/3.$$

This total residual has a compound logistic-normal distribution. Multiplying the total residual by the factor

$$\sqrt{\frac{\pi^2/3}{\mathbf{a}_2'\boldsymbol{\Psi}_{(L)}\mathbf{a}_2 + \pi^2/3}} \approx \frac{1}{\sqrt{1 + 0.30\mathbf{a}_2'\boldsymbol{\Psi}_{(L)}\mathbf{a}_2}},$$

we obtain a residual with the variance of the logistic distribution. Since the logistic and normal distributions are very similar,

$$\text{Pr}\left(\frac{\xi}{\sqrt{1 + 0.30\mathbf{a}_2'\boldsymbol{\Psi}_{(L)}\mathbf{a}_2}} \le \frac{\mathbf{x}'\boldsymbol{\beta} + \mathbf{a}_1'\mathbf{w}_{(L)}}{\sqrt{1 + 0.30\mathbf{a}_2'\boldsymbol{\Psi}_{(L)}\mathbf{a}_2}}\right) \approx F\left(\frac{\mathbf{x}'\boldsymbol{\beta} + \mathbf{a}_1'\mathbf{w}_{(L)}}{\sqrt{1 + 0.30\mathbf{a}_2'\boldsymbol{\Psi}_{(L)}\mathbf{a}_2}}\right),$$

where F is the logistic cumulative distribution function. Note that Zeger *et al.* (1988) use $\frac{(15/16)^2}{\pi^2/3} \approx 0.35$ in place of 0.30.

Other link functions are generally not preserved under marginalization. A useful discussion of conditional and marginal effects for different models is given by Ritz and Spiegelman (2004).

4.8.2 Marginal covariance structures

In general, the covariance structure marginal with respect to the latent variables can be derived using the relation

$$\text{Cov}(\mathbf{y}_{(L)}|\mathbf{X}_{(L)}, \mathbf{w}_{(L)}) = \text{E}_{\boldsymbol{\zeta}}[\text{Cov}(\mathbf{y}_{(L)}|\boldsymbol{\nu}_{(L)})] + \text{Cov}_{\boldsymbol{\zeta}}[\text{E}(\mathbf{y}_{(L)}|\boldsymbol{\nu}_{(L)})].$$

The marginal covariance between different level-1 units i and i' (omitting higher-level subscripts) becomes

$$\text{Cov}(y_i, y_{i'}|\mathbf{X}_{(L)}, \mathbf{w}_{(L)}) = \text{E}_{\boldsymbol{\zeta}}[\text{Cov}(y_i, y_{i'}|\nu_i, \nu_{i'})] + \text{Cov}_{\boldsymbol{\zeta}}[\text{E}(y_i|\nu_i), \text{E}(y_{i'}|\nu_{i'})]$$
$$= \text{Cov}_{\boldsymbol{\zeta}}[g^{-1}(\nu_i), g^{-1}(\nu_{i'})], \qquad (4.31)$$

the covariance between the conditional expectations. Note that the first term

Table 4.2 *Nomenclature for bivariate normal latent response correlations*

	Continuous	Dichotomous	Ordinal	Censored
Continuous	Pearson			
Dichotomous	Biserial	Tetrachoric		
Ordinal	Polyserial	Polychoric	Polychoric	
Censored	Tobitserial[?]	Bitobit[?]	Polytobit[?]	Tobit

[?]We may have invented these terms

above is zero because $\text{Cov}(y_i, y_{i'}|\nu_i, \nu_{i'}) = 0$ due to conditional independence given the latent variables. All pairs of units having the same two values of the linear predictor will have the same marginal covariances and correlations. However, with the exception of an identity link, the 'intraclass' correlation between the observed responses will differ between clusters with different cluster-specific covariates and between pairs of units within clusters if there are lower-level covariates (such as time in longitudinal or panel data).

For <u>continuous responses and latent responses</u>, the marginal covariance structure is

$$\boldsymbol{\Omega}_{(L)} \equiv \text{Cov}(\mathbf{y}^*_{(L)}|\mathbf{X}_{(L)}, \mathbf{w}_{(L)}) = \mathbf{A}_{2(L)}\boldsymbol{\Psi}_{(L)}\mathbf{A}'_{2(L)} + \boldsymbol{\Theta}, \qquad (4.32)$$

where $\boldsymbol{\Theta}$ is the typically diagonal covariance matrix of the 'errors' $\boldsymbol{\epsilon}$ (see Display 4.1). This yields the correlation structure

$$\boldsymbol{\rho}_{(L)} \equiv \text{Cor}(\mathbf{y}^*_{(L)}|\mathbf{X}_{(L)}, \mathbf{w}_{(L)}) = [\text{Diag}(\boldsymbol{\Omega}_{(L)})]^{-\frac{1}{2}}\boldsymbol{\Omega}_{(L)}[\text{Diag}(\boldsymbol{\Omega}_{(L)})]^{-\frac{1}{2}}. \quad (4.33)$$

Consider the special case of multinormal latent responses. The possible combinations of response types and corresponding names given to the latent response correlations are presented in Table 4.2. Note that historically these names have referred to bivariate correlations without conditioning on explanatory variables but will here be used generally.

For a <u>Poisson response with a log link</u>, the marginal variance becomes

$$\begin{aligned}
\text{Var}(y|\mathbf{x}, \mathbf{w}_{(L)}) &= \mathbf{E}_{\boldsymbol{\zeta}}[\text{Var}(y|\mathbf{x}, \mathbf{w}_{(L)}, \boldsymbol{\zeta}_{(L)})] + \text{Var}_{\boldsymbol{\zeta}}[\mathbf{E}(y|\mathbf{x}, \mathbf{w}_{(L)}, \boldsymbol{\zeta}_{(L)})] \\
&= \mathbf{E}_{\boldsymbol{\zeta}}[\exp(\mathbf{x}'\boldsymbol{\beta} + \mathbf{a}'_1\mathbf{w}_{(L)} + \mathbf{a}'_2\boldsymbol{\zeta}_{(L)})] \\
&\quad + \text{Var}_{\boldsymbol{\zeta}}[\exp(\mathbf{x}'\boldsymbol{\beta} + \mathbf{a}'_1\mathbf{w}_{(L)} + \mathbf{a}'_2\boldsymbol{\zeta}_{(L)})] \\
&= \mathbf{E}(y|\mathbf{x}, \mathbf{w}_{(L)}) + \exp(\mathbf{x}'\boldsymbol{\beta} + \mathbf{a}'_1\mathbf{w}_{(L)})^2\text{Var}[\exp(\mathbf{a}'_2\boldsymbol{\zeta}_{(L)})].
\end{aligned}$$

For normal $\boldsymbol{\zeta}$, it follows from a general result for the variance of log-normal random variables (e.g. Johnson *et al.*, 1994, p. 212) that

$$\text{Var}(\exp(\mathbf{a}'_2\boldsymbol{\zeta}_{(L)})) = \exp(\mathbf{a}'_2\boldsymbol{\Psi}_{(L)}\mathbf{a}_2)[\exp(\mathbf{a}'_2\boldsymbol{\Psi}_{(L)}\mathbf{a}_2) - 1],$$

so that

$$\text{Var}(y|\mathbf{x}, \mathbf{w}_{(L)}) = \mathbf{E}(y|\mathbf{x}, \mathbf{w}_{(L)})\left\{1 + \mathbf{E}(y|\mathbf{x}, \mathbf{w}_{(L)})[\exp(\mathbf{a}'_2\boldsymbol{\Psi}_{(L)}\mathbf{a}_2) - 1]\right\}.$$

Without latent variables, the marginal variance reduces to $E(y|\mathbf{x}, \mathbf{w}_{(L)})$, the variance function of the Poisson distribution. The latent variables therefore lead to an increase in the variance that is proportional to the expectation squared. In contrast, the usual quasi-likelihood approach sets the variance function equal to $\phi^* E(y|\mathbf{x}, \mathbf{w}_{(L)})$ (see Section 2.3.1), thus assuming that the variance is proportional to the expectation.

For <u>dichotomous responses</u>, the marginal variance is

$$\mathrm{Var}(y|\mathbf{x}, \mathbf{w}_{(L)}) = \Pr(y = 1|\mathbf{x}, \mathbf{w}_{(L)})[1 - \Pr(y = 1|\mathbf{x}, \mathbf{w}_{(L)})],$$

which for a probit link has the simple form

$$\Phi\left(\frac{\mathbf{x}'\boldsymbol{\beta} + \mathbf{a}_1'\mathbf{w}_{(L)}}{\sqrt{\mathbf{a}_2'\boldsymbol{\Psi}_{(L)}\mathbf{a}_2 + 1}}\right)\left[1 - \Phi\left(\frac{\mathbf{x}'\boldsymbol{\beta} + \mathbf{a}_1'\mathbf{w}_{(L)}}{\sqrt{\mathbf{a}_2'\boldsymbol{\Psi}_{(L)}\mathbf{a}_2 + 1}}\right)\right].$$

Note that the relationship between mean and variance is the same as for Bernoulli models without latent variables. This is as expected since overdispersion is not possible for dichotomous responses.

4.9 Reduced form distribution and likelihood

The conditional distribution of the observed responses $\mathbf{y}$ given the explanatory variables $\mathbf{X}$ is called the *reduced form distribution*. There are two ways of deriving the reduced form, via latent variable integration or latent response integration. The first approach is based on the specification of conditional independence of the observed responses given the latent variables. The second rests on the specification of multivariate normality of the latent responses marginal with respect to the latent variables.

4.9.1 Latent variable integration

The reduced form distribution is the distribution of the responses marginal to the latent variables but conditional on the explanatory variables. If the latent variables are discrete, this is obtained by summing the joint probabilities of the responses and class membership over the classes giving a finite mixture (see Section 3.4). If the latent variables are continuous, the latent variables are integrated out giving an infinite mixture. Thus the latent variable distribution is often referred to as the *mixing distribution*. When integrating over the latent variables at the different levels, we will exploit conditional independence among the units at a given level given the latent variables at all higher levels.

Two-level random intercept model

For simplicity, consider the two-level random intercept model

$$\nu_{ij} = \mathbf{x}_{ij}'\boldsymbol{\beta} + \zeta_j^{(2)},$$

and let the conditional probability (or probability density) of the response y_{ij} be denoted $g^{(1)}(y_{ij}|\mathbf{x}_{ij}, \zeta_j^{(2)}; \boldsymbol{\vartheta})$ where $\boldsymbol{\vartheta}$ is the vector of fundamental parameters. The marginal distribution of the responses $\mathbf{y}_j^{(2)}$ for the jth level-2 unit given the matrix of covariates $\mathbf{X}_{j(2)}$ for that unit is

$$g^{(2)}(\mathbf{y}_j^{(2)}|\mathbf{X}_{j(2)};\boldsymbol{\vartheta}) = \int h(\zeta_j^{(2)}) \prod_{i=1}^{n_j} g^{(1)}(y_{ij}|\mathbf{x}_{ij},\zeta_j^{(2)};\boldsymbol{\vartheta})\mathrm{d}\zeta_j^{(2)},$$

where $h(\cdot)$ is the density of the random intercept and the product is over all level-1 units i within the jth level-2 unit. This product represents the joint probability (density) of the responses given the random intercept since the responses are conditionally independent given the random intercept.

General model

We will use the notation of the GRC formulation, denoting the conditional distribution of a level-1 unit i as $g^{(1)}(y_{(1)}|\mathbf{X}_{(1)},\zeta^{(2^+)};\boldsymbol{\vartheta})$, where $\zeta^{(l^+)}$ is the vector of disturbances at levels l and above. The multivariate distribution of the latent variables at level l will be denoted $h^{(l)}(\zeta^{(l)})$. The conditional distribution of the responses of a level-l unit, conditional on the latent variables at levels $l+1$ and above, is a function of the distributions of the level-$(l-1)$ units within the unit:

$$g^{(l)}(\mathbf{y}_{(l)}|\mathbf{X}_{(l)},\zeta^{([l+1]^+)};\boldsymbol{\vartheta}) = \int h^{(l)}(\zeta^{(l)}) \prod g^{(l-1)}(\mathbf{y}_{(l-1)}|\mathbf{X}_{(l-1)},\zeta^{(l^+)};\boldsymbol{\vartheta})\mathrm{d}\zeta^{(l)}.$$

This recursive relationship can be used to build up the likelihood, increasing l from 2 to $L-1$. The reduced form distribution of the responses of a level-L unit then is

$$g^{(L)}(\mathbf{y}_{(L)}|\mathbf{X}_{(L)};\boldsymbol{\vartheta}) = \int h^{(L)}(\zeta^{(L)}) \prod g^{(L-1)}(\mathbf{y}_{(L-1)}|\mathbf{X}_{(L-1)},\zeta^{(L)};\boldsymbol{\vartheta})\mathrm{d}\zeta^{(L)},$$

and the reduced form distribution of all responses is the product

$$g(\mathbf{y}|\mathbf{X};\boldsymbol{\vartheta}) = \prod g^{(L)}(\mathbf{y}_{(L)}|\mathbf{X}_{(L)};\boldsymbol{\vartheta}). \tag{4.34}$$

Latent variable integration using Gauss-Hermite quadrature and other methods is discussed in Section 6.3.

4.9.2 Latent response integration

We will assume that the latent responses (marginal w.r.t the *latent disturbances* ζ) have a multivariate normal distribution. It follows that that univariate, bivariate etc. latent response distributions (marginal w.r.t other *latent responses* y^*) are also normal.

Consider now the ith latent response y_i^* underlying the observed response y_i. y_i could for instance represent the response on the ith item (for a subject) in a measurement model or the response of the ith unit (in a cluster) in a random effects model. We can write the latent response model as (omitting higher-level subscripts)

$$y_i^* = \mu_i + \xi_i,$$

where the mean μ_i is given in (4.28), the covariance matrix $\mathbf{\Omega}_{(L)}$ of the vector of total residuals $\boldsymbol{\xi}_{(L)}$ in (4.32) and the corresponding correlation matrix $\boldsymbol{\rho}_{(L)}$ in (4.33).

Univariate observed response distribution

For an ordinal or dichotomous response y_i, the marginal probabilities of the response categories given the explanatory variables can be expressed as

$$\Pr(y_i = a_s) = \frac{1}{\sqrt{\omega_{ii}}} \int_{\kappa_{s-1}/\sqrt{\omega_{ii}}}^{\kappa_s/\sqrt{\omega_{ii}}} \phi\left(\frac{\mu_i + \xi_i}{\sqrt{\omega_{ii}}}\right) d\xi_i, \qquad (4.35)$$

where ϕ is the standard normal density and ω_{ii} is the ith diagonal element of $\mathbf{\Omega}_{(L)}$. The integration limits are just the *reduced form thresholds*

$$\tau_{is} = \frac{\kappa_s}{\sqrt{\omega_{ii}}}.$$

For a left-censored continuous response the probability is as above with $\tau_{i,s-1} = -\infty$ and κ_s equal to the censoring limit. Here we have kept the mean structure separate from the threshold structure; the mean structure is taken to be the part of the linear predictor that is constant across the response categories (e.g. Muthén, 1984). Another possibility would be to write the integral as

$$\begin{aligned}
\Pr(y_i = a_s) &= \frac{1}{\sqrt{\omega_{ii}}} \int_{\tau_{i,s-1}^*}^{\tau_{is}^*} \phi\left(\frac{\xi_i}{\sqrt{\omega_{ii}}}\right) d\xi_i \\
&= \Phi(\tau_{is}^*) - \Phi(\tau_{is-1}^*),
\end{aligned}$$

where

$$\tau_{is}^* = \frac{\kappa_s - \mu_i}{\sqrt{\omega_{ii}}}.$$

Bivariate observed response distribution

For two ordinal or dichotomous responses y_i and $y_{i'}$, having S and T categories respectively, the joint response probabilities are

$$\Pr(y_i = a_s, y_{i'} = b_t) = \frac{1}{\sqrt{\omega_{ii} \omega_{i'i'}}} \int_{\tau_{s-1}}^{\tau_s} \int_{\tau_{t-1}}^{\tau_t} \varphi\left(\frac{\mu_i + \xi_i}{\sqrt{\omega_{ii}}}, \frac{\mu_{i'} + \xi_{i'}}{\sqrt{\omega_{i'i'}}}; \rho_{ii'}\right) d\xi_i d\xi_{i'},$$

where $\varphi(.,.; \rho_{ii'})$ is the bivariate standard normal density with correlation $\rho_{ii'}$ between latent responses y_i^* and $y_{i'}^*$.

For censored responses, the response distribution for a left censored response (at κ_{is}) and an observed continuous response is

$$\Pr(y_i^* < \kappa_{is}, y_{i'}^* = y_{i'}) = \frac{1}{\sqrt{\omega_{ii} \omega_{i'i'}}} \int_{-\infty}^{\kappa_{is}} \varphi\left(\frac{\mu_i + \xi_i}{\sqrt{\omega_{ii}}}, \frac{\mu_j + \xi_{i'}}{\sqrt{\omega_{i'i'}}}; \rho_{ii'}\right) d\xi_i.$$

The bivariate (and trivariate etc.) distributions of all combinations between dichotomous, ordinal, classified and censored responses involve similar integrals.

Univariate and bivariate response distributions form the basis for the limited information estimation method to be discussed in Section 6.7. Integrating over all the latent responses for a particular level-L unit gives the reduced form distribution for that unit $g^{(L)}(\mathbf{y}^{(L)}|\mathbf{X}^{(L)};\boldsymbol{\vartheta})$. A popular method for performing high-dimensional latent variable integration is by simulation; see Section 6.3.4.

4.9.3 The likelihood

The marginal likelihood $f(\boldsymbol{\vartheta};\mathbf{y}|\mathbf{X})$ is proportional to the reduced form distribution of all units and is considered as a function of the parameters for given values of the responses.

$$f(\boldsymbol{\vartheta};\mathbf{y}|\mathbf{X}) \propto g(\mathbf{y}|\mathbf{X};\boldsymbol{\vartheta}) = \prod g^{(L)}(\mathbf{y}^{(L)}|\mathbf{X}^{(L)};\boldsymbol{\vartheta}).$$

4.10 Reduced form parameters

The parameters of the reduced form for the latent variables and the reduced form for the linear predictor could be referred to as reduced form parameters. However, this term is usually reserved for parameters with the following properties:

1. the reduced form parameters are functions of the fundamental parameters $\boldsymbol{\vartheta}$

2. the reduced form parameters completely characterize the reduced form distribution

3. the reduced form distribution depends on the fundamental parameters only through the reduced form parameters.

In the special case of multivariate normal latent variables and conditionally normal observed responses, the reduced form distribution for a top-level unit is multivariate normal. The distribution is in this case completely characterized by the first and second order moments (the mean structure and covariance structure). It follows that in the above list of properties, 'reduced form distribution' can be replaced by 'first and second order moments'. In Section 4.9.2 we showed that the reduced form distribution for probit models with multivariate normal latent variables is completely characterized by the mean and threshold structure and tetrachoric correlations. These quantities can therefore replace 'reduced form distribution' in the above list of properties in this case.

To illustrate these ideas for the case of conditionally normal observed responses, we will now derive the reduced form parameters for a one-factor model with four items:

$$\begin{bmatrix} y_1 \\ y_2 \\ y_3 \\ y_4 \end{bmatrix} = \begin{bmatrix} \lambda_1 \\ \lambda_2 \\ \lambda_3 \\ \lambda_4 \end{bmatrix} \eta + \begin{bmatrix} \epsilon_1 \\ \epsilon_2 \\ \epsilon_3 \\ \epsilon_4 \end{bmatrix},$$

where $\eta \sim N(0, \psi)$, $\epsilon_i \sim N(0, \theta_{ii})$, and $\text{Cov}(\epsilon_i, \epsilon_{i'}) = 0$. (No intercepts are specified because the responses have been mean-centered.) All in all there are 9 unknown parameters placed in the column vector $\boldsymbol{\vartheta}$

$$\boldsymbol{\vartheta} = [\lambda_1, \lambda_2, \lambda_3, \lambda_4, \psi, \theta_{11}, \theta_{22}, \theta_{33}, \theta_{44}]'. \tag{4.36}$$

From the model implication $\text{E}(y_i) = 0$ and the specification of normality it follows that all information is contained in the second order moments of the y_i:

$$\boldsymbol{\Omega}(\boldsymbol{\vartheta}) \equiv \text{Cov}(\mathbf{y}) = \begin{bmatrix} \lambda_1^2 \psi + \theta_{11} & & & \\ \lambda_2 \psi \lambda_1 & \lambda_2^2 \psi + \theta_{22} & & \\ \lambda_3 \psi \lambda_1 & \lambda_3 \psi \lambda_2 & \lambda_3^2 \psi + \theta_{33} & \\ \lambda_4 \psi \lambda_1 & \lambda_4 \psi \lambda_2 & \lambda_4 \psi \lambda_3 & \lambda_4^2 \psi + \theta_{44} \end{bmatrix}.$$

There are in total 10 nonredundant variances and covariances which are placed in the vector of reduced form parameters $\mathbf{m}(\boldsymbol{\vartheta})$, given as

$$\begin{aligned} \mathbf{m}(\boldsymbol{\vartheta}) &= \text{vech}(\boldsymbol{\Omega}(\boldsymbol{\vartheta})) \\ &= [\lambda_1^2 \psi + \theta_{11}, \lambda_2 \psi \lambda_1, \lambda_3 \psi \lambda_1, \lambda_4 \psi \lambda_1, \lambda_2^2 \psi + \theta_{22}, \lambda_3 \psi \lambda_2, \lambda_4 \psi \lambda_2, \\ &\quad \lambda_3^2 \psi + \theta_{33}, \lambda_4 \psi \lambda_3, \lambda_4^2 \psi + \theta_{44}]. \end{aligned} \tag{4.37}$$

In general, the situation may be more complicated since elements of the covariance matrix may depend on covariates and be written as polynomials in these covariates. The corresponding reduced form parameters in this case are the set of unique coefficients (up to a multiplicative constant) of these polynomials. For example, consider a random coefficient model with a single covariate x_{ij}

$$y_{ij} = \beta_1 + \beta_2 x_{ij} + \eta_{1j} + \eta_{2j} x_{ij} + \epsilon_{ij},$$

having a random slope η_{2j} which is regressed on a random intercept η_{1j},

$$\eta_{2j} = b_{21} \eta_{1j} + \zeta_{2j},$$
$$\eta_{1j} = \zeta_{1j},$$

where $\begin{bmatrix} \zeta_{1j} \\ \zeta_{2j} \end{bmatrix} \sim N_2 \left(\begin{bmatrix} 0 \\ 0 \end{bmatrix}, \begin{bmatrix} \psi_{11} & \\ \psi_{21} & \psi_{22} \end{bmatrix} \right)$ and $\epsilon_{ij} \sim N(0, \theta)$. The conditional variances of the responses become

$$\text{Var}[y_{ij}|x_{ij}] = \psi_{11} + 2b_{21} \psi_{11} x_{ij} + (b_{21}^2 \psi_{11} + \psi_{22}) x_{ij}^2 + \theta,$$

and the covariances

$$\text{Cov}[y_{ij}, y_{i'j}|x_{i'j}, x_{i'j}] = \psi_{11} + b_{21} \psi_{11}(x_{ij} + x_{i'j}) + (b_{21}^2 \psi_{11} + \psi_{22}) x_{ij} x_{i'j}.$$

If $i = 1, 2$, the reduced form parameters are $\psi_{11} + \theta$, $b_{21} \psi_{11}$, $b_{21}^2 \psi_{11} + \psi_{22}$, (the reduced form parameters of the variance for $i = 1$), $\psi_{11} + \theta$, $b_{21} \psi_{11}$, $b_{21}^2 \psi_{11} + \psi_{22}$ (the reduced form parameters of the variance for $i = 2$), ψ_{11}, $b_{21} \psi_{11}$, $b_{21}^2 \psi_{11} + \psi_{22}$ (the reduced form parameters of the covariance between $i = 1$ and $i = 2$), β_1, β_2 (the reduced form parameters of the mean for $i = 1$)

and β_1, β_2 (the reduced form parameters of the mean for $i = 2$). The 6 nonredundant reduced form parameters can then be assembled in

$$\mathbf{m}(\boldsymbol{\vartheta}) \;=\; [\beta_1, \beta_2, \psi_{11}, \psi_{11} + \theta, b_{21}\psi_{11}, b_{21}^2\psi_{11} + \psi_{22}].$$

4.11 Summary and further reading

We have introduced a general model framework unifying multilevel, structural equation, latent class and longitudinal models. The framework accommodates factors and random coefficients at different levels, regression structures among them and a wide range of response processes and flexible specifications of latent variable distributions. The framework is essentially the generalized linear latent and mixed model (GLLAMM) framework discussed in Rabe-Hesketh *et al.* (2004a) for the case of continuous latent variables.

Development of the GLLAMM framework has been in parallel with the development of the Stata program `gllamm` available from `www.gllamm.org`. The program can estimate all the models discussed in this chapter except (currently) models including both discrete and continuous latent variables and discrete latent variable models with more general structural models than that given in equation (4.15). Furthermore, the multivariate normal distribution is the only continuous latent variable distribution currently available. A relatively nontechnical treatment of the model framework with details of using `gllamm` to estimate the models given in Rabe-Hesketh *et al.* (2004b). Except where stated otherwise, all models described in the Application Part have been estimated using `gllamm`.

Several other more or less general model frameworks with latent variables have been suggested, some of which are mentioned here. Muthén's general model framework (e.g. Muthén, 2001, 2002) includes latent traits as well as latent classes and handles continuous, dichotomous and ordinal responses. Other seminal contributions to multilevel structural equation modeling with continuous responses include Goldstein and McDonald (1988), McDonald and Goldstein (1989). Brief discussions can be found in Raudenbush and Bryk (2002), Hox (2002) and de Boeck and Wilson (2004). Fox (2001) considers multilevel item response models from a Bayesian perspective. Skrondal (1996) considers latent trait and multilevel models with continuous, censored, dichotomous and ordinal responses. In the single level setting, Sammel *et al.* (1997) and Moustaki and Knott (2000) discuss latent trait models with continuous, dichotomous and ordinal responses. Bartholomew and Knott (1999) and Moustaki (1996) discuss both latent trait and latent class models for continuous, dichotomous and polytomous response processes. Arminger and Küsters (1988, 1989) discuss models with continuous, dichotomous, ordinal and polytomous responses as well as counts. Hagenaars (1993) and Vermunt (1997) cover latent class models with structural equations.

We have derived the reduced form distribution using two approaches: latent variable integration and latent response integration. An advantage of latent variable integration is that it handles all response types. A disadvantage is that conditional independence of the responses given the latent variables must be specified. However, this disadvantage may be partly overcome by inducing dependence using additional latent variables. An advantage of latent response integration is that it is easy to relax the conditional independence assumption, but a disadvantage is that it is confined to response models with a latent response formulation, not for instance models with a logit link or Poisson distribution.

CHAPTER 5

Identification and equivalence

5.1 Introduction

The statistical models considered in this book are fairly complex, going beyond merely description or exploratory analysis of data. A basic idea is to construct structural statistical models purporting to represent the main features of the 'data generating mechanism'; the empirical process having generated the observed data. In this setting the issues of identification and equivalence of statistical models become fundamental, since the structural parameters of scientific interest differ from the reduced form parameters. Broadly speaking, identification and equivalence concern the prospects of making inferences regarding the structural model based on the reduced form distribution for the observed variables.

A parametric statistical model is said to be identified if there is a unique value of the parameter vector ϑ or *parameter point* that can generate a given reduced form distribution $g(\mathbf{y}|\mathbf{X}, \vartheta)$. If a model is not identified there are several sets of parameters that could have produced the reduced form distribution. The model parameters are in this sense arbitrary, a situation that is detrimental for scientific inference. A crucial implication is that consistent parameter estimation is precluded (e.g. Gabrielsen, 1978).

The related concept of equivalence concerns the prospects of distinguishing empirically between different statistical models. If two models are merely reparameterizations, producing identical reduced form distributions, they are equivalent and equally well compatible with any data. Equivalence is of course detrimental when the models represent different and perhaps contradictory substantive data generating mechanisms. On the other hand, we can sometimes take advantage of equivalence to simplify estimation problems as is shown in Section 5.3.2.

In Section 5.2, we present some useful definitions of identification. Analytic investigation of identification proceeds by studying the properties of the mappings between reduced form parameters, which are presumed to be identified, and the fundamental parameters. This approach is illustrated for a number of models. Finally, empirical identification is considered.

In Section 5.3 we first present definitions of equivalence and then discuss the analytic approach to equivalence. This involves investigating whether there is a one-to-one transformation between parameterizations generating the same reduced form distribution. After illustrating this analytic approach using some examples, we also consider empirical equivalence.

Our treatment is fairly informal, focusing on how analytic investigation of identification and equivalence can proceed in practice, and we will refer to the literature for deeper insight. We will in particular consider methods that are straightforward to implement in software for computer algebra such as Mathematica (Wolfram, 2003) or Maple (Maple 9 Learning Guide, 2003).

Although the reader should appreciate the importance of identification and equivalence, he or she may wish to skip some of the more technical parts of this chapter.

5.2 Identification

5.2.1 Definitions

The following definitions are useful:

- Two parameter points $\boldsymbol{\vartheta}^1, \boldsymbol{\vartheta}^2$ are called *observationally equivalent* if they imply the same reduced form distribution for the observed random variables; $g(\mathbf{y}|\mathbf{X}; \boldsymbol{\vartheta}^1) = g(\mathbf{y}|\mathbf{X}; \boldsymbol{\vartheta}^2)$.

- A parameter vector $\boldsymbol{\vartheta} \in \mathcal{A}$ is *globally identified* if for any parameter point $\boldsymbol{\vartheta}^1 \in \mathcal{A}$ there is no other observationally equivalent point $\boldsymbol{\vartheta}^2 \in \mathcal{A}$.

- A parameter point $\boldsymbol{\vartheta}^0 \in \mathcal{A}$ is *locally identified* if there exists an open neighborhood of $\boldsymbol{\vartheta}^0$ containing no other $\boldsymbol{\vartheta}$ which is observationally equivalent to $\boldsymbol{\vartheta}^0$.

It should be noted that local identification everywhere in the parameter space $\mathcal{A}$ is a necessary but not sufficient condition for global identification, see Bechger *et al.* (2001, p.362) for an example. Also note that the models considered in this book typically imply nonlinear moment structures. It follows that local identification at one point in $\mathcal{A}$ does not imply local identification everywhere in $\mathcal{A}$ and that parameter points can often be found which are not locally identified (see Section 5.2.4 for an example). Hence, many of the models considered in this book are likely not to be globally identifiable and we must resort to the weaker notion of local identification (see also McDonald, 1982).

5.2.2 Methods for analytical investigation of local identification

Recall the unidimensional factor model

$$y_{ij} = \beta_i + \lambda_i \eta_j + \epsilon_{ij}, \quad \eta_j \sim N(\gamma, \psi), \quad \epsilon_{ij} \sim N(0, \theta_{ii}),$$

from Section 3.3.2. By considering a linear transformation of the factor, $f_j = a\eta_j + c$, we can write the model as

$$
\begin{aligned}
y_{ij} &= (\beta_i - \lambda_i c/a) + (\lambda_i/a)f_j + \epsilon_{ij}, \quad f_j \sim N(a\gamma + c, a^2\psi), \quad \epsilon_{ij} \sim N(0, \theta_{ii}) \\
&= \beta_i^* + \lambda_i^* f_j + \epsilon_{ij}, \quad f_j \sim N(\gamma^*, \psi^*), \quad \epsilon_{ij} \sim N(0, \theta_{ii}),
\end{aligned}
$$

where

$$\beta_i^* = \beta_i - \lambda_i c/a, \quad \lambda_i^* = \lambda_i/a, \quad \gamma^* = a\gamma + c, \quad \psi^* = a^2\psi.$$

Therefore, different parameter points generate the same reduced form distribution and the model is not identified. In order to identify the model we could fix the mean and variance of the factor for instance to zero and one, respectively. Although it was easy to demonstrate that the above model was not identified, it is not straightforward to show that the suggested parameter restrictions render the model identified. In fact, it can be shown that the model is not identified with fewer than three items.

Under suitable assumptions, a necessary and sufficient condition for local identification at a parameter point is that the (theoretical) information matrix is nonsingular at the point (e.g. Rothenberg, 1971). In principle this condition can be used for investigating identification, but this approach is usually not feasible in practice because the information matrix is usually analytically intractable in complex models.

In the special case where reduced form parameters exist (that completely characterize the reduced form distribution), as for the normal case, the standard approach to identification instead focuses on the mappings between fundamental and reduced form parameters (see Section 4.10). This approach, which yields necessary and sufficient conditions for local identification, appears to be due to Wald (1950); see Fisher (1966) for a survey. Dupačová and Wold (1982) applied this idea to conventional structural equation models with latent variables.

In this chapter we extend the mapping approach beyond normal responses to models with dichotomous and/or ordinal responses generated from normal latent responses crossing thresholds, i.e. models with probit links. This is possible since reduced form parameters exist in this case that completely characterize the reduced form distribution. For models where there are no reduced form parameters that completely characterize the distribution, it may nevertheless be useful to consider the mapping between fundamental parameters and the reduced form parameters of the first and second order moments, the idea being that identification relying on higher order moments is likely to be fragile. For example, for dichotomous or ordinal responses, identification is likely to be fragile for logit models, if the analogous probit model is not identified (e.g. Rabe-Hesketh and Skrondal, 2001).

The fundamental parameters $\vartheta \in \mathcal{A}$ produce reduced form parameters $\mathbf{m} \in \mathcal{A}'$ via mappings

$$m_s = h_s(\vartheta), \qquad 1 \leq s \leq S,$$

where $h_s(\cdot)$, $1 \leq s \leq S$, are continuously differentiable known functions.

The probability distribution of the observed variables depends on the fundamental parameters $\vartheta \in \mathcal{A}$ only through the S-dimensional reduced form parameter vector $\mathbf{m}$,

$$g(\mathbf{y}|\mathbf{X}; \vartheta) = g^*(\mathbf{y}|\mathbf{X}; h_1(\vartheta), \ldots, h_S(\vartheta)) = g^*(\mathbf{y}|\mathbf{X}; \mathbf{m}) \quad \text{for all } \vartheta \in \mathcal{A},$$

where g^* is the distribution in terms of the reduced form parameters. Identification of ϑ can therefore be investigated by considering characteristics of mappings from ϑ to $\mathbf{m}$.

Consider a particular fundamental parameter vector $\boldsymbol{\vartheta}^0$ generating the reduced form parameter vector $\mathbf{m}^0$,

$$m_s^0 = h_s(\boldsymbol{\vartheta}^0), \quad 1 \leq s \leq S.$$

Then $\boldsymbol{\vartheta}^0$ is identifiable if and only if $\boldsymbol{\vartheta}^0$ is the unique solution of the equations

$$m_s^0 = h_s(\boldsymbol{\vartheta}), \quad 1 \leq s \leq S. \tag{5.1}$$

Hence, the identification of $\boldsymbol{\vartheta}^0$ depends solely on the properties of the mappings $h_s(\cdot)$. The identification problem therefore reduces to the question of uniqueness of solutions to systems of equations so that we can use classical results of calculus.

It is evident that a necessary but not sufficient condition for identification is that there are at least as many elements in the reduced form parameter vector as there are unknown parameters; $v \leq S$. In order to derive stronger identification results, it is useful to define the Jacobian of the mapping

$$J(\boldsymbol{\vartheta}) = \left[\frac{\partial h_s}{\partial \vartheta_j}, \quad 1 \leq s \leq S, 1 \leq j \leq v, \right].$$

A parameter vector $\boldsymbol{\vartheta}^0$ is a *regular point* if there is an open neighborhood of $\boldsymbol{\vartheta}^0$ in which the Jacobian has constant rank. If we know nothing about $\boldsymbol{\vartheta}^0$ except that $\boldsymbol{\vartheta}^0 \in \mathcal{A}$ it makes sense to assume that it is a regular point, since almost all points in $\mathcal{A}$ are regular points. However, we will encounter a case where $\boldsymbol{\vartheta}^0$ is not a regular point in Section 5.2.4.

If $\boldsymbol{\vartheta}^0$ is a regular point, the system of equations (5.1) has the unique solution $\boldsymbol{\vartheta}^0$ if and only if the rank of the Jacobian is equal to the number of fundamental parameters v. The analysis of identification in this chapter will therefore rely on the following Lemma:

Lemma 1: If $\boldsymbol{\vartheta}^0$ is a regular point of $J(\boldsymbol{\vartheta})$, then $\boldsymbol{\vartheta}^0$ is locally identified if and only if Rank $\left[J(\boldsymbol{\vartheta}^0) \right] = v$.

5.2.3 Applying the Jacobian method for local identification

One-factor model with four continuous items

We return to the one-factor model with four continuous items introduced in Section 4.10, whose parameter vector $\boldsymbol{\vartheta}$ and reduced form parameters $\mathbf{m}(\boldsymbol{\vartheta})$

were given in (4.36) and (4.37), respectively. The 10×9 Jacobian becomes

$$
\mathbf{J}(\boldsymbol{\vartheta}) = \begin{bmatrix}
2\psi\lambda_1 & 0 & 0 & 0 & \lambda_1^2 & 1 & 0 & 0 & 0 \\
\psi\lambda_2 & \psi\lambda_1 & 0 & 0 & \lambda_1\lambda_2 & 0 & 0 & 0 & 0 \\
\psi\lambda_3 & 0 & \psi\lambda_1 & 0 & \lambda_1\lambda_3 & 0 & 0 & 0 & 0 \\
\psi\lambda_4 & 0 & 0 & \psi\lambda_1 & \lambda_1\lambda_4 & 0 & 0 & 0 & 0 \\
0 & 2\psi\lambda_2 & 0 & 0 & \lambda_2^2 & 0 & 1 & 0 & 0 \\
0 & \psi\lambda_3 & \psi\lambda_2 & 0 & \lambda_2\lambda_3 & 0 & 0 & 0 & 0 \\
0 & \psi\lambda_4 & 0 & \psi\lambda_2 & \lambda_2\lambda_4 & 0 & 0 & 0 & 0 \\
0 & 0 & 2\psi\lambda_3 & 0 & \lambda_3^2 & 0 & 0 & 1 & 0 \\
0 & 0 & \psi\lambda_4 & \psi\lambda_3 & \lambda_3\lambda_4 & 0 & 0 & 0 & 0 \\
0 & 0 & 0 & 2\psi\lambda_4 & \lambda_4^2 & 0 & 0 & 0 & 1
\end{bmatrix}, \qquad (5.2)
$$

where the element in row r and column c is $\partial m_r(\boldsymbol{\vartheta})/\partial\vartheta_c$. The rank of this Jacobian, Rank[$\mathbf{J}(\boldsymbol{\vartheta})$], is 8 which is one less than the number of parameters, so the model is not locally identified.

Let $\mathbf{N}(\boldsymbol{\vartheta})$ be a vector satisfying the equation $\mathbf{J}(\boldsymbol{\vartheta})\mathbf{N}(\boldsymbol{\vartheta}) = \mathbf{0}$. Such a vector is called a basis for the nullspace of the Jacobian $\mathbf{J}(\boldsymbol{\vartheta})$. In the present example the vector

$$
\left[-\frac{\lambda_1}{2\psi}, -\frac{\lambda_2}{2\psi}, -\frac{\lambda_3}{2\psi}, -\frac{\lambda_4}{2\psi}, 1, 0, 0, 0, 0 \right]'
$$

represents a basis for the nullspace. We see that linear dependencies are evident among factor loadings λ_i and the factor variance ψ, but not involving the error variances θ_{ii}. Importantly, it follows that the θ_{ii} are locally identified and may be consistently estimated, although the model as a whole is not identified. This is often called *partial identification*. It seems to be a good idea to either fix a factor loading or the factor variance to a constant in order to identify the factor variance and the loadings.

First consider whether the factor model is identified after 'anchoring', fixing one of the factor loadings to an arbitrary nonzero constant, typically $\lambda_1 = 1$. We let a star be appended to the 8 remaining unknown parameters. The 10×8 Jacobian is obtained by setting $\lambda_1 = 1$ in (5.2) and omitting the first column since this contains derivatives with respect to λ_1 which is no longer a parameter of the model. The resulting Jacobian is

$$
\mathbf{J}(\boldsymbol{\vartheta}^*) = \begin{bmatrix}
0 & 0 & 0 & 1 & 1 & 0 & 0 & 0 \\
\psi^* & 0 & 0 & \lambda_2^* & 0 & 0 & 0 & 0 \\
0 & \psi^* & 0 & \lambda_3^* & 0 & 0 & 0 & 0 \\
0 & 0 & \psi^* & \lambda_4^* & 0 & 0 & 0 & 0 \\
2\psi^*\lambda_2^* & 0 & 0 & (\lambda_2^*)^2 & 0 & 1 & 0 & 0 \\
\psi^*\lambda_3^* & \psi^*\lambda_2^* & 0 & \lambda_2^*\lambda_3^* & 0 & 0 & 0 & 0 \\
\psi^*\lambda_4^* & 0 & \psi^*\lambda_2^* & \lambda_2^*\lambda_4^* & 0 & 0 & 0 & 0 \\
0 & 2\psi^*\lambda_3^* & 0 & (\lambda_3^*)^2 & 0 & 0 & 1 & 0 \\
0 & \psi^*\lambda_4^* & \psi^*\lambda_3^* & \lambda_3^*\lambda_4^* & 0 & 0 & 0 & 0 \\
0 & 0 & 2\psi^*\lambda_4^* & (\lambda_4^*)^2 & 0 & 0 & 0 & 1
\end{bmatrix},
$$

with $\text{Rank}[\mathbf{J}(\boldsymbol{\vartheta}^*)] = 8$, so the model is locally identified after anchoring if $\boldsymbol{\vartheta}^*$ is a regular point.

We then investigate whether the factor model is identified after 'factor standardization' where we instead fix the factor variance to an arbitrary nonzero constant, typically $\psi = 1$. To distinguish the parameters from those under anchoring we put a bar above the symbol. Setting $\psi = 1$ in (5.2) and omitting the fifth column, the Jacobi matrix becomes

$$
\mathbf{J}(\bar{\boldsymbol{\vartheta}}) =
\begin{bmatrix}
2\bar{\lambda}_1 & 0 & 0 & 0 & 1 & 0 & 0 & 0 \\
\bar{\lambda}_2 & \bar{\lambda}_1 & 0 & 0 & 0 & 0 & 0 & 0 \\
\bar{\lambda}_3 & 0 & \bar{\lambda}_1 & 0 & 0 & 0 & 0 & 0 \\
\bar{\lambda}_4 & 0 & 0 & \bar{\lambda}_1 & 0 & 0 & 0 & 0 \\
0 & 2\bar{\lambda}_2 & 0 & 0 & 0 & 1 & 0 & 0 \\
0 & \bar{\lambda}_3 & \bar{\lambda}_2 & 0 & 0 & 0 & 0 & 0 \\
0 & \bar{\lambda}_4 & 0 & \bar{\lambda}_2 & 0 & 0 & 0 & 0 \\
0 & 0 & 2\bar{\lambda}_3 & 0 & 0 & 0 & 1 & 0 \\
0 & 0 & \bar{\lambda}_4 & \bar{\lambda}_3 & 0 & 0 & 0 & 0 \\
0 & 0 & 0 & 2\bar{\lambda}_4 & 0 & 0 & 0 & 1
\end{bmatrix},
$$

with $\text{Rank}[\mathbf{J}(\bar{\boldsymbol{\vartheta}})] = 8$, so the model is locally identified if $\bar{\boldsymbol{\vartheta}}$ is a regular point.

Structural equation model with six continuous items and correlated errors

We now consider a structural equation model for panel data with three latent variables discussed by Wheaton *et al.* (1977). The model is recursive in that one latent variable serves as explanatory variable whereas the other two represent the response variables at two panel waves or occasions. Each latent variable is measured by two items, where one item serves as anchor for each factor. An important feature is that errors for repeated measures of the same item are correlated, making investigation of identification by means of pen and paper quite complex (see also Jöreskog and Sörbom, 1989, p.173-174). A path diagram of the model is shown in Figure 5.1 where the ϵ_{ij} are represented by small circles.

The measurement part of the model (corresponding to equation (3.31)) is:

$$
\begin{bmatrix}
y_{1j} \\
y_{2j} \\
y_{3j} \\
y_{4j} \\
y_{5j} \\
y_{6j}
\end{bmatrix}
=
\begin{bmatrix}
1 & 0 & 0 \\
\lambda_{21} & 0 & 0 \\
0 & 1 & 0 \\
0 & \lambda_{42} & 0 \\
0 & 0 & 1 \\
0 & 0 & \lambda_{63}
\end{bmatrix}
\begin{bmatrix}
\eta_{1j} \\
\eta_{2j} \\
\eta_{3j}
\end{bmatrix}
+
\begin{bmatrix}
\epsilon_{1j} \\
\epsilon_{2j} \\
\epsilon_{3j} \\
\epsilon_{4j} \\
\epsilon_{5j} \\
\epsilon_{6j}
\end{bmatrix},
$$

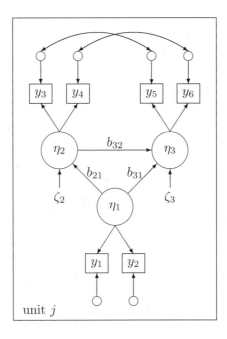

Figure 5.1 *Path diagram of model by Wheaton* et al. *(1977)*

where

$$
\begin{bmatrix} \epsilon_{1j} \\ \epsilon_{2j} \\ \epsilon_{3j} \\ \epsilon_{4j} \\ \epsilon_{5j} \\ \epsilon_{6j} \end{bmatrix} \sim N_6 \left(\begin{bmatrix} 0 \\ 0 \\ 0 \\ 0 \\ 0 \\ 0 \end{bmatrix}, \begin{bmatrix} \theta_{11} & & & & & \\ 0 & \theta_{22} & & & & \\ 0 & 0 & \theta_{33} & & & \\ 0 & 0 & 0 & \theta_{44} & & \\ 0 & 0 & \theta_{53} & 0 & \theta_{55} & \\ 0 & 0 & 0 & \theta_{64} & 0 & \theta_{66} \end{bmatrix} \right).
$$

The structural part (corresponding to equation (3.32)) is specified as

$$
\begin{bmatrix} \eta_{1j} \\ \eta_{2j} \\ \eta_{3j} \end{bmatrix} = \begin{bmatrix} 0 & 0 & 0 \\ b_{21} & 0 & 0 \\ b_{31} & b_{32} & 0 \end{bmatrix} \begin{bmatrix} \eta_{1j} \\ \eta_{2j} \\ \eta_{3j} \end{bmatrix} + \begin{bmatrix} \zeta_{1j} \\ \zeta_{2j} \\ \zeta_{3j} \end{bmatrix},
$$

where

$$
\begin{bmatrix} \zeta_{1j} \\ \zeta_{2j} \\ \zeta_{3j} \end{bmatrix} \sim N_3 \left(\begin{bmatrix} 0 \\ 0 \\ 0 \end{bmatrix}, \begin{bmatrix} \psi_{11} & & \\ 0 & \psi_{22} & \\ 0 & 0 & \psi_{33} \end{bmatrix} \right).
$$

This model has 17 parameters and implies 21 nonredundant second order moments. We obtain $\mathrm{Rank}[\mathbf{J}(\boldsymbol{\vartheta})]=17$, so the model is locally identified if $\boldsymbol{\vartheta}$ is a regular point.

Due to the panel design it might be tempting to also let the disturbances ζ_2

and ζ_3 be correlated, e.g. $\psi_{32} \neq 0$. However, we then obtain $\text{Rank}[\mathbf{J}(\boldsymbol{\vartheta})] = 17$ for 18 parameters, demonstrating that this model is not locally identified.

One-factor model with four dichotomous items

Consider the following one-factor model for four continuous underlying or latent response variables:

$$
\begin{bmatrix} y_{1j}^* \\ y_{2j}^* \\ y_{3j}^* \\ y_{4j}^* \end{bmatrix} = \begin{bmatrix} \beta_1 \\ \beta_2 \\ \beta_3 \\ \beta_4 \end{bmatrix} + \begin{bmatrix} 1 \\ \lambda_2 \\ \lambda_3 \\ \lambda_4 \end{bmatrix} \eta_j + \begin{bmatrix} \epsilon_{1j} \\ \epsilon_{2j} \\ \epsilon_{3j} \\ \epsilon_{4j} \end{bmatrix},
$$

where $\eta_j \sim N(0, \psi)$, $\epsilon_{ij} \sim N(0, \theta_{ii})$ and $\text{Cov}(\epsilon_{ij}, \epsilon_{i'j}) = 0$. Note that we have 'anchored' the factor by imposing $\lambda_1 = 1$.

There are 12 unknown parameters

$$
\boldsymbol{\vartheta} = [\beta_1, \beta_2, \beta_3, \beta_4, \lambda_2, \lambda_3, \lambda_4, \psi, \theta_{11}, \theta_{22}, \theta_{33}, \theta_{44}]',
$$

and the covariance matrix of the latent response variables becomes

$$
\boldsymbol{\Omega}(\boldsymbol{\vartheta}) = \begin{bmatrix} \psi + \theta_{11} & & & \\ \lambda_2\psi & \lambda_2^2\psi + \theta_{22} & & \\ \lambda_3\psi & \lambda_3\psi\lambda_2 & \lambda_3^2\psi + \theta_{33} & \\ \lambda_4\psi & \lambda_4\psi\lambda_2 & \lambda_4\psi\lambda_3 & \lambda_4^2\psi + \theta_{44} \end{bmatrix}.
$$

The present model is identical to the one-factor model for four continuous items including intercepts β_i, with the crucial difference that the underlying or latent response variables are not observed. Instead, the latent response variables are related to observed dichotomous responses via threshold functions

$$
y_{ij} = \begin{cases} 1 & \text{if } y_{ij}^* > 0 \\ 0 & \text{otherwise.} \end{cases}
$$

The marginal probability $\text{Pr}(y_{ij} = 1)$ becomes

$$
\text{Pr}(y_{ij}=1) = \frac{1}{\sqrt{\lambda_i^2\psi+\theta_{ii}}} \int_0^\infty \phi\left(\frac{\beta_i+\xi_{ij}}{\sqrt{\lambda_i^2\psi+\theta_{ii}}}\right) d\xi_{ij} = \Phi\left(\frac{\beta_i}{\sqrt{\lambda_i^2\psi+\theta_{ii}}}\right),
$$

where ϕ is the standard normal density, $\xi_{ij} = \lambda_i\eta_j + \epsilon_{ij}$, and Φ is the cumulative standard normal distribution. The means μ_i of the latent responses y_{ij}^* underlying the items i are

$$
\mu_i = \frac{\beta_i}{\sqrt{\lambda_i^2\psi+\theta_{ii}}},
$$

and are identified from the marginal probabilities. The joint response probabilities $p_{stuv} \equiv \text{Pr}(y_1 = s, y_2 = t, y_3 = u, y_4 = v)$ can be expressed as

$$
p_{stuv} = \frac{1}{\sqrt{(\psi+\theta_{11})(\lambda_2^2\psi+\theta_{22})(\lambda_3^2\psi+\theta_{33})(\lambda_4^2\psi+\theta_{44})}} \int_{-\tau_{s-1}}^{-\tau_s} \int_{-\tau_{t-1}}^{-\tau_t} \int_{-\tau_{u-1}}^{-\tau_u} \int_{-\tau_{v-1}}^{-\tau_v} \times
$$

$$
\varphi\left(\frac{\beta_1+\xi_1}{\sqrt{\psi+\theta_{11}}}, \frac{\beta_2+\xi_2}{\sqrt{\lambda_2^2\psi+\theta_{22}}}, \frac{\beta_3+\xi_3}{\sqrt{\lambda_3^2\psi+\theta_{33}}}, \frac{\beta_4+\xi_4}{\sqrt{\lambda_4^2\psi+\theta_{44}}}; \mathbf{R}\right) d\xi_1 d\xi_2 d\xi_3 d\xi_4
$$

where $\varphi(\cdot, \cdot, \cdot, \cdot; \mathbf{R})$ is the four dimensional standard normal density with tetrachoric correlation matrix

$$\mathbf{R}(\boldsymbol{\vartheta}) = \text{diag}(\boldsymbol{\Omega}(\boldsymbol{\vartheta}))^{-1/2}\boldsymbol{\Omega}(\boldsymbol{\vartheta})\text{diag}(\boldsymbol{\Omega}(\boldsymbol{\vartheta}))^{-1/2}$$

$$= \begin{bmatrix} 1 \\ \frac{\lambda_2\psi}{\sqrt{\lambda_2^2\psi+\theta_{22}}\sqrt{\psi+\theta_{11}}} & 1 \\ \frac{\lambda_3\psi}{\sqrt{\lambda_3^2\psi+\theta_{33}}\sqrt{\psi+\theta_{11}}} & \frac{\lambda_3\psi\lambda_2}{\sqrt{\lambda_3^2\psi+\theta_{33}}\sqrt{\lambda_2^2\psi+\theta_{22}}} & 1 \\ \frac{\lambda_4\psi}{\sqrt{\lambda_4^2\psi+\theta_{44}}\sqrt{\psi+\theta_{11}}} & \frac{\lambda_4\psi\lambda_2}{\sqrt{\lambda_4^2\psi+\theta_{44}}\sqrt{\lambda_2^2\psi+\theta_{22}}} & \frac{\lambda_4\psi\lambda_3}{\sqrt{\lambda_4^2\psi+\theta_{44}}\sqrt{\lambda_3^2\psi+\theta_{33}}} & 1 \end{bmatrix}.$$

The tetrachoric correlation matrix is well known to be identified.

All in all there are 10 reduced form parameters, 4 means and 6 nonredundant tetrachoric correlations, which are assembled in

$$\mathbf{m}(\boldsymbol{\vartheta}) = \left[\frac{\beta_1}{\sqrt{\psi+\theta_{11}}}, \frac{\beta_2}{\sqrt{\lambda_2^2\psi+\theta_{22}}}, \frac{\beta_3}{\sqrt{\lambda_3^2\psi+\theta_{33}}}, \frac{\beta_4}{\sqrt{\lambda_4^2\psi+\theta_{44}}}, \right.$$

$$\frac{\lambda_2\psi}{\sqrt{\lambda_2^2\psi+\theta_{22}}\sqrt{\psi+\theta_{11}}}, \frac{\lambda_3\psi}{\sqrt{\lambda_3^2\psi+\theta_{33}}\sqrt{\psi+\theta_{11}}}, \frac{\lambda_4\psi}{\sqrt{\lambda_4^2\psi+\theta_{44}}\sqrt{\psi+\theta_{11}}},$$

$$\frac{\lambda_3\psi\lambda_2}{\sqrt{\lambda_3^2\psi+\theta_{33}}\sqrt{\lambda_2^2\psi+\theta_{22}}}, \frac{\lambda_4\psi\lambda_2}{\sqrt{\lambda_4^2\psi+\theta_{44}}\sqrt{\lambda_2^2\psi+\theta_{22}}},$$

$$\left. \frac{\lambda_4\psi\lambda_3}{\sqrt{\lambda_4^2\psi+\theta_{44}}\sqrt{\lambda_3^2\psi+\theta_{33}}} \right].$$

In this case we need not proceed to obtain the rank of the Jacobian since there are more unknown parameters, 12, than identified reduced form parameters, 10. Thus it is obvious that the model is not identified.

Consider now fixing the variances of the errors ϵ_{ij} to 1, e.g. $\theta_{11}=\theta_{22}=\theta_{33}=\theta_{44}=1$. There are now 8 unknown parameters

$$\boldsymbol{\vartheta} = [\beta_1, \beta_2, \beta_3, \beta_4, \lambda_2, \lambda_3, \lambda_4, \psi]'$$

and the 10 identified reduced form parameters become

$$\mathbf{m}(\boldsymbol{\vartheta}) = \left[\frac{\beta_1}{\sqrt{\psi+1}}, \frac{\beta_2}{\sqrt{\lambda_2^2\psi+1}}, \frac{\beta_3}{\sqrt{\lambda_3^2\psi+1}}, \frac{\beta_4}{\sqrt{\lambda_4^2\psi+1}}, \right.$$

$$\frac{\lambda_2\psi}{\sqrt{\lambda_2^2\psi+1}\sqrt{\psi+1}}, \frac{\lambda_3\psi}{\sqrt{\lambda_3^2\psi+1}\sqrt{\psi+1}}, \frac{\lambda_4\psi}{\sqrt{\lambda_4^2\psi+1}\sqrt{\psi+1}},$$

$$\frac{\lambda_3\psi\lambda_2}{\sqrt{\lambda_3^2\psi+1}\sqrt{\lambda_2^2\psi+1}}, \frac{\lambda_4\psi\lambda_2}{\sqrt{\lambda_4^2\psi+1}\sqrt{\lambda_2^2\psi+1}},$$

$$\left. \frac{\lambda_4\psi\lambda_3}{\sqrt{\lambda_4^2\psi+1}\sqrt{\lambda_3^2\psi+1}} \right]. \tag{5.3}$$

The 10×8 Jacobian $\mathbf{J}(\boldsymbol{\vartheta})$ becomes complicated and too huge to be presented in this case, but the main point is that $\text{Rank}[\mathbf{J}(\boldsymbol{\vartheta})]$ is 8, so the model is locally identified if $\boldsymbol{\vartheta}$ is a regular point.

Coull and Agresti model for four dichotomous responses

Coull and Agresti (2000) suggested a multivariate binomial logit-normal (BLN) model. In their first example they specified a simple model with a separate intercept for each of four occasions and no other covariates. We will here consider the probit-normal version of their model, denoted the BPN model in Rabe-Hesketh and Skrondal (2001):

$$
\begin{bmatrix} y_{1j}^* \\ y_{2j}^* \\ y_{3j}^* \\ y_{4j}^* \end{bmatrix}
=
\begin{bmatrix} \beta_1 \\ \beta_2 \\ \beta_3 \\ \beta_4 \end{bmatrix}
+
\begin{bmatrix} \eta_{1j} \\ \eta_{2j} \\ \eta_{3j} \\ \eta_{4j} \end{bmatrix}
+
\begin{bmatrix} \epsilon_{1j} \\ \epsilon_{2j} \\ \epsilon_{3j} \\ \epsilon_{4j} \end{bmatrix},
$$

where

$$
\begin{bmatrix} \eta_{1j} \\ \eta_{2j} \\ \eta_{3j} \\ \eta_{4j} \end{bmatrix}
\sim N_4 \left(
\begin{bmatrix} 0 \\ 0 \\ 0 \\ 0 \end{bmatrix},
\begin{bmatrix} \sigma^2 & & & \\ \rho_1\sigma^2 & \sigma^2 & & \\ \rho_1\sigma^2 & \rho_1\sigma^2 & \sigma^2 & \\ \rho_2\sigma^2 & \rho_2\sigma^2 & \rho_2\sigma^2 & \sigma^2 \end{bmatrix}
\right),
$$

$\epsilon_{ij} \sim N(0,1)$ and $\mathrm{Cov}(\epsilon_{ij}, \epsilon_{i'j}) = 0$. The latent response variables are related to observed dichotomous responses via threshold functions

$$
y_{ij} = \begin{cases} 1 & \text{if } y_{ij}^* > 0 \\ 0 & \text{otherwise.} \end{cases}
$$

The model is not identified since the variances of the latent response errors are not identified as shown above for the one-factor model with dichotomous items.

Rabe-Hesketh and Skrondal (2001) suggested that σ^2 be fixed at a positive value to ensure identification. Actually, σ^2 cannot be fixed to any positive value (Rabe-Hesketh and Skrondal, 2001, p. 1258) but we will for simplicity impose $\sigma^2 = 1$ here. There are then 6 unknown parameters

$$
\boldsymbol{\vartheta} = [\beta_1, \beta_2, \beta_3, \beta_4, \rho_1, \rho_2]',
$$

implying the tetrachoric correlation matrix

$$
\mathbf{R}(\boldsymbol{\vartheta}) =
\begin{bmatrix}
1 & & & \\
\frac{\rho_1}{2} & 1 & & \\
\frac{\rho_1}{2} & \frac{\rho_1}{2} & 1 & \\
\frac{\rho_2}{2} & \frac{\rho_2}{2} & \frac{\rho_2}{2} & 1
\end{bmatrix},
$$

where the off-diagonal elements are identified. From the marginal probabilities

$$
\Pr(y_{ij}=1) = \Phi(\frac{\beta_i}{\sqrt{2}}),
$$

we can identify

$$
\mu_i = \frac{\beta_i}{\sqrt{2}}.
$$

The identified reduced form parameters are placed in

$$\mathbf{m}(\boldsymbol{\vartheta}) = \left[\frac{\beta_1}{\sqrt{2}}, \frac{\beta_2}{\sqrt{2}}, \frac{\beta_3}{\sqrt{2}}, \frac{\beta_4}{\sqrt{2}}, \frac{\rho_1}{2}, \frac{\rho_1}{2}, \frac{\rho_2}{2}, \frac{\rho_1}{2}, \frac{\rho_2}{2}, \frac{\rho_2}{2} \right], \tag{5.4}$$

and the 10×6 Jacobian becomes

$$\mathbf{J}(\boldsymbol{\vartheta}) = \begin{bmatrix} \frac{1}{\sqrt{2}} & 0 & 0 & 0 & 0 & 0 \\ 0 & \frac{1}{\sqrt{2}} & 0 & 0 & 0 & 0 \\ 0 & 0 & \frac{1}{\sqrt{2}} & 0 & 0 & 0 \\ 0 & 0 & 0 & \frac{1}{\sqrt{2}} & 0 & 0 \\ 0 & 0 & 0 & 0 & \frac{1}{2} & 0 \\ 0 & 0 & 0 & 0 & \frac{1}{2} & 0 \\ 0 & 0 & 0 & 0 & 0 & \frac{1}{2} \\ 0 & 0 & 0 & 0 & \frac{1}{2} & 0 \\ 0 & 0 & 0 & 0 & 0 & \frac{1}{2} \\ 0 & 0 & 0 & 0 & 0 & \frac{1}{2} \end{bmatrix},$$

where $\text{Rank}[\mathbf{J}(\boldsymbol{\vartheta})] = 6$, so the model is locally identified. Also note that $\boldsymbol{\vartheta}$ is not involved in the Jacobian, so the model is locally identified throughout parameter space.

One-factor model with three ordinal items

Consider the following one-factor model for three continuous underlying or latent response variables:

$$\begin{bmatrix} y_{1j}^* \\ y_{2j}^* \\ y_{3j}^* \end{bmatrix} = \begin{bmatrix} \beta_1 \\ \beta_2 \\ \beta_3 \end{bmatrix} + \begin{bmatrix} \lambda_1 \\ \lambda_2 \\ \lambda_3 \end{bmatrix} \eta_j + \begin{bmatrix} \epsilon_{1j} \\ \epsilon_{2j} \\ \epsilon_{3j} \end{bmatrix},$$

where $\eta_j \sim \mathrm{N}(0,1)$, $\epsilon_{ij} \sim \mathrm{N}(0,\theta_{ii})$ and $\text{Cov}(\epsilon_{ij}, \epsilon_{i'j}) = 0$. The latent response variables are related to observed trichotomous (three-category) responses via threshold functions with constant thresholds across items

$$y_{ij} = \begin{cases} 0 & \text{if} \quad y_{ij}^* \leq 0 \\ 1 & \text{if} \quad 0 < y_{ij}^* \leq \kappa_2 \\ 2 & \text{if} \quad \kappa_2 < y_{ij}^*. \end{cases}$$

There are 10 unknown parameters

$$\boldsymbol{\vartheta} = [\kappa_2, \beta_1, \beta_2, \beta_3, \lambda_1, \lambda_2, \lambda_3, \theta_{11}, \theta_{22}, \theta_{33}]'.$$

The marginal probabilities are

$$\Pr(y_{ij}=0) = \Phi\left(\frac{\beta_i}{\sqrt{\lambda_i^2 + \theta_{ii}}}\right), \tag{5.5}$$

$$\Pr(y_{ij}=1) = \Phi\left(\frac{\beta_i - \kappa_2}{\sqrt{\lambda_i^2 + \theta_{ii}}}\right) - \Phi\left(\frac{\beta_i}{\sqrt{\lambda_i^2 + \theta_{ii}}}\right),$$

and

$$\Pr(y_{ij}=2) \quad = \quad 1 - \Phi(\frac{\beta_i - \kappa_2}{\sqrt{\lambda_i^2+\theta_{ii}}}). \tag{5.6}$$

From (5.5) we can then identify the means for $i=1,2,3$

$$\mu_i \quad = \quad \frac{\beta_i}{\sqrt{\lambda_i^2+\theta_{ii}}},$$

and from (5.6) the threshold

$$\tau_i \quad = \quad \frac{\kappa_2}{\sqrt{\lambda_i^2+\theta_{ii}}},$$

can be identified.

The polychoric correlation matrix becomes

$$\mathbf{R}(\vartheta) = \begin{bmatrix} 1 & & \\ \frac{\lambda_2\lambda_1}{\sqrt{\lambda_2^2+\theta_{22}}\sqrt{\lambda_1^2+\theta_{11}}} & 1 & \\ \frac{\lambda_3\lambda_1}{\sqrt{\lambda_3^2+\theta_{33}}\sqrt{\lambda_1^2+\theta_{11}}} & \frac{\lambda_3\lambda_2}{\sqrt{\lambda_3^2+\theta_{33}}\sqrt{\lambda_2^2+\theta_{22}}} & 1 \end{bmatrix},$$

where the nonredundant polychoric correlations are identified. The 9 reduced form parameters (means, thresholds and polychoric correlations) are placed in

$$\mathbf{m}(\vartheta) = \left[\frac{\beta_1}{\sqrt{\lambda_1^2+\theta_{11}}}, \frac{\beta_2}{\sqrt{\lambda_2^2+\theta_{22}}}, \frac{\beta_3}{\sqrt{\lambda_3^2+\theta_{33}}}, \frac{\kappa_2}{\sqrt{\lambda_1^2+\theta_{11}}}, \frac{\kappa_2}{\sqrt{\lambda_2^2+\theta_{22}}}, \frac{\kappa_2}{\sqrt{\lambda_3^2+\theta_{33}}}, \right.$$
$$\left. \frac{\lambda_2\lambda_1}{\sqrt{\lambda_2^2+\theta_{22}}\sqrt{\lambda_1^2+\theta_{11}}}, \frac{\lambda_3\lambda_1}{\sqrt{\lambda_3^2+\theta_{33}}\sqrt{\lambda_1^2+\theta_{11}}}, \frac{\lambda_3\lambda_2}{\sqrt{\lambda_3^2+\theta_{33}}\sqrt{\lambda_2^2+\theta_{22}}} \right].$$

Since there are more unknown parameters, 10, than identified reduced form parameters, 9, the model is obviously not identified.

To obtain identification, Muraki (1990) considered the special case of the above model where the error variances of the items are fixed to one ($\theta_{11} = \theta_{22} = \theta_{33} = 1$). There are now 7 unknown parameters

$$\vartheta = [\kappa_2, \beta_1, \beta_2, \beta_3, \lambda_1, \lambda_2, \lambda_3]',$$

and the 9 reduced form parameters are placed in

$$\mathbf{m}(\vartheta) \quad = \quad \left[\frac{\beta_1}{\sqrt{\lambda_1^2+1}}, \frac{\beta_2}{\sqrt{\lambda_2^2+1}}, \frac{\beta_3}{\sqrt{\lambda_3^2+1}}, \frac{\kappa_2}{\sqrt{\lambda_1^2+1}}, \frac{\kappa_2}{\sqrt{\lambda_2^2+1}}, \frac{\kappa_2}{\sqrt{\lambda_3^2+1}}, \right.$$
$$\left. \frac{\lambda_2\lambda_1}{\sqrt{\lambda_2^2+1}\sqrt{\lambda_1^2+1}}, \frac{\lambda_3\lambda_1}{\sqrt{\lambda_3^2+1}\sqrt{\lambda_1^2+1}}, \frac{\lambda_3\lambda_2}{\sqrt{\lambda_3^2+1}\sqrt{\lambda_2^2+1}} \right].$$

The 9×7 Jacobian has $\text{Rank}[\mathbf{J}(\vartheta)]=7$, so the model is locally identified if ϑ is a regular point.

Skrondal (1996) pointed out that the constraints imposed by Muraki (1990) were unnecessarily restrictive. He showed that it suffices to fix one of the error variances to obtain local identification. Without loss of generality, we fix the error variance of the first item ($\theta_{11}=1$). There are now 9 unknown parameters

$$\boldsymbol{\vartheta} = [\kappa_2, \beta_1, \beta_2, \beta_3, \lambda_1, \lambda_2, \lambda_3, \theta_{22}, \theta_{33}]',$$

and the 9 identified reduced form parameters are placed in

$$\mathbf{m}(\boldsymbol{\vartheta}) = \left[\frac{\beta_1}{\sqrt{\lambda_1^2+1}}, \frac{\beta_2}{\sqrt{\lambda_2^2+\theta_{22}}}, \frac{\beta_3}{\sqrt{\lambda_3^2+\theta_{33}}}, \frac{\kappa_2}{\sqrt{\lambda_1^2+1}}, \frac{\kappa_2}{\sqrt{\lambda_2^2+\theta_{22}}}, \frac{\kappa_2}{\sqrt{\lambda_3^2+\theta_{33}}}, \right.$$
$$\left. \frac{\lambda_2\lambda_1}{\sqrt{\lambda_2^2+\theta_{22}}\sqrt{\lambda_1^2+1}}, \frac{\lambda_3\lambda_1}{\sqrt{\lambda_3^2+\theta_{33}}\sqrt{\lambda_1^2+1}}, \frac{\lambda_3\lambda_2}{\sqrt{\lambda_3^2+\theta_{33}}\sqrt{\lambda_2^2+\theta_{22}}} \right].$$

The 9×9 Jacobian, which is too huge to be presented, has Rank$[\mathbf{J}(\boldsymbol{\vartheta})]=9$, so the model is locally identified as shown by Skrondal (1996) if $\boldsymbol{\vartheta}$ is a regular point. We will use this parametrization in investigating the life-satisfaction of Americans in Section 10.4.

Consider now the model where the error variances are constrained to one as in Muraki (1990), but the threshold is permitted to vary across items. The threshold functions become

$$y_{ij} = \begin{cases} 0 & \text{if} \quad y_{ij}^* \leq 0 \\ 1 & \text{if} \quad 0 < y_{ij}^* \leq \kappa_{i2} \\ 2 & \text{if} \quad \kappa_{i2} < y_{ij}^* \end{cases}$$

where we note that the thresholds have index i. There are now 9 unknown parameters

$$\boldsymbol{\vartheta} = [\kappa_{12}, \kappa_{22}, \kappa_{32}, \beta_1, \beta_2, \beta_3, \lambda_1, \lambda_2, \lambda_3]',$$

and the 9 reduced form parameters are placed in

$$\mathbf{m}(\boldsymbol{\vartheta}) = \left[\frac{\beta_1}{\sqrt{\lambda_1^2+1}}, \frac{\beta_2}{\sqrt{\lambda_2^2+1}}, \frac{\beta_3}{\sqrt{\lambda_3^2+1}}, \frac{\kappa_{12}}{\sqrt{\lambda_1^2+1}}, \frac{\kappa_{22}}{\sqrt{\lambda_2^2+1}}, \frac{\kappa_{32}}{\sqrt{\lambda_3^2+1}}, \right.$$
$$\left. \frac{\lambda_2\lambda_1}{\sqrt{\lambda_2^2+1}\sqrt{\lambda_1^2+1}}, \frac{\lambda_3\lambda_1}{\sqrt{\lambda_3^2+1}\sqrt{\lambda_1^2+1}}, \frac{\lambda_3\lambda_3}{\sqrt{\lambda_3^2+1}\sqrt{\lambda_2^2+1}} \right].$$

The 9×9 Jacobian has Rank$[\mathbf{J}(\boldsymbol{\vartheta})]=9$, so the model is locally identified if $\boldsymbol{\vartheta}$ is a regular point.

5.2.4 Regular points and local identification

Two-factor model with four continuous items

Consider the following two-factor model with four continuous items where both factors are 'anchored':

$$
\begin{bmatrix} y_{1j} \\ y_{2j} \\ y_{3j} \\ y_{4j} \end{bmatrix}
=
\begin{bmatrix} 1 & 0 \\ \lambda_{21} & 0 \\ 0 & 1 \\ 0 & \lambda_{42} \end{bmatrix}
\begin{bmatrix} \eta_{1j} \\ \eta_{2j} \end{bmatrix}
+
\begin{bmatrix} \epsilon_{1j} \\ \epsilon_{2j} \\ \epsilon_{3j} \\ \epsilon_{4j} \end{bmatrix},
$$

where $\begin{bmatrix} \eta_{1j} \\ \eta_{2j} \end{bmatrix} \sim N_2\left(\begin{bmatrix} 0 \\ 0 \end{bmatrix}, \begin{bmatrix} \psi_{11} & \\ \psi_{21} & \psi_{22} \end{bmatrix} \right)$, $\epsilon_{ij} \sim N(0, \theta_{ii})$, and $\mathrm{Cov}(\epsilon_{ij}, \epsilon_{i'j}) = 0$. A path diagram of this model is given in the left panel of Figure 5.2.

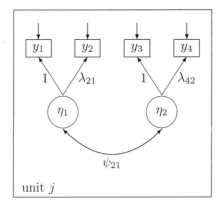 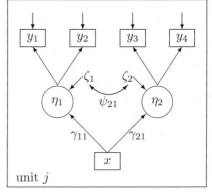

Figure 5.2 *Two-factor models with and without exogenous variable*

There are 9 unknown parameters placed in the column vector $\boldsymbol{\vartheta}$

$$
\boldsymbol{\vartheta} = [\lambda_{21}, \lambda_{42}, \psi_{11}, \psi_{21}, \psi_{22}, \theta_{11}, \theta_{22}, \theta_{33}, \theta_{44}]',
$$

implying the covariance matrix

$$
\boldsymbol{\Omega}(\boldsymbol{\vartheta}) =
\begin{bmatrix}
\psi_{11} + \theta_{11} & & & \\
\lambda_{21}\psi_{11} & \lambda_{21}^2\psi + \theta_{22} & & \\
\psi_{21} & \psi_{21}\lambda_{21} & \psi_{22} + \theta_{33} & \\
\lambda_{42}\psi_{21} & \lambda_{42}\psi_{21}\lambda_{21} & \lambda_{42}\psi_{22} & \lambda_{42}^2\psi_{22} + \theta_{44}
\end{bmatrix}.
$$

The 10 reduced form parameters (nonredundant variances and covariances) are placed in

$$
\begin{aligned}
\mathbf{m}(\boldsymbol{\vartheta}) = \ & [\psi_{11}+\theta_{11}, \lambda_{21}\psi_{11}, \psi_{21}, \lambda_{42}\psi_{21}, \lambda_{21}^2\psi_{11}+\theta_{22}, \psi_{21}\lambda_{21}, \lambda_{42}\psi_{21}\lambda_{21}, \\
& \psi_{22}+\theta_{33}, \lambda_{42}\psi_{22}, \lambda_{42}^2\psi_{22}+\theta_{44}].
\end{aligned}
$$

The 10×9 Jacobian becomes

$$
\mathbf{J}(\boldsymbol{\vartheta}) =
\begin{bmatrix}
0 & 0 & 1 & 0 & 0 & 1 & 0 & 0 & 0 \\
\psi_{11} & 0 & \lambda_{21} & 0 & 0 & 0 & 0 & 0 & 0 \\
0 & 0 & 0 & 1 & 0 & 0 & 0 & 0 & 0 \\
0 & \psi_{21} & 0 & \lambda_{42} & 0 & 0 & 0 & 0 & 0 \\
2\lambda_{21}\psi_{11} & 0 & \lambda_{21}^2 & 0 & 0 & 0 & 1 & 0 & 0 \\
\psi_{21} & 0 & 0 & \lambda_{21} & 0 & 0 & 0 & 0 & 0 \\
\lambda_{42}\psi_{21} & \lambda_{21}\psi_{21} & 0 & \lambda_{21}\lambda_{42} & 0 & 0 & 0 & 0 & 0 \\
0 & 0 & 0 & 0 & 1 & 0 & 0 & 1 & 0 \\
0 & \psi_{22} & 0 & 0 & \lambda_{42} & 0 & 0 & 0 & 0 \\
0 & 2\lambda_{42}\psi_{22} & 0 & 0 & \lambda_{42}^2 & 0 & 0 & 0 & 1
\end{bmatrix},
$$

with $\text{Rank}[\mathbf{J}(\boldsymbol{\vartheta})] = 9$, so the model is locally identified if $\boldsymbol{\vartheta}$ is a regular point.

Consider now the special case where the factors are uncorrelated, $\psi_{21} = 0$. Substituting this restriction in the above Jacobian, the rank becomes 7. It is thus evident that $\boldsymbol{\vartheta}^0$ with $\psi_{21} = 0$ is not a regular point since the rank of the Jacobian in this case is not constant in the neighborhood of $\boldsymbol{\vartheta}^0$. Furthermore, the model is not locally identified for $\psi_{21} = 0$ since the rank of the Jacobian is 7 for 8 unknown parameters $(\lambda_{21}, \lambda_{42}, \psi_{11}, \psi_{22}, \theta_{11}, \theta_{22}, \theta_{33}, \theta_{44})$. We are thus in the somewhat unusual situation where a model becomes locally identified by relaxing (not imposing) a parameter restriction. The above model nicely illustrates that Lemma 1 can only be applied if the parameter point is a *regular point*.

Two-factor model with four continuous items and exogenous variable

Consider now the extension of the above model where the factors are regressed on a covariate x_j

$$
\begin{bmatrix} \eta_{1j} \\ \eta_{2j} \end{bmatrix} = \begin{bmatrix} \gamma_{11} \\ \gamma_{21} \end{bmatrix} x_j + \begin{bmatrix} \zeta_{1j} \\ \zeta_{2j} \end{bmatrix},
$$

where $\begin{bmatrix} \zeta_{1j} \\ \zeta_{2j} \end{bmatrix} \sim \mathrm{N}_2 \left(\begin{bmatrix} 0 \\ 0 \end{bmatrix}, \begin{bmatrix} \psi_{11} & \\ \psi_{21} & \psi_{22} \end{bmatrix} \right)$. A path diagram of this model is given in the right panel of Figure 5.2.

The unknown parameters are

$$
\boldsymbol{\vartheta} = [\gamma_{11}, \gamma_{21}, \lambda_{21}, \lambda_{42}, \psi_{11}, \psi_{21}, \psi_{22}, \theta_{11}, \theta_{22}, \theta_{33}, \theta_{44}]'.
$$

We obtain the regression/mean structure

$$
\begin{aligned}
\mathrm{E}(y_{1j}|x_j) &= \gamma_{11}x_j \\
\mathrm{E}(y_{2j}|x_j) &= \lambda_{21}\gamma_{11}x_j \\
\mathrm{E}(y_{3j}|x_j) &= \gamma_{21}x_j \\
\mathrm{E}(y_{4j}|x_j) &= \lambda_{42}\gamma_{21}x_j,
\end{aligned}
$$

and the vector of reduced form parameters becomes

$$
\begin{aligned}
\mathbf{m}(\boldsymbol{\vartheta}) = \; & [\gamma_{11}, \lambda_{21}\gamma_{11}, \gamma_{21}, \lambda_{42}\gamma_{21}, \psi_{11}+\theta_{11}, \lambda_{21}\psi_{11}, \psi_{21}, \lambda_{42}\psi_{21}, \lambda_{21}^2\psi_{11}+\theta_{22}, \\
& \psi_{21}\lambda_{21}, \lambda_{42}\psi_{21}\lambda_{21}, \psi_{22}+\theta_{33}, \lambda_{42}\psi_{22}, \lambda_{42}^2\psi_{22}+\theta_{44}].
\end{aligned}
$$

The 14×11 Jacobian becomes

$$
\mathbf{J}(\boldsymbol{\vartheta}) =
\begin{bmatrix}
1 & 0 & 0 & 0 & 0 & 0 & 0 & 0 & 0 & 0 & 0 \\
\lambda_{21} & 0 & \gamma_{11} & 0 & 0 & 0 & 0 & 0 & 0 & 0 & 0 \\
0 & 1 & 0 & 0 & 0 & 0 & 0 & 0 & 0 & 0 & 0 \\
0 & \lambda_{42} & 0 & \gamma_{21} & 0 & 0 & 0 & 0 & 0 & 0 & 0 \\
0 & 0 & 0 & 0 & 1 & 0 & 0 & 1 & 0 & 0 & 0 \\
0 & 0 & \psi_{11} & 0 & \lambda_{21} & 0 & 0 & 0 & 0 & 0 & 0 \\
0 & 0 & 0 & 0 & 0 & 1 & 0 & 0 & 0 & 0 & 0 \\
0 & 0 & 0 & \psi_{21} & 0 & \lambda_{42} & 0 & 0 & 0 & 0 & 0 \\
0 & 0 & 2\,\lambda_{21}\,\psi_{11} & 0 & \lambda_{21}^{2} & 0 & 0 & 0 & 1 & 0 & 0 \\
0 & 0 & \psi_{21} & 0 & 0 & \lambda_{21} & 0 & 0 & 0 & 0 & 0 \\
0 & 0 & \lambda_{42}\,\psi_{21} & \lambda_{21}\,\psi_{21} & 0 & \lambda_{21}\,\lambda_{42} & 0 & 0 & 0 & 0 & 0 \\
0 & 0 & 0 & 0 & 0 & 0 & 1 & 0 & 0 & 1 & 0 \\
0 & 0 & 0 & \psi_{22} & 0 & 0 & \lambda_{42} & 0 & 0 & 0 & 0 \\
0 & 0 & 0 & 2\,\lambda_{42}\,\psi_{22} & 0 & 0 & \lambda_{42}^{2} & 0 & 0 & 0 & 1
\end{bmatrix}
$$

with $\mathrm{Rank}[\mathbf{J}(\boldsymbol{\vartheta})] = 11$, so the model is locally identified if $\boldsymbol{\vartheta}$ is a regular point. Consider once again the special case where $\psi_{21} = 0$; the factors are uncorrelated. Substituting this restriction in the above Jacobian the rank remains 11, so $\psi_{21} = 0$ no longer implies that $\boldsymbol{\vartheta}$ is an irregular point, in contrast to the case without a covariate.

5.2.5 Empirical identification

Analytical identification proceeds in terms of unknown true parameters $\boldsymbol{\vartheta}$. A useful complement to this 'theoretical' approach is 'empirical' identification which is instead based on properties of estimated parameters. Although less stringent than the analytical method since it is based on estimated parameters instead of theoretical parameters, the empirical method has some advantages as compared to the former approach: First, empirical investigation is based on the estimated information matrix, a natural byproduct of maximum likelihood estimation. Second, empirical identification is more general in the sense that it does not rest on the existence of globally identified reduced form parameters that completely characterize the reduced form distribution. Third, it can be argued that empirical identification assesses identification where it matters, at the parameter estimates. For instance, inferences are expected to be problematic for the two-factor model in Section 5.2.4 if $\widehat{\psi}_{21} \approx 0$. Fourth, empirical identification addresses problems that may be inherent in the sample on which inferences must be based. Collinearity among predictor variables in linear regression is an example of an empirical identification problem.

Inspired by Wiley (1973) and McDonald and Krane (1977), we suggest the following definition:

- A model is *empirically identified* for a sample if the estimated information matrix at the maximum likelihood solution $\widehat{\boldsymbol{\vartheta}}$ is nonsingular.

Note that this condition is simply an empirical counterpart of the condition based on the theoretical information matrix (e.g. Rothenberg, 1971) mentioned earlier.

A measure of how close a matrix is to singularity is the *condition number*, defined as the square root of the ratio of the largest to the smallest eigenvalue. In practice, we say that a model is *empirically underidentified* if the condition number is 'large', exceeding some threshold. When a model is empirically underidentified, standard errors and intercorrelations of parameter estimates will be high. We would for example expect this scenario when there is collinearity among predictor variables and for the two-factor model in Section 5.2.4 when $\widehat{\psi}_{21} \approx 0$.

The binomial logit-normal (BLN) model

We now consider the BLN model discussed by Coull and Agresti (2000), the logit version of the BPN models discussed above. Rabe-Hesketh and Skrondal (2001) argued that the BLN model is not identified from the first and second order moments and is therefore likely to be empirically underidentified since information in higher order moments is likely to be scarce.

Estimating the model without constraining σ^2, the condition number is 179.5 which is extremely large (the smallest eigenvalue was less than 0.004) and indicates that the observed information matrix is nearly singular. Thus, the BLN model appears to be empirically unidentified. We also estimated the model constraining σ^2 equal to its maximum likelihood estimate of 4.06, giving a condition number of 5.2.

Inverting the estimated information matrices, we obtained the estimated covariance matrices of the parameters estimates. As can be seen in Table 5.1, the estimated standard errors decrease substantially when σ is fixed. The correlations of the parameter estimates are shown in Table 5.2. For the unconstrained model, the parameter estimates are highly intercorrelated, most correlations approaching ± 1, the smallest correlation (in absolute value) being -0.79, whereas the highest correlation for the constrained model is 0.19.

Having demonstrated empirical underidentification, we now investigate if this is due to the scarce information in the higher order moments. For a range of values of σ, we computed the other parameters to preserve the means and correlations implied by the maximum likelihood solution. The models with these different sets of parameter values imply identical first and second order moments of the latent responses but different higher order moments. The deviance of these models is plotted against σ in Figure 5.3 where σ increases from 1.35, the lowest value consistent with the correlations of the latent responses, to 8. The deviance hardly changes at all although the higher order moments were deliberately ignored in determining the other parameters for each value of σ. This provides direct evidence for the scarcity of information in the higher order moments of the latent responses. Note that the curve in Figure 5.3 represents an upper bound for the deviance corresponding to the profile likelihood (see Section 8.3.5) for σ since the other parameters are not

Table 5.1 *Parameter estimates, standard errors and deviance for constrained and unconstrained versions of the BLN model (20 quadrature points per dimension)*

	Est	Standard Error Unconstrained	Standard Error Constrained
β_1	-4.04	6.85	0.39
β_2	-4.42	7.34	0.41
β_3	-4.69	7.77	0.42
β_4	-4.56	7.57	0.42
σ	4.06	7.99	–
ρ_1	0.43	0.28	0.10
ρ_2	-0.25	0.20	0.12
Deviance	6.28		

Source: Rabe-Hesketh and Skrondal (2001)

Table 5.2 *Correlation matrices of parameter estimates for the BLN model; above the diagonal: constrained model; below the diagonal: unconstrained model*

	β_1	β_2	β_3	β_4	σ	ρ_1	ρ_2
β_1	1	0.190	0.187	-0.083	–	0.050	-0.025
β_2	0.997	1	0.185	-0.080	–	0.062	-0.033
β_3	0.997	0.998	1	-0.077	–	0.069	-0.037
β_4	0.997	0.997	0.997	1	–	0.014	-0.045
σ	-0.998	-0.998	-0.999	-0.998	1	–	–
ρ_1	0.941	0.942	0.942	0.941	-0.942	1	-0.106
ρ_2	-0.814	-0.815	-0.815	-0.815	0.815	-0.788	1

Source: Rabe-Hesketh and Skrondal (2001)

estimated by maximum likelihood. Estimating the model with σ fixed at 8.2, for example, gives a deviance of only 6.53.

5.3 Equivalence

5.3.1 Definitions

- Two statistical models $\mathcal{M}_1$ and $\mathcal{M}_2$ with $\vartheta_A \in \mathcal{A}$ and $\vartheta_B \in \mathcal{B}$ are *globally equivalent* if they are reparameterizations in the sense that there exist one-to-one transformations between ϑ_A and ϑ_B throughout $\mathcal{A}$ and $\mathcal{B}$ making them observationally equivalent.

As was the case for identification in nonlinear moment structures, the prospects

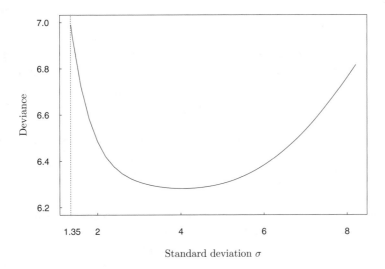

Figure 5.3 *Deviance for different values of the standard deviation σ. The other parameters have been computed to preserve the first and second order moments implied by the maximum likelihood solution. (Source: Rabe-Hesketh and Skrondal, 2001)*

for global equivalence seem to be bleak, and we resort to a notion of local equivalence:

- Two statistical models $\mathcal{M}_1$ and $\mathcal{M}_2$ with locally identified parameter points $\boldsymbol{\vartheta}_A^1 \in \mathcal{A}$ and $\boldsymbol{\vartheta}_B^2 \in \mathcal{B}$ are *locally equivalent* if they are reparameterizations in the sense that there exists a one-to-one transformation between $\boldsymbol{\vartheta}_A$ and $\boldsymbol{\vartheta}_B$ in open neighborhoods of the points making them observationally equivalent.

5.3.2 Analytical investigation of equivalence

We propose the following approach for models that are completely characterized by lower order moments. Consider the reduced form parameters $m_s = h_s(\boldsymbol{\vartheta}_A)$ and $m_s^* = h_s^*(\boldsymbol{\vartheta}_B)$ of two potentially equivalent models with fundamental parameter vectors $\boldsymbol{\vartheta}_A$ and $\boldsymbol{\vartheta}_B$. If the models are equivalent it follows that for each parameter point $\boldsymbol{\vartheta}_A^0$ there is a point $\boldsymbol{\vartheta}_B^0$ such that

$$h_s(\boldsymbol{\vartheta}_A^0) = h_s^*(\boldsymbol{\vartheta}_B^0).$$

We then investigate whether and under which conditions one-to-one transformation between the fundamental parameters of the two models can be found by solving for $\boldsymbol{\vartheta}_A$ in terms of $\boldsymbol{\vartheta}_B$ and for $\boldsymbol{\vartheta}_B$ in terms of $\boldsymbol{\vartheta}_A$.

For the special case where we want to investigate local equivalence for two

submodels $\mathcal{M}_1$ and $\mathcal{M}_2$ of a common underidentified model $\mathcal{M}_{12}$, we consider
the following approach suggested by Luijben (1991). Let $\boldsymbol{\vartheta}^{12}$ denote the vector of v fundamental parameters of $\mathcal{M}_{12}$. Define $\mathcal{M}_0$ as the model where the
restrictions on $\mathcal{M}_{12}$ leading to $\mathcal{M}_1$ as well as $\mathcal{M}_2$ are imposed. Also let $\boldsymbol{\vartheta}_R^{12}$,
assumed to be a regular point of $\mathcal{M}_{12}$, be a restricted version of $\boldsymbol{\vartheta}^{12}$ where
the restrictions leading to $\mathcal{M}_0$ are imposed. Under the restrictions yielding
$\mathcal{M}_0$, the parameter vectors of $\mathcal{M}_1$ and $\mathcal{M}_2$ are assumed to be locally identi-
fied and regular points. Define the Jacobian $\mathbf{J}(\boldsymbol{\vartheta}^{12})\,|_{\boldsymbol{\vartheta}^{12}=\boldsymbol{\vartheta}_R^{12}}$ as the Jacobian
$\mathbf{J}(\boldsymbol{\vartheta}^{12})$ of the mappings from fundamental parameters $\boldsymbol{\vartheta}^{12}$ to the reduced form
parameter vector $\mathbf{m}^{12}$ with the restrictions in $\boldsymbol{\vartheta}_R^{12}$ inserted.

The analysis of equivalence can in the present setting rely on the following
Lemma:

Lemma 2: $\mathcal{M}_1$ and $\mathcal{M}_2$ are *locally equivalent* if and only if

$$\text{Rank}\left[\mathbf{J}(\boldsymbol{\vartheta}^{12})\,|_{\boldsymbol{\vartheta}^{12}=\boldsymbol{\vartheta}_R^{12}}\right] < v.$$

As for the Jacobian strategy for identification, this approach for investigating
equivalence is confined to models where there are reduced form parameters
completely characterizing the reduced form distribution.

Unfortunately, we cannot use Lemma 2 when no common underidentified
model is known in which the two possibly equivalent submodels are nested.
Bekker *et al.* (1994) point out that Luijben's approach is quite restrictive and
consider investigation of local equivalence for the general case. Unfortunately,
the derived conditions appear to be extremely difficult to evaluate in practice.

One-factor model with four continuous items

Let us once more return to the one-factor model with four continuous items
introduced in Section 4.10. On pages 139-140 we demonstrated that there were
two locally identified models, one with anchoring ($\lambda_1 = 1$) and another with
factor standardization ($\psi = 1$).

We proceed by equating the reduced form parameters produced by the two
parametrizations:

$$
\begin{aligned}
\mathbf{m}(\bar{\boldsymbol{\vartheta}}) &= [\bar{\lambda}_1^2 + \bar{\theta}_{11}, \bar{\lambda}_2\bar{\lambda}_1, \bar{\lambda}_3\bar{\lambda}_1, \bar{\lambda}_4\bar{\lambda}_1, \bar{\lambda}_2^2 + \bar{\theta}_{22}, \bar{\lambda}_3\bar{\lambda}_2, \bar{\lambda}_4\bar{\lambda}_2, \\
&\quad\; \bar{\lambda}_3^2 + \bar{\theta}_{33}, \bar{\lambda}_4\bar{\lambda}_3, \bar{\lambda}_4^2 + \bar{\theta}_{44}] \\
&= [\psi^* + \theta_{11}^*, \lambda_2^*\psi^*, \lambda_3^*\psi^*, \lambda_4^*\psi^*, (\lambda_2^*)^2\psi^* + \theta_{22}^*, \lambda_3^*\psi^*\lambda_2^*, \lambda_4^*\psi^*\lambda_2^*, \\
&\quad\; (\lambda_3^*)^2\psi^* + \theta_{33}^*, \lambda_4^*\psi^*\lambda_3^*, (\lambda_4^*)^2\psi^* + \theta_{44}^*] = \mathbf{m}(\boldsymbol{\vartheta}^*).
\end{aligned}
$$

Solving for $\boldsymbol{\vartheta}^*$ gives us unique solutions

$$
\left\{\theta_{11}^* = \bar{\theta}_{11}, \theta_{22}^* = \bar{\theta}_{22}, \theta_{33}^* = \bar{\theta}_{33}, \theta_{44}^* = \bar{\theta}_{44}, \lambda_2^* = \frac{\bar{\lambda}_2}{\bar{\lambda}_1}, \lambda_3^* = \frac{\bar{\lambda}_3}{\bar{\lambda}_1}, \right.
$$

$$
\left. \lambda_2^* = \frac{\bar{\lambda}_4}{\bar{\lambda}_1}, \psi^* = \bar{\lambda}_1^2 \right\}.
$$

The standardized model can apparently generate parameters throughout the
parameter space of the anchored model as long as $\bar{\lambda}_1 \neq 0$.

Solving for $\bar{\boldsymbol{\vartheta}}$, we obtain unique solutions, subject to determination of the sign of the factor loadings $\bar{\lambda}_i$,

$$\left\{\bar{\theta}_{11} = \theta_{11}^*, \bar{\theta}_{22} = \theta_{22}^*, \bar{\theta}_{33} = \theta_{33}^*, \bar{\theta}_{44} = \theta_{44}^*, \bar{\lambda}_1 = \pm\sqrt{\psi^*}, \bar{\lambda}_2 = \pm\lambda_2^*\sqrt{\psi^*},\right.$$
$$\left.\bar{\lambda}_3 = \pm\lambda_3^*\sqrt{\psi^*}, \bar{\lambda}_4 = \pm\lambda_4^*\sqrt{\psi^*}\right\}.$$

It seems that the anchored model can generate parameters throughout the parameter space of the standardized model if $\psi^* \geq 0$.

Note that both models are nested in the nonrestricted model. Hence, we can apply Lemma 2 to investigate whether the models are locally equivalent. Substituting the restrictions $\lambda_1 = 1$ (anchoring) and $\psi = 1$ (factor standardization) in the Jacobian for the nonidentified model given in (5.2) produces

$$\mathbf{J}(\boldsymbol{\vartheta}^{12})\,|_{\boldsymbol{\vartheta}^{12}=\boldsymbol{\vartheta}_R^{12}} = \begin{bmatrix} 2 & 0 & 0 & 0 & 1 & 1 & 0 & 0 & 0 \\ \lambda_2 & 1 & 0 & 0 & \lambda_2 & 0 & 0 & 0 & 0 \\ \lambda_3 & 0 & 1 & 0 & \lambda_3 & 0 & 0 & 0 & 0 \\ \lambda_4 & 0 & 0 & 1 & \lambda_4 & 0 & 0 & 0 & 0 \\ 0 & 2\lambda_2 & 0 & 0 & \lambda_2^2 & 0 & 1 & 0 & 0 \\ 0 & \lambda_3 & \lambda_2 & 0 & \lambda_2\lambda_3 & 0 & 0 & 0 & 0 \\ 0 & \lambda_4 & 0 & \lambda_2 & \lambda_2\lambda_4 & 0 & 0 & 0 & 0 \\ 0 & 0 & 2\lambda_3 & 0 & \lambda_3^2 & 0 & 0 & 1 & 0 \\ 0 & 0 & \lambda_4 & \lambda_3 & \lambda_3\lambda_4 & 0 & 0 & 0 & 0 \\ 0 & 0 & 0 & 2\lambda_4 & \lambda_4^2 & 0 & 0 & 0 & 1 \end{bmatrix}.$$

The rank of this Jacobian, $\mathrm{Rank}\left[J(\boldsymbol{\vartheta}^{12})\,|_{\boldsymbol{\vartheta}^{12}=\boldsymbol{\vartheta}_R^{12}}\right]$, is 8 which is one less the number of parameters, so the models are locally equivalent under the assumptions stated above.

Although equivalent parametrizations, anchoring is often regarded as preferable to standardization since it ensures factorial invariance (see Section 3.3.2).

BPN model and restricted dichotomous one-factor model

We now investigate whether the identified BPN model presented on page 144 is equivalent to the restricted version of the one-factor model with four dichotomous indicators introduced on page 142. In this case we are not aware of any common underidentified model in which the two models are nested and use of Lemma 2 is precluded.

We therefore proceed by substituting the restrictions $\lambda_1 = \lambda_2 = \lambda_3 = 1$ into (5.3), obtaining the reduced form parameters

$$\mathbf{m}(\boldsymbol{\vartheta}) = \left[\frac{\beta_1}{\sqrt{\psi+1}}, \frac{\beta_2}{\sqrt{\psi+1}}, \frac{\beta_3}{\sqrt{\psi+1}}, \frac{\beta_4}{\sqrt{\lambda_4^2\psi+1}}, \frac{\psi}{\psi+1}, \frac{\psi}{\psi+1}, \frac{\lambda_4\psi}{\sqrt{\lambda_4^2\psi+1}\sqrt{\psi+1}},\right.$$
$$\left.\frac{\psi}{\psi+1}, \frac{\lambda_4\psi}{\sqrt{\lambda_4^2\psi+1}\sqrt{\psi+1}}, \frac{\lambda_4\psi}{\sqrt{\lambda_4^2\psi+1}\sqrt{\psi+1}}, \frac{\lambda_4\psi}{\sqrt{\lambda_4^2\psi+1}\sqrt{\psi+1}}\right].$$

Equating these with (5.4), but letting the intercepts now be denoted b_i to distinguish the parameters, we solve for the parameters of the restricted one-factor model in terms of the BPN parameters and obtain:

$$
\left\{
\beta_1 = \frac{b_1 \sqrt{1 - \frac{\rho_1}{-2+\rho_1}}}{\sqrt{2}}, \beta_2 = \frac{b_2 \sqrt{1 - \frac{\rho_1}{-2+\rho_1}}}{\sqrt{2}}, \beta_3 = \frac{b_3 \sqrt{1 - \frac{\rho_1}{-2+\rho_1}}}{\sqrt{2}},
\right.
$$

$$
\left.
\beta_4 = \frac{b_4 \sqrt{1 - \frac{\rho_2{}^2}{-2\rho_1+\rho_2{}^2}}}{\sqrt{2}}, \lambda_4 = \frac{\pm\sqrt{-2+\rho_1}\,\rho_2}{\sqrt{\rho_1}\sqrt{-2\rho_1+\rho_2{}^2}}, \psi = \frac{-\rho_1}{-2+\rho_1}
\right\}.
$$

The solutions are unique apart from the sign of λ_4 and the BPN model can apparently generate parameters throughout the parameter space of the restricted one-factor model.

Solving for the BPN parameters in terms of the parameters of the restricted one-factor model, we get

$$
\left\{
b_1 = \frac{\sqrt{2}\,\beta_1}{\sqrt{1+\psi}}, b_2 = \frac{\sqrt{2}\,\beta_2}{\sqrt{1+\psi}}, b_3 = \frac{\sqrt{2}\,\beta_3}{\sqrt{1+\psi}}, b_4 = \frac{\sqrt{2}\,\beta_4}{\sqrt{1+\psi\,\lambda_4{}^2}},
\right.
$$

$$
\left.
\rho_1 = \frac{2\,\psi}{1+\psi}, \rho_2 = \frac{2\,\psi\,\lambda_4}{\sqrt{1+\psi}\,\sqrt{1+\psi\,\lambda_4{}^2}}
\right\}.
$$

The factor variance ψ is obviously nonnegative, $\psi \geq 0$. From $\rho_1 = 2\psi/(1+\psi)$ it follows that $\rho_1 \geq 0$; the restricted one factor model cannot generate negative ρ_1 for the BPN model. The BPN model and the restricted one factor model are thus not globally equivalent.

Importantly, as long as the restriction $\rho_1 \geq 0$ is reasonable we can make use of the equivalence to greatly simplify estimation of the BPN model. Instead of having to integrate over four random effects to obtain the marginal likelihood, we need only evaluate a one-dimensional integral (Rabe-Hesketh and Skrondal, 2001).

Another and perhaps more prominent example of lack of global equivalence concerns a multivariate linear model with a compound symmetric residual covariance matrix and a random intercept model. The former specifies variances equal to A and covariances equal to B. For the random intercept model the *implied* marginal variances and covariances become $\psi + \theta_{ii}$ and ψ, respectively. Note that the covariances are necessarily non-negative in the random intercept model since $\psi \geq 0$ whereas the covariances B need not be positive under compound symmetry (see also Lindsey, 1999).

5.3.3 Empirical equivalence

Analogously to the case for identification, we now consider 'empirical' investigation of equivalence:

- Two models are *empirically equivalent* for a sample if there are one-to-one functions relating the parameter estimates across models such that almost identical likelihoods are produced.

As for empirical identification, investigation of empirical equivalence is less formal than 'theoretical' equivalence since it is based on parameter estimates. However, we must in practice resort to empirical equivalence for models that are not completely characterized by say their first and second order moments, since this renders the theoretical approach unfeasible.

BLN and restricted dichotomous one-factor models

We now consider empirical equivalence between the identified BLN model and the restricted one-factor logit model. The deviances of these models were 6.28 and 6.58, respectively. The deviances are so close since, as expected, the implied first and second order moments are nearly identical for both models: The means are respectively estimated as $\widehat{\mu}_1 = -0.91, -0.87$, $\widehat{\mu}_2 = -0.99, -0.96$, $\widehat{\mu}_3 = -1.05, -1.03$, $\widehat{\mu}_4 = -1.03, -0.97$ and the correlations of the latent responses are estimated as $\widehat{\rho}_{12} = \widehat{\rho}_{13} = \widehat{\rho}_{22} = 0.36, 0.34$ and $\widehat{\rho}_{14} = \widehat{\rho}_{24} = \widehat{\rho}_{34} = -0.21, -0.21$.

We can transform the parameter estimates of the one-factor model into estimates for the BLN model using the equations

$$b_r = \beta_r \frac{\sqrt{\sigma^2 + \pi^2/3}}{\sqrt{\psi + \pi^2/3}}, \quad r = 1, 2, 3$$

$$b_4 = \beta_4 \frac{\sqrt{\sigma^2 + \pi^2/3}}{\sqrt{\lambda_4^2 \psi + \pi^2/3}}$$

$$\rho_1 = \frac{\psi(\sigma^2 + \pi^2/3)}{\sigma^2(\psi + \pi^2/3)}$$

$$\rho_2 = \frac{\lambda_4 \psi(\sigma^2 + \pi^2/3)}{\sigma^2 \sqrt{(\psi + \pi^2/3)(\lambda_4^2 \psi + \pi^2/3)}},$$

where we substitute the maximum likelihood estimate of 4.06 for σ. The resulting estimates $(-3.87, -4.27, -4.55, -4.30, 0.41, -0.25)$ are close to the maximum likelihood estimates for the BLN model $(-4.04, -4.42, -4.69, -4.56, 0.43, -0.25)$.

5.4 Summary and further reading

We have defined identification and equivalence and shown how both properties can be investigated analytically as well as empirically. Jacobians, their rank and a basis for their nullspace are easily obtained using computer algebra. For simplicity, we have imposed parameter restrictions by direct substitution in the Jacobians. Alternatively, we could have augmented the Jacobian used in this chapter with a Jacobian of a restriction matrix (e.g. Rothenberg, 1971).

Our discussion has been confined to identification of parametric models. A more daunting task is 'nonparametric' identification, where the models are characterized by constraints on functions (generally not parameterized). Identification then concerns whether there are more than one set of functions generating the same distribution for the observations. Nonparametric identification for proportional hazards models with latent variables has been extensively studied in econometrics (e.g. Brinch, 2001; Van den Berg, 2001).

We have only considered identification for models where the fundamental parameters ϑ are unknown constants. In the Bayesian setting, Chechile (1977) discusses the notion of 'posterior-probabilistically identified' and demonstrates that models may be identified in this sense although they are not 'likelihood-identified'. An example where identification is achieved through the prior distributions is discussed in Knorr-Held and Best (2001).

It is worth noting that the use of Markov Chain Monte Carlo (MCMC) methods or other simulation methods (see Chapter 6) may be dangerous from the point of view of identification. As stated by Keane (1992, p.193), this is because "simulation error will generate contours where the true objective function is flat and will generate a nonsingular Hessian when the true Hessian is singular". Keane illustrated the danger by referring to Horowitz *et al.* (1982) who did not discover that a particular multinomial probit model was nonidentified.

Equivalence is not only an issue for the latent variable models discussed in this book. For instance, MacCallum *et al.* (1993) point out that the multidimensional scaling models for three-way proximity data suggested by Tucker (1972) and Carroll and Chang (1972) are equivalent, although they represent widely different representations of individual differences in judgment tasks. Equivalence may also involve different types of models, for instance latent class and Rasch models (e.g. Lindsay *et al.*, 1991; Heinen, 1996) or factor models for continuous responses and latent profile models (e.g. Bartholomew, 1987, 1993; Molenaar and von Eye, 1994).

A modern and comprehensive discussion of identification and equivalence for parametric models, including formal definitions, assumptions and theorems, is provided by Bekker *et al.* (1994). Useful treatments of identification include Koopmans and Reiersøl (1950), Wald (1950), Anderson and Rubin (1956), Fisher (1966), Geraci (1976), Rothenberg (1971), Dupacóvá and Wold (1982), Hsiao (1983), Rabe-Hesketh and Skrondal (2001) and Bechger *et al.* (2001). Contributions to the equivalence literature include Stelzl (1986), Breckler (1990), Jöreskog and Sörbom (1990), Luijben (1991), McCallum *et al.* (1993), Hershberger (1994), Raykov and Penev (1999), Rabe-Hesketh and Skrondal (2001) and Bechger *et al.* (2002).

The identification problem has been given ample attention in econometrics where complex structural models have been used for a long time. This stands in contrast to biometrics, where much simpler models have traditionally been used. The equivalence problem appears to have attracted most interest in psychometrics. However, identification and equivalence should definitely be given more attention throughout statistics due to the increasing popularity of highly structured models.

CHAPTER 6

Estimation

6.1 Introduction

In this chapter we describe a number of estimation methods that have been proposed for latent variable models belonging to the general model framework presented in Chapter 4. We believe that a relatively nontechnical overview of different methods is useful, since some of the methods are alien outside particular methodological disciplines. The estimation methods are sketched in more or less detail, referring to the pertinent literature for technical details. An incomplete overview of software implementing the different estimation methods is provided in an appendix.

We also consider the strengths and weaknesses of different methods. The estimation methods turn out to be quite heterogeneous according to criteria such as generality of the accommodated model class, robustness, computational efficiency, treatment of missing data and performance of the estimators.

So far in this book we have treated latent variables as random and parameters as fixed which is the most common approach. Alternatively, both latent variables and parameters can be treated as fixed, either for theoretical reasons or for computational convenience. In contrast, Bayesians treat both latent variables and parameters as random variables. Although often viewed as a fundamentally different statistical paradigm, this approach is currently also often adopted for practical reasons. Recognizing these different perspectives is important for delineating different kinds of estimation methods.

Random latent variables and fixed parameters

When latent variables are treated as random and parameters as fixed, inference is usually based on the *marginal likelihood*, the likelihood of the data given the latent variables, integrated (or summed in the discrete case) over the latent variable distribution. In the case of continuous latent variables the likelihood generally does not have a closed form. In Section 6.3, we will hence describe several more or less accurate approximate methods of integration, including numerical and Monte Carlo integration (simulated likelihood). Different methods for maximizing likelihoods, including the EM and Newton-Raphson algorithms, are reviewed in Section 6.4.

In Section 6.5 we discuss *nonparametric maximum likelihood estimation* (NPMLE), where we relax the assumption of normal latent variables. The idea of *restricted maximum likelihood* (REML) is briefly described in Section 6.6. For some models the dimensionality of integration can be considerably reduced by using a *limited information* approach described in Section 6.7. Section 6.8.3 describes *penalized quasi-likelihood* (PQL), an approximate method that avoids integration, and Section 6.9 discusses the algorithmically similar *generalized estimating equations* (GEE). GEE is very different from the other approaches considered in this book since dependence among the responses is not explicitly modeled using latent variables, but instead treated as a nuisance. Furthermore, the regression parameters are no longer interpretable as conditional or cluster-specific effects, but as marginal or population averaged effects.

Fixed latent variables and parameters

When latent variables are construed as unknown *fixed parameters* instead of random variables, integration is avoided. The fixed effects approach can be viewed as conditional on the effects in the sample. In this case it is irrelevant whether the clusters can realistically be considered a random sample from a population. We describe two fixed effects approaches to estimation in Section 6.10. In *joint maximum likelihood* (JML) estimation the latent variables and model parameters are jointly estimated, whereas the latent variables are loosely speaking 'conditioned away' in *conditional maximum likelihood* (CML) estimation.

Random latent variables and parameters

The Bayesian approach described in Section 6.11 treats both latent variables and parameters as random and bases inference on their *posterior distribution* given the observed data. In Section 6.11.5, we describe the popular Markov chain Monte Carlo (MCMC) method for sampling from the posterior distribution and estimating parameters by their posterior means.

6.2 Maximum likelihood: Closed form marginal likelihood

The integral involved in the marginal likelihood can in some instances be explicitly solved and expressed in closed form, the canonical examples being the LISREL model and the linear mixed model. In these cases multivariate normal latent variables and multivariate normal responses given the latent variables produce *multivariate normal marginal distributions*.

Estimation of linear mixed models is discussed in Section 6.8.1. For the LISREL model introduced in Section 3.5, the model-implied covariance matrix was shown to be

$$\boldsymbol{\Sigma} \;=\; \boldsymbol{\Lambda}(\mathbf{I}-\mathbf{B})^{-1}\boldsymbol{\Psi}(\mathbf{I}-\mathbf{B})^{-1}\boldsymbol{\Lambda}' + \boldsymbol{\Theta}.$$

Since the mean structure is often not of interest in these models we let the n-dimensional response vector $\mathbf{y}_j$ have zero expectation in this section. The likelihood can then be expressed as

$$f(\boldsymbol{\vartheta}; \mathbf{Y}) = \prod_{j=1}^{J}(2\pi)^{-\frac{n}{2}}|\boldsymbol{\Sigma}^{-1}|\exp(-\mathbf{y}_j'\boldsymbol{\Sigma}^{-1}\mathbf{y}_j).$$

The empirical covariance matrix $\mathbf{S}$ of $\mathbf{y}$ is the sufficient statistic for the parameters structuring $\boldsymbol{\Sigma}$. Since $\mathbf{S}$ has a Wishart distribution, it can be shown (e.g. Jöreskog, 1967) that instead of maximizing the likelihood we can equivalently minimize the fitting function

$$F_{\mathrm{ML}} = \log|\boldsymbol{\Sigma}| + \mathrm{tr}(\mathbf{S}\boldsymbol{\Sigma}^{-1}) - \log|\mathbf{S}| - n,$$

with respect to the unknown free parameters $\boldsymbol{\Lambda}$, $\boldsymbol{\Psi}$ and $\boldsymbol{\Theta}$. F_{ML} is nonnegative and only zero if there is a perfect fit in the sense that the fitted $\boldsymbol{\Sigma}$ equals $\mathbf{S}$. The fitting function also provides an estimated information matrix for the maximum likelihood estimates.

Browne (1984) suggested a general family of weighted least squares (WLS) fit functions for covariance structures,

$$F_{\mathrm{WLS}} = [\boldsymbol{\sigma}-\mathbf{s}]'\,\mathbf{W}^{-1}\,[\boldsymbol{\sigma}-\mathbf{s}], \tag{6.1}$$

where $\boldsymbol{\sigma}$ and $\mathbf{s}$ are vectors containing the nonredundant elements in the model-implied and empirical covariance matrix, respectively, and $\mathbf{W}$ is positive definite weight matrix. The maximum likelihood estimator is obtained by using $\widehat{\boldsymbol{\Sigma}}$, the covariance matrix implied by the parameter estimates, as weight matrix. WLS methods are also useful for limited information estimation of models without closed form likelihoods, a topic we will discuss in Section 6.7.

Generalized linear random intercept models may also have closed form likelihoods. Specifically, combining Poisson distributed responses given the mean $\exp(\nu)$ with a gamma distribution for the mean produces a *negative binomial* marginal model (e.g. Greenwood and Yule, 1920; Hausman *et al.*, 1984). For dichotomous responses it is well known that the *beta binomial* model where probabilities are assumed to be beta distributed has a closed form likelihood (e.g. Skellam, 1948; Williams, 1975). Here, a regression model is often specified for the marginal logits (e.g. Heckman and Willis, 1977). Unlike the negative binomial model, this is not a generalized linear mixed model since it cannot be specified by including an additive random intercept in the linear predictor. Unfortunately, these useful results are not applicable for the common situation where covariates vary within clusters (e.g. Neuhaus and Jewell, 1990).

6.3 Maximum likelihood: Approximate marginal likelihood

The reduced form distribution of the responses given the explanatory variables was derived using latent variable integration in Section 4.9.1. Regarded as a function of the fundamental parameters for given responses, this is the marginal likelihood $f(\boldsymbol{\vartheta}; \mathbf{y}, \mathbf{X})$.

For the general model, the marginal likelihood is

$$f(\boldsymbol{\vartheta}; \mathbf{y}, \mathbf{X}) = \prod g^{(L)}(\mathbf{y}_{(L)}),$$

where the product is over all top-level clusters. Let $g^{(l)}(\mathbf{y}_{(l)}|\boldsymbol{\zeta}^{(l+)})$ be the joint conditional probability (density) of the responses for a level-l unit, given the latent variables at levels l and above, $\boldsymbol{\zeta}^{(l+)} = (\boldsymbol{\zeta}^{(l)\prime}, \ldots, \boldsymbol{\zeta}^{(L)\prime})'$. Starting from $l=2$, we can recursively evaluate the integrals

$$g^{(l)}(\mathbf{y}_{(l)}|\boldsymbol{\zeta}^{([l+1]^+)}) = \int h^{(l)}(\boldsymbol{\zeta}^{(l)}) \prod g^{(l-1)}(\mathbf{y}_{(l-1)}|\boldsymbol{\zeta}^{(l^+)}) \mathrm{d}\boldsymbol{\zeta}^{(l)}, \qquad (6.2)$$

up to level L. Here we have simplified the notation by setting

$$g^{(l)}(\mathbf{y}_{(l)}|\boldsymbol{\zeta}^{([l+1]^+)}) \equiv g^{(l)}(\mathbf{y}_{(l)}|\mathbf{X}_{(L)}, \boldsymbol{\zeta}^{([l+1]^+)}; \boldsymbol{\vartheta})$$

and will continue to do so in the remainder of the chapter.

We will describe some integration methods in detail for the two-level random intercept model and indicate how they are extended for the general model. Setting $\eta_j^{(2)} = \zeta_j$, the random intercept model is given by

$$\nu_{ij} = \mathbf{x}_{ij}'\boldsymbol{\beta} + \zeta_j.$$

The joint density of the responses for the jth level-2 unit is

$$g^{(2)}(\mathbf{y}_{j(2)}) = \int_{-\infty}^{\infty} h(\zeta_j) \prod_i g^{(1)}(y_{ij}|\zeta_j) \mathrm{d}\zeta_j. \qquad (6.3)$$

Unfortunately, there are in general no closed forms for the integrals involved. There are several approaches to approximating the integrals:

- Laplace approximation,

- Numerical integration using quadrature or adaptive quadrature,

- Monte Carlo integration,

which are described in Sections 6.3.1 to 6.3.3. Section 6.3.4 describes a tailor-made simulation approach for multivariate normal latent responses, based on latent response integration instead of latent variable integration (see Section 4.9).

6.3.1 Laplace approximation

For a unidimensional integral, the Laplace approximation can be written as

$$\int_{-\infty}^{\infty} \exp[f(x)]\mathrm{d}x \approx \int_{-\infty}^{\infty} \exp[f(\tilde{x}) - (x - \tilde{x})^2/2\sigma^2)]\mathrm{d}x$$

$$= \int_{-\infty}^{\infty} \exp[f(\tilde{x})]\sqrt{2\pi}\sigma\phi(x; \tilde{x}, \sigma^2)\mathrm{d}x$$

$$= \exp[f(\tilde{x})]\sqrt{2\pi}\sigma, \qquad (6.4)$$

where $\phi(x; \tilde{x}, \sigma^2)$ is a normal density with mean $\tilde{x}$ and variance σ^2, $\tilde{x}$ is the mode of $f(x)$ and hence of $\exp[f(x)]$ and

$$\sigma^2 = -\left(\frac{\partial^2 f(\tilde{x})}{\partial x^2}\right)^{-1},$$

minus the inverse of the second derivative of $f(x)$ with respect to x, evaluated at the mode $\tilde{x}$.

The approximation is derived by expanding $f(x)$ as a second order Taylor series around its mode so that the first order term vanishes (see inside brackets in the first line of (6.4)). The approximation is exact if the integrand is proportional to a normal density with mean $\tilde{x}$ and variance σ^2 since $f(x)$ is in this case quadratic in x.

For a random intercept model, we need to evaluate the integral in (6.3). The integrand (corresponding to $\exp[f(x)]$),

$$h(\zeta_j) \prod_i g^{(1)}(y_{ij}|\zeta_j), \qquad (6.5)$$

is the product of the 'prior' density of ζ_j and the joint probability (density) of the responses given ζ_j. After normalization with respect to ζ_j, this integrand is therefore just the 'posterior' density of ζ_j given the observed responses for cluster j (see also Section 7.2). In the Laplace approximation, $\tilde{x}$ therefore corresponds to the *posterior mode* $\tilde{\zeta}_j$ and σ corresponds to the curvature of the posterior at the mode, σ_j. The approximation becomes

$$\ln g^{(2)}(\mathbf{y}_{j(2)}) \approx \ln(\sqrt{2\pi}\sigma_j) + \ln h(\tilde{\zeta}_j) + \sum_i \ln g^{(1)}(y_{ij}|\tilde{\zeta}_j)$$

$$= \ln(\sigma_j/\sqrt{\psi}) - \tilde{\zeta}_j^2/(2\psi) + \sum_i \ln g^{(1)}(y_{ij}|\tilde{\zeta}_j). \qquad (6.6)$$

This approximation is good whenever the posterior density of the random intercept is approximately normal. It is well known that this is the case for large sample sizes (cluster sizes in this setting). This asymptotic normality is sometimes referred to as a Bayesian central limit theorem (e.g. Carlin and Louis, 2000). The posterior also becomes more normal as the conditional response probabilities become more normal, e.g. Poisson with large mean or binomial with large denominator. In this case the posterior mode approaches the posterior mean and σ_j the posterior standard deviation.

In penalized quasi-likelihood (PQL) methods (e.g. Schall, 1991; McGilchrist, 1994; Breslow and Clayton, 1993), the first term in (6.6) is ignored and the remaining terms are maximized with respect to the fixed effects parameters β (for known variance parameters). It is important to note that this does not correspond to maximum likelihood. Instead, the penalized quasi log-likelihood

$$-\zeta_j^2/(2\psi) + \sum_i \ln g^{(1)}(y_{ij}|\zeta_j)$$

is jointly maximized. This is accomplished by maximization with respect to β

and then with respect to ζ_j for given $\boldsymbol{\beta}$, since $\tilde{\zeta}_j$ maximizes (6.5) and the log of (6.5) differs from the above only by the constant $\ln(\sqrt{2\pi\psi})$. An alternative derivation of the PQL approach is discussed in Section 6.8.3.

Lee and Nelder (1996, 2001) define the *hierarchical likelihood*, or *h*-likelihood, as the joint distribution of the responses and latent variables treating the latent variables as observed. The log of the *h*-likelihood is therefore

$$\ell_h = -\zeta_j^2/(2\psi) + \sum_i \ln g^{(1)}(y_{ij}|\zeta_j) - \ln(\sqrt{2\pi\psi}). \tag{6.7}$$

Maximizing the *h*-likelihood with respect to $\boldsymbol{\beta}$ and ζ_j (for fixed ψ) leads to the same estimates as penalized quasi-likelihood. The merits of the *h*-likelihood are that it does not require integration and allows flexible specification of latent variable distributions. However, in the context of missing data problems, Little and Rubin (1983; 2002, p. 124) argue that the approach does not "generally share the optimal properties of ML estimation except under trivial asymptotics in which the proportion of missing data goes to zero as the sample size increases". In latent variable models, the missing data are the realizations of the latent variables so that the proportion of missing data goes to zero only if the cluster sizes go to infinity; see also Clayton (1996a). A useful discussion follows Lee and Nelder (1996).

For linear mixed models, the likelihood equations for the *h*-likelihood are the famous 'mixed model equations' for $\boldsymbol{\beta}$ and $\boldsymbol{\zeta}$ (Henderson, 1975; Harville, 1977). For given random effects covariance matrix $\boldsymbol{\Psi}$, the estimator for $\boldsymbol{\beta}$ is the maximum marginal likelihood estimator and the estimator for $\boldsymbol{\zeta}$ is the empirical Bayes predictor or best linear unbiased predictor (BLUP) discussed in Section 7.3.1.

Stiratelli *et al.* (1984) derive the same estimating equations for random effects logistic regression with dichotomous responses by maximizing the posterior distribution for $\boldsymbol{\beta}$ and ζ_j under a diffuse prior for $\boldsymbol{\beta}$ (so that the posterior is essentially the *h*-likelihood).

The Laplace approximation is based on a second order Taylor series expansion of $f(x)$. Fourth order Laplace approximations are more accurate if the posterior density is not normal and have been used to correct the bias of parameter estimates obtained using second order Laplace (Breslow and Lin, 1995; Lin and Breslow, 1996). Approximate maximum likelihood using a sixth order Laplace approximation, known as LaPlace6, was proposed by Raudenbush *et al.* (2000).

In small simulation studies of dichotomous responses, the sixth order Laplace approximation performed considerably better than PQL, somewhat better than 20-point Gauss Hermite quadrature (Raudenbush and Yang, 1998; Raudenbush *et al.*, 2000) and similarly to 7-point adaptive quadrature (Raudenbush *et al.*, 2000). However, Laplace6 (as implemented in HLM) was considerably faster than adaptive quadrature (as implemented in SAS NLMIXED). As far as we are aware, this method has so far only been used for generalized linear mixed models with nested random effects.

6.3.2 Numerical integration

Gauss-Hermite quadrature

Quadrature approximates an integral by a weighted sum of the integrand evaluated at a set of values or locations of the variable being integrated out. The locations and weights are referred to as *quadrature rules*. Gauss-Hermite quadrature rules are designed to evaluate integrals of the form

$$\int_{-\infty}^{\infty} \exp(-x^2) f(x) dx \approx \sum_{r=1}^{R} p_r^* f(a_r^*)$$

exactly with R points if $f(x)$ is a $(2R-1)$th degree polynomial. Since the 'weight function' $\exp(-x^2)$ is proportional to a normal density, we can use the rule to integrate out the normally distributed latent variable in (6.3). We first change the variable of integration to a standard normal variable $v_j = \zeta_j / \sqrt{\psi}$ so that the integral in (6.3) becomes

$$g^{(2)}(\mathbf{y}_{j(2)}) = \int_{-\infty}^{\infty} \phi(v_j) \prod_i g^{(1)}(y_{ij} | \sqrt{\psi} v_j) dv_j, \tag{6.8}$$

where $\phi(\cdot)$ is the standard normal density

$$\phi(v_j) = \frac{1}{\sqrt{2\pi}} \exp(-v_j^2/2).$$

Applying the Gauss-Hermite quadrature rule to this integral gives

$$\int_{-\infty}^{\infty} \phi(v_j) \prod_i g^{(1)}(y_{ij} | \sqrt{\psi} v_j) dv_j \approx \sum_r p_r \prod_i g^{(1)}(y_{ij} | \sqrt{\psi} a_r), \tag{6.9}$$

where $p_r \equiv p_r^* / \sqrt{\pi}$ and $a_r \equiv \sqrt{2} a_r^*$.

The multivariate integrals in (6.2) required for the general model can be evaluated using *cartesian product quadrature*. Here we change the variables of integration to independent standard normally distributed latent variables $\mathbf{v}^{(l)}$ so that

$$\boldsymbol{\zeta}^{(l)} = \mathbf{Q}^{(l)} \mathbf{v}^{(l)}, \tag{6.10}$$

where $\mathbf{Q}^{(l)}$ is the Cholesky decomposition of the covariance matrix $\boldsymbol{\Psi}^{(l)}$ of $\boldsymbol{\zeta}^{(l)}$. The integrals over the M_l latent variables at level l then become

$$g^{(l)}(\mathbf{y}_{(l)} | \mathbf{v}^{([l+1]^+)})$$

$$= \int_{-\infty}^{\infty} \phi(v_1^{(l)}) \ldots \int_{-\infty}^{\infty} \phi(v_{M_l}^{(l)}) \prod g^{(l-1)}(\mathbf{y}_{(l-1)} | \mathbf{v}^{(l^+)}) dv_{M_l}^{(l)} \ldots dv_1^{(l)}$$

$$\approx \sum_{r_1=1}^{R_1^{(l)}} p_{r_1} \ldots \sum_{r_{M_l}=1}^{R_{M_l}^{(l)}} p_{r_{M_l}} \prod g^{(l-1)}(\mathbf{y}_{(l-1)} | a_{r_1}, \ldots, a_{r_{M_l}}, \mathbf{v}^{([l+1]^+)}), \tag{6.11}$$

where $\mathbf{v}^{(l^+)} = (\mathbf{v}^{(l)\prime}, \ldots, \mathbf{v}^{(L)\prime})'$.

The latent variables are therefore evaluated at a rectangular grid of points as shown in the top left panel of Figure 6.1. A different number of quadrature points $R_m^{(l)}$ can be used for each latent variable $v_m^{(l)}$ requiring a total of $\prod_{m=1}^{M_l} R_m^{(l)}$ evaluations of the integrand.

An alternative to cartesian product quadrature rules is *spherical quadrature* rules which are specifically designed for integrating over multivariate normal densities (Stroud, 1971). As the name suggests, the quadrature points lie on (hyper)spheres (circles in two dimensions) instead of rectangles as shown in the bottom left panel of Figure 6.1. Importantly, spherical rules require fewer points than cartesian rules to obtain a given precision. However, for some dimensionalities and required levels of precision, no spherical rules are currently available. This is perhaps the reason why spherical rules have not been used much for latent variable models, exceptions being Clarkson and Zhan (2002) and Rabe-Hesketh *et al.* (2005).

Gaussian quadrature was used for probit item response (IRT) models within a Fisher scoring algorithm (see Section 6.4.2) by Bock and Lieberman (1970) and within an EM algorithm (see Section 6.4.1) by Bock and Aitkin (1981). Butler and Moffitt (1982) suggested quadrature for random intercept probit regression models. Gaussian quadrature has also been used for models with other links, for instance the one-parameter logistic IRT model (Thissen, 1982), and generalized linear mixed models (e.g. Hedeker and Gibbons, 1994, 1996a).

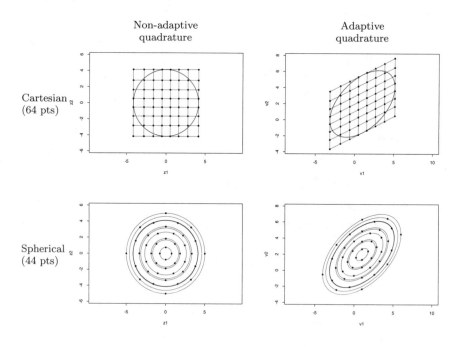

Figure 6.1 *Locations for nonadaptive and adaptive integration using cartesian and spherical product rules, where $\mu_1 = 1$, $\mu_2 = 2$, $\tau_1 = \tau_2 = 1$ and the posterior correlation is 0.5 (Source: Rabe-Hesketh* et al., *2005a)*

Gaussian quadrature works well if the product in equation (6.9) is well approximated by a low degree polynomial. However, in practice a large number of quadrature points is often required to approximate the likelihood (e.g. Crouch and Spiegelman, 1990). This will be the case for instance if there are a large number of level-1 units in a cluster j. The product in (6.9) can then have a very sharp peak and be poorly approximated by a polynomial. If insufficient quadrature points are used, the peak could be located between adjacent quadrature points a_r and a_{r+1} so that a substantial part of the likelihood contribution of cluster j is lost (see upper panel of Figure 6.2). These problems have been pointed out by Lesaffre and Spiessens (2001) for dichotomous responses and Albert and Follmann (2000) for counts. Note that the quadrature approximation can fail even for small cluster sizes for counts, since the individual $g^{(1)}(y_{ij}|v_j)$ can have sharp peaks. The approximation can also be poor for large random effects variances. Small simulation studies show that Gauss Hermite quadrature performs better than PQL (Raudenbush and Yang, 1998; Raudenbush *et al.*, 2000), but worse than adaptive quadrature (Rabe-Hesketh *et al.*, 2005a).

Adaptive quadrature

To overcome the problems with ordinary quadrature, adaptive quadrature essentially shifts and scales the quadrature locations to place them under the peak of the integrand. As discussed in Section 6.3.1, the integrand

$$\phi(v_j) \prod_i g^{(1)}(y_{ij}|v_j)$$

is proportional to the posterior density and can often be well approximated by a normal density $\phi(v_j; \mu_j, \tau_j^2)$ with some cluster-specific mean μ_j and variance τ_j^2. Instead of treating the *prior* density as the 'weight function' when applying the quadrature rule as in (6.9), we therefore rewrite the integral as

$$g^{(2)}(\mathbf{y}_{j(2)}) = \int_{-\infty}^{\infty} \phi(v_j; \mu_j, \tau_j^2) \left[\frac{\phi(v_j) \prod_i g^{(1)}(y_{ij}|\sqrt{\psi}v_j)}{\phi(v_j; \mu_j, \tau_j^2)} \right] dv_j, \qquad (6.12)$$

and treat the normal density approximating the *posterior* density as the weight function.

Changing the variable of integration from v_j to $z_j = (v_j - \mu_j)/\tau_j$ and applying the standard quadrature rule yields

$$
\begin{aligned}
g^{(2)}(\mathbf{y}_{j(2)}) &= \int_{-\infty}^{\infty} \frac{\phi(z_j)}{\tau_j} \left[\frac{\phi(\tau_j z_j + \mu_j) \prod_i g^{(1)}(y_{ij}|\sqrt{\psi}(\tau_j z_j + \mu_j))}{\frac{1}{\sqrt{2\pi}\tau_j} \exp(-z_j^2/2)} \right] \tau_j \, dz_j \\
&\approx \sum_{r=1}^{R} p_r \left[\frac{\phi(\tau_j a_r + \mu_j) \prod_i g^{(1)}(y_{ij}|\sqrt{\psi}(\tau_j a_r + \mu_j))}{\frac{1}{\sqrt{2\pi}\tau_j} \exp(-a_r^2/2)} \right] \\
&= \sum_{r=1}^{R} \pi_{jr} \prod_i g^{(1)}(y_{ij}|\sqrt{\psi}\alpha_{jr}),
\end{aligned}
$$

where

$$\alpha_{jr} \equiv \tau_j a_r + \mu_j,$$

and

$$\pi_{jr} \equiv \sqrt{2\pi}\, \tau_j\, \exp(a_r^2/2)\, \phi(\tau_j a_r + \mu_j)\, p_r.$$

The term in square brackets will be well approximated by a low-degree polynomial if the posterior density is approximately normal so that the numerator is approximately proportional to the denominator. We would therefore expect the method to require fewer quadrature points than nonadaptive quadrature and to work well with large cluster sizes. The superiority of adaptive quadrature can be seen in Figure 6.2 which illustrates for $R = 5$ how adaptive quadrature translates and scales the locations so that they lie directly under the integrand.

When there are several latent variables, the posterior covariances must also be taken into account; see Naylor and Smith (1988) and Rabe-Hesketh *et al.* (2005a) for details. For two latent variables with a posterior correlation of 0.5, the second column of Figure 6.1 shows how adaptive quadrature transforms the locations to fit more closely the elliptical contours of the (approximately) bivariate normal posterior.

Naylor and Smith (1982) take the mean μ_j and variance τ_j^2 of the normal density approximating the posterior density to be the posterior mean and variance. Unfortunately, these posterior moments are not known exactly but must themselves be obtained using adaptive quadrature. Integration is therefore iterative. Using starting values $\mu_j^0 = 0$ and $\tau_j^0 = 1$ to define α_{jr}^0 and π_{jr}^0, the posterior means and variances are updated in the kth iteration using

$$g^{(2)k}(\mathbf{y}_{j(2)}) = \sum_{r=1}^{R} \pi_{jr}^{k-1} \prod_{i=1}^{n_j} g^{(1)}(y_{ij}|\sqrt{\psi}\alpha_{jr}^{k-1})$$

$$\mu_j^k = \sum_{r=1}^{R} \alpha_{jr}^{k-1} \left[\frac{\pi_{jr}^{k-1} \prod_{i=1}^{n_j} g^{(1)}(y_{ij}|\sqrt{\psi}\alpha_{jr}^{k-1})}{g^{(2)k}(\mathbf{y}_{j(2)})} \right]$$

$$(\tau_j^k)^2 = \sum_{r=1}^{R} (\alpha_{jr}^{k-1})^2 \left[\frac{\pi_{jr}^{k-1} \prod_{i=1}^{n_j} g^{(1)}(y_{ij}|\sqrt{\psi}\alpha_{jr}^{k-1})}{g^{(2)k}(\mathbf{y}_{j(2)})} \right] - (\mu_j^k)^2,$$

and this is repeated until convergence. A similar iterative algorithm is described in Naylor and Smith (1988).

An alternative to computing the posterior moments μ_j and τ_j^2 is to use the mode and the curvature at the mode (Liu and Pierce, 1994) as in the first order Laplace approximation described in Section 6.3.1. In this case, adaptive quadrature with $R = 1$ quadrature point is equivalent to the first order Laplace approximation. An advantage of using the mode and curvature at the mode instead of the posterior moments is that computing the former does not require numerical integration. However, the approach is not easily generalized to multilevel models which led Rabe-Hesketh *et al.* (2005a) to

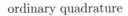

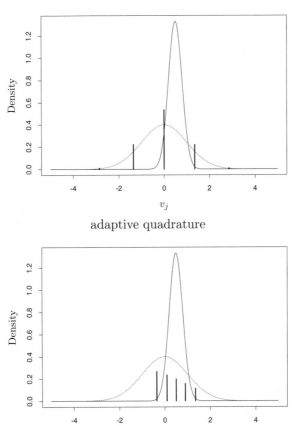

Figure 6.2 *Prior (dotted curve) and posterior (solid curve) densities and quadra-*
ture weights (bars) for ordinary quadrature and adaptive quadrature. Note that the
integrand is proportional to the posterior density. (Source: Rabe-Hesketh et al.*, 2002)*

adopt the Naylor and Smith approach for models belonging to our general
model framework.

Adaptive quadrature has been used by Pinheiro and Bates (1995) for two-
level nonlinear mixed models, Bock and Schilling (1997) for exploratory factor
analysis with dichotomous responses and Rabe-Hesketh *et al.* (2002, 2005a)
for generalized linear latent and mixed models. Adaptive quadrature, as implemented in the Stata program `gllamm`, is used in most applications in this
book involving continuous latent variables.

Monte Carlo experiments have been carried out for two-level models with
dichotomous responses with varying cluster sizes and intraclass correlations to

compare the performance of adaptive and ordinary quadrature (Rabe-Hesketh *et al.* 2005a). The performance of adaptive quadrature was excellent, requiring fewer quadrature points than ordinary quadrature. For combinations of large cluster sizes and high intraclass correlations ordinary quadrature sometimes failed, whereas adaptive quadrature worked well with a sufficient number of points.

6.3.3 Monte Carlo Integration

Let φ be a vector of random variables with distribution $h(\varphi)$. Assume that we require the expectation of a function $f(\varphi)$ over φ,

$$\mathrm{E}[f(\varphi)] = \int_{-\infty}^{\infty} h(\varphi)f(\varphi)\mathrm{d}\varphi.$$

Monte Carlo integration approximates the expectation by the mean of $f(\varphi)$ over simulated values of φ. Different versions of Monte Carlo integration arise according to how the simulation proceeds.

Crude Monte Carlo integration

In this case independent samples $\varphi^{(r)}$, $r = 1, \ldots, R$, are drawn from $h(\varphi)$, providing the simulator

$$\mathrm{E}[f(\varphi)] \approx \bar{f} \equiv \frac{1}{R} \sum_{r=1}^{R} f(\varphi^{(r)}).$$

By a strong law of large numbers, $\bar{f}$ converges to $\mathrm{E}(f(\varphi))$ with probability 1 as $R \to \infty$. By a central limit theorem, $\bar{f}$ is approximately normally distributed when R is large with mean $\mathrm{E}(f(\varphi))$ and variance

$$\mathrm{Var}(\bar{f}) = \frac{1}{R(R-1)} \sum_{r=1}^{R} [f(\varphi^{(r)}) - \bar{f}]^2.$$

Letting $\varphi = v_j$, $h(\varphi) = \phi(\varphi)$ and $f(\varphi) = \prod_i g^{(1)}(y_{ij}|\sqrt{\psi}v_j)$, we see that the likelihood for the random intercept model,

$$\int_{-\infty}^{\infty} \phi(v_j) \prod_i g^{(1)}(y_{ij}|\sqrt{\psi}v_j)\, \mathrm{d}v_j,$$

takes the form of $\mathrm{E}[f(\varphi)]$. Monte Carlo integration of the likelihood can then proceed by sampling from $\phi(v_j)$ and evaluating the mean of $\prod_i g^{(1)}(y_{ij}|\sqrt{\psi}v_j)$.

Unlike the Laplace approximation, which improves only as the cluster sizes n_j increase, we can improve the precision simply by increasing the number of simulations R. Furthermore, in contrast to quadrature or the Laplace approximation, we can assess the accuracy of the approximation by estimating the variance of the simulator.

Importance sampling

Crude Monte Carlo integration can be improved by using importance sampling to reduce the sampling variance. A judiciously chosen importance density $g(\varphi)$ is used to simulate $E[f(\varphi)]$ when it is either difficult to sample φ from $h(\varphi)$ or $h(\varphi)$ is not smooth. The integral is then written as

$$E[f(\varphi)] = \int_{-\infty}^{\infty} g(\varphi)\frac{h(\varphi)f(\varphi)}{g(\varphi)}d\varphi,$$

where $g(\varphi)$ is a density for which (a) it is easy to draw φ, (b) the support is the same as for $f(\varphi)$, (c) it is easy to evaluate $\frac{h(\varphi)f(\varphi)}{g(\varphi)}$ given φ and (d) $\frac{h(\varphi)f(\varphi)}{g(\varphi)}$ is bounded and smooth in the parameters over the support of φ. The importance sampling simulator is

$$E[f(\varphi)] \approx \frac{1}{R}\sum_{r=1}^{R}\frac{h(\varphi^{(r)})f(\varphi^{(r)})}{g(\varphi^{(r)})},$$

where $\varphi^{(r)}$ is a draw from the importance density $g(\varphi)$.

Considering a random intercept model, an importance sampler can be constructed as in (6.12)

$$\int_{-\infty}^{\infty} \phi(v_j;\mu_j,\tau_j^2)\left[\frac{\phi(v_j)\prod_i g^{(1)}(y_{ij}|v_j)}{\phi(v_j;\mu_j,\tau_j^2)}\right]dv_j,$$

where the importance density $\phi(v_j;\mu_j,\tau_j^2)$ is the normal density approximating the posterior density. Samples are drawn from $\phi(v_j;\mu_j,\tau_j^2)$ to compute the mean of the term in brackets. Note that this method is analogous to adaptive quadrature, which can be viewed as a deterministic version of importance sampling as pointed out by Pinheiro and Bates (1995). The multivariate extension of this importance sampler has been used for generalized linear mixed models by Kuk (1999) and Skaug (2002).

An interesting alternative to Monte Carlo integration is quasi-Monte Carlo where samples are drawn deterministically; see Shaw (1988) and Fang and Wang (1994, Chapter 2).

6.3.4 A tailored simulator: GHK

Some simulators are tailored for specific models, for instance the Geweke-Hajivassiliou-Keane (GHK) (Geweke, 1989; Hajivassiliou and Ruud, 1994; Keane, 1994) and Stern simulators (Stern, 1992) for models with multinormal latent responses, such as multinomial probit and multivariate probit models.

For these models the likelihood contributions are probabilities, say p, of the form

$$p = \Pr\left[(\tau_1^- < \epsilon_1 \le \tau_1^+), (\tau_2^- < \epsilon_2 \le \tau_2^+), \ldots, (\tau_S^- < \epsilon_S \le \tau_S^+)\right],$$

where $\epsilon = (\epsilon_1, \epsilon_2, \ldots, \epsilon_S)'$ is an S-dimensional multivariate normal vector with mean zero and covariance matrix $\boldsymbol{\Sigma}$. In the *multivariate* probit case, $\boldsymbol{\epsilon}$ would represent residuals of the latent responses $\mathbf{y}^*$ and in the *multinomial* probit case differences between utility residuals. The integration limits or thresholds τ_s would typically depend on covariates. For instance in multivariate probit models for ordinal or dichotomous responses, $\tau_s = \kappa_s - \mathbf{x}'\boldsymbol{\beta}$, where κ_s are the parameters of the threshold model (see Section 2.4).

Note that the probability p is an integral over a rectangular region of the *latent response distribution*, not over the latent variable distribution. In contrast, other methods described in this section use latent variable integration (see Section 4.9 for a discussion of latent response versus latent variable integration). Latent response integration is also used in the limited information method to be discussed in Section 6.7.

Here we briefly describe the popular GHK simulator. First, we exploit the fact that the probability p can be expressed as a product of sequentially conditioned univariate normal distributions. Defining

$$Q_1 \equiv \Pr\left[(\tau_1^- < \epsilon_1 \leq \tau_1^+)\right],$$

$$Q_2 \equiv \Pr\left[(\tau_2^- < \epsilon_2 \leq \tau_2^+) \mid (\tau_1^- < \epsilon_1 \leq \tau_1^+)\right],$$

$$\vdots$$

$$Q_S \equiv \Pr\left[(\tau_S^- < \epsilon_S \leq \tau_S^+) \mid (\tau_{S-1}^- < \epsilon_{S-1} \leq \tau_{S-1}^+), \dots, (\tau_1^- < \epsilon_1 \leq \tau_1^+)\right],$$

the probability can be written as

$$p = \prod_{s=1}^{S} Q_s.$$

It is easy to calculate $Q_1 = \Phi(\tau_1^+/\sigma_{11}) - \Phi(\tau_1^-/\sigma_{11})$, where $\Phi(\cdot)$ denotes the univariate cumulative standard normal distribution function and σ_{11} is the standard deviation of ϵ_1. However, each Q_s, $s = 2, \dots, S$ is a *conditional* probability that ϵ_s lies within an interval, given that the other ϵ_t (which are correlated with ϵ_s) lie within specific intervals, which is difficult to evaluate.

We therefore orthogonalize the residuals $\boldsymbol{\epsilon}$ using a lower diagonal Cholesky decomposition of the covariance matrix $\boldsymbol{\Sigma}$, $\boldsymbol{\Sigma} = \mathbf{C}\mathbf{C}'$, with elements c_{sm}, $c_{sm} = 0$ if $m > s$. We can then write $\boldsymbol{\epsilon} = \mathbf{C}\mathbf{u}$ where $\mathbf{u}$ is an orthogonal vector, having *independent* standard normal components u_s.

The algorithm then proceeds as follows:

1. For replication $r = 1, \dots, R$:

(a) For $s = 1$:

- Evaluate Q_{1r}:

$$Q_{1r} = \Phi(\tau_1^+/c_{11}) - \Phi(\tau_1^-/c_{11}).$$

- Simulate u_{1r} from a doubly-truncated standard normal distribution, with truncation points at τ_1^-/c_{11} and τ_1^+/c_{11}, so that $\epsilon_{1r} = c_{11}u_{1r}$ fulfills the condition $\tau_1^- < \epsilon_{1r} \leq \tau_1^+$ for the conditional probabilities Q_2 to Q_S.

(b) For $s = 2$:

- Evaluate Q_{2r}:
  Treating u_{1r} as known, Q_{2r} becomes an easily evaluated *unconditional* probability

  $$Q_{2r} = \Phi([\tau_2^+ - c_{21}u_{1r}]/c_{22}) - \Phi([\tau_2^- - c_{21}u_{1r}]/c_{22}).$$

  The first term follows from the equivalence

  $$\epsilon_{2r} = c_{21}u_{1r} + c_{22}u_{2r} \leq \tau_2^+ \iff u_{2r} \leq [\tau_2^+ - c_{21}u_{1r}]/c_{22},$$

  and similarly for the second term.
- Simulate u_{2r} from a doubly-truncated standard normal distribution truncated at $[\tau_2^- - c_{21}u_{1r}]/c_{22}$ and $[\tau_2^+ - c_{21}u_{1r}]/c_{22}$ so that ϵ_{2r} satisfies the conditions in the conditional probabilities Q_3 to Q_S.

(c) For $s = 3, \ldots, S$:

- Evaluate Q_{sr} sequentially, treating u_{1r} to $u_{s-1,r}$ as known:

  $$Q_{sr} = \Phi([\tau_s^+ - \sum_{m=1}^{s-1} c_{sm}u_{mr}]/c_{ss}) - \Phi([\tau_s^- - \sum_{m=1}^{s-1} c_{sm}u_{mr}]/c_{ss}),$$

  where the first term follows from

  $$\epsilon_{sr} = \sum_{m=1}^{s} c_{sm}u_{mr} \leq \tau_s^+ \iff u_{sr} \leq [\tau_s^+ - \sum_{m=1}^{s-1} c_{sm}u_{mr}]/c_{ss}.$$

- Simulate u_{sr} from a doubly-truncated standard normal distribution truncated at $[\tau_s^- - \sum_{m=1}^{s-1} c_{sm}u_{mr}]/c_{ss}$ and $[\tau_s^+ - \sum_{m=1}^{s-1} c_{sm}u_{mr}]/c_{ss}$ so that ϵ_{sr} satisfies the conditions in the conditional probabilities Q_{s+1} to Q_S. (This step is not needed for $s = S$.)

2. After R replications: The required simulated probability is obtained as

$$\check{p} = \frac{1}{R} \sum_{r=1}^{R} \prod_{s=1}^{S} Q_{sr}.$$

Note that $\check{p}$ is an unbiased simulator of p, with $\mathrm{E}(\check{p}) = p$, where the expectation is over imagined repeats of the simulation. However, the simulated log-likelihood is a sum of terms of the form $\log(\check{p})$ and these will generally be biased because $\mathrm{E}[\log(\check{p})] \neq \log(p)$, due to the nonlinear log transformation. The bias can be reduced by increasing the number of replications R.

The GHK simulator is typically used in conjunction with gradient methods (to be discussed in Section 6.4.2) to obtain *maximum simulated likelihood* (MSL) estimators. We refer to Train (2003) for a detailed discussion of the properties of MSL and related estimators.

In econometrics the GHK simulator is very popular for models with multi-normal latent responses, for instance probit panel (longitudinal) models (e.g. Keane, 1994; Geweke et al., 1994). This is probably because the simulator has been shown to outperform other simulators although it is relatively easy to implement. In contrast to the crude Monte Carlo approach originally proposed by Lerman and Manski (1981), the GHK simulator is a continuous and differentiable function of the parameters and produces simulated probabilities that are unbiased and bounded in the (0,1) interval. Furthermore, GHK is more statistically efficient than other simulators (e.g. Hajivassiliou *et al.*, 1996). We refer to Stern (1997), Train (2003) and Cappellari and Jenkins (2003) for relatively nontechnical discussions of the GHK and related simulators.

6.4 Maximizing the likelihood

There are several methods for maximizing the likelihood, the most common being the Expectation-Maximization (EM) algorithm and Newton-Raphson or Fisher scoring algorithms to be described in Sections 6.4.2 and 6.4.1. Each of the integration methods introduced above may be combined with the maximization methods to be described.

6.4.1 EM algorithm

The Expectation-Maximization (EM) algorithm is an iterative algorithm for maximum likelihood estimation in incomplete data problems. The algorithm was given its name by Dempster *et al.* (1977) who presented the general theory for the algorithm and a number of examples. Orchard and Woodbury (1972) first noted the general applicability of the underlying idea, calling it the 'missing information principle', although applications of the EM algorithm date back at least to McKendrick (1926).

Perhaps the most prominent application is estimation when there are missing data on random variables whose realizations would otherwise be observed (e.g. Little and Rubin, 2002). Another application, which is more important in the present setting, is in the estimation of latent variable models. In this case the realizations of latent variables are interpreted as missing data (e.g. Becker *et al.*, 1997).

The motivating idea behind the EM algorithm is as follows: rather than performing one complex estimation, the observed data is augmented with latent data that permits estimation to proceed in a sequence of simple estimation steps.

The complete data $\mathbf{C} = \{\mathbf{y}, \mathbf{X}, \boldsymbol{\zeta}\}$ consist of two parts: the incomplete data $\mathbf{y}$ and $\mathbf{X}$ that are observable and the unobservable or latent data $\boldsymbol{\zeta}$. The complete data log-likelihood, imagining that the latent data were observed, is denoted $\ell_h(\boldsymbol{\vartheta}|\mathbf{C})$. Here we used the h subscript since the log-likelihood is just the h-log-likelihood of Lee and Nelder (1996). In general, the complete data log-likelihood itself is involved at each iteration of the EM algorithm, which takes the following form at the $(k+1)$th step :

E-step: Evaluate the posterior expectation

$$Q(\boldsymbol{\vartheta}|\boldsymbol{\vartheta}^k) \equiv \mathrm{E}_{\boldsymbol{\zeta}}[\ell_h(\boldsymbol{\vartheta}|\mathbf{C})|\mathbf{y}, \mathbf{X}; \boldsymbol{\vartheta}^k],$$

the conditional expectation of the complete data log-likelihood with respect to the latent variables, given the incomplete data and the estimates $\boldsymbol{\vartheta}^k$ from the previous iteration, i.e. an expectation over the posterior density of $\boldsymbol{\zeta}$.

M-step: Maximize $Q(\boldsymbol{\vartheta}|\boldsymbol{\vartheta}^k)$ with respect to $\boldsymbol{\vartheta}$ to produce an updated estimate $\boldsymbol{\vartheta}^{k+1}$.

This can sometimes be accomplished analytically, but usually requires iterative algorithms such as gradient methods (see Section 6.4.2).

We now consider implementation of the EM algorithm for two-level latent variable models with multinormal latent variables. As in Section 6.3.2 it is convenient to change the variables of integration to independent standard normal latent variables $\mathbf{v}_j = (v_{j1}, v_{j2}, \ldots, v_{jM})'$ using the Cholesky decomposition $\boldsymbol{\zeta}_j = \mathbf{Q}\mathbf{v}_j$ (where $\mathbf{Q}$ depends on $\boldsymbol{\vartheta}$). The complete data log-likelihood, treating the orthogonalized latent variables $\mathbf{v}_j$ as observed, can then be expressed as

$$
\begin{aligned}
\ell_h(\boldsymbol{\vartheta}|\mathbf{C}) &= \ln \prod_j \left\{ \prod_i g(y_{ij}|\mathbf{Q}\mathbf{v}_j) \prod_m \phi(v_{jm}) \right\} \\
&= \sum_j \left\{ \sum_i \ln g(y_{ij}|\mathbf{Q}\mathbf{v}_j) + \sum_m \ln \phi(v_{jm}) \right\} \\
&= \sum_j \ell_h^j(\boldsymbol{\vartheta}|\mathbf{C}),
\end{aligned}
\tag{6.13}
$$

where $\ell_h^j(\boldsymbol{\vartheta}|\mathbf{C})$ is a cluster-contribution to the complete data log-likelihood.

E-step: Evaluate

$$
\begin{aligned}
Q(\boldsymbol{\vartheta}|\boldsymbol{\vartheta}^k) &= \mathrm{E}_{\boldsymbol{\zeta}}[\ell_h(\boldsymbol{\vartheta}|\mathbf{C})|\mathbf{y};\boldsymbol{\vartheta}^k] \\
&= \sum_j \int \ell_h^j(\boldsymbol{\vartheta}|\mathbf{C})\omega(\mathbf{v}_j|\mathbf{y}_{j(2)};\boldsymbol{\vartheta}^k)\mathrm{d}\mathbf{v}_j,
\end{aligned}
$$

where $\omega(\mathbf{v}_j|\mathbf{y}_{j(2)};\boldsymbol{\vartheta}^k)$ is the posterior density of the latent variables $\mathbf{v}_j$ for cluster j given the observed responses $\mathbf{y}_{j(2)}$ for that cluster. Using Bayes theorem, the posterior becomes

$$
\omega(\mathbf{v}_j|\mathbf{y}_{j(2)};\boldsymbol{\vartheta}^k) = \frac{\prod_i g(y_{ij}|\mathbf{Q}^k\mathbf{v}_j;\boldsymbol{\vartheta}^k)\prod_m \phi(v_{jm})}{\int \prod_i g(y_{ij}|\mathbf{Q}^k\mathbf{v}_j;\boldsymbol{\vartheta}^k)\prod_m \phi(v_{jm})\,\mathrm{d}\mathbf{v}_j},
\tag{6.14}
$$

where the k superscript in $\mathbf{Q}^k$ denotes that this matrix depends on $\boldsymbol{\vartheta}^k$. Using (6.13) and (6.14), $Q(\boldsymbol{\vartheta}|\boldsymbol{\vartheta}^k)$ simplifies to

$$
\begin{aligned}
Q(\boldsymbol{\vartheta}|\boldsymbol{\vartheta}^k) &= \sum_j \frac{1}{A_j^k} \int_{-\infty}^{\infty} \left\{ \sum_i \ln g(y_{ij}|\mathbf{Q}\mathbf{v}_j;\boldsymbol{\vartheta}) + \sum_m \ln \phi(v_{jm}) \right\} \\
&\quad \times \prod_i g(y_{ij}|\mathbf{Q}^k\mathbf{v}_j;\boldsymbol{\vartheta}^k)\prod_m \phi(v_{jm})\,\mathrm{d}\mathbf{v}_j,
\end{aligned}
$$

where
$$A_j^k = \int_{-\infty}^{\infty} \prod_i g(y_{ij}|\mathbf{Q}^k\mathbf{v}_j; \boldsymbol{\vartheta}^k) \prod_m \phi(v_{jm}) \, d\mathbf{v}_j.$$

Note that A_j^k does not depend on the unknown parameters $\boldsymbol{\vartheta}$ but only on the values $\boldsymbol{\vartheta}^k$ obtained in the previous iteration.

The E-step is complicated since the integral cannot in general be solved analytically, and several approximate methods based on numerical integration or simulation have been suggested. Monte Carlo integration has been suggested (e.g. Wei and Tanner, 1990) to yield Monte Carlo Expectation Maximization (MCEM) algorithms. In the present context, we consider vector draws $\mathbf{d}_{jr} = (d_{jr1}, d_{jr2}, \ldots, d_{jrM})'$ for the independent normally distributed random variables in $\mathbf{v}_j$ with replications $r = 1, 2, \ldots, R$. This provides the Monte Carlo integration approximation

$$Q^{MC}(\boldsymbol{\vartheta}|\boldsymbol{\vartheta}^k) = \sum_j \sum_r c_{jr}^{MC} \left\{ \sum_i \ln g(y_{ij}|\mathbf{Q}\mathbf{d}_{jr}; \boldsymbol{\vartheta}) + \sum_m \ln \phi(d_{jrm}) \right\},$$

with weights that do not depend on the unknown parameters $\boldsymbol{\vartheta}$,

$$c_{jr}^{MC} = \frac{\sum_i g(y_{ij}|\mathbf{Q}^k\mathbf{d}_{jr}; \boldsymbol{\vartheta}^k)}{\sum_r \sum_i g(y_{ij}|\mathbf{Q}^k\mathbf{d}_{jr}; \boldsymbol{\vartheta}^k)},$$

satisfying $\sum_r c_{jr}^{MC} = 1$. As an alternative to using crude Monte Carlo integration, Meng and Schilling (1996) suggest using Gibbs sampling (see Section 6.11.5). A problematic feature of MCEM is the inability to quantify the Monte Carlo error introduced at each step of the algorithm (e.g. McCulloch, 1997; Hobert, 2000). If the number of replications R is too small the E-step will be swamped by Monte Carlo error, whereas an unnecessarily large R is wasteful. In fact, Booth et al. (2001) point out that MCEM does not converge in the usual sense unless R increases with k.

Since the latent variable distribution is specified as normal, Gauss-Hermite quadrature can alternatively be used yielding

$$Q^{GH}(\boldsymbol{\vartheta}|\boldsymbol{\vartheta}^k) = \sum_j \sum_{r_1} c_{r_1}^{GH} \cdots \sum_{r_M} c_{r_M}^{GH} \left\{ \sum_i \ln g(y_{ij}|\mathbf{Q}\mathbf{a}_r; \boldsymbol{\vartheta}) + \sum_m \ln \phi(a_{rm}) \right\},$$

with weights that do not depend on $\boldsymbol{\vartheta}$,

$$c_{r_m}^{GH} = \frac{p_{r_m} \sum_i g(y_{ij}|\mathbf{Q}^k\mathbf{a}_r; \boldsymbol{\vartheta}^k)}{\sum_r p_{r_m} \sum_i g(y_{ij}|\mathbf{Q}^k\mathbf{a}_r; \boldsymbol{\vartheta}^k)},$$

satisfying $\sum_{r_m} c_{r_m}^{GH} = 1$. Here, $\mathbf{a}_r = (a_{r1}, a_{r2}, \ldots, a_{rM})'$ and $p_{r_m} = p_r$ denote quadrature locations and weights, respectively. Bock and Schilling (1997) suggest improving the quadrature approximation by using adaptive quadrature.

M-step: Maximize $Q(\boldsymbol{\vartheta}|\boldsymbol{\vartheta}^k)$ with respect to $\boldsymbol{\vartheta}$.

For MCEM, this is equivalent to solving the equation

$$\frac{\partial Q^{MC}(\boldsymbol{\vartheta}|\boldsymbol{\vartheta}^k)}{\partial \boldsymbol{\vartheta}} = \sum_j \sum_r c_{jr}^{MC} \frac{\partial \sum_i \ln g(y_{ij}|\mathbf{Qd}_{jr};\boldsymbol{\vartheta})}{\partial \boldsymbol{\vartheta}} = \mathbf{0}.$$

This is a weighted score function for $\boldsymbol{\vartheta}$ in a generalized linear model for expanded data y_{ij}, d_{jr} with known weights c_{jr}^{MC}. $Q(\boldsymbol{\vartheta}|\boldsymbol{\vartheta}^k)$ can thus simply be maximized by weighted maximum likelihood, using standard software.

For Gauss-Hermite quadrature the M-step amounts to solving

$$\frac{\partial Q^{GH}(\boldsymbol{\vartheta}|\boldsymbol{\vartheta}^k)}{\partial \boldsymbol{\vartheta}} = \sum_j \sum_{r_1} c_{r_1}^{GH} \cdots \sum_{r_M} c_{r_M}^{GH} \frac{\partial \sum_i \ln g(y_{ij}|\mathbf{Qa}_r;\boldsymbol{\vartheta})}{\partial \boldsymbol{\vartheta}} = \mathbf{0},$$

which can also be maximized with an appropriately weighted maximum likelihood algorithm. See Aitkin (1999a) for a suggested implementation of this algorithm for two-level models and Vermunt (2004) for higher-level models.

Usually, the E-step is the demanding step but in some situations the M-step is more difficult. One simplification may be the use of the ECM algorithm (Meng and Rubin, 1993) which replaces each M-step with a sequence of conditional maximization steps with subsets of $\boldsymbol{\vartheta}$ being fixed at their previous values. Other modifications are discussed in Little and Rubin (2002). Another possibility is to use simulation in the M-step, proceeding by crude Monte Carlo integration or by means of more elaborate approaches such as the Metropolis algorithm (see Section 6.11.5) used by McCulloch (1994, 1997) or importance sampling as suggested by Booth and Hobert (1999).

It should be noted that EM works best if the complete data distribution is of *regular exponential family* form. In this case it can be shown (e.g. Tanner, 1996) that the E-step consists of estimating the complete data sufficient statistics by their posterior expectations. Given these estimates, the likelihood equations for the M-step then take the same form as for complete data, so that standard software can be used.

It is instructive to consider estimation of a conventional exploratory factor model, where implementation of the EM algorithm is extremely simple. Expected sufficient statistics, expressed in closed form, are obtained in the E-step whereas the M-step requires only elementary linear algebra.

Example: EM exploratory factor analysis

Consider the exploratory factor model introduced in Section 3.3.3,

$$\mathbf{y}_j = \boldsymbol{\Lambda}\boldsymbol{\eta}_j + \boldsymbol{\epsilon}_j,$$

where $\mathbf{y}_j$ are mean-centered responses, $\boldsymbol{\eta}_j \sim \mathrm{N}_M(\mathbf{0},\mathbf{I})$, $\boldsymbol{\epsilon}_j \sim \mathrm{N}_n(\mathbf{0},\boldsymbol{\Theta})$, $\boldsymbol{\eta}_j$ and $\boldsymbol{\epsilon}_j$ independent, $\boldsymbol{\Theta}$ diagonal and $\boldsymbol{\Lambda}$ unstructured.

It follows from the exploratory factor model that

$$\mathbf{y}_j \sim \mathrm{N}_n(\mathbf{0},\boldsymbol{\Sigma}),$$

where
$$\mathbf{\Sigma} \equiv \mathrm{Cov}(\mathbf{y}_j) \;=\; \mathbf{\Lambda}\mathbf{\Lambda}' + \mathbf{\Theta}.$$

We also obtain
$$\mathbf{y}_j | \boldsymbol{\eta}_j \sim \mathrm{N}_n(\mathbf{\Lambda}\boldsymbol{\eta}_j, \mathbf{\Theta}).$$

Importantly, if the common factors were observed, the exploratory factor model would just become a multivariate regression model with response $\mathbf{y}_j$, standardized covariates $\boldsymbol{\eta}_j$, regression parameters $\mathbf{\Lambda}$ and residual covariance matrix $\mathbf{\Theta}$. The complete data log-likelihood thus equals that of multivariate regression under normality, with the sufficient statistics

$$\mathbf{S}_{yy} = \frac{1}{N} \sum_{j=1}^{N} \mathbf{y}_j \mathbf{y}_j',$$

$$\mathbf{S}_{y\eta} = \frac{1}{N} \sum_{j=1}^{N} \mathbf{y}_j \boldsymbol{\eta}_j',$$

and

$$\mathbf{S}_{\eta\eta} = \frac{1}{N} \sum_{j=1}^{N} \boldsymbol{\eta}_j \boldsymbol{\eta}_j'.$$

The conditional distribution of the factors, given the responses, becomes

$$\boldsymbol{\eta}_j | \mathbf{y}_j \sim \mathrm{N}_M(\widetilde{\boldsymbol{\eta}}_j, \mathbf{\Upsilon}),$$

where
$$\widetilde{\boldsymbol{\eta}}_j \equiv \mathrm{E}(\boldsymbol{\eta}_j | \mathbf{y}_j) \;=\; \mathbf{F}\mathbf{y}_j,$$
$$\mathbf{\Upsilon} \equiv \mathrm{Cov}(\boldsymbol{\eta}_j | \mathbf{y}_j) \;=\; \mathbf{I} - \mathbf{F}\mathbf{\Lambda},$$

and $\mathbf{F}$ is the 'factor scoring matrix' (see page 227) for the regression method of factor scoring,

$$\mathbf{F} \;\equiv\; \mathbf{\Lambda}'\mathbf{\Sigma}^{-1} \;=\; \mathbf{\Lambda}'(\mathbf{\Lambda}\mathbf{\Lambda}' + \mathbf{\Theta})^{-1}.$$

E-step: the conditional expectations (over the factors $\boldsymbol{\eta}_j$) of the sufficient statistics, given the responses $\mathbf{y}$ and parameters $\boldsymbol{\vartheta}^k$, become

$$\mathrm{E}_{\boldsymbol{\eta}}(\mathbf{S}_{yy} | \mathbf{y}, \boldsymbol{\vartheta}^k) = \frac{1}{N} \sum_{j=1}^{N} \mathrm{E}_{\boldsymbol{\eta}}(\mathbf{y}_j \mathbf{y}_j' | \mathbf{y}, \boldsymbol{\vartheta}^k) = \mathbf{S}_{yy},$$

$$\mathrm{E}_{\boldsymbol{\eta}}(\mathbf{S}_{y\eta} | \mathbf{y}, \boldsymbol{\vartheta}^k) \;=\; \frac{1}{N} \sum_{j=1}^{N} \mathrm{E}_{\boldsymbol{\eta}}(\mathbf{y}_j \boldsymbol{\eta}_j' | \mathbf{y}, \boldsymbol{\vartheta}^k) \;=\; \mathbf{S}_{yy} \mathbf{F}^{k\prime},$$

and

$$\mathrm{E}_{\boldsymbol{\eta}}(\mathbf{S}_{\eta\eta} | \mathbf{y}, \boldsymbol{\vartheta}^k) \;=\; \frac{1}{N} \sum_{j=1}^{N} \mathrm{E}_{\boldsymbol{\eta}}(\boldsymbol{\eta}_j \boldsymbol{\eta}_j' | \mathbf{y}, \boldsymbol{\vartheta}^k) \;=\; \mathbf{F}^k \mathbf{S}_{yy} \mathbf{F}^{k\prime} + \mathbf{\Upsilon}^k.$$

Here,
$$\mathbf{F}^k \equiv \mathbf{\Lambda}^{k\prime}(\mathbf{\Lambda}^k \mathbf{\Lambda}^{k\prime} + \mathbf{\Theta}^k)^{-1}$$

and

$$\Upsilon^k \equiv \mathbf{I} - \mathbf{F}^k \mathbf{\Lambda}^k.$$

M-step: express the standard multivariate regression estimators in terms of the expected sufficient statistics from the E-step (instead of the usual sufficient statistics):

$$\mathbf{\Lambda}^{k+1} = \mathrm{E}_{\boldsymbol{\eta}}(\mathbf{S}_{y\eta}|\mathbf{y},\boldsymbol{\vartheta}^k)\{\mathrm{E}_{\boldsymbol{\eta}}(\mathbf{S}_{\eta\eta}|\mathbf{y},\boldsymbol{\vartheta}^k)\}^{-1}$$
$$= \mathbf{S}_{yy}\mathbf{F}^{k\prime}(\mathbf{F}^k\mathbf{S}_{yy}\mathbf{F}^{k\prime} + \Upsilon^k)^{-1},$$

$$\mathbf{\Theta}^{k+1} = \mathrm{diag}\left(\mathrm{E}_{\boldsymbol{\eta}}(\mathbf{S}_{yy}|\mathbf{y},\boldsymbol{\vartheta}^k) - \right.$$
$$\left. \mathrm{E}_{\boldsymbol{\eta}}(\mathbf{S}_{y\eta}|\mathbf{y},\boldsymbol{\vartheta}^k)[\mathrm{E}_{\boldsymbol{\eta}}(\mathbf{S}_{\eta\eta}|\mathbf{y},\boldsymbol{\vartheta}^k)]^{-1}\mathrm{E}_{\boldsymbol{\eta}}(\mathbf{S}_{y\eta}|\mathbf{y},\boldsymbol{\vartheta}^k)'\right)$$
$$= \mathrm{diag}(\mathbf{S}_{yy} - \mathbf{S}_{yy}\mathbf{F}^{k\prime}(\mathbf{F}^k\mathbf{S}_{yy}\mathbf{F}^{k\prime} + \Upsilon^k)^{-1}\mathbf{F}^k\mathbf{S}_{yy})$$
$$= \mathrm{diag}(\mathbf{S}_{yy} - \mathbf{\Lambda}^{k+1}\mathbf{F}^k\mathbf{S}_{yy}).$$

The idea of using the EM algorithm for estimation of factor models was due to Dempster *et al.* (1977). Elaboration and extension to confirmatory factor models (see Section 3.3.3) was presented by Rubin and Thayer (1982) and Schoenberg and Richtand (1984). Liu *et al.* (1998) show that the algorithm presented above is a special case of the so-called parameter extended EM (PX-EM) algorithm. Chen (1981) describes estimation of the conventional MIMIC model (see Section 3.5) using the EM algorithm.

Example continued: EM exploratory factor analysis

The EM algorithm can alternatively be implemented in a slightly different way, which may have more intuitive appeal. Noting that the above equations for $\mathbf{\Lambda}^{k+1}$ and $\mathbf{\Theta}^{k+1}$ correspond to linear regression of $\mathbf{y}_j$ on $\widetilde{\boldsymbol{\eta}}_j^k = \mathbf{F}^k\mathbf{y}_j$, we may proceed with the following iterative algorithm:

E-step: Impute the missing $\boldsymbol{\eta}_j$ by $\widetilde{\boldsymbol{\eta}}_j^k$. This is an example of the 'regression method' for factor scoring to be described in Section 7.3.1.

M-step: Estimate $\mathbf{\Lambda}_i^{k+1}$ (row i of $\mathbf{\Lambda}$) and θ_{ii}^{k+1} by OLS regression of y_{ij} (the ith component of $\mathbf{y}_j$) on $\widetilde{\boldsymbol{\eta}}_j^k$. This simplification arises since the model has a diagonal residual covariance matrix $\mathbf{\Theta}$.

This formulation illustrates that the EM algorithm can be viewed as a formalization of an intuitive approach to handling missing data: (1) impute missing values by predicted values, (2) estimate parameters treating imputed values as data, (3) re-impute the missing values treating the new estimates as true parameters, (4) re-estimate the parameters, and so on until convergence. It is important to keep in mind that this imputation based approach only works if the complete data likelihood equations are linear in the missing data. Otherwise, the approach may yield severely biased estimates.

Moving outside the exponential family, the numerical inaccuracy of the E-step is liable to produce artificial modes for the function to be maximized in the M-step (e.g. Meng and Rubin, 1992). This led Meng and Schilling (1996) to use data augmentation, generating latent responses $\mathbf{y}^*$ to modify the E-step for the probit factor model of Bock and Aitkin (1981). Although the model for $\mathbf{y}$ is not a member of the exponential family, Meng and Schilling exploit the fact that the complete data distribution of the latent responses $\mathbf{y}^*$ is in exponential family form. See also Section 6.11.5 for an example of data augmentation in probit models.

The EM algorithm has been used for a wide range of latent variable models. For instance, factor and MIMIC models for continuous responses have been estimated by Dempster *et al.* (1981) and Chen (1981), respectively, exploratory probit factor models by Bock and Aitkin (1981) and the one-parameter logistic IRT model by Thissen (1982). Linear mixed models were considered by Strenio *et al.* (1983) and Raudenbush and Bryk (2002) and generalized linear mixed models by Aitkin *et al.* (1981), Aitkin (1999a) and Vermunt (2004). The EM algorithm is the most popular method for estimating latent class and finite mixture models (e.g. Goodman, 1974).

An often mentioned advantage of the EM algorithm is ease of implementation as compared to other optimization methods. Although this is certainly true in many settings, it should be evident from the formulae derived above that this argument does not appear to have much force in the context of complex latent variable models. Theoretical advantages include the fact that each iteration increases the likelihood and that if the sequence $\boldsymbol{\vartheta}^k$ converges, it converges to a local maximum or saddle point.

An important disadvantage of the EM algorithm is that convergence can be very slow whenever there is a large fraction of missing information. Another disadvantage is that an estimated information matrix is not a direct byproduct of maximization, in contrast to the case for gradient methods such as Newton-Raphson. One possible approach is to augment EM with a final Newton-Raphson step after convergence. Procedures for obtaining the information matrix within the EM algorithm have been suggested by Louis (1982), Meng and Rubin (1991) and Oakes (1999) among others.

6.4.2 Gradient methods

In order to maximize the log-likelihood, we must solve the likelihood equations

$$\frac{\partial \ell(\boldsymbol{\vartheta})}{\partial \boldsymbol{\vartheta}} = \mathbf{0}.$$

Gradient methods are iterative where we let the parameters in the kth iteration be denoted $\boldsymbol{\vartheta}^k$.

Newton-Raphson and Fisher Scoring

The *Newton-Raphson algorithm* can be derived by considering an approximation of the derivatives of the log-likelihood using a first order Taylor series expansion around the current parameter estimates $\boldsymbol{\vartheta}^k$:

$$\frac{\partial \ell(\boldsymbol{\vartheta})}{\partial \boldsymbol{\vartheta}} \approx \frac{\partial \ell(\boldsymbol{\vartheta}^k)}{\partial \boldsymbol{\vartheta}} + \frac{\partial^2 \ell(\boldsymbol{\vartheta}^k)}{\partial \boldsymbol{\vartheta} \partial \boldsymbol{\vartheta}'}(\boldsymbol{\vartheta} - \boldsymbol{\vartheta}^k)$$

$$= \mathbf{g}(\boldsymbol{\vartheta}^k) + \mathbf{H}(\boldsymbol{\vartheta}^k)(\boldsymbol{\vartheta} - \boldsymbol{\vartheta}^k), \tag{6.15}$$

where $\mathbf{g}(\boldsymbol{\vartheta}^k)$ is the V-dimensional gradient vector and $\mathbf{H}(\boldsymbol{\vartheta}^k)$ is the Hessian, the $v \times v$ matrix of second derivatives of the log-likelihood with respect to the parameters, evaluated at $\boldsymbol{\vartheta}^k$. The updated parameters $\boldsymbol{\vartheta}^{k+1}$ are the parameters for which this first order Taylor expansion is zero, i.e.,

$$\mathbf{g}(\boldsymbol{\vartheta}^k) + \mathbf{H}(\boldsymbol{\vartheta}^k)(\boldsymbol{\vartheta}^{k+1} - \boldsymbol{\vartheta}^k) = \mathbf{0}$$

so that

$$\boldsymbol{\vartheta}^{k+1} = \boldsymbol{\vartheta}^k - \mathbf{H}(\boldsymbol{\vartheta}^k)^{-1}\mathbf{g}(\boldsymbol{\vartheta}^k).$$

Note that the Taylor expansion is exact if the log-likelihood is quadratic in the parameters, in which case the maximum is found in a single iteration. The canonical example is the standard linear regression model.

The *Fisher scoring algorithm* is similar to the Newton-Raphson algorithm but the negative of Fisher's information matrix $\mathbf{I}(\boldsymbol{\vartheta}^k)$ is used instead of the Hessian, i.e.

$$\boldsymbol{\vartheta}^{k+1} = \boldsymbol{\vartheta}^k + \mathbf{I}(\boldsymbol{\vartheta}^k)^{-1}\mathbf{g}(\boldsymbol{\vartheta}^k),$$

where

$$\mathbf{I}(\boldsymbol{\vartheta}^k) = -\mathrm{E}\left(\mathbf{H}(\boldsymbol{\vartheta}^k)\right).$$

An advantage of the Newton-Raphson and Fisher scoring algorithms compared with EM is that they provide estimates of the standard errors for the maximum likelihood estimates $\hat{\boldsymbol{\vartheta}}$. In the case of Fisher scoring, the inverse information is used, whereas the inverse of $-\mathbf{H}(\hat{\boldsymbol{\vartheta}})$, the 'observed information', is used in the Newton-Raphson algorithm (see also Section 8.3).

Quasi-Newton methods

Newton-Raphson and Fisher scoring require second order derivatives of the log-likelihood with respect to the parameters. Computing these analytically is often difficult and computing them numerically can be very slow. Thus, *quasi-Newton algorithms* only requiring gradients have been proposed.

A useful algorithm was described by Berndt, Hall, Hall and Hausman (1974). Their BHHH or BH^3 algorithm is based on the fact that, under correct model specification, the information matrix equals the covariance matrix of the gradients,

$$\mathbf{I}(\boldsymbol{\vartheta}^k) = -\mathrm{E}\left(\mathbf{H}(\boldsymbol{\vartheta}^k)\right) = \mathrm{E}\left(\mathbf{g}(\boldsymbol{\vartheta}^k)\mathbf{g}(\boldsymbol{\vartheta}^k)'\right).$$

From a law of large numbers it follows that a consistent estimator of this covariance matrix is given by

$$\mathbf{I}_{\mathrm{BH}^3}(\boldsymbol{\vartheta}^k) \equiv \sum_{z=1}^{J} \mathbf{g}^z(\boldsymbol{\vartheta}^k)\mathbf{g}^z(\boldsymbol{\vartheta}^k)',$$

where $\mathbf{g}^z(\boldsymbol{\vartheta}^k)$ are score vectors, the top-level cluster contributions to the gradients,

$$\mathbf{g}(\boldsymbol{\vartheta}^k) = \sum_{z=1}^{Z} \mathbf{g}^z(\boldsymbol{\vartheta}^k).$$

The BH^3 algorithm uses this estimator in Fisher-Scoring, giving

$$\boldsymbol{\vartheta}^{k+1} = \boldsymbol{\vartheta}^k + [\mathbf{I}_{\text{BH}^3}(\boldsymbol{\vartheta}^k)]^{-1}\mathbf{g}(\boldsymbol{\vartheta}^k).$$

A merit of the BH^3 algorithm is that only gradients are required; neither Hessians nor Fisher information matrices must be computed. Little and Rubin (2002) claim that the performance of the BH^3 algorithm can be erratic because the accuracy of the approximation to the information matrix depends on the validity of the model, but our experience suggests that the algorithm works well even with 'bad' starting values.

Other examples of quasi-Newton algorithms include Davidon-Fletcher-Powell (DFP) and Broyden-Fletcher-Goldfarb-Shanno (BFGS) which involve different approximations of the Hessian $\mathbf{H}(\boldsymbol{\vartheta}^k)$. Although the approximations work well for optimization, caution should be exercised in basing estimated covariance matrices for the estimated parameters on these approximations (e.g. Thisted, 1987).

The Newton-Raphson algorithm has been used for latent class models by Haberman (1989), for generalized linear mixed models by Pan and Thompson (2003) and for generalized linear latent and mixed models by Rabe-Hesketh *et al.* (2002, 2004a). Pan and Thompson used analytical first and second order derivatives of the log-likelihood whereas Rabe-Hesketh *et al.* used numerical derivatives. McCulloch (1997) and Kuk and Cheng (1997) discuss Monte-Carlo Newton Raphson algorithms for generalized linear mixed models.

Fisher-scoring was used by Longford (1987) for linear mixed models. The BH^3 algorithm was used in latent variable modeling by Arminger and Küsters (1989), Skrondal (1996) and Hedeker and Gibbons (1994, 1996a), among others. Davidon-Fletcher-Powell (DFP) was introduced for factor models by Jöreskog (1967). Skaug (2002) used a quasi-Newton method with line search for generalized linear mixed models. Here first order derivatives were obtained by automatic differentiation, that is, the code for evaluating the derivatives was generated automatically by a computer program from the code to evaluate the log-likelihood.

6.5 Nonparametric maximum likelihood estimation

Estimation of models with discrete latent variables is straightforward using EM or gradient methods since the likelihood is a *finite mixture* so that no integration is involved. The discrete distribution is characterized by a finite set of locations e_c, $c = 1, \ldots, C$ and probabilities or masses π_c (at these locations). If the number of masses C of the discrete distribution is chosen to maximize the likelihood, the nonparametric maximum likelihood estimator can be achieved (e.g. Simar, 1976; Laird, 1978; Lindsay, 1983); see Section 4.4.2. Attempting to add an additional mass-point would then either result in one estimated probability approaching zero or two estimated locations nearly coinciding.

In this section we briefly describe methods for finding the number of masses of the nonparametric maximum likelihood estimator (NPMLE). A common approach is to start estimation with a large number of mass-points and omit points that either merge with other points or whose mass approaches zero during maximization of the likelihood (e.g. Butler and Louis, 1992). Another approach is to introduce mass-points one by one using the concept of a directional derivative (e.g. Simar, 1976; Jewell, 1982; Böhning, 1982; Lindsay, 1983; Rabe-Hesketh et al., 2003a), referred to as the Gateaux derivative by Heckman and Singer (1984).

Consider a model with a single latent variable with maximized log-likelihood $\ell(\widehat{\boldsymbol{\vartheta}}^C, \widehat{\boldsymbol{\pi}}^C, \widehat{\mathbf{e}}^C)$ for C masses. To determine whether this is the NPMLE solution, we consider changing the discrete mass-point distribution along the path $([1 - \lambda]\widehat{\boldsymbol{\pi}}^C, \lambda)'$ with locations $(\widehat{\mathbf{e}}^C, e^{C+1})'$, where $\lambda = 0$ corresponds to the current solution and $\lambda = 1$ places unit mass at a new location e^{C+1}. The directional derivative is then defined as

$$
\Delta(e^{C+1}) = \lim_{\lambda \to 0} \frac{\ell(\widehat{\boldsymbol{\vartheta}}^C, ([1 - \lambda]\widehat{\boldsymbol{\pi}}^C, \lambda)', (\widehat{\mathbf{e}}^C, e^{C+1})') - \ell(\widehat{\boldsymbol{\vartheta}}^C, \widehat{\boldsymbol{\pi}}^C, \widehat{\mathbf{e}}^C)}{\lambda}.
$$

(6.16)

According to the general mixture maximum likelihood theorem (Lindsay, 1983; Böhning, 1982), the NPMLE has been found if and only if $\Delta(e^{C+1}) \leq 0$ for all e^{C+1}.

Rabe-Hesketh et al. (2003a) suggested searching for a new location e^{C+1} over a fine grid spanning a wide range of values and to terminate the algorithm, if for a small value of λ the numerator of (6.16) is negative for all locations. This approach is similar to the algorithm proposed by Simar (1976), adapted by Heckman and Singer (1984), Follmann and Lambert (1989), among others. Algorithms for finding the NPMLE (both the number of masses and parameter estimates) are described in Lindsay (1995) and Böhning (2000).

The important merit of NPMLE is that we need not assume a parametric distribution for the latent variables, potentially making inferences more robust. However, for generalized linear models with covariate measurement error, simulations indicate that inference based on multivariate normal latent variables is fairly robust to misspecification (e.g. Thoresen and Laake, 2000). Although Schafer (2001) and Rabe-Hesketh et al. (2003a) found that estimates assuming the conventional (misspecified) model were biased, these estimates had a smaller root mean squared error than the unbiased NPML estimates. Little is known about the performance of NPMLE for models with a large number of latent variables. For models with categorical responses, boundary solutions where locations approach $\pm \infty$ can pose problems.

Nonparametric maximum likelihood estimation has been used for survival or duration models (e.g. Heckman and Singer, 1984; Holmås, 2002), item response models (e.g. de Leeuw and Verhelst, 1986; Lindsay *et al.*, 1991), generalized linear models with covariate measurement error (e.g. Roeder *et al.*, 1996; Aitkin and Rocci, 2002; Rabe-Hesketh *et al.*, 2003a), random coefficient models (e.g. Davies and Pickles, 1987; Aitkin, 1999a) and meta-analysis (e.g. Aitkin, 1999b). We use NPMLE in many applications in this book, for instance in Sections 9.5, 11.2, 11.3.3 and 14.2.

6.6 Restricted/Residual maximum likelihood (REML)

It is instructive to initially consider the simple linear regression model

$$y_i = \mathbf{x}_i'\boldsymbol{\beta} + \epsilon_i, \quad \epsilon_i \sim N(0, \theta), \quad i = 1, \ldots, N$$

where $\boldsymbol{\beta}$ contains P regression parameters. The maximum likelihood estimator of the residual variance is

$$\widehat{\theta} = \frac{1}{N} \sum_{i=1}^{N} (y_i - \mathbf{x}_i'\widehat{\boldsymbol{\beta}})^2.$$

It is well known that this estimator is downward biased with 'bias factor' $\frac{N-P}{N}$, i.e., $E(\widehat{\theta}) = \frac{N-P}{N}\theta$. $\widehat{\theta}$ would be unbiased if the regression parameters $\boldsymbol{\beta}$ were known, but is biased when based on $\widehat{\boldsymbol{\beta}}$ since $\mathbf{x}_i'\widehat{\boldsymbol{\beta}}$ 'fits the data more closely' than $\mathbf{x}_i'\boldsymbol{\beta}$. Thus, the bias-corrected estimator,

$$\widehat{\theta} = \frac{1}{N-P} \sum_{i=1}^{N} (y_i - \mathbf{x}_i'\widehat{\boldsymbol{\beta}})^2,$$

is typically used instead.

The same bias issue applies for latent variable models where estimates of variance parameters are expected to be biased downwards. For instance, for a two-level random intercept model Raudenbush and Bryk (2002) point out that the maximum likelihood estimator of the random intercept variance ψ is biased with approximate bias factor $\frac{J-P}{J}$.

To address this problem Patterson and Thompson (1971) suggested so-called restricted or residual maximum likelihood method (REML). Maximum likelihood is in this case not applied directly to the responses $\mathbf{y}$ but instead to linear functions or 'error contrasts' of the responses, say $\mathbf{A}\mathbf{y}$. Importantly, $\mathbf{A}$ is specified as orthogonal to $\mathbf{X}$ so $\mathbf{A}\mathbf{y}$ 'sweeps out' the fixed effects from the model. It follows that REML itself does not produce estimates of the fixed effects $\boldsymbol{\beta}$.

REML can alternatively be derived from a Bayesian perspective (see Section 6.11). Specifically, a flat prior is specified for $\boldsymbol{\beta}$, whereas the variance and covariance parameters are regarded as fixed. The latter parameters can be estimated using maximum marginal likelihood, after integrating out the $\boldsymbol{\beta}$ and latent variables. Empirical Bayes (see Section 7.3.1) is employed to 'score' the latent variables and estimate the $\boldsymbol{\beta}$ parameters (e.g. Harville, 1977; Dempster *et al.*, 1981; Laird and Ware, 1982).

Although developed for linear mixed models with purely continuous responses, approximate REML methods based on penalized quasi-likelihoods have also been suggested for generalized linear mixed models (e.g. Schall, 1991; Breslow and Clayton, 1993; McGilchrist, 1994; Stiratelli *et al.*, 1984). Longford (1993, p.236) suggested another approach, adding a penalty term to the marginal log-likelihood.

There does not seem to be a clear winner when contrasting the performance of REML and maximum likelihood (ML). The standard argument in favor of REML over ML is that unbiased estimators of variance and covariance parameters are produced. It should be noted, however, that REML is biased for unbalanced designs. Moreover, the bias of ML will be important only if there are few clusters compared to the number of fixed effects. In this case the utility of latent variable modeling itself may be questionable, so the performance of REML versus ML becomes a secondary issue. Furthermore, the mean squared error is often used as optimality criterion instead of bias. Interestingly, the mean squared error may be larger for REML (e.g. Corbeil and Searle, 1976), as was also indicated by simulations conducted by Busing (1993) and van der Leeden and Busing (1994). A disadvantage of using REML is that deviance testing is precluded for fixed parameters, since REML itself does not provide estimates of fixed effects. On the other hand, for the special case of balanced mixed ANOVA models, REML estimates of variances and covariances are identical to classical ANOVA moment estimators. It follows that the REML estimator in this particular case has minimal variance properties and does not rely on any normality assumption. Finally, it has been argued that REML is less sensitive to outliers than ML (Verbyla, 1993).

6.7 Limited information methods

In this section we consider models for conditionally (given the latent variables) multivariate normal latent responses $\mathbf{y}_j^*$ with constant cluster size, $n_j = n$, $j = 1, \ldots, J$. We will consider two-level models here, although higher-level models are accommodated if the number of level-1 units in the highest-level units is constant, as in multivariate growth curve models. We furthermore assume that the latent variables ζ are multivariate normal so that the marginal distribution (w.r.t the latent variables) of the latent responses is multivariate normal. In this case the reduced form parameters are the parameters characterizing the marginal mean and covariance structure; see Section 4.9.2. It follows from multivariate normality that the univariate and bivariate distributions (marginal w.r.t the other latent responses) are also normal.

In the limited information approach we first use the univariate and bivariate distributions to estimate 'empirical' or unrestricted versions of the reduced-form parameters. For instance, in a model without covariates this would be the unrestricted means and covariances (tetrachoric correlations in the dichotomous case) of the $\mathbf{y}_j^*$, which would generally not satisfy the model-implied constraints. We then estimate the structural parameters using a weighted least squares fitting function as in (6.1), minimizing the distance between the model-implied and estimated (unrestricted) reduced form parameters. For continuous responses where $\mathbf{y}_j = \mathbf{y}_j^*$, the univariate and bivariate distributions contain all information about the reduced form parameters. In contrast, for coarsened responses (such as dichotomous) this is no longer the case giving rise to the term 'limited information'.

In the present context, the idea of using univariate and bivariate information appears to be due to Christofferson (1975) and extended by Muthén in a series of papers (e.g. Muthén, 1978, 1981, 1982, 1983, 1984, 1988a, 1989bc). Since this limited information approach is little known outside psychometrics, we will provide a somewhat detailed sketch of it.

Consider the model introduced by Muthén (1983, 1984), which generalizes the structural equation model presented on page 78 to latent responses $\mathbf{y}^*$. The structural model was given in equation (3.33),

$$\boldsymbol{\eta}_j = \boldsymbol{\alpha} + \mathbf{B}\boldsymbol{\eta}_j + \boldsymbol{\Gamma}\mathbf{x}_j + \boldsymbol{\zeta}_j,$$

and the response model is

$$\mathbf{y}_j^* = \boldsymbol{\nu} + \boldsymbol{\Lambda}\boldsymbol{\eta}_j + \mathbf{K}\mathbf{x}_j + \boldsymbol{\epsilon}_j.$$

For continuous responses, the observed responses simply equal the latent responses. Dichotomous, ordinal and censored observed responses are related to the latent responses via threshold functions as described in Section 2.4. The reduced form becomes

$$\mathbf{y}_j^* = \boldsymbol{\nu} + \boldsymbol{\Lambda}(\mathbf{I} - \mathbf{B})^{-1}[\boldsymbol{\alpha} + \boldsymbol{\Gamma}\mathbf{x}_j + \boldsymbol{\zeta}_j] + \mathbf{K}\mathbf{x}_j + \boldsymbol{\epsilon}_j,$$

with expectation structure

$$E(\mathbf{y}_j^*|\mathbf{x}_j) = \underbrace{\boldsymbol{\nu} + \boldsymbol{\Lambda}(\mathbf{I} - \mathbf{B})^{-1}\boldsymbol{\alpha}}_{\boldsymbol{\Pi}_0} + \underbrace{[\boldsymbol{\Lambda}(\mathbf{I} - \mathbf{B})^{-1}\boldsymbol{\Gamma} + \mathbf{K}]}_{\boldsymbol{\Pi}_1}\mathbf{x}_j,$$

and covariance structure

$$\boldsymbol{\Omega} \equiv \mathrm{Cov}(\mathbf{y}_j^*|\mathbf{x}_j) = \boldsymbol{\Lambda}(\mathbf{I} - \mathbf{B})^{-1}\boldsymbol{\Psi}(\mathbf{I} - \mathbf{B})^{-1\prime}\boldsymbol{\Lambda}' + \boldsymbol{\Theta}.$$

For simplicity we assume from now on that the observed responses are dichotomous. In this case the diagonal elements of either $\boldsymbol{\Theta}$ or $\boldsymbol{\Omega}^*$ are typically fixed to one for identification. Following Muthén, we use the second parametrization in this section so that the covariance structure becomes

$$\boldsymbol{\Omega}^* = \mathrm{diag}(\boldsymbol{\Omega})^{-\frac{1}{2}}\,\boldsymbol{\Omega}\,\mathrm{diag}(\boldsymbol{\Omega})^{-\frac{1}{2}}.$$

We furthermore assume that the thresholds are set to zero for identification. In order to simplify the subsequent development we define the augmented covariate vector $\mathbf{z}_j = (1, \mathbf{x}_j')'$, including a one for the intercepts in addition to the covariates. The expectation structure can then be written as

$$E(\mathbf{y}_j^*|\mathbf{z}_j) = \boldsymbol{\Pi}\mathbf{z}_j,$$

where $\boldsymbol{\Pi} = (\boldsymbol{\Pi}_0, \boldsymbol{\Pi}_1)$ is the reduced form regression matrix including intercepts. Following Muthén, we specify multinormal distributions for ζ_j and ϵ_j, and obtain

$$\mathbf{y}_j^* | \mathbf{z}_j \sim N_n(\mathbf{\Pi} \mathbf{z}_j, \mathbf{\Omega}^*).$$

We can write the *univariate* distribution of a variable or response i, y_{ij}^*, $i = 1, 2, \ldots, n$, as

$$y_{ij}^* | \mathbf{z}_j \sim N(\boldsymbol{\pi}_i \mathbf{z}_j, 1),$$

where $\boldsymbol{\pi}_i$ is the ith row of $\mathbf{\Pi}$. The *bivariate* distribution of two responses y_{ij}^* and $y_{i'j}^*$, $i, i' = 1, 2, \ldots, n$, $i \neq i'$, can be written as

$$y_{ij}^*, y_{i'j}^* | \mathbf{z}_j \sim N_2 \left(\left[\begin{array}{c} \boldsymbol{\pi}_i \mathbf{z}_j \\ \boldsymbol{\pi}_{i'} \mathbf{z}_j \end{array} \right], \left[\begin{array}{cc} 1 & \\ \omega_{ii'}^* & 1 \end{array} \right] \right),$$

where $\omega_{ii'}^*$ is the ii'th residual correlation element in $\mathbf{\Omega}^*$.

We now spell out the three-stage limited information approach developed by Muthén. Note that $\mathbf{\Pi}$ and $\mathbf{\Omega}^*$ will stand for unrestricted versions of the reduced form parameters.

Stage 1: The first estimation stage produces limited information maximum likelihood estimates for the reduced form intercepts and regression parameters $\mathbf{\Pi}$ from univariate information. A univariate probit regression is specified for each i,

$$\Pr(y_{ij} = 1 | \mathbf{z}_{ij}) = \Phi_1(\boldsymbol{\pi}_i \mathbf{z}_j),$$

where $\Phi_1(\cdot)$ is the standard normal cumulative distribution function. The log-likelihood contribution for cluster j and variable i becomes

$$\ell_i^j(\boldsymbol{\pi}_i) \equiv y_{ij} \ln \Phi_1(\boldsymbol{\pi}_i \mathbf{z}_j) + (1 - y_{ij}) \ln[1 - \Phi_1(\boldsymbol{\pi}_i \mathbf{z}_j)].$$

For each i, the univariate log-likelihood $\sum_{j=1}^{J} \ell_i^j(\boldsymbol{\pi}_i)$ is then maximized, producing consistent estimates $\widehat{\boldsymbol{\pi}}_i$. The gradients for cluster j are assembled in

$$\mathbf{g}_1^j = \left[\frac{\partial \ell_1^j}{\partial \boldsymbol{\pi}_1}, \frac{\partial \ell_2^j}{\partial \boldsymbol{\pi}_2}, \ldots, \frac{\partial \ell_n^j}{\partial \boldsymbol{\pi}_n} \right]'.$$

Stage 2a: Conditional on the first stage reduced form intercept and regression parameter estimates $\widehat{\mathbf{\Pi}}$, the reduced form residual correlations in $\mathbf{\Omega}^*$ are estimated by limited information 'pseudo' maximum likelihood based on bivariate information for each pair of responses ii' $i > i'$,

$$\Pr(y_{ij} = 1, y_{i'j} = 1 | \mathbf{z}_{ij}, \mathbf{z}_{i'j}) = \Phi_2(\boldsymbol{\pi}_i \mathbf{z}_j, \boldsymbol{\pi}_{i'} \mathbf{z}_j, \omega_{ii'}^*),$$
$$\Pr(y_{ij} = 0, y_{i'j} = 1 | \mathbf{z}_{ij}, \mathbf{z}_{i'j}) = \Phi_2(-\boldsymbol{\pi}_i \mathbf{z}_j, \boldsymbol{\pi}_{i'} \mathbf{z}_j, -\omega_{ii'}^*),$$
$$\Pr(y_{ij} = 1, y_{i'j} = 0 | \mathbf{z}_{ij}, \mathbf{z}_{i'j}) = \Phi_2(\boldsymbol{\pi}_i \mathbf{z}_j, -\boldsymbol{\pi}_{i'} \mathbf{z}_j, -\omega_{ii'}^*),$$
$$\Pr(y_{ij} = 0, y_{i'j} = 0 | \mathbf{z}_{ij}, \mathbf{z}_{i'j}) = \Phi_2(-\boldsymbol{\pi}_i \mathbf{z}_j, -\boldsymbol{\pi}_{i'} \mathbf{z}_j, \omega_{ii'}^*),$$

where $\Phi_2(\mu_1, \mu_2, \rho)$ is the bivariate standard normal cumulative distribution function with means μ_1 and μ_2 and correlation ρ and $\omega_{ii'}^*$ is the residual correlation. The corresponding bivariate log-likelihood contribution, *given* the estimates $\widehat{\pi}_i$ and $\widehat{\pi}_{i'}$ from stage 1, becomes

$$
\begin{aligned}
\ell_{ii'}^j(\omega_{ii'}^*|\widehat{\pi}_i,\widehat{\pi}_{i'}) \;=\; & y_{ij}y_{i'j}\ln\Phi_2(\widehat{\pi}_i\mathbf{z}_j,\widehat{\pi}_{i'}\mathbf{z}_j,\omega_{ii'}^*) \;+ \\
& y_{ij}(1-y_{i'j})\ln\Phi_2(\widehat{\pi}_i\mathbf{z}_j,-\widehat{\pi}_{i'}\mathbf{z}_j,-\omega_{ii'}^*) \;+ \\
& (1-y_{ij})y_{i'j}\ln\Phi_2(-\widehat{\pi}_i\mathbf{z}_j,\widehat{\pi}_{i'}\mathbf{z}_j,-\omega_{ii'}^*) \;+ \\
& (1-y_{ij})(1-y_{i'j})\ln\Phi_2(-\widehat{\pi}_i\mathbf{z}_j,-\widehat{\pi}_{i'}\mathbf{z}_j,\omega_{ii'}^*).
\end{aligned}
$$

For each pair ii', the 'pseudo log-likelihood' (in the sense of Parke, 1986) $\sum_{j=1}^J \ell_{ii'}^j(\omega_{ii'}^*|\widehat{\pi}_i,\widehat{\pi}_{i'})$ is then maximized, producing the consistent estimate $\widehat{\omega}_{ii'}^*$. The gradients for cluster j, evaluated at the maximum pseudolikelihood, are assembled in

$$
\mathbf{g}_2^j = \left[\frac{\partial\ell_{21}^j}{\partial\omega_{21}^*}, \frac{\partial\ell_{31}^j}{\partial\omega_{31}^*}, \frac{\partial\ell_{32}^j}{\partial\omega_{32}^*}, \ldots, \frac{\partial\ell_{n,n-1}^j}{\partial\omega_{n,n-1}^*}\right]'.
$$

For later use, we also define

$$
\mathbf{g}^j = [\mathbf{g}_1^{j'}, \mathbf{g}_2^{j'}]',
$$

$$
\mathbf{g} = \sum_{j=1}^N \mathbf{g}^j,
$$

and put the gradients of the bivariate log-likelihoods with respect to the reduced form intercepts and regression parameters, evaluated at the maximum pseudo-likelihood, in the vector

$$
\mathbf{g}_{21}^j = \left[\frac{\partial\ell_{21}^j}{\partial\pi_1}, \frac{\partial\ell_{21}^j}{\partial\pi_2}, \frac{\partial\ell_{31}^j}{\partial\pi_1}, \frac{\partial\ell_{31}^j}{\partial\pi_3}, \ldots, \frac{\partial\ell_{n,n-1}^j}{\partial\pi_{n-1}}, \frac{\partial\ell_{n,n-1}^j}{\partial\pi_n}\right]'.
$$

Stage 2b: Let the nonredundant elements of the reduced form parameters $\mathbf{\Pi}$ and $\mathbf{\Omega}^*$ be assembled in the vector $\boldsymbol{\sigma}$. The estimated asymptotic covariance matrix of the estimated reduced form parameters, $\widetilde{\mathrm{Cov}}(\widehat{\boldsymbol{\sigma}})$, is then derived based on marginal information from stages 1 and 2a (e.g. Lee, 1982; Muthén, 1984).

Expanding the gradient $\mathbf{g}(\widehat{\boldsymbol{\sigma}})$ of the maximum likelihood estimates around the true value $\bar{\boldsymbol{\sigma}}$ using the mean value theorem gives

$$
\mathbf{0} \;=\; \mathbf{g}(\widehat{\boldsymbol{\sigma}}) \;=\; \mathbf{g}(\bar{\boldsymbol{\sigma}}) + \frac{\partial\mathbf{g}(\boldsymbol{\sigma}^*)}{\partial\boldsymbol{\sigma}}(\widehat{\boldsymbol{\sigma}}-\bar{\boldsymbol{\sigma}}),
$$

where $\boldsymbol{\sigma}^*$ is some point between $\widehat{\boldsymbol{\sigma}}$ and $\bar{\boldsymbol{\sigma}}$. Multiplying by $J^{\frac{1}{2}}$, we obtain

$$
J^{\frac{1}{2}}(\widehat{\boldsymbol{\sigma}}-\bar{\boldsymbol{\sigma}}) \;=\; \left(-J^{-1}\frac{\partial\mathbf{g}(\boldsymbol{\sigma}^*)}{\partial\boldsymbol{\sigma}}\right)^{-1} \times J^{-\frac{1}{2}}\mathbf{g}(\bar{\boldsymbol{\sigma}}).
$$

Regarding the first term, using a law of large numbers gives

$$
J^{-1}\frac{\partial\mathbf{g}(\boldsymbol{\sigma}^*)}{\partial\boldsymbol{\sigma}} = J^{-1}\sum_{j=1}^J\frac{\partial\mathbf{g}^j(\boldsymbol{\sigma}^*)}{\partial\boldsymbol{\sigma}} \xrightarrow{\ \mathrm{P}\ } \lim_{J\to\infty}J^{-1}\sum_{j=1}^J\mathrm{E}\left(\frac{\partial\mathbf{g}^j(\bar{\boldsymbol{\sigma}})}{\partial\boldsymbol{\sigma}}\right) \equiv \mathbf{A},
$$

where $\xrightarrow{\text{P}}$ denotes convergence in probability. Note that $\mathbf{A}$ is partitioned as

$$\mathbf{A} = \begin{bmatrix} \mathbf{A}_{11} & \mathbf{0} \\ \mathbf{A}_{21} & \mathbf{A}_{22} \end{bmatrix},$$

where $\mathbf{A}_{12} = \mathbf{0}$, since the derivatives of the gradients of the univariate likelihoods with respect to the correlations are zero because no correlations are involved there.

For the second term, using a multivariate central limit theorem gives

$$J^{-\frac{1}{2}}\mathbf{g}(\bar{\sigma}) = J^{-\frac{1}{2}}\sum_{j=1}^{J}\mathbf{g}^{j}(\bar{\sigma}) \quad \xrightarrow{\text{D}} \quad N(\mathbf{0}, \mathbf{V}),$$

where $\xrightarrow{\text{D}}$ denotes convergence in distribution and

$$\mathbf{V} \equiv \lim_{J\to\infty} \text{Cov}(J^{-\frac{1}{2}}\mathbf{g}(\bar{\sigma})) = \lim_{J\to\infty} J^{-1}\sum_{j=1}^{J} \text{E}\left(\mathbf{g}^{j}(\bar{\sigma})\mathbf{g}^{j}(\bar{\sigma})'\right)$$

since $\sum_{j=1}^{J}\text{E}(\mathbf{g}^{j}(\bar{\sigma})) = \mathbf{0}$, the expected scores are zero at the maximum. It follows that

$$J^{\frac{1}{2}}(\hat{\sigma} - \sigma) \quad \xrightarrow{\text{D}} \quad N(\mathbf{0}, \mathbf{A}^{-1}\mathbf{V}\mathbf{A}^{-1\prime}).$$

Under correct model specification, it follows from the 'information matrix equality' that

$$\lim_{J\to\infty} J^{-1}\sum_{j=1}^{J}\text{E}\left(\frac{\partial\mathbf{g}^{j}(\sigma)}{\partial\sigma}\right) = -\lim_{J\to\infty} J^{-1}\sum_{j=1}^{J}\text{E}\left(\mathbf{g}^{j}(\sigma)\mathbf{g}^{j}(\sigma)'\right).$$

Using this result, we can estimate the matrices in $\mathbf{A}$ using the gradients obtained in the previous stages:

$$\widehat{\mathbf{A}}_{11} = -J^{-1}\sum_{j=1}^{J}\mathbf{g}_{1}^{j}\mathbf{g}_{1}^{j\prime},$$

$$\widehat{\mathbf{A}}_{22} = -J^{-1}\sum_{j=1}^{J}\mathbf{g}_{2}^{j}\mathbf{g}_{2}^{j\prime},$$

and

$$\widehat{\mathbf{A}}_{21} = -J^{-1}\sum_{j=1}^{J}\mathbf{g}_{21}^{j}\mathbf{g}_{1}^{j\prime}.$$

The covariance matrix $\mathbf{V}$ is estimated by the empirical covariance of the gradients

$$\widehat{\mathbf{V}} = J^{-1}\sum_{j=1}^{J}\mathbf{g}^{j}\mathbf{g}^{j\prime}.$$

The asymptotic covariance matrix of the estimated reduced form parameters is finally estimated as

$$\widehat{\mathbf{W}} \equiv \widehat{\mathrm{Cov}}(\widehat{\boldsymbol{\sigma}}) = \widehat{\mathbf{A}}^{-1}\widehat{\mathbf{V}}\widehat{\mathbf{A}}^{-1\prime}.$$

Stage 3: The reduced form parameters are regarded as functions $\boldsymbol{\sigma}(\boldsymbol{\vartheta})$ of the fundamental parameters $\boldsymbol{\vartheta}$. A consistent estimator $\widehat{\boldsymbol{\vartheta}}$ is obtained by fitting $\boldsymbol{\sigma}(\boldsymbol{\vartheta})$ to $\widehat{\boldsymbol{\sigma}}$, minimizing the weighted least squares (WLS) criterion

$$\mathrm{F}(\boldsymbol{\vartheta}) = \frac{1}{2}\left[\boldsymbol{\sigma}(\boldsymbol{\vartheta})-\widehat{\boldsymbol{\sigma}}\right]'\widehat{\mathbf{W}}^{-1}\left[\boldsymbol{\sigma}(\boldsymbol{\vartheta})-\widehat{\boldsymbol{\sigma}}\right]. \tag{6.17}$$

Defining

$$\boldsymbol{\Delta} \equiv \frac{\partial\boldsymbol{\sigma}(\widehat{\boldsymbol{\vartheta}})}{\partial\boldsymbol{\vartheta}},$$

a model-based estimator of the asymptotic covariance matrix of $\widehat{\boldsymbol{\vartheta}}$ is

$$\widehat{\mathrm{Cov}}(\widehat{\boldsymbol{\vartheta}}) = J^{-1}(\boldsymbol{\Delta}'\widehat{\mathbf{W}}^{-1}\boldsymbol{\Delta})^{-1}.$$

This estimator is consistent if the model is correctly specified. A large sample chi-square distributed test statistic of absolute fit against the estimated reduced form parameters is obtained as $2J\,F(\widehat{\boldsymbol{\vartheta}})$.

We refer to Küsters (1987) and Muthén and Satorra (1996) for technical details. Olsson (1979) provides details regarding the first two stages for polychoric correlations, Olsson *et al.* (1982) for polyserial correlations and Muthén (1989c) for tobit correlations. An overview of different latent response correlations, denoted 'polytobiserial' by Küsters (1987), was presented in Table 4.2.

Importantly, an analogy to robust normal theory estimation (see Satorra (1990) for a review) was suggested by Muthén (1993) for dichotomous responses and Muthén *et al.* (1997) for the general Muthén model. The 'robust' asymptotic covariance matrix of $\widehat{\boldsymbol{\vartheta}}$ is obtained as

$$\widehat{\mathrm{Cov}}(\widehat{\boldsymbol{\vartheta}}) = J^{-1}(\boldsymbol{\Delta}'\widetilde{\mathbf{W}}^{-1}\boldsymbol{\Delta})^{-1}\boldsymbol{\Delta}'\widetilde{\mathbf{W}}^{-1}\widehat{\mathbf{W}}\widetilde{\mathbf{W}}^{-1}\boldsymbol{\Delta}(\boldsymbol{\Delta}'\widetilde{\mathbf{W}}^{-1}\boldsymbol{\Delta})^{-1}.$$

Muthén (1993) suggested simply using $\widetilde{\mathbf{W}}=\mathbf{I}$ in the above expression as well as in the WLS criterion (6.17), effectively simplifying the latter to unweighted least squares. Muthén *et al.* (1997) instead specified $\widetilde{\mathbf{W}}$ as a diagonal matrix with estimated variances of $\widehat{\boldsymbol{\sigma}}$ as elements. The resulting $\widetilde{\mathbf{W}}$ is then used in the 'robust' covariance matrix and the fit criterion now becomes diagonally weighted least squares. A beneficial feature of these approaches is that $\widehat{\mathbf{W}}$ need not be inverted, which can be problematic for 'large' models and/or small samples and/or highly skewed dichotomous responses. Satorra and Bentler (e.g. Satorra and Bentler, 1994; Satorra, 1992) furthermore propose 'robust' tests for absolute fit. Muthén *et al.* (1997) also discuss the connections to GEE, a methodology discussed in Section 6.9.

The limited information methodology developed by Muthén and others has many merits. It handles a general model framework, although only models with multinormal latent responses. Thus, models with for instance a logit link or Poisson distribution cannot be accommodated. The methodology is very computationally efficient, reducing a possibly high dimensional integration problem to a series of univariate and bivariate integrations, which is especially valuable for models with many latent variables. The approach also appears to be remarkably efficient, producing estimates that are very close to maximum likelihood. An important limitation is that missing data can only be handled by using multiple-group models so that only a few missing data patterns can be handled in practice. Monte Carlo experiments (e.g. Muthén and Kaplan, 1992) have shown that the method can perform poorly for complex models if the sample size is small.

6.8 Maximum quasi-likelihood

We first discuss the iteratively reweighted least squares (IRLS) algorithm for generalized linear models and the iterative generalized least squares (IGLS) algorithm for multivariate linear models. This is not only of interest in itself but also as a precursor for quasi-likelihood, marginal and penalized quasi-likelihood (MQL and PQL) as well as generalized estimating equations (GEE).

6.8.1 Iteratively reweighted least squares

Consider a generalized linear model (see Section 2.2) with log-likelihood

$$\ell = \sum_i [y_i\theta_i - b(\theta_i)]/\phi + c(y_i, \phi).$$

The likelihood equations in this case are

$$\frac{\partial \ell}{\partial \beta_p} = \sum_i \left[y_i \frac{\partial \theta_i}{\partial \mu_i} \frac{\partial \mu_i}{\partial \beta_p} - \frac{\partial b(\theta_i)}{\partial \theta_i} \frac{\partial \theta_i}{\partial \mu_i} \frac{\partial \mu_i}{\partial \beta_p} \right] / \phi = 0,$$

for $p = 1, \ldots, P$. For the cumulant function $b(\cdot)$,

$$\frac{\partial b(\theta_i)}{\partial \theta_i} = \mu_i,$$

$$\frac{\partial^2 b(\theta_i)}{\partial \theta_i^2} = \frac{\partial \mu_i}{\partial \theta_i} = \left[\frac{\partial \theta_i}{\partial \mu_i} \right]^{-1} = V(\mu_i).$$

Substituting these expressions into the likelihood equations,

$$\frac{\partial \ell}{\partial \beta_p} = \sum_i \frac{\partial \mu_i}{\partial \beta_p} [\phi V(\mu_i)]^{-1} [y_i - \mu_i] = 0. \qquad (6.18)$$

This can be further simplified using the relations $\mu_i = g^{-1}(\nu_i)$ and $\nu_i = \mathbf{x}_i'\boldsymbol{\beta}$,

$$\frac{\partial \mu_i}{\partial \beta_p} = \frac{\partial \mu_i}{\partial \nu_i} x_{pi} = \frac{\partial g^{-1}(\nu_i)}{\partial \nu_i} x_{pi} = \frac{x_{pi}}{g'(\mu_i)},$$

where $g'(\mu_i)$ is the first derivative of $g(\cdot)$ evaluated at μ_i.

Substituting these expressions into the likelihood equations,

$$\frac{\partial \ell}{\partial \beta_p} = \sum_i \frac{x_{pi}}{g'(\mu_i)\phi V(\mu_i)}[y_i - \mu_i] = 0. \tag{6.19}$$

For a linear model (identity link and normal distribution) with possibly heteroscedastic residual variance $\phi_i = \sigma_i^2$, $V(\mu_i) = 1$, $g'(\mu_i) = 1$, so that the likelihood equations are linear in β,

$$\frac{\partial \ell}{\partial \beta_p} = \sum_i \frac{x_{pi}}{\sigma_i^2}[y_i - \mathbf{x}_i'\beta] = 0, \tag{6.20}$$

and can be solved by *weighted least squares* with weights $1/\sigma_i^2$,

$$\hat{\beta} = (\mathbf{X}'\mathbf{V}^{-1}\mathbf{X})^{-1}\mathbf{X}'\mathbf{V}^{-1}\mathbf{y}, \tag{6.21}$$

where $\mathbf{V}$ is a diagonal matrix with diagonal elements equal to σ_i^2.

Iteratively reweighted least squares (IRLS) is a procedure which linearizes the likelihood equations in each iteration so that estimates for the next iteration can be found by weighted least squares. Let the estimates from the 'current' iteration be denoted β^k and the corresponding mean μ_i^k. A *working variate* z_i^k is defined as

$$z_i^k = g(\mu_i^k) + [y_i - \mu_i^k]g'(\mu_i^k)$$

so that

$$y_i - \mu_i^k = [z_i^k - \mathbf{x}_i'\beta^k]/g'(\mu_i^k).$$

We now show that the estimates can be updated by weighted least squares (as if the model were linear) with weights given by $1/\sigma_i^2 = [g'(\mu_i^k)^2\phi V(\mu_i^k)]^{-1}$. Substituting these weights into the weighted least squares equations (6.20),

$$0 = \sum_i \frac{x_{pi}}{g'(\mu_i^k)^2\phi V(\mu_i^k)}[z_i - \mathbf{x}_i'\beta] = \sum_i \frac{x_{pi}}{g'(\mu_i^k)\phi V(\mu_i^k)}\frac{[z_i - \mathbf{x}_i'\beta]}{g'(\mu_i^k)}$$

$$= \sum_i \frac{x_{pi}}{g'(\mu_i^k)\phi V(\mu_i^k)}[y_i - \mu_i],$$

we obtain the original likelihood equations (6.19), except that the denominators of the first terms (the 'weights') are held fixed at the estimates from the previous iteration k. Solving these equations by weighted least squares gives estimates β^{k+1}, leading to new weights, then new estimates using 'reweighted' least squares and so on, iterating until convergence. Note that for generalized linear models, the iteratively reweighted least squares algorithm is identical to Fisher scoring.

Another way of conceptualizing the algorithm is by approximating the generalized linear model by a linear model using a first order Taylor series expansion. Let $h(\nu_i) \equiv g^{-1}(\nu_i) = \mu_i$ and $h'(\nu_i)$ be the first derivative evaluated at ν_i. In the kth iteration, y_i is approximated as

$$y_i = h(\nu_i^k) + \mathbf{x}_i'(\beta - \beta^k)h'(\nu_i^k) + \epsilon,$$

where $\text{Var}(\epsilon) = \phi V(\mu_i)$. Rearranging terms,

$$y_i - h(\nu_i^k) + \mathbf{x}_i'\boldsymbol{\beta}^k h'(\nu_i^k) = \mathbf{x}_i'\boldsymbol{\beta} h'(\nu_i^k) + \epsilon.$$

Multiplying by $1/h'(\nu_i^k) = g'(\mu_i^k)$, we obtain

$$[y_i - \mu_i^k]g'(\mu_i^k) + \mathbf{x}_i'\boldsymbol{\beta}^k = z_i^k = \mathbf{x}_i'\boldsymbol{\beta} + g'(\mu_i^k)\epsilon,$$

a linear model for z_i^k with mean $\mathbf{x}_i'\boldsymbol{\beta}$ and variance $g'(\mu^k)^2 \phi V(\mu)$.

Note that the details of the algorithm depend only on the link and variance functions of the generalized linear model. If we wish to specify an arbitrary link and variance function, we can employ the same algorithm to estimate the parameters even if the specification does not correspond to any statistical model. This idea of estimating parameters without specifying a model is known as *quasi-likelihood* (Wedderburn, 1974) and the corresponding equations as quasi-score equations or *estimating equations*. McCullagh (1983) showed that quasi-likelihood estimators have similar properties to maximum likelihood estimators such as consistency and asymptotic normality with covariance matrix given by the same formula as for maximum likelihood.

6.8.2 Iterative generalized least squares

In multivariate linear models with known residual covariance matrix $\mathbf{V}_{(D)}$, the parameters can be estimated by generalized least squares (GLS) where the diagonal matrix in (6.21) is replaced by the (nondiagonal) covariance matrix. Since the residual covariance matrix is generally not known, we must use iterative methods such as iterative generalized least squares (IGLS). Writing a multilevel linear mixed model for the response vector of the entire sample as

$$\mathbf{y} = \mathbf{X}\boldsymbol{\beta} + \boldsymbol{\Lambda}_{(D)}\boldsymbol{\zeta}_{(D)} + \boldsymbol{\epsilon},$$

and letting $\mathbf{V}_{(D)}^k$ be the 'current' estimate of the covariance matrix of the total residual $\boldsymbol{\xi} = \boldsymbol{\Lambda}_{(D)}\boldsymbol{\zeta}_{(D)} + \boldsymbol{\epsilon}$, the regression parameters can be updated using GLS

$$\boldsymbol{\beta}^{k+1} = (\mathbf{X}'(\mathbf{V}_{(D)}^k)^{-1}\mathbf{X})^{-1}\mathbf{X}(\mathbf{V}_{(D)}^k)^{-1}\mathbf{y}. \tag{6.22}$$

Using these updated estimates, the variance parameters $\boldsymbol{\Psi}^{(l)}$ are estimated from the residuals $\mathbf{r}^{k+1} = \mathbf{y} - \mathbf{X}'\boldsymbol{\beta}^{k+1}$, giving a new estimate of the covariance matrix $\mathbf{V}_{(D)}^{k+1}$. Specifically, the matrix of cross-products $\mathbf{r}^{k+1}\mathbf{r}^{k+1\prime}$ is formed and its expectation equated to $\mathbf{V}_{(D)}^{k+1}$. The expectation of the vectorized matrix of cross-products can be written as a linear regression with variance parameters as coefficients. For instance, for a two-level random intercept model (omitting the superscripts),

$$
E[\text{vec}(\mathbf{rr}')] = E
\begin{pmatrix}
r_{11}^2 \\
r_{21}r_{11} \\
r_{12}^2 \\
r_{12}r_{11} \\
r_{12}r_{21} \\
\vdots \\
r_{n_J J}^2
\end{pmatrix}
=
\begin{pmatrix}
\psi + \theta \\
\psi \\
\psi + \theta \\
0 \\
0 \\
\vdots \\
\psi + \theta
\end{pmatrix}
= \psi
\begin{pmatrix}
1 \\
1 \\
1 \\
0 \\
0 \\
\vdots \\
1
\end{pmatrix}
+ \theta
\begin{pmatrix}
1 \\
0 \\
1 \\
0 \\
0 \\
\vdots \\
1
\end{pmatrix}.
$$

The variance and covariance parameters are then estimated by generalized least squares where the covariance matrix of $\text{vec}(\mathbf{rr}')$ is derived from the previous estimate $\mathbf{V}_{(D)}^k$.

The IGLS algorithm hence iterates between updating the parameters of the fixed and random parts of the model. The resulting estimates are maximum likelihood estimates under normality of $\boldsymbol{\zeta}_{(D)}$ and $\boldsymbol{\epsilon}$ (Goldstein, 1986). Goldstein (1986) shows how inversion of the very high-dimensional covariance matrix $\mathbf{V}_{(D)}^k$ can be simplified, exploiting its block diagonal structure.

After convergence, the standard errors of the estimated regression parameters are estimated from the last GLS step treating the covariance matrix $V_{(D)}$ as known. These standard errors are generally correct since the estimates of the fixed part are uncorrelated with the estimates of the random part. An important exception is the situation where responses are missing at random with missingness depending on observed responses, not just covariates. Consider for instance a linear random intercept model for longitudinal data without covariates. If the probability of dropout increases with the magnitude of the observed response prior to dropout, a larger random intercept variance (i.e. a higher intraclass correlation) would imply a larger fixed intercept (since the imputed values for those who dropped out would be higher), making the two estimates positively correlated. See also Verbeke and Molenberghs (2000, Chapter 21).

For details of the IGLS algorithm we refer to Chapter 2 and Appendix 2.1 in Goldstein (2003). A slight modification of IGLS to restricted iterative generalized least squares (RIGLS) leads to restricted maximum likelihood (REML) estimates (Goldstein, 1989). Yang *et al.* (1999) propose an extension of the IGLS algorithm for multilevel structural equation models (see also Rabe-Hesketh *et al.*, 2001d).

6.8.3 Marginal and penalized quasi-likelihood

Marginal quasi-likelihood (MQL) and penalized quasi-likelihood (PQL) have been derived in a number of ways (see Section 6.3.1 and McCulloch and Searle, 2001). Here we give a summary based on the description in Goldstein (2003).

MQL and PQL are based on approximating generalized linear mixed models by linear mixed models so that the IGLS algorithm can be applied (which no longer corresponds to maximum likelihood). This linearization method is analogous to iteratively reweighted least squares described in Section 6.8.1.

In generalized linear mixed models the conditional expectation of the response is (retaining only the i subscript for level 1 units) $\mu_i = h(\nu_i)$, where $h(\cdot)$ is the inverse link function. The model for y_i is linearized by expanding $h(\nu_i)$ as a first order Taylor series around a known 'current' value of the linear predictor from iteration k,

$$\nu_i^k = \mathbf{x}_i'\boldsymbol{\beta}^k + \sum_{l=1}^{L} \mathbf{z}_i^{(l)\prime}\boldsymbol{\zeta}^{(l)k}, \tag{6.23}$$

giving

$$y_i \approx h(\nu_i^k) + \mathbf{x}_i'[\boldsymbol{\beta} - \boldsymbol{\beta}^k]h'(\nu_i^k) + \sum_l \mathbf{z}_i^{(l)}[\boldsymbol{\zeta}^{(l)} - \boldsymbol{\zeta}^{(l)k}]h'(\nu_i^k) + \epsilon_i. \tag{6.24}$$

Here ϵ_i is a heteroscedastic error term with variance $\phi V(\mu_i)$ corresponding to the chosen distribution family. Note that this expression is linear in the unknown parameters $\boldsymbol{\beta}$. The sum of the terms involving known current values $\boldsymbol{\beta}^k$ and $\boldsymbol{\zeta}^{(l)k}$ is treated as an offset o_i,

$$o_i = h(\nu_i^k) - h'(\nu_i^k)\mathbf{x}_i'\boldsymbol{\beta}^k - \sum_l h'(\nu_i^k)\mathbf{z}_i^{(l)}\boldsymbol{\zeta}^{(l)k},$$

and the terms involving latent variables $\boldsymbol{\zeta}^{(l)}$ contribute to the total residual ξ_i,

$$\xi_i = \sum_l h'(\nu_i^k)\mathbf{z}_i^{(l)}\boldsymbol{\zeta}^{(l)} + \epsilon_i,$$

giving

$$y_i = o_i + h'(\nu_i^k)\mathbf{x}'(\boldsymbol{\beta}) + \xi_i.$$

Multiplying $\mathbf{x}_i$ by $h'(\nu_i^k)$, we can therefore obtain $\boldsymbol{\beta}^{k+1}$ using generalized least squares as in (6.22) where $V_{(D)}^k$ is the covariance matrix of the total residuals for all units for the current estimates.

There are several variants of this algorithm (Goldstein, 1991, 2003; Longford, 1993, 1994). In marginal quasi-likelihood (MQL), $\boldsymbol{\zeta}^{(l)k}$ is set to zero in (6.23) and (6.24). In penalized quasi-likelihood (PQL), the expansion is improved by setting the latent variables equal to the posterior modes based on the linearized model, $\boldsymbol{\zeta}^{(l)k} = \widetilde{\boldsymbol{\zeta}}^{(l)k}$. The difference between MQL and PQL is hence in the offset used. Since MQL sets the latent variables to zero, the fixed effects estimates are essentially marginal effects. These are attenuated relative to the required conditional effects as discussed in Section 4.8.1.

The estimator for the latent variables is more easily expressed in terms of the 'working variate'

$$z_i = (y_i - o_i)/h'(\nu_i),$$

so that the linear approximation can be written using the GF formulation (see Section 4.2.2) as

$$\mathbf{z}_{(L)} = \mathbf{X}_{(L)}\boldsymbol{\beta} + \boldsymbol{\Lambda}_{(L)}\boldsymbol{\zeta}_{(L)} + g'(\mu_i)\boldsymbol{\epsilon}$$

and the latent variables are updated using

$$\zeta_{(L)}^{k+1} = \Psi_{(L)}\Lambda_{(L)}'\mathbf{V}_{(L)}^{-1}(\mathbf{z}_{(L)} - \mathbf{X}_{(L)}\beta^k),$$

where $\mathbf{V}_{(L)}$ is a block of $\mathbf{V}_{(D)}$ for a top-level unit. The parameters of the random part are estimated from the residuals $r_i = y_i - h(\nu_i)$ as in Section 6.8.2. The algorithm therefore iterates between updating β for given $\zeta_{(L)}^k$ and Ψ^k and updating Ψ and $\zeta_{(L)}$ given current values of the other parameters.

The algorithms have been improved considerably by using a second order Taylor expansion in the latent variables (Goldstein and Rasbash, 1996). In the case of PQL, this improves both the offset and the variance of the total residual. In the case of MQL, the offset is not affected since the random part is set to zero. For more details see Appendix 4.1 in Goldstein (2003).

The PQL approach has been used for generalized linear mixed models by Goldstein (1991), Schall (1991), Breslow and Clayton (1993), Longford (1993), Wolfinger and O'Connell (1993), Engel and Keen (1994) and McGilchrist (1994) among others. The algorithm is computationally very efficient since numerical integration is avoided. Furthermore, the approach can be used for models with crossed random effects (e.g. Breslow and Clayton, 1993; and, in linear mixed models, Goldstein, 1987), and (spatially or temporally) autocorrelated random effects (e.g. Breslow and Clayton, 1993; Langford et al., 1999), as well as multiple membership models (e.g. Hill and Goldstein, 1998; Rasbash and Browne, 2001); see also Section 3.2.6. See Section 6.11.5 for an alternative approach to the analysis of models with crossed random effects. However, PQL has not been used for models with factor loadings or structural equations.

These methods work well when the conditional distribution of the responses given the random effects is close to normal, for example with a Poisson distribution if the mean is 5 or greater (Breslow, 2003), or 7 or greater (McCulloch and Searle, 2001) or if the responses are proportions with large binomial denominators. The methods also work well if the conditional joint distributions of the responses belonging to each cluster are nearly normal or, equivalently, if the posterior distribution of the random effects is nearly normal. Even for dichotomous responses, this becomes increasingly the case as the cluster sizes increase.

However, both MQL and PQL perform poorly for dichotomous responses with small cluster sizes (e.g. Rodriguez and Goldman, 1995, 2001; Breslow and Lin, 1995; Lin and Breslow, 1996; Breslow et al., 1998; Goldstein and Rasbash, 1996; Browne and Draper, 2004; McCulloch and Searle, 2001; Breslow, 2003). In such situations, PQL is a better approximation than MQL and second order expansions of the random part (MQL-2 or PQL-2) yield better results than first order expansions (MQL-1 or PQL-1). However, Rodriquez and Goldman (2001) found that estimates of both fixed and random parameters were attenuated even for PQL-2, in contrast to ML and Gibbs sampling, for 'large' random effects variances. They point out that it is hazardous to use methods that work well for 'small' random effects variances since the degree of within-cluster dependence is rarely known in advance. Unfortunately,

MQL-2 and PQL-2 are sometimes numerically unstable, a problem reported by Rodriquez and Goldman (2001) for one of their examples.

The standard errors for $\widehat{\boldsymbol{\beta}}$ do not take into account the imprecision in the estimates of $\widehat{\boldsymbol{\Psi}}$. This can result in large biases since the fixed effects estimates are generally correlated with the variance estimates in generalized linear mixed models. This is obvious in PQL where the offset used to estimate the fixed effects depends on the variance. Another drawback of marginal and penalized quasi-likelihood is that no likelihood is provided, precluding for instance the use of likelihood ratio testing, model selection criteria such as AIC, BIC etc. and likelihood based diagnostics (see Chapter 8).

6.9 Generalized Estimating Equations (GEE)

Estimation using 'Generalized Estimating Equations' (GEE) was initially advocated in a series of papers by Liang, Zeger and their colleagues (see Liang and Zeger, 1986; Zeger and Liang, 1986; Zeger et al., 1988). GEE can be considered as a generalization of the quasi-likelihood method described in Section 6.8.1 to *multivariate* regression models. This methodology is popular for dependent responses, for instance repeated measurements or clustered data.

The marginal expectation (with respect to any latent variables) of the responses is modeled using a generalized linear model. For a two-level model with n_j observations for cluster j,

$$g[\mathrm{E}(y_{ij}|\mathbf{x}_{ij})] \;=\; \mathbf{x}'_{ij}\boldsymbol{\beta}.$$

The regression parameters $\boldsymbol{\beta}$ then represent *marginal* or *population averaged effects*. Importantly, these effects differ from conditional or cluster-specific effects (see Figure 1.6 and Section 3.6.5).

We now consider the marginal variances and covariances of the responses given the covariates. The variances are assumed to be $\phi V(\mu_{ij})$, corresponding to the specified generalized linear model (see Section 2.2). Combining these variances with a *working correlation matrix* $\mathbf{R}_j(\boldsymbol{\alpha})$ structured by parameters $\boldsymbol{\alpha}$, the covariance matrix becomes

$$\mathbf{V}_j = \mathbf{B}_j^{1/2}\mathbf{R}_j(\boldsymbol{\alpha})\mathbf{B}_j^{1/2}\phi,$$

where $\mathbf{B}_j$ is a diagonal matrix with elements $b''(\theta_{ij}) = V(\mu_{ij})$.

The quasi-score equation in (6.18) can now be generalized to *generalized estimating equations* of the form

$$S_{\boldsymbol{\beta}}(\boldsymbol{\beta},\boldsymbol{\alpha}) \;=\; \sum_j \frac{\partial \boldsymbol{\mu}'_j}{\partial \boldsymbol{\beta}}\mathbf{V}_j^{-1}[\mathbf{y}_j - \boldsymbol{\mu}_j] \;=\; \mathbf{0},$$

which depend not only on the marginal effects $\boldsymbol{\beta}$ but also on the dependence parameters $\boldsymbol{\alpha}$. Here,

$$\frac{\partial \boldsymbol{\mu}_j}{\partial \boldsymbol{\beta}} = \mathbf{B}_j\boldsymbol{\Delta}_j\mathbf{X}_j,$$

where $\boldsymbol{\Delta}_j$ is a diagonal matrix with elements $\frac{\partial \theta_{ij}}{\partial \nu_{ij}}$.

Liang and Zeger (1986) propose iterating between (1) estimation of $\boldsymbol{\beta}$ (for given $\boldsymbol{\alpha}$ and ϕ) solving the generalized estimating equations and (2) estimation of $\boldsymbol{\alpha}$ and ϕ (for given $\boldsymbol{\beta}$) based on Pearson residuals r_{ij}^P,

$$r_{ij}^P = [y_{ij} - b'(\widehat{\theta}_{ij})]/[b''(\widehat{\theta}_{ij})]^{\frac{1}{2}}.$$

They propose a moment estimator for the overdispersion parameter ϕ,

$$\widehat{\phi} = \frac{1}{N} \sum_{j=1}^{N} \frac{1}{n_j} \sum_{i=1}^{n_j} (r_{ij}^P)^2,$$

and the following moment estimators for $\boldsymbol{\alpha}$ for different correlation structures $\mathbf{R}_j(\boldsymbol{\alpha})$:

- 'Independence'

    - Correlation structure: $\mathrm{Cor}(y_{ij}, y_{i'j}) = 0$.

- 'Exchangeable'

    - Correlation structure: $\mathrm{Cor}(y_{ij}, y_{i'j}) = \alpha$, $i \neq i'$,

    - Estimator: $\widehat{\alpha} = \frac{1}{N} \sum_{j=1}^{N} \frac{1}{n_j(n_j-1)} \sum_{i \neq i'} r_{ij}^P r_{i'j}^P$.

- 'AR(1)'

    - Correlation structure: $\mathrm{Cor}(y_{ij}, y_{i+t,j}) = \alpha^t$, $t = 0, 1, \ldots, n_j - i$,

    - Estimator: $\widehat{\alpha} = \frac{1}{N} \sum_{j=1}^{N} \frac{1}{(n_j-1)} \sum_{i \leq n_j - 1} r_{ij}^P r_{i+1,j}^P$.

- 'Unstructured'

    - Correlation structure: $\mathrm{Cor}(y_{ij}, y_{i'j}) = \alpha_{ii'}$, $i \neq i'$,

    - Estimator: $\widehat{\alpha}_{ii'} = \frac{1}{N} \sum_{j=1}^{N} r_{ij}^P r_{i'j}^P$.

Liang and Zeger (1986) showed that the estimated *marginal* effects $\widehat{\boldsymbol{\beta}}$ are asymptotically normal and consistent as the number of clusters increases. Importantly, these estimates are 'robust' in the sense that they are consistent for misspecified correlation structures, assuming that the mean structure is correctly specified. Consistent estimates of the covariance matrix of the estimated marginal effects are typically obtained by means of the so called sandwich estimator described in Section 8.3.3.

Instead of using the above moment estimators of $\boldsymbol{\alpha}$, Prentice (1988) suggested adding a second set of estimating equations for the correlation parameters $\boldsymbol{\alpha}$ (and possibly ϕ). Define the vector of products of Pearson residuals $\mathbf{u}_j = (r_{1j}^P r_{2j}^P, r_{1j}^P r_{3j}^P, \ldots, r_{n_{j-1},j}^P r_{n_j,j}^P)'$, and the diagonal matrix $\mathbf{W}_j = \mathrm{diag}\{\mathrm{Var}(r_{1j}^P r_{2j}^P), \mathrm{Var}(r_{1j}^P r_{2j}^P), \ldots, \mathrm{Var}(r_{n_{j-1},j}^P r_{n_j,j}^P)\}$. The estimating equations for $\boldsymbol{\alpha}$ can then be expressed as

$$S_{\boldsymbol{\alpha}}(\boldsymbol{\beta}, \boldsymbol{\alpha}) = \sum_j \frac{\partial \mathrm{E}(\mathbf{u}_j)'}{\partial \boldsymbol{\alpha}} \mathbf{W}_j^{-1} [\mathbf{u}_j - \mathrm{E}(\mathbf{u}_j)] = \mathbf{0}.$$

These approaches based on Pearson correlations are problematic for categorical responses where the Pearson correlation is in general not a suitable measure of association. For the special case of dichotomous responses this makes some sense, but a problem is that the admissible range of the correlation depends on the marginal probabilities (e.g. Lord and Novick, 1968; Bishop *et al.*, 1975). A more natural measure of association for categorical data is the odds-ratio, and a parametrization based on marginal odds-ratios was proposed by Lipsitz *et al.* (1991). The odds-ratios are typically structured to simplify the working correlation matrix, for instance by specifying a common odds-ratio. Log-linear models may also be specified letting the odds-ratio depend on covariates. In general, the specification of the working correlation matrix entails a trade-off between simplicity and loss off efficiency due to misspecification (e.g. Fitzmaurice, 1995).

Zhao and Prentice (1990) and Liang *et al.* (1992) proposed extending the above first order estimating equations (GEE-1) to second order estimating equations (GEE-2). Here, a joint estimating equation

$$S_{\boldsymbol{\beta},\boldsymbol{\alpha}}(\boldsymbol{\beta},\boldsymbol{\alpha}) = \sum_j \frac{\partial[\boldsymbol{\mu}_j, \mathrm{E}(\mathbf{u}_j)]'}{\partial[\boldsymbol{\beta},\boldsymbol{\alpha}]} \left(\begin{array}{cc} \mathbf{V}_j & \\ \mathrm{Cov}(\mathbf{u}_j,\mathbf{y}_j) & \mathbf{W}_j \end{array} \right)^{-1} \left(\begin{array}{c} \mathbf{y}_j - \boldsymbol{\mu}_j \\ \mathbf{u}_j - \mathrm{E}(\mathbf{u}_j) \end{array} \right) = \mathbf{0}$$

is simultaneously solved for $\boldsymbol{\beta}$ and $\boldsymbol{\alpha}$. The major merit of GEE-2 as compared to GEE-1 is in efficiency gain, primarily for $\boldsymbol{\alpha}$. However, the robustness of GEE-1 is lost since GEE-2 rests on correct specification of the dependence structure. Moreover, obtaining the required estimate of $\mathrm{Cov}(\mathbf{u}_j,\mathbf{y}_j)$ is quite involved.

When marginal odds-ratios are used to represent dependence, Carey *et al.* (1993) suggest estimating $\boldsymbol{\alpha}$ using logistic regression with an offset. This implementation of GEE is called 'alternating logistic regressions' (ALR). ALR preserves the robustness of GEE-1 regarding $\widehat{\boldsymbol{\beta}}$ while producing estimates $\widehat{\boldsymbol{\alpha}}$ that are almost as efficient as using the more complex GEE-2 methodology.

Since explicit integration is avoided, the GEE methodology is definitely an important contribution to the estimation of models for longitudinal and clustered data. We use GEE for longitudinal data on respiratory infection in Section 9.2 where it is also compared to random effects modeling. Interestingly, GEE has recently been extended to factor models (Reboussin and Liang, 1998), where the dependence structure is of primary interest.

A rather severe limitation is that missing data can apparently only be handled under the restrictive assumption of missing completely at random MCAR (Liang and Zeger, 1986), since the estimating equations will otherwise be biased (e.g. Rotnitzky and Wypij, 1994). However, it is often not recognized that missingness may actually depend on covariates but not on observed responses (Little, 1995). Robins *et al.* (1994) suggest combining estimating equations with inverse probability weighting, yielding consistent estimators if the missing data mechanism is correctly specified.

Another limitation is that it is in general difficult to assess model adequacy in GEE (e.g. Albert, 1999); likelihood based diagnostics are for instance not available. The use of GEE should furthermore be reserved to problems where marginal or population averaged effects are of interest and avoided in analyses of etiology. This is because causal processes must operate at the cluster or individual level, not the population level. Population averaged effects are therefore merely descriptive and largely determined by the degree of heterogeneity in the population. Finally, Lindsey and Lambert (1998) and Crouchley and Davies (1999) point out that the estimated regression parameters are no longer consistent if there are endogenous covariates such as 'baseline' (initial) responses in longitudinal data.

6.10 Fixed effects methods

The main focus in this book, including the discussion previously in this chapter, has been on latent variables as random variables. In this section, we depart from this interpretation and instead consider latent variables as *unknown fixed parameters*.

6.10.1 Joint maximum likelihood

At first sight, it may appear natural to attempt the simultaneous estimation of the fundamental parameters $\boldsymbol{\vartheta}$ and the 'latent scores' or values attained by the latent variables $\boldsymbol{\zeta}$ by maximizing the likelihood of the responses given the latent variables. For the two-level random intercept model, the log-likelihood

$$\ln \prod_{ij} g^{(1)}(y_{ij}|\mathbf{x}_{ij}, \zeta_j; \boldsymbol{\vartheta})$$

is maximized with respect to *both* $\boldsymbol{\beta}$ and the ζ_j, $j = 1, \ldots, J$. Thus, the random intercepts are simply treated as fixed parameters and estimated alongside $\boldsymbol{\beta}$. Note that this likelihood differs from the h-likelihood in equation (6.7) on page 164, the joint likelihood of the responses and latent variables, which is also jointly maximized with respect to both parameters and latent scores.

In a three-level random intercept model estimation of the level-three intercepts would require constraints on the level-2 intercepts, for instance that they add to zero for each level-3 cluster. It is therefore more natural to omit all higher-level latent variables that have lower-level counterparts. Hence we assume that the model only includes latent variables at level 2.

A fundamental problem may arise due to the fact that the number of latent scores to be estimated increases with the number of level-2 clusters J. This is the well-known problem of estimating 'structural parameters' in the presence of 'incidental parameters'. In our context the fundamental parameters are regarded as structural and the latent scores as incidental. The basic problem is that maximum likelihood estimators of the structural parameters are not necessarily consistent when incidental parameters are present (Neyman and Scott, 1948; see also Lancaster, 2000). Zellner (1971, p. 114-115) demonstrates how inconsistency comes about in a simple example.

Importantly, there is no incidental parameter problem for linear models with normally distributed residuals or loglinear models with a Poisson distribution. In these models the estimates coincide with those from conditional maximum likelihood to be described in Section 6.10.2. See Cameron and Trivedi (1998) for further discussion.

For common factor models with continuous responses, Anderson and Rubin (1956) demonstrated that the likelihood presented by Lawley (1942) did not have a maximum. As a solution to this problem McDonald (1979) suggested using a maximum likelihood ratio (MLR) estimator, where the alternative is that $\mathbf{\Psi}$ is any positive definite matrix. It turns out that the resulting estimates of the structural parameters equal those obtained by ML in the random factor case, whereas the estimators for the factor scores are inconsistent, being given by the Kestelman expressions for 'indeterminate' factor scores (Kestelman, 1952; Guttmann, 1955).

Turning to dichotomous responses, the ramifications for the logistic regression model with cluster-specific intercepts, $n=2$ level-1 units and one dichotomous covariate were established by Andersen (1973) (see also Chamberlain, 1980 and Breslow and Day, 1980). Specifically, Andersen demonstrated that the joint maximum likelihood estimator is inconsistent for $J \to \infty$, since $\widehat{\boldsymbol{\beta}}$ converges in probability to $2\boldsymbol{\beta}$. Simulations performed by Katz (2001) suggest that joint maximum likelihood is safe if $n_j = n > 20$ and might be acceptable if $8 < n < 16$. For the Rasch model, simulations indicate that $\widehat{\boldsymbol{\beta}}$ converges to $\frac{n}{n-1}\boldsymbol{\beta}$, where n is the number of items. Haberman (1977) proved that the bias vanishes if $J \to \infty$, $n \to \infty$ and $\frac{J}{n} \to \infty$. For fixed effects probit regression Heckman (1981b) conducted a Monte Carlo experiment and found modest bias, always towards zero.

Numerical problems are common in the case of dichotomous responses. If all responses for a cluster are zero (one), the joint maximum likelihood estimate $\widehat{\zeta}_j$ will diverge to $-\infty$ (∞). Moreover, if an item i is failed (passed) by all units in the Rasch model, $\widehat{\beta}_i$ tends to $-\infty$ (∞). Note that the latter is a problem in marginal maximum likelihood estimation as well. For the two-parameter logistic IRT model it was observed by Wright (1977) that arbitrary upper bounds on the discrimination parameters typically must be imposed in order to prevent the estimates from diverging. Another important limitation of joint maximum likelihood is that cluster-level covariates cannot be included.

6.10.2 Conditional maximum likelihood

Rasch (1960) suggested using conditional maximum likelihood estimation in the context of the one-parameter logistic IRT model or Rasch model,

$$\Pr(y_{ij}=1) = \frac{\exp(\beta_i + \zeta_j)}{1 + \exp(\beta_i + \zeta_j)},$$

discussed in Section 3.3.4. Importantly, this approach circumvents the incidental parameter problem.

The conditional maximum likelihood approach requires the existence of sufficient statistics for the parameters ζ_j. The likelihood is then maximized conditional on these sufficient statistics.

Consider a logistic regression model where there are fixed cluster-specific intercepts ζ_j, so the linear predictor becomes $\nu_{ij} = \mathbf{x}'_{ij}\boldsymbol{\beta}+\zeta_j$. For dichotomous responses, $y_{ij} = 0, 1$, the model becomes

$$\Pr(y_{ij}=1|\mathbf{x}_{ij}) = \frac{\exp(\mathbf{x}'_{ij}\boldsymbol{\beta}+\zeta_j)}{1+\exp(\mathbf{x}'_{ij}\boldsymbol{\beta}+\zeta_j)}. \tag{6.25}$$

The Rasch model is the special case of the fixed effect logistic regression where $\mathbf{x}'_{ij}\boldsymbol{\beta} = \beta_i$.

The responses within a cluster j are independently distributed due to the cluster-specific intercepts, with joint probability

$$\Pr(\mathbf{y}_j) = \prod_{i=1}^{n_j}\Pr(y_{ij}=1)^{y_{ij}}[1-\Pr(y_{ij}=1)]^{1-y_{ij}}.$$

Substituting (6.25) for $\Pr(y_{ij}=1)$ and reexpressing, the joint probability can alternatively be written as

$$\Pr(\mathbf{y}_j) = \exp\left[\zeta_j\sum_{i=1}^{n_j}y_{ij} + \boldsymbol{\beta}'\sum_{i=1}^{n_j}\mathbf{x}_{ij}y_{ij} + a(\zeta_j,\boldsymbol{\beta})\right],$$

where

$$a(\zeta_j,\boldsymbol{\beta}) = \ln\prod_{i=1}^{n_j}[1+\exp(\mathbf{x}'_{ij}\boldsymbol{\beta}+\zeta_j)].$$

It follows from the theory of exponential family distributions that $\sum_{i=1}^{n_j}y_{ij}$ is a minimal sufficient statistic for the cluster-specific parameter ζ_j. Consequently, the conditional distribution $\Pr(\mathbf{y}_j|\sum_{i=1}^{n_j}y_{ij})$ does not depend on ζ_j. The idea of conditional maximum likelihood is to estimate the 'structural parameters' by maximizing $\prod_j\Pr(\mathbf{y}_j|\sum_{i=1}^{n_j}y_{ij})$, which does not contain the 'incidental parameters' ζ_j.

The joint probability can also be expressed as

$$\Pr(\mathbf{y}_j) = \frac{\exp[\zeta_j\sum_{i=1}^{n_j}y_{ij} + \boldsymbol{\beta}'\sum_{i=1}^{n_j}\mathbf{x}_{ij}y_{ij}]}{\prod_{i=1}^{n_j}[1+\exp(\mathbf{x}'_{ij}\boldsymbol{\beta}+\zeta_j)]},$$

and the probability of $\tau_j = \sum_{i=1}^{n_j}y_{ij}$ as

$$\Pr\left(\sum_{i=1}^{n_j}y_{ij}=\tau_j\right) = \sum_{\mathbf{d}_j\in B(\tau_j)}\frac{\exp[\zeta_j\tau_j + \boldsymbol{\beta}'\sum_{i=1}^{n_j}\mathbf{x}_{ij}d_{ij}]}{\prod_{i=1}^{n_j}[1+\exp(\mathbf{x}'_{ij}\boldsymbol{\beta}+\zeta_j)]}, \quad \tau_j = 0, 1, \ldots, n_j,$$

where $B(\tau_j) = \{\mathbf{d}_j = (d_{1j}, \ldots, d_{n_j,j}) : d_{ij} = 0 \text{ or } 1, \sum_{i=1}^{n_j}d_{ij} = \tau_j\}$. The number of elements in $B(\tau_j)$ is $\binom{n_j}{\tau_j}$, which increases rapidly with the cluster size. The conditional probability $\Pr(\mathbf{y}_j|\sum_{i=1}^{n_j}y_{ij} = \tau_j)$ can now be obtained by dividing $\Pr(\mathbf{y}_j)$ by $\Pr(\sum_{i=1}^{n_j} y_{ij} = \tau_j)$,

$$\Pr\left(\mathbf{y}_j \,\middle|\, \sum_{i=1}^{n_j} y_{ij} = \tau_j\right) = \frac{\exp[\boldsymbol{\beta}' \sum_{i=1}^{n_j} \mathbf{x}_{ij} y_{ij}]}{\sum_{\mathbf{d}_j \in B(\tau_j)} \exp[\boldsymbol{\beta}' \sum_{i=1}^{n_j} \mathbf{x}_{ij} d_{ij}]}.$$

It is worth noting that when $\tau_j = 1$ this probability has the same form as the multinomial logit model (see Section 2.4.3) and a partial likelihood contribution for Cox regression (see Section 2.5).

The conditional likelihood $l_C(\boldsymbol{\beta}|\mathbf{X})$ becomes

$$l_C(\boldsymbol{\beta}|\mathbf{X}) = \prod_{j=1}^{N} \frac{\exp[\boldsymbol{\beta}' \sum_{i=1}^{n_j} \mathbf{x}_{ij} y_{ij}]}{\sum_{\mathbf{d}_j \in B(\tau_j)} \exp[\boldsymbol{\beta}' \sum_{i=1}^{n_j} \mathbf{x}_{ij} d_{ij}]},$$

which does not depend on the incidental parameters ζ_j. Note that clusters with $\tau_j = 0$ or $\tau_j = n_j$, having only zero or unit responses, do not contribute to the likelihood since their conditional probabilities become 1. Hence, there may be a considerable loss of data, especially when the clusters are small.

It is instructive to investigate the conditional likelihood for the special case of clusters of size 2; e.g. $n_j = 2$. Here, the only situation contributing information is $\tau_j = 1$, e.g. $(y_{1j} = 0, y_{2j} = 1)$ and $(y_{1j} = 1, y_{2j} = 0)$. The conditional probability of the former becomes

$$\begin{aligned}
\Pr(y_{1j}=0, y_{2j}=1 | y_{1j}+y_{2j}=1) &= \frac{\Pr(y_{1j}=0, y_{2j}=1)}{\Pr(y_{1j}=0, y_{2j}=1) + \Pr(y_{1j}=1, y_{2j}=0)} \\
&= \frac{\exp[\boldsymbol{\beta}'(\mathbf{x}_{2j} - \mathbf{x}_{1j})]}{1 + \exp[\boldsymbol{\beta}'(\mathbf{x}_{2j} - \mathbf{x}_{1j})]}.
\end{aligned}$$

The resulting conditional likelihood thus reduces to an (unconditional) logistic likelihood for dichotomous responses $y_j' = 1$ if $(y_{1j} = 0, y_{2j} = 1)$ and $y_j' = 0$ if $(y_{1j} = 1, y_{2j} = 0)$ (discarding concordant responses) with covariates $\mathbf{x}_{2j} - \mathbf{x}_{1j}$. Importantly, it follows that elements of $\boldsymbol{\beta}$ pertaining to covariates that do not vary over units cannot be estimated by conditional maximum likelihood, since the corresponding elements of $\mathbf{x}_{2j} - \mathbf{x}_{1j}$ become zero for all j. In a (matched) case-control study the responses are $(y_{1j} = 0, y_{2j} = 1)$ if y_{2j} and y_{1j} represent the case and control, respectively. It follows from the above result that the conditional likelihood now reduces to an (unconditional) likelihood for a logistic model with constant response $y_j' = 1$ and covariates $\mathbf{x}_{2j} - \mathbf{x}_{1j}$. See also Holford et al. (1978).

Andersen (1970) demonstrates that conditional maximum likelihood yields consistent and asymptotically normal estimators under weak regularity conditions. For the exponential family he also shows that the estimates are asymptotically efficient under 'S-ancillarity' (e.g. Barndorff-Nielsen, 1978). Andersen (1973) demonstrates that this condition holds for the Rasch model.

An advantage of conditional maximum likelihood is that we need not make distributional assumptions regarding the latent variables. Importantly, this implies that inference based on this approach is likely to be more 'robust' than using random effects. If the random effects model is true we would expect a loss of efficiency using the conditional maximum likelihood estimator, but Andersen (1973) demonstrates that the loss is small. Another important merit is that correlations between the latent variable and covariates is unproblematic, a feature that has attracted a lot of attention in econometrics (e.g. Hausman, 1978). In contrast, random effects models usually assume that the random effects and the covariates are uncorrelated. However, in random effects models this assumption can be relaxed for some covariates by including the cluster means of these covariates as additional predictors in the model; see page 52.

Unfortunately, relying on conditional maximum likelihood severely limits the types of models that can be estimated. The sufficient statistic required to construct a conditional likelihood only exists for simple models with cluster-specific intercepts. Furthermore, it is required that the models belong to the exponential family, having canonical links. For continuous responses, this is the standard identity link, for dichotomous responses the logit link (in contrast to for instance the probit), for counts the log link and for unordered polytomous responses the multinomial logit (e.g. Andersen, 1973; Chamberlain, 1980). Even simple models with cluster-specific intercepts cannot be estimated if they include cluster-specific covariates. As discussed on pages 81 to 84 the estimated regression parameters for within-cluster covariates therefore reflect only the within-cluster effects. In contrast, random effects estimates are a weighted average of within-cluster and between-cluster effects (equal to the within-cluster effects in the case of balanced data). Another drawback compared to random effects models is that we cannot have a structural model where cluster-specific fixed effects are regressed on covariates or other cluster-specific effects.

6.11 Bayesian methods

6.11.1 Introduction

In the Bayesian approach there is no distinction between latent variables and parameters; all are considered random quantities. Let $\mathbf{D}$ denote the observed data and $\mathbf{L}$ denote parameters as well as latent variables (including missing data). Inference requires setting up a joint probability distribution $\Pr(\mathbf{D}, \mathbf{L})$ over all random quantities. The joint distribution comprises two parts: the *likelihood* $\Pr(\mathbf{D}|\mathbf{L})$ and the *prior* distribution $\Pr(\mathbf{L})$. Specifying $\Pr(\mathbf{D}|\mathbf{L})$ and $\Pr(\mathbf{L})$ gives a *full probability model*, where

$$\Pr(\mathbf{D}, \mathbf{L}) = \Pr(\mathbf{D}|\mathbf{L})\Pr(\mathbf{L}).$$

Having observed $\mathbf{D}$, Bayes theorem is used to obtain the *posterior distribution* of $\mathbf{L}$ given $\mathbf{D}$,

$$\Pr(\mathbf{L}|\mathbf{D}) = \frac{\Pr(\mathbf{D}|\mathbf{L})\Pr(\mathbf{L})}{\int \Pr(\mathbf{D}|\mathbf{L})\Pr(\mathbf{L}) \, \mathrm{d}\mathbf{L}}.$$

Loosely speaking, the posterior distribution updates prior 'knowledge' (represented by the prior distribution) with information in the observed data (represented by the likelihood). Note that the posterior distribution is proportional to the product of the prior distribution and the likelihood.

All the information regarding the unknown quantities is contained in their posterior distribution given the data **D**. However, it is difficult if not impossible to comprehend a posterior distribution with possibly thousands of dimensions and Bayesians thus typically summarize the information in the posterior. A preferred summary is the posterior expectation of the parameters, since these 'estimates' minimize the posterior expectation of a quadratic loss function (mean squared error of estimates). The expectation also has the merit that its value for a subset of parameters is invariant with respect to marginalization over the remaining parameters. In contrast, the posterior modes are not invariant under marginalization. Other features of the posterior are also used for Bayesian inference, including moments, quantiles and highest posterior density regions. All these quantities can be expressed as posterior expectations of functions of **L**.

6.11.2 Bayes modal or modal a posteriori (MAP)

Before the advent of Markov Chain Monte Carlo (MCMC) methods (see below), Bayes modal or modal a posteriori (MAP) methods were often used to approximate expectations, since modes are often easier to approximate numerically. Lindley and Smith (1972) suggested that inference could be based on the *joint* posterior mode which can be obtained by using standard optimization methods. However, posterior expectations of subsets of parameters are generally better approximated by the mode after marginalization over the other parameters (e.g. O'Hagan, 1976). Inference regarding structural parameters $\boldsymbol{\theta}$ would thus preferably proceed by considering the mode of marginal posteriors with incidental parameters and latent variables integrated out. Another problem with the use of joint posterior modes is that an incidental parameter problem may arise, invalidating the asymptotic normality of Bayes modal predictions. Thus, care must be exercised to ensure that Bayes modal predictions coincide with Bayes (mean) predictions in large sample situations. In the context of IRT models, the joint posterior mode has been used by Swaminathan and Gifford (1982, 1985, 1986) for the one, two, and three-parameter logistic models respectively.

6.11.3 Hierarchical Bayesian models

Latent variable models can be viewed as *hierarchical* Bayesian models. This is because the prior distributions of some of the parameters, namely the latent variables, depend on further parameters (the variances and covariances of the latent variables) known as *hyperparameters*. The distributions of these hyperparameters are known as *hyperpriors*. At stage one the distribution $\Pr(\mathbf{D}|\boldsymbol{\zeta},\boldsymbol{\vartheta}_1)$ of the responses is specified conditional on the latent variables $\boldsymbol{\zeta}$ and parameters $\boldsymbol{\vartheta}_1$. At stage two, the prior distribution $\Pr(\boldsymbol{\vartheta}_1)$ of the parameters and the prior distribution $\Pr(\zeta|\vartheta_2)$ of the latent variables are specified, the latter depending on hyperparameters ϑ_2. At stage 3, a hyperprior $\Pr(\vartheta_2)$ for the hyperparameters is specified. In generalized linear mixed models ϑ_1 would be regression and possibly dispersion and threshold parameters, whereas ϑ_2 would be variance and covariance parameters of the random effects. The posterior distribution of the parameters given the data can be written as

$$\Pr(\zeta, \vartheta_1, \vartheta_2|\mathbf{D}) \ \propto \ \Pr(\mathbf{D}|\zeta, \vartheta_1)\Pr(\vartheta_1)\Pr(\zeta|\vartheta_2)\Pr(\vartheta_2). \qquad (6.26)$$

Thus, the term 'hierarchical' does not refer to the data structure but to this sequential model specification. Bayesian hierarchical linear models are presented by Lindley (1971), Lindley and Smith (1972) and Smith (1973) and applied in Novick *et al.* (1973). A 'frequentist' latent variable model can be viewed as 'empirical Bayesian' because the parameters (variances and covariances) of the 'prior' distribution for the latent variables are estimated instead of assuming a 'hyperprior' distribution.

6.11.4 Prior distributions

There appear to be four different motivations for using prior (and hyperprior) distributions, the first 'truly' Bayesian and the others 'pragmatic' Bayesian. True Bayesians would specify prior distributions reflecting prior beliefs or knowledge regarding the parameters. For instance, factor loadings in independent clusters factor models are expected to be positive and this prior belief could be represented by appropriate prior distributions. Another example would be a prior for a treatment effect in a clinical trial based on elicited expert opinion (e.g. Spiegelhalter *et al.*, 1994).

Second, a prior can be used to ensure that estimates are confined to the permitted parameter space, for instance to avoid 'Heywood cases' where unique factors have negative variances (e.g. Martin and McDonald, 1975). In latent class modeling priors are sometimes used to prevent boundary solutions where conditional response probabilities approach zero or one with corresponding logit parameters approaching $-\infty$ and ∞.

Third, priors can aid identification. For instance, a measurement model without replicate measures could be identified by specifying a prior for the measurement error variance. This approach may be preferable to the conventional approach of treating the variance as a known parameter, since the prior can reflect parameter uncertainty. In the same vain, priors have been used as a cure for the problem of excessive standard errors in the three-parameter logistic IRT model (Wainer and Thissen, 1982).

Finally, perhaps the most prevalent pragmatic reason for using priors is that Markov chain Monte Carlo (MCMC) methods (described below) can then be used for estimating complex models for which other methods perform badly or are unfeasible, for instance generalized linear mixed models with crossed random effects. The prior is typically specified as 'noninformative' (also denoted as flat, vague or diffuse) to minimize its effect on statistical inference. The likelihood component of the posterior then dominates so that the posterior becomes nearly proportional to the likelihood. This pragmatic approach is for instance reflected in modern books on Bayesian modeling such as Congdon (2001) and the examples accompanying the popular BUGS software (Spiegelhalter *et al.*, 1996bc).

6.11.5 Markov chain Monte Carlo

The aim of MCMC is to draw parameters and latent variables $\varphi \equiv \mathbf{L}$ from the posterior distribution to obtain the expectation of a function $f(\varphi)$. Drawing *independent* samples from $h(\varphi)$ as in crude Monte Carlo integration described in Section 6.3.3 may not be feasible. However, consistent estimators of the expectation $\mathrm{E}[f(\varphi)]$ can be obtained from *dependent* samples as long as the samples are drawn throughout the support of $h(\varphi)$ in correct proportions. This can be accomplished by using a Markov chain with the target distribution $h(\varphi)$ as its stationary distribution, leading to *Markov chain Monte Carlo* (MCMC).

Let $\{\varphi^{(0)}, \varphi^{(1)}, \ldots\}$ be a sequence of random variables. In a *first order homogenous Markov chain* the next state $\varphi^{(r+1)}$ is sampled from a distribution $P(\varphi^{(r+1)}|\varphi^{(r)})$, which only depends on the current state $\varphi^{(r)}$ and neither on the 'history' of the chain $\{\varphi^{(0)}, \varphi^{(1)}, \ldots, \varphi^{(r-1)}\}$ nor r. Importantly, the chain will gradually 'forget' its initial state and eventually converge to a unique stationary distribution. To obtain the required distribution we discard the states up to the 'time' when we believe that stationarity has been reached, known as the 'burn-in' period. Sometimes several chains with different initial states are used to monitor convergence to a stationary distribution. After the burn-in, we need to run the chain sufficiently long for the sample averages of $f(\varphi)$ to reliably estimate the required expectations; determining how long is not trivial since the draws are dependent.

There are several ways of constructing a Markov chain with the target distribution $h(\varphi)$ as stationary distribution. We start with the most complex algorithm and gradually proceed to the simpler.

The Metropolis-Hastings algorithm

From $\varphi^{(r)}$, the next state $\varphi^{(r+1)}$ is obtained as follows:

Step 1: Sample a candidate point $\check{\varphi}$ from some proposal distribution $q(\check{\varphi}|\varphi^{(r)})$. For instance, $q(\check{\varphi}|\varphi^{(r)})$ could be multivariate normal with mean $\varphi^{(r)}$ and fixed covariance matrix.

Step 2: Accept the candidate point with probability

$$\alpha(\varphi^{(r)}, \check{\varphi}) = \min\left(1, \frac{h(\check{\varphi})q(\varphi^{(r)}|\check{\varphi})}{h(\varphi^{(r)})q(\check{\varphi}|\varphi^{(r)})}\right).$$

Roughly speaking, the algorithm proceeds in the following way: In Step 1 we sample from a convenient but incorrect distribution. In Step 2 we correct for this in a correct but rather nonintuitive way.

If the candidate point is accepted, the next state becomes $\varphi^{(r+1)} = \check{\varphi}$; if the candidate is rejected, the chain does not move and $\varphi^{(r+1)} = \varphi^{(r)}$. Use of the Metropolis-Hastings algorithm thus requires the ability to draw from the proposal distribution and calculate the fraction involved in obtaining the acceptance probability. Importantly, the target distribution need not be normalized so that the denominator of posterior distributions may be omitted.

The stationary distribution will be $h(\varphi)$ whatever proposal distribution $q(\check{\varphi}|\varphi^{(r)})$ is used. However, the rate of convergence will of course depend on how close the proposal distribution is to the target distribution.

The Metropolis algorithm

This algorithm is the special case of the Metropolis-Hastings algorithm where only symmetric proposal distributions where $q(\check{\varphi}|\varphi^{(r)}) = q(\varphi^{(r)}|\check{\varphi})$, such as the multivariate normal or t distributions, are considered. In this case the acceptance probability simplifies to

$$\alpha(\varphi^{(r)}, \check{\varphi}) = \min\left(1, \frac{h(\check{\varphi})}{h(\varphi^{(r)})}\right),$$

which does not depend on the proposal distribution. The *random walk Metropolis* algorithm arises as the special case where $q(\check{\varphi}|\varphi^{(r)}) = q(|\check{\varphi} - \varphi^{(r)}|)$.

The single components Metropolis-Hastings algorithm

Consider now the partitioning of the vector φ into k components, which can be blocks and not necessarily scalars, $\varphi = \{\varphi_1, \varphi_2, \ldots, \varphi_k\}$. Instead of updating the entire vector φ, single-component methods update the k components φ_i one at a time. Let $\varphi_i^{(r)}$ denote the state of component φ_i at the end of iteration r. Define $\varphi_{-i}^{(r)} = \{\varphi_1^{(r+1)}, \ldots, \varphi_{i-1}^{(r+1)}, \varphi_{i+1}^{(r)}, \ldots, \varphi_k^{(r)}\}$ as φ without its ith element, after completing step $i - 1$ of iteration $r + 1$. We have presumed a fixed *updating order*, although different types of random order are possible. It may also be beneficial to update highly dependent components more frequently than others (e.g. Zeger and Karim, 1991).

In the single components Metropolis-Hastings algorithm the candidate $\check{\varphi}_i$ is drawn from the proposal distribution $q(\check{\varphi}_i|\varphi_i^{(r)}, \varphi_{-i}^{(r)})$. The candidate is accepted with probability

$$\alpha(\varphi_{-i}^{(r)}, \varphi_i^{(r)}, \check{\varphi}_i) = \min\left(1, \frac{h(\check{\varphi}_i|\varphi_{-i}^{(r)})q(\varphi_i^{(r)}|\check{\varphi}_i, \varphi_{-i}^{(r)})}{h(\varphi_i^{(r)}|\varphi_{-i}^{(r)})q(\check{\varphi}_i|\varphi_i^{(r)}, \varphi_{-i}^{(r)})}\right).$$

Here, $h(\varphi_i|\varphi_{-i})$ is the *full conditional distribution* for φ_i under $h(\varphi)$, the distribution of the ith component of φ conditioning on all the remaining components. Importantly, $\alpha(\varphi_{-i}, \varphi_i, \check{\varphi}_i)$ simplifies when $h(\varphi)$ derives from a conditional independence model.

The Gibbs sampler

The basic idea of the Gibbs sampler is to utilize the fact that conditional distributions of (blocks of) random variables may be relatively simple in spite of a complicated joint distribution. It is a special case of the single components Metropolis-Hastings algorithm where the proposal distribution $q(\check{\varphi}_i|\varphi_i^{(r)}, \varphi_{-i}^{(r)})$ for updating the ith component of φ is the full conditional distribution $h(\check{\varphi}_i|\varphi_{-i}^{(r)})$. Note that substituting $h(\check{\varphi}_i|\varphi_{-i}^{(r)})$ for $q(\check{\varphi}_i|\varphi_i^{(r)}, \varphi_{-i}^{(r)})$ in $\alpha(\varphi_{-i}^{(r)}, \varphi_i^{(r)}, \check{\varphi}_i)$ produces an acceptance probability

$$\alpha(\varphi_{-i}^{(r)}, \varphi_i^{(r)}, \check{\varphi}_i) = 1,$$

so candidates are always accepted in Gibbs sampling. Hence, the target distribution is simulated by performing a random walk on the vector φ, altering one of its components at a time. The optimal scenario for the Gibbs sampler is where the components of φ are independent in the target distribution, in which case each iteration produces a new independent draw of φ. If the components are highly correlated, convergence can be improved by orthogonalizing the components.

Straightforward use of the Gibbs sampler requires that samples can be drawn from the full conditional distributions derived from the target distribution. When this is impossible, the more complicated Metropolis-Hastings algorithm may be used. Alternatively, Gilks and Wild (1992) suggested using adaptive rejection sampling for the common case of univariate and log-concave full conditional distributions (see also Dellaportas and Smith, 1993). This approach is implemented in the BUGS software (Spiegelhalter et al., 1996a). For continuous conditional distributions that are difficult to simulate from, Albert and Chib (1993) suggest simulating from discretized versions.

Example: Gibbs sampling for random intercept probit model

Consider the random intercept probit model

$$\Pr(y_{ij} = 1|\mathbf{x}_{ij}) = \Phi(\mathbf{x}_{ij}'\boldsymbol{\beta} + \zeta_j),$$

where $\zeta_j \sim N(0, \psi)$. It is useful to express this model as a latent response model

$$y_{ij}^* = \mathbf{x}_{ij}'\boldsymbol{\beta} + \zeta_j + \epsilon_{ij},$$

where $\epsilon_{ij} \sim N(0, 1)$, independent of ζ_j. The latent responses are related to observed responses y_{ij} via the threshold function

$$y_{ij} = \begin{cases} 1 & \text{if } y_{ij}^* > 0 \\ 0 & \text{if } y_{ij}^* \leq 0. \end{cases}$$

For the regression coefficients $\boldsymbol{\beta}$ we consider the conjugate multivariate normal prior

$$\boldsymbol{\beta} \sim N_p(\boldsymbol{\beta}_0, \boldsymbol{\Sigma}_{\boldsymbol{\beta}}),$$

with $\boldsymbol{\beta}_0$ and $\boldsymbol{\Sigma}_{\boldsymbol{\beta}}$ presumed known. A conjugate inverse-gamma (IG) density is specified for ψ,

$$\Pr(\psi) = \frac{b^a}{\Gamma(a)} \psi^{-(a+1)} \exp^{-\frac{b}{\psi}}, \quad \psi > 0,$$

where $a > 0$ and $b > 0$ are known shape and scale parameters, respectively, and $\Gamma(\cdot)$ is the gamma function. It is assumed that ψ and β are independent. The posterior distribution $\Pr(\mathbf{y}^*, \boldsymbol{\zeta}, \psi, \boldsymbol{\beta} | \mathbf{y}, \mathbf{X})$ is the distribution of the unknown parameters ($\boldsymbol{\beta}$ and ψ) and latent variables $\boldsymbol{\zeta}$ and latent responses $\mathbf{y}^*$, given the data $\mathbf{X}$ and $\mathbf{y}$. The posterior can be expressed as

$$\Pr(\mathbf{y}^*, \boldsymbol{\zeta}, \psi, \boldsymbol{\beta} | \mathbf{y}, \mathbf{X}) \propto [\Pr(\mathbf{y} | \mathbf{y}^*) \Pr(\mathbf{y}^* | \boldsymbol{\zeta}, \boldsymbol{\beta}, \mathbf{X}) \Pr(\boldsymbol{\zeta} | \psi)] \Pr(\psi) \Pr(\boldsymbol{\beta}). \quad (6.27)$$

Here, the normalizing constant is the marginal distribution of the observed responses $\mathbf{y}$. The joint density of $\mathbf{y}$, $\mathbf{y}^*$, and $\boldsymbol{\zeta}$ (given $\boldsymbol{\beta}$ and ψ) in square brackets, referred to as the 'augmented complete data likelihood', takes the form

$$\Pr(\mathbf{y}, \mathbf{y}^*, \boldsymbol{\zeta} | \boldsymbol{\beta}, \psi, \mathbf{X}) = \prod_{j=1}^{J} \left\{ \prod_{i=1}^{n_j} [I(\mathbf{y}_{ij}^* > 0) I(\mathbf{y}_{ij} = 1)][I(\mathbf{y}_{ij}^* \leq 0) I(\mathbf{y}_{ij} = 0)] \right.$$
$$\left. \times \phi(\mathbf{y}_{ij}^*; \, x_{ij}'\boldsymbol{\beta} + \zeta_j, 1) \right\} \phi(\zeta_j; \, 0, \psi),$$

where $I(\cdot)$ is the indicator function. Substituting this expression into the posterior in (6.27), it is evident that the normalizing constant does not have a closed form, making it difficult to simulate directly from the posterior distribution.

Fortunately, the Gibbs sampler can instead be applied since the full conditional distributions for the latent responses, parameters and latent variables are simple:

1. Independent truncated normal full conditionals of y_{ij}^*, $j = 1, \ldots, J$ $i = 1, \ldots, n_j$:

$$\Pr(y_{ij}^* | \mathbf{X}, \mathbf{y}, \mathbf{y}^*, \boldsymbol{\zeta}, \psi, \boldsymbol{\beta}) = \phi^-(y_{ij}^*; \, x_{ij}'\boldsymbol{\beta} + \zeta_j, 1)^{y_{ij}} \phi^+(y_{ij}^*; \, x_{ij}'\boldsymbol{\beta} + \zeta_j, 1)^{1-y_{ij}},$$

where $\phi^-(\cdot)$ is a left-truncated normal density equal to 0 when $y_{ij}^* \leq 0$ and $\phi^+(\cdot)$ is a right-truncated normal density equal to 0 when $y_{ij}^* > 0$. To simulate y_{ij}^* from the truncated normals, we can simulate from a normal density and discard the draws falling outside the permitted interval. To avoid this 'waste' of simulations, we can alternatively first generate a random uniform $(0,1)$ variate u_{ij}. Then if $y_{ij} = 1$ we calculate $y_{ij}^* = \Phi^{-1}\left[1 - \Phi(x_{ij}'\boldsymbol{\beta} + \zeta_j)[1 - u_{ij}]\right]$ and if $y_{ij} = 0$ we calculate $y_{ij}^* = \Phi^{-1}\left[[1 - \Phi(x_{ij}'\boldsymbol{\beta} + \zeta_j)]u_{ij}\right]$.
   Importantly, having simulated y_{ij}^* the full conditionals of the other random variables become independent of y_{ij}.

2. Multinormal full conditional $\Pr(\boldsymbol{\beta} | \psi, \boldsymbol{\zeta}, \mathbf{y}^*, \mathbf{X}, \mathbf{y})$ of $\boldsymbol{\beta}$ with mean $\bar{\boldsymbol{\mu}}_{\boldsymbol{\beta}} = \bar{\boldsymbol{\Sigma}}_{\boldsymbol{\beta}}[\boldsymbol{\Sigma}_{\boldsymbol{\beta}}^{-1}\boldsymbol{\beta}_0 + \sum_{j=1}^{J} \mathbf{X}_j'(\mathbf{y}_j^* - \mathbf{1}_{n_j}\zeta_j)]$ and covariance matrix $\bar{\boldsymbol{\Sigma}}_{\boldsymbol{\beta}} = (\boldsymbol{\Sigma}_{\boldsymbol{\beta}}^{-1} + \mathbf{X}'\mathbf{X})^{-1}$.

3. Independent normal full conditionals of ζ_j, $j = 1, \ldots, J$:

$$\Pr(\zeta_j | \boldsymbol{\beta}, \psi, \mathbf{y}^*, \mathbf{X}, \mathbf{y}) \; = \; \phi(\zeta_j; \; (\psi + n_j)^{-1} \sum_{i=1}^{n_j} (y_{ij}^* - x_{ij}'\boldsymbol{\beta}), \; (\psi + n_j)^{-1}).$$

The mean of the conditional is a special case of the so-called empirical Bayes predictor derived in (7.5). Note that the full conditional would not be multinormal if we had not augmented the data with y_{ij}^*.

4. Inverse-gamma full conditional of ψ:

$$\Pr(\psi | \boldsymbol{\beta}, \boldsymbol{\zeta}, \mathbf{y}^*, \mathbf{X}, \mathbf{y}) \; = \; \frac{(b + \frac{1}{2}\sum_{j=1}^{J}\zeta_j^2)^{a + \frac{J}{2}}}{\Gamma(a + \frac{J}{2})} \psi^{-(a + 1 + \frac{J}{2})}$$

$$\times \; \exp[-\psi^{-1}(b + \frac{1}{2}\sum_{j=1}^{J}\zeta_j^2)].$$

Gibbs sampling simply proceeds by sampling from the full conditional distributions above from some starting values and iterating to a stationary distribution.

Note that the straightforward application of the Gibbs sampler in our example rests on a judiciously chosen set-up, involving the probit specification and data augmentation with latent responses (e.g. Tanner and Wong, 1987). More complex procedures must be invoked in other cases, for instance for the random intercept logit model where Zeger and Karim (1991) suggested using rejection sampling, Spiegelhalter et al. (1996a) adaptive rejection sampling (Gilks and Wild, 1992) and Browne and Draper (2004) a hybrid Metropolis-Gibbs approach.

Alternating imputation posterior algorithm

A special kind of MCMC algorithm has been suggested by Clayton and Rasbash (1999) for generalized linear models with crossed random effects. Their algorithm is based on the imputation posterior (IP) algorithm of Tanner and Wong (1987) which iterates between an 'imputation' (I) step and a 'posterior' (P) step. The algorithm is similar to Gibbs sampling except that, in the P-step, the whole parameter vector is sampled from its conditional posterior distribution given the latent variables, instead of single components.

The algorithm can be outlined as follows:

- I-step: Draw a sample $\boldsymbol{\zeta}^r$ from the posterior distribution of $\boldsymbol{\zeta}$ given $\mathbf{y}$, $\boldsymbol{\vartheta}^{r-1}$ and $\mathbf{X}$ (data augmentation).

- P-step: Draw a sample $\boldsymbol{\vartheta}^r$ from the posterior distribution of $\boldsymbol{\vartheta}$ given $\boldsymbol{\zeta}^r$, $\mathbf{y}$ and $\mathbf{X}$.

As in Gibbs sampling, the algorithm is run until the stationary distribution has been reached (for a burn in period) and the parameters are estimated by their mean

$$E(\boldsymbol{\vartheta}|\mathbf{y},\mathbf{X}) \approx \bar{\boldsymbol{\vartheta}} \equiv \frac{1}{R}\sum_{r=1}^{R}\boldsymbol{\vartheta}^{r}.$$

In the I-step, the usual empirical Bayesian posterior distribution of the latent variables for fixed parameters is used (see Section 7.2) but with parameters set equal to $\boldsymbol{\vartheta}^{r}$ instead of the maximum likelihood estimates. In the P-step, the random effects drawn in the previous iteration are treated as fixed offsets. The posterior distribution of the parameters is then approximated by a multivariate normal distribution with mean given by the maximum likelihood estimates $\widehat{\boldsymbol{\vartheta}}^{r}$ (treating $\boldsymbol{\zeta}^{r}$ as offsets) and covariance matrix $\widehat{\boldsymbol{\Sigma}}^{r}$ derived from the Hessian. This approximate 'sampling distribution' approximates the true Bayesian posterior if uniform priors are assumed for all parameters $\boldsymbol{\vartheta}$. The variance of the parameter estimates is then estimated by (using Rao-Blackwellization)

$$\text{Var}(\boldsymbol{\vartheta}|\mathbf{y},\mathbf{X}) \approx \frac{1}{R}\sum_{r=1}^{R}\widehat{\boldsymbol{\Sigma}}^{r} + \frac{1}{R}\sum_{r=1}^{R}(\boldsymbol{\vartheta}^{r}-\bar{\boldsymbol{\vartheta}})(\boldsymbol{\vartheta}^{r}-\bar{\boldsymbol{\vartheta}})',$$

the sum of within and between-imputation variances. Clayton and Rasbash (1999) point out that this Rao-Blackwellization cannot be used in Gibbs sampling where *individual* parameters are sampled from their conditionals (not the entire parameter vector $\boldsymbol{\vartheta}$ as here), since the conditional *covariances* (the off-diagonal elements of $\widehat{\boldsymbol{\Sigma}}^{r}$) are in this case not available.

Note that Clayton and Rasbash (1999) argue that ideally, the P-step should consist of two parts: (1) use REML to obtain an approximate posterior for the variance and covariance parameters and draw samples from this posterior and (2) approximate the posterior of the fixed parameters by a multivariate normal distribution with mean and variance from the ML solution setting the variance parameters equal to the draws from (1).

Clayton and Rasbash (1999) use this algorithm to estimate models with crossed random effects. In their application women were artificially inseminated on several occasions using sperm from different donors and the response was success or failure for each attempt. Since sperm from each donor was also used to inseminate different women, the woman-specific random effects are crossed with donor-specific random effects. In their Alternating Imputation-Posterior (AIP) algorithm, Clayton and Rasbash therefore alternated between donor and woman 'wings' of the IP algorithm. In the donor wing, one iteration of IP is carried out, treating the woman-specific effects as offsets and drawing a new sample of donor-specific random effects. In the woman wing, the donor-specific random effects are treated as offsets in one iteration of IP to update the woman-specific random effects. The wings are alternated until convergence.

Obtaining separate estimates of the model parameters from the two wings allows convergence of the algorithm to be assessed. The final estimates are averages over both wings. Ecochard and Clayton (2001) suggest running both wings in parallel. In Section 11.4 we use a modified version of the AIP algorithm for disease mapping with spatially correlated random effects where quadrature methods cannot be used.

6.11.6 Advantages and disadvantages of MCMC

MCMC methods have been used for a variety of latent variable models, including generalized linear mixed models (e.g. Zeger and Karim, 1991; Clayton, 1996b), multilevel models (e.g. Browne, 1998), covariate measurement error models (e.g. Richardson and Gilks, 1993), disease mapping (e.g. Mollié, 1996), multilevel factor models (e.g. Goldstein and Browne, 2002) and multilevel item response models (e.g. Ansari and Jedidi, 2000; Fox and Glas, 2001). Numerous applications can be found in Congdon (2001) and Spiegelhalter et al. (1996bc). We use MCMC for a meta-analysis in Section 9.5 and for disease mapping in Section 11.4.

An important merit of MCMC is that the approach can be used to estimate complex models for which other methods are either unfeasible or work poorly. Another advantage is that any characteristics of the posterior distribution can be investigated based on stationary simulated values, for instance posterior means and percentiles of ranks of random effects in institutional comparisons (e.g. Goldstein and Spiegelhalter, 1996).

There are a number of more or less controversial issues that must be settled in using MCMC methods. First, the *burn-in*, the number of initial iterates to discard because of dependence on the starting values, must somehow be determined. Unfortunately, it is considerably more difficult to monitor convergence to a distribution than to a point (Gelman and Rubin, 1996). A popular approach is to use an arbitrary large number, or to run a number of chains with different initial states to assess convergence. However, recommendations regarding the *number of chains* that one should rely on have been conflicting, including several long chains, one very long chain and several short chains (the latter seems to be misguided).

Another problem is deciding when to *stop* the chain to ensure acceptable precision of the estimates. It can be particularly hard to judge convergence of the estimates when there is slow mixing, that is, when the chain moves slowly "through bottlenecks of the target distribution" (Gelman and Rubin, 1996). When mixing is poor, the chain has to be run for a very long time to obtain accurate estimates.

Although the Bayesian approach can be useful for identification, implementation via MCMC makes it hard to discover lack of identification (e.g. Keane, 1992). One reason is that a flat posterior would not be detected as a natural byproduct of estimation, in contrast to maximization using gradient methods. Inadequate mixing of the chain could moreover falsely indicate that an unidentified parameter has been estimated with reasonable precision.

Another problem concerns the specification of noninformative priors for variance parameters in random effects models (e.g. Natarajan and Kass, 2000;

Hobert, 2000). For instance, Hobert (2000) points out that the prior used by Zeger and Karim (1991) in Gibbs sampling can lead to an improper posterior distribution (not integrating to one) although all full conditionals are proper. Unfortunately, this problem may not be apparent from the Gibbs output. Moreover, the prior is not noninformative in any formal sense even if a proper posterior is obtained, meaning that the prior is actually driving the inferences. This problem is shared by 'diffuse' proper priors which are typically proper conjugate priors that are nearly improper. Furthermore, these priors may lead to Gibbs samplers which converge very slowly. Hobert (2000) concludes that choosing a prior for a variance parameter is currently a real dilemma for a Bayesian with no prior information. Hobert also raises concerns about the theoretical properties of estimated standard errors.

6.12 Summary

The advantage of full maximum likelihood through explicit integration is (1) consistency if data are missing at random (MAR) and (2) the availability of a likelihood for likelihood based inference. Accuracy can be improved and assessed by increasing the number of quadrature points or Monte Carlo replications. Furthermore, the methods are applicable for the general model framework. The drawback is computational inefficiency. The Gauss-Hermite methods in particular become computationally demanding as the number of latent variables increases. The 6th order Laplace approximation by Raudenbush *et al.* (2000) appears to be very efficient and may be sufficiently accurate in many situations.

Muthén's limited information approach is an excellent alternative for a general class of models with multinormal latent responses. However, cluster sizes must be (nearly) constant with either few missing data or few missing data patterns. The estimation method is computationally extremely efficient and appears to produce estimates that are very close to maximum likelihood, except for complex models with small samples. Surprisingly, this approach has received scant attention in the multilevel modeling literature. A related limited information method is the 'pseudo-likelihood' approach; see e.g. le Cessie and van Houwelingen (1994) and Geys *et al.* (2002).

MQL and PQL are also very computationally efficient whatever the number of latent variables. Unfortunately, these methods sometimes produce severely biased estimates. Unlike methods based on integration, the accuracy cannot be improved gradually, making it difficult to assess the reliability. This can be rectified by using parametric bootstrapping for bias correction as suggested by Kuk (1995) (see also Goldstein, 2003). MQL and PQL are currently confined to generalized linear mixed models.

MCMC methods allow estimation of a very wide range of models and have become increasingly popular. However, this flexibility can lead to specification of overly complicated models that may not be identified and where it is difficult to assess the impact of the prior distributions. These problems are often exacerbated by inadequate description of the model and lack of tables with estimates of *all* model parameters and their standard errors. Furthermore, convergence can be slow and difficult to monitor.

For some applications where latent variable models are for some reason not considered appropriate, GEE or fixed effects methods may be useful.

Unfortunately, there is a paucity of simulation studies comparing the performance of different estimation methods, assessing the effects of different 'factors' such as cluster size, intraclass correlation, sample size etc. on performance in the systematic way proposed by Skrondal (2000). Useful overviews of different estimation methods for latent variable models are given in Breslow (2003), McCulloch and Searle (2001), Rabe-Hesketh *et al.* (2002) and Goldstein (2003).

Appendix: Some software and references

For each of the main methods discussed in this chapter, we now list some references as well as software implementing the method. Note that omission of software is not informative about its quality. We do not provide addresses or links to the software since such information is quickly out of date and can easily be found on the internet.

- Closed form marginal likelihood (Section 6.2)

  - Useful references: Browne and Arminger (1995) for structural equation modeling with latent variables (other methods are also discussed) and Verbeke and Molenberghs (2000) for linear mixed models.
  - Software for structural equation modeling:

    * AMOS (Arbuckle and Wothke, 1999, 2003)
    * EQS (Bentler, 1995)
    * LISREL (Jöreskog and Sörbom, 1994; Jöreskog et al., 2001)
    * MECOSA (Arminger et al., 1996)
    * MX (Neale et al., 2002)

  - Software for linear mixed models (also REML):

    * PROC MIXED in SAS (Verbeke and Molenberghs, 1997)
    * lme in S-PLUS and R (Pinheiro and Bates, 2000)
    * Linear Mixed Models in SPSS (SPSS, 2001)
    * MLwiN (Rasbash et al., 2004)
    * HLM (Raudenbush et al., 2004)

- Laplace approximation

  - Useful references: Tierney and Kadane (1986), Tanner (1996, Section 3.2), and Raudenbush et al. (2000).
  - Software: HLM for sixth order Laplace in multilevel generalized linear mixed models (Raudenbush et al., 2004).

- Gauss-Hermite quadrature

  - Useful references: Stroud and Secrest (1966), and Davis and Rabinowitz (1984).
  - Software:

    * aML for multilevel and multiprocess models (Lillard and Panis, 2000)
    * BILOG-MG for binary logistic item-response models (e.g. Du Toit, 2003)
    * EGRET for two-level random intercept models (EGRET for Windows User Manual, 2000)
    * gllamm for generalized linear latent and mixed models in Stata (Rabe-Hesketh et al., 2004b, 2005b)
    * LIMDEP for two-level random intercept models (Greene, 2002a)
    * MIXNO for two-level multinomial logit models (Hedeker, 1999)

- * MIXOR for two-level ordinal logistic and probit regression (Hedeker and Gibbons, 1996a)
- * MIXREG for two-level linear mixed models with autocorrelated errors (Hedeker and Gibbons, 1996b)
- * MIXSUR and MIXPREG for counts and discrete-time durations (by Hedeker)
- * MULTILOG for multinomial logit item-response models (e.g. Du Toit, 2003)
- * PARSCALE for ordinal probit and logit item response models (e.g. Du Toit, 2003)
- * SABRE for two-level generalized linear mixed models (Francis *et al.*, 1996)
- * Stata's xt commands for two-level random intercept models (Stata Cross-Sectional Time-Series, 2003; StataCorp, 2003)
- * TESTFACT for exploratory multidimensional probit factor models (Bock *et al.*, 1999; Du Toit, 2003)

• Adaptive quadrature

  - Useful references: Pinheiro and Bates (1995), Bock and Schilling (1997), Evans and Swartz (2000), and Rabe-Hesketh *et al.* (2002, 2005a).
  - Software:

    * gllamm for generalized linear latent and mixed models in Stata (Rabe-Hesketh *et al.*, 2002, 2004b, 2005b)
    * NLMIXED for two-level generalized linear mixed models in SAS (Wolfinger, 1999)
    * TESTFACT for exploratory multidimensional probit factor models (Bock *et al.*, 1999; Du Toit, 2003)

• Monte Carlo integration:

  - Useful references: Train (2003), Cappellari and Jenkins (2003), and Gourieroux and Montfort (1996).
  - Software:

    * mvprobit for multivariate probit regression in Stata (Cappellari and Jenkins, 2003)
    * NLOGIT for multinomial logit and probit and other discrete choice models with random effects (Greene, 2002b)
    * DCM in Ox (Eklöf and Weeks, 2003)
    * Mixed logit estimation routine for panel data in GAUSS (Train *et al.*, 1999)

• EM algorithm:

  - Useful references: Tanner (1996), Schafer (1997), McLachlan and Krishnan (1997) and Little and Rubin (2002).
  - Software (some examples):

  * HLM for multilevel generalized linear mixed models (Raudenbush *et al.*, 2004)
  * Latent GOLD for latent class and related models with many different response types (Vermunt and Magidson, 2000, 2003a)

- Gradient methods

  - Useful references: Judge *et al.* (1985), Fletcher (1987), Gould *et al.* (2003), Thisted (1987), and Everitt (1987).
  - Software for maximizing arbitrary likelihood:

    * Stata's ml command (Gould *et al.*, 2003)
    * GAUSS's add-on application CML (Schoenberg, 1996)

- Limited information:

  - Useful references: Skrondal (1996), Küsters (1987) and Muthén and Satorra (1995).
  - Software:

    * Mplus (Muthén and Muthén, 1998, 2003)
    * MECOSA (Arminger *et al.*, 1996),

- (Marginal and penalized) quasi-likelihood:

  - Useful references: Goldstein (2003) and Breslow (2003).
  - Software:

    * HLM for multilevel generalized linear mixed models (Raudenbush *et al.*, 2004)
    * MLwiN for generalized linear mixed models (Rasbash *et al.*, 2004)
    * GLIMMIX for generalized linear mixed models in SAS (SAS/Stat User's Guide, version 8, 2000)
    * GLMM for generalized linear mixed models in Genstat (Payne, 2002)
    * glmmPQL for generalized linear mixed models in S and R (Venables and Ripley, 2002)

- Hierarchical-likelihood (h-likelihood):

  - Useful references: Lee and Nelder (1996, 2001)
  - Software:

    * HG procedures in Genstat (Payne, 2002)

- Generalized Estimating Equations:

  - Useful references: Pickles (1998), Hardin and Hilbe (2002), and Molenberghs (2002).
  - Software:

    * Stata's xtgee command (Stata Cross-Sectional Time-Series, 2003)
    * GENMOD in SAS (SAS/Stat User's Guide, version 8, 2000)
    * GEE in Genstat (Payne, 2002)

- Joint maximum likelihood

  - Useful references: Hambleton and Swaminathan (1985) for IRT; Lancaster (2000) on the incidental parameter problem.

- Conditional maximum likelihood:

  - Useful references: Clayton and Hills (1993), Breslow and Day (1980), Hamerle and Rönning (1995) and Cameron and Trivedi (1998).
  - Software:

    * EGRET (EGRET for Windows User Manual, 2000)
    * Stata's `clogit`, `xtpoisson`, etc. commands (Stata Cross-Sectional Time-Series, 2003)
    * LOGISTIC in SAS (SAS/Stat User's Guide, version 8, 2000)

- Bayes:

  - Useful references: Gelman *et al.* (2003) and Carlin and Louis (2000) on 'pragmatic' Bayesian statistics; Casella and George (1992) on the Gibbs sampler; Chib and Greenberg (1995) on Metropolis-Hastings; Rubin (1991), Gelman and Rubin (1996), Gilks *et al.* (1996), and Gilks (1998), on different MCMC methods; Zeger and Karim (1991), Albert (1992), Albert and Chib (1993), Dellaportas and Smith (1993), Clayton (1996b), Arminger and Muthén (1998) and Fox and Glas (2001) on MCMC for generalized linear mixed models, structural equation models and IRT models.
  - Software:

    * BUGS and WinBUGS for general Bayesian and hierarchical Bayesian models (Spiegelhalter *et al.*, 1996abc; see also Congdon, 2001)
    * GLMMGibbs for generalized linear mixed models by Gibbs sampling in R (Myles and Clayton, 2001)
    * MLwiN for generalized linear mixed models and multilevel factor models (Browne, 2004; see also Goldstein and Browne, 2002)
    * Routine for mixed logits with bounded distributions in Gauss (Train, 2002)

CHAPTER 7

Assigning values to latent variables

7.1 Introduction

In this chapter we discuss methods for assigning values to latent variables for individual clusters. The clusters could for instance be subjects in measurement and longitudinal modeling or schools in multilevel modeling. For continuous latent variables we will refer to this as *latent scoring* (factor scoring or random effects scoring) and for discrete latent variables as *classification*. Note that this terminology does not distinguish between different response types; factor scoring would for instance include scoring for IRT models with dichotomous responses.

Sometimes scoring and classification are the main aims of latent variable modeling, canonical examples being ability scoring based on IRT models and medical diagnosis based on latent class models. Other examples include disease mapping, small area estimation and assessments of institutional performance.

In the previous chapter we considered estimation of the fundamental parameters ϑ. Here we assume that these parameters have been estimated as $\widehat{\vartheta}$, yielding the structural parameter estimates $\widehat{\theta} = \mathbf{h}(\widehat{\vartheta})$. Sometimes, for instance in educational testing, the parameters are estimated using data from a large *calibration sample* which does not include the clusters to be scored. Advantages of this approach are that the parameter estimates are very precise. However, transporting the estimates across different populations can be problematic.

Unlike the previous chapter, we focus exclusively on frequentist methods, treating the estimated structural parameters as known. In the Bayesian approach both parameters and latent variables are treated as random variables, so there is no fundamental distinction between parameter estimation and latent scoring or classification. In the empirical Bayesian approach, inference regarding the latent variables is based on the conditional posterior distribution given the parameters (with estimates plugged in). The empirical Bayesian posterior distribution is discussed in Section 7.2.

We consider three methods of assigning values to latent variables that are motivated from general statistical principles. *Prediction* using empirical Bayes (EB) (also called expected a posteriori, EAP) is discussed in Section 7.3 and prediction using empirical Bayes modal (EBM) (also called modal a posteriori, MAP) in Section 7.4. *Estimation* using maximum likelihood (ML) is treated in Section 7.5. Empirical Bayes is the most common approach for latent scoring, whereas empirical Bayes modal is the most common approach for classification.

For each of the scoring methods different notions of variability of latent scores are contrasted. This is especially useful for empirical Bayes prediction where there is confusion in the literature regarding the meaning of different types of variances.

As well as discussing scoring and classification for the general model, we also investigate the special case of models without a structural part (e.g. factor models and random effects models) with multivariate normal latent variables and responses, henceforth for brevity referred to as the '*linear case*'. It turns out that many familiar scoring methods can be recognized as special cases of the general approaches, providing a deeper motivation for and understanding of these methods. The closed form expressions for the 'linear case' are also helpful for discussing concepts such as 'shrinkage'. To aid interpretation, the expressions are also presented for the special case of a two-level random intercept model. For more complex linear models, we find it instructive to substitute numerical values into the formulae in order to 'see' what happens in concrete examples. In Section 7.6 we demonstrate how the three scoring methods are related in the 'linear case'.

Since ad hoc methods are often used for latent scoring of hypothetical constructs, such approaches are briefly discussed in Section 7.7. In Section 7.8 we explore the wide range of uses of latent scoring and classification. A list of software implementing different methods for assigning values to latent variables is provided in an appendix.

Our discussion is confined to continuous latent variables unless otherwise indicated. We also confine our explicit treatment to disturbances or residuals ζ (with zero means) instead of the η, which in addition to the disturbances may be composed of regressions on observed covariates as well as on other latent variables. This is to simplify notation and because the disturbances are often of interest in their own right. Since the structural equations are linear, scores for the corresponding η can be obtained by substituting the scores for the disturbances into the reduced form for the latent variables (4.18). If there is no structural model, we simply have $\zeta = \eta$.

7.2 Posterior distributions

Frequentists often turn to Bayesian principles when assigning values to latent variables. The reason for this is that the latent variables can be interpreted as random 'parameters' (e.g. random effects) with a 'prior' distribution $h(\zeta; \theta)$, making the models appear similar to Bayesian models. An important difference is that the fully Bayesian approach discussed in Section 6.11 would also assume a prior for the structural parameters θ in addition to the prior for the disturbances ζ. In this case the priors for the parameters of the prior for ζ, e.g. the variances and covariances of the latent variables, would be referred to as *hyperpriors*. A Bayesian would base parameter estimation as well as latent scoring on posterior distributions given the responses $\mathbf{y}$. The relevant posterior for scoring would be marginal with respect to $\boldsymbol{\theta}$, whereas the posterior for parameter estimation would be marginal with respect to $\boldsymbol{\zeta}$.

When it comes to latent scoring, frequentists typically adopt an empirical-Bayesian approach. They estimate the parameters $\boldsymbol{\theta}$ by maximum likelihood (or another method) but rely on the conditional posterior distribution of the latent variables, given the estimated parameters $\widehat{\boldsymbol{\theta}}$, for prediction. For convenience we will in the sequel use Bayesian terminology, which would be technically correct if $\widehat{\boldsymbol{\theta}}$ were not estimated model parameters but fixed constants.

We have three different sources of information concerning the disturbances $\boldsymbol{\zeta}$. The first piece of information is the prior distribution $h(\boldsymbol{\zeta};\boldsymbol{\theta})$ representing our a priori knowledge about the latent variables, typically specified as multivariate normal when they are continuous. The second piece of information is provided by the observed responses $\mathbf{y}$ 'measuring' the latent variables. The third piece of information, which may not always be available, are covariates $\mathbf{X}$ in the response and/or structural models.

Note that it is not always clear whether one should use covariate information in models for latent scoring and classification. For instance, most people would agree that covariates such as age, gender and ethnicity should be used in diagnosis of heart disease if this reduces the risk of misdiagnosis. In contrast, use of such covariate information might be considered 'politically incorrect' or unfair in educational testing, even if it improves the quality of ability assessment (see also Section 9.4).

A natural way of combining the sources of information is through the *posterior distribution* $\omega(\boldsymbol{\zeta}|\mathbf{y},\mathbf{X};\widehat{\boldsymbol{\theta}})$ of $\boldsymbol{\zeta}$, the distribution of $\boldsymbol{\zeta}$ updated with or *given* the data $\mathbf{y}$ and $\mathbf{X}$. Thus the posterior provides a natural setting for inference concerning latent scores (see also Bartholomew, 1981). Using Bayes theorem, we obtain

$$\omega(\boldsymbol{\zeta}|\mathbf{y},\mathbf{X};\widehat{\boldsymbol{\theta}}) = \frac{\Pr(\mathbf{y},\boldsymbol{\zeta}|\mathbf{X};\widehat{\boldsymbol{\theta}})}{\Pr(\mathbf{y}|\mathbf{X};\widehat{\boldsymbol{\theta}})}.$$

In the general L-level model, the linear predictors of each level-1 unit in a top-level cluster z will generally depend on several elements of the vector of latent variables for that cluster $\boldsymbol{\zeta}_{z(L)}$. For example, in a three-level random intercept model, as shown in Display 3.2 on page 59, each response depends on both a level-2 and level-3 random intercept. It follows that a given response provides information on more than a single latent variable so that the latent variables are generally dependent under the posterior distribution. It is therefore useful to consider the joint posterior distribution of $\boldsymbol{\zeta}_{z(L)}$, given all responses $\mathbf{y}_{z(L)}$ and all covariates $\mathbf{X}_{z(L)}$ for a top-level cluster z,

$$\omega(\boldsymbol{\zeta}_{z(L)}|\mathbf{y}_{z(L)},\mathbf{X}_{z(L)};\widehat{\boldsymbol{\theta}}) = \frac{\Pr(\mathbf{y}_{z(L)},\boldsymbol{\zeta}_{z(L)}|\mathbf{X}_{z(L)};\widehat{\boldsymbol{\theta}})}{\Pr(\mathbf{y}_{z(L)}|\mathbf{X}_{z(L)};\widehat{\boldsymbol{\theta}})}.$$

To fix ideas, consider a simple two-level random intercept model. For a given cluster j, the joint density of the random intercept $\zeta_{j(2)}$ and the responses $\mathbf{y}_j$ in the numerator above can be written as $h(\zeta_{j(2)};\boldsymbol{\theta})\prod g^{(1)}(y_{ij}|\mathbf{x}_{ij},\zeta_{j(2)};\boldsymbol{\theta})$.

Here, we have utilized that the y_{ij} are conditionally independent given $\zeta_{j(2)}$. The marginal density of the responses in the denominator is just the integral of this joint density with respect to the random intercept so that the posterior density becomes

$$\omega(\zeta_{j(2)}|\mathbf{y}_{j(2)}, \mathbf{X}_{j(2)}; \widehat{\boldsymbol{\theta}}) = \frac{h(\zeta_{j(2)}; \widehat{\boldsymbol{\theta}}) \prod g^{(1)}(y_{ij}|\mathbf{x}_{ij}, \zeta_{j(2)}; \widehat{\boldsymbol{\theta}})}{\int h(\zeta_{j(2)}; \widehat{\boldsymbol{\theta}}) \prod g^{(1)}(y_{ij}|\mathbf{x}_{ij}, \zeta_{j(2)}; \widehat{\boldsymbol{\theta}}) \, d\zeta_{j(2)}}.$$

Observe that the posterior cannot in general be expressed in closed form because the integral in the denominator does not have an analytical solution. Since the denominator is just the likelihood contribution of the jth level-2 unit, the problem is the same as that of evaluating the marginal likelihood discussed in Section 6.3.

Writing the general L-level model as a two-level model in terms of $\boldsymbol{\eta}_{(L)}$ and $\boldsymbol{\zeta}_{(L)}$ as shown in Section 4.2.2, the expression above also applies in the L-level case,

$$\omega(\boldsymbol{\zeta}_{z(L)}|\mathbf{y}_{z(L)}, \mathbf{X}_{z(L)}; \widehat{\boldsymbol{\theta}})$$
$$= \frac{h(\boldsymbol{\zeta}_{z(L)}; \widehat{\boldsymbol{\theta}}) \prod g^{(1)}(y_{ij\ldots z}|\mathbf{x}_{ij\ldots z}, \boldsymbol{\zeta}_{z(L)}; \widehat{\boldsymbol{\theta}})}{\int h(\boldsymbol{\zeta}_{z(L)}; \widehat{\boldsymbol{\theta}}) \prod g^{(1)}(y_{ij\ldots z}|\mathbf{x}_{ij\ldots z}, \boldsymbol{\zeta}_{z(L)}; \widehat{\boldsymbol{\theta}}) \, d\boldsymbol{\zeta}_{z(L)}}, \qquad (7.1)$$

where the products are now over all level-1 units within the zth level-L unit. However, the integral in the denominator has a very high dimensionality and numerical integration should make use of the conditional independence structure to evaluate the integral recursively as in equation (6.2) on page 162.

In the 'linear case', it follows from standard results on conditional multivariate normal densities (e.g. Anderson, 2003) that the posterior density is *multivariate normal* (see equations (7.3) and (7.7) for the expectation vector and covariance matrix). For other response types, it follows from the *Bayesian central limit theorem* (e.g. Carlin and Louis, 2000) that the posterior density tends to multinormality as the number of units in the cluster increases.

Finally, consider a *discrete* random intercept model with prior probabilities π_{cj} for locations $\zeta^{(2)} = e_c$, $c = 1, \ldots, C$. The posterior probability that the intercept equals e_c is given by

$$\omega(e_c|\mathbf{y}_{j(2)}, \mathbf{X}_{j(2)}; \widehat{\boldsymbol{\theta}}) = \frac{\pi_{cj} \prod g^{(1)}(y_{ij}|\mathbf{x}_{ij}, e_c; \widehat{\boldsymbol{\theta}})}{\sum_{k=1}^{C} \pi_{kj} \prod g^{(1)}(y_{ij}|\mathbf{x}_{ij}, e_k; \widehat{\boldsymbol{\theta}})}.$$

In medical diagnosis using a single test result y_j, these probabilities are called the 'positive predictive value' if $y_j = 1$ (positive test result) and $c = 2$ (disease present) and the 'negative predictive value' if $y_j = 0$ (negative test result) and $c = 1$ (disease absent). In this context the prior probability is the prevalence of disease which represents the physician's knowledge of the patient diagnosis before seeing the test result. If the test is useful, the posterior probabilities will be substantially closer to zero or one than the prior probabilities. See Section 9.3 for an application to diagnosis of myocardial infarction (heart attack).

7.3 Empirical Bayes (EB)

7.3.1 Empirical Bayes prediction

Empirical Bayes prediction is undoubtedly the most widely used method for both factor and random effects scoring. Empirical Bayes predictors (see Efron and Morris, 1973, 1975; Morris, 1983) of the latent variables $\boldsymbol{\zeta}_{(L)}$ are their posterior means with parameter estimates $\widehat{\boldsymbol{\theta}}$ plugged in,

$$\widetilde{\boldsymbol{\zeta}}_{z(L)}^{\text{EB}} = \mathrm{E}(\boldsymbol{\zeta}_{z(L)}|\mathbf{y}_{z(L)}, \mathbf{X}_{z(L)}; \widehat{\boldsymbol{\theta}}).$$

Whenever the prior distribution is parametric, the predictor is denoted 'parametric empirical Bayes'.

The reason for the term 'empirical Bayes' is that, as noted earlier, Bayesian principles are adapted to a frequentist setting by plugging in estimated model parameters. The Bayesian would obtain the posterior distribution of the latent variables, marginal to $\boldsymbol{\theta}$, instead of simply plugging in estimates for $\boldsymbol{\theta}$. It is evident that the Bayes and empirical Bayes approaches are based on different philosophical underpinnings, prompting Lindley (1969) to remark:

> "there is no one less Bayesian than an empirical Bayesian."

Unfortunately, ambiguity still prevails in the terminology for Bayes and Empirical Bayes prediction. This is reflected in the term 'expected a posteriori' predictor (EAP) predominant in the psychometric literature (e.g. Bock and Aitkin, 1981; Bock and Mislevy, 1982; Bock, 1983, 1985; Muraki and Engelhard Jr., 1985) which seems to imply true Bayes prediction.

Despite the theoretical differences between Bayes and empirical Bayes inference, whenever $\widehat{\boldsymbol{\theta}}$ is consistent, the effect of substituting estimates for parameters is expected to be small if the likelihood dominates the (hyper)prior of $\boldsymbol{\theta}$, as in large samples and/or vague (hyper)priors. Little is known, however, about the consequences in the small or moderate sample situation. A reassuring theoretical result is provided by Deely and Lindley (1981) who point out that the empirical Bayes predictor is a first order approximation to the Bayes predictor.

The empirical Bayes predictor can be justified by considering the *summed quadratic loss function* defined as the unweighted sum of the squared errors of a predictor $\boldsymbol{\zeta}_{(L)}$. Dropping the z subscript,

$$\mathrm{L}^{\text{EB}}(\boldsymbol{\zeta}_{(L)}, \widetilde{\boldsymbol{\zeta}}_{(L)}) = (\boldsymbol{\zeta}_{(L)} - \widetilde{\boldsymbol{\zeta}}_{(L)})'(\boldsymbol{\zeta}_{(L)} - \widetilde{\boldsymbol{\zeta}}_{(L)}).$$

Treating the parameters as known, the empirical Bayes predictor minimizes the expected posterior loss:

$$\int (\boldsymbol{\zeta}_{(L)} - \widetilde{\boldsymbol{\zeta}}_{(L)})'(\boldsymbol{\zeta}_{(L)} - \widetilde{\boldsymbol{\zeta}}_{(L)})\,\omega(\boldsymbol{\zeta}_{(L)}|\mathbf{y}_{(L)}, \mathbf{X}_{(L)}; \widehat{\boldsymbol{\theta}})\,\mathrm{d}\boldsymbol{\zeta}_{(L)}. \tag{7.2}$$

Note that this can be given the intuitively pleasing interpretation as proportional to a *posterior mean squared error of prediction*, where the expectation is taken over the posterior distribution.

Searle *et al.* (1992, p.262) show that the empirical Bayes predictor also minimizes the mean squared error of prediction over the sampling distribution of **y** *if the parameters are treated as known*. For frequentists, this result might be more useful than the (empirical) Bayesian justification in terms of posterior loss. McCulloch and Searle (2001, p.257-258) emphasize that substitution of estimated parameters in the empirical Bayes predictor is purely pragmatic and has no statistical rationale.

It is clear that the mean squared error loss function is meaningless for truly discrete latent variables where predictions must coincide with the locations of the latent classes. Empirical Bayes prediction should therefore not be used in this case. However, empirical Bayes prediction can be used in nonparametric maximum likelihood estimation (NPMLE), where the discrete distribution is interpreted as a nonparametric estimator of a possibly continuous latent variable distribution. Empirical Bayes prediction based on NPMLE has been used by Clayton and Kaldor (1987), Laird (1982) and Rabe-Hesketh *et al.* (2003a), among others. The latter paper found that EB predictions based on NPMLE outperformed predictions based on normality for a skewed latent variable distribution. See Section 11.4 for an application of empirical Bayes prediction based on NPMLE for disease mapping.

For a two-level random intercept model with random intercept $\eta_j^{(2)} = \zeta_j^{(2)}$,

$$\nu_{ij} = \mathbf{x}'_{ij}\boldsymbol{\beta} + \zeta_j^{(2)},$$

the empirical Bayes predictor becomes

$$\widetilde{\zeta}_{j(2)}^{EB} = \frac{\int \zeta_j^{(2)} h(\zeta_j^{(2)}; \widehat{\boldsymbol{\theta}}) \prod_i g^{(1)}(y_{ij}|\mathbf{x}_{ij}, \zeta_j^{(2)}; \widehat{\boldsymbol{\theta}}) \, d\zeta_j^{(2)}}{\int h(\zeta_j^{(2)}; \widehat{\boldsymbol{\theta}}) \prod_i g^{(1)}(y_{ij}|\mathbf{x}_{ij}, \zeta_j^{(2)}; \widehat{\boldsymbol{\theta}}) \, d\zeta_j^{(2)}}.$$

In the general L-level model, the empirical Bayes prediction of the latent variable $\zeta^{(l)}$ at level l can be obtained recursively as

$$\mathrm{E}\left(\zeta^{(l)}|\boldsymbol{\zeta}^{([l+1]+)}\right) = \frac{\int \zeta^{(l)} h(\zeta^{(l)}) \prod g^{(l-1)} \, d\zeta^{(l)}}{\int h(\zeta^{(l)}) \prod g^{(l-1)} \, d\zeta^{(l)}},$$

and

$$\mathrm{E}\left(\zeta^{(l)}|\boldsymbol{\zeta}^{([l+k+1]+)}\right) = \frac{\int \mathrm{E}\left(\zeta^{(l)}|\boldsymbol{\zeta}^{([l+k]+)}\right) h(\zeta^{(l+k)}) \prod g^{(l+k-1)} \, d\zeta^{(l+k)}}{\int h(\zeta^{(l+k)}) \prod g^{(l+k-1)} \, d\zeta^{(l+k)}},$$

where $k = 1, \ldots, L-l$ and we have written $g^{(l)}$ for $g^{(l)}(\mathbf{y}_{(l)}|\mathbf{X}_{(l)}, \boldsymbol{\zeta}^{([l+1]+)}; \widehat{\boldsymbol{\theta}})$ to simplify notation. In general, it is impossible to obtain empirical Bayes predictions using analytical integration. Any of the numerical integration methods discussed in Section 6.3 can be used, see for instance page 167 for adaptive quadrature. For discrete distributions, the integrals are simply replaced by sums.

The 'linear case': We now consider the 'linear case' with no structural model, i.e. $\boldsymbol{\eta}_{z(L)} = \boldsymbol{\zeta}_{z(L)}$, specified as in (4.3),

$$\mathbf{y}_{z(L)} = \mathbf{X}_{z(L)}\boldsymbol{\beta} + \boldsymbol{\Lambda}_{z(L)}\boldsymbol{\zeta}_{z(L)} + \boldsymbol{\epsilon}_{z(L)}.$$

Here, the latent variables $\boldsymbol{\zeta}_{z(L)}$ are multivariate normal with estimated covariance matrix $\widehat{\boldsymbol{\Psi}}_{(L)}$ and the disturbances $\boldsymbol{\epsilon}_{z(L)}$ are multivariate normal with estimated diagonal covariance matrix $\widehat{\boldsymbol{\Theta}}_{z(L)}$ with elements $\widehat{\theta}_{ii}$ (not to be confused with the vector $\widehat{\boldsymbol{\theta}}$ of *all* estimated parameters).

The empirical Bayes predictor can in this case be expressed as

$$\widetilde{\boldsymbol{\zeta}}_{z(L)}^{\text{EB}} = \widehat{\boldsymbol{\Psi}}_{(L)}\widehat{\boldsymbol{\Lambda}}'_{z(L)}\widehat{\boldsymbol{\Sigma}}_{z(L)}^{-1}\left(\mathbf{y}_{z(L)} - \mathbf{X}_{z(L)}\widehat{\boldsymbol{\beta}}\right), \qquad (7.3)$$

where

$$\widehat{\boldsymbol{\Sigma}}_{z(L)} \equiv \widehat{\boldsymbol{\Lambda}}_{z(L)}\widehat{\boldsymbol{\Psi}}_{(L)}\widehat{\boldsymbol{\Lambda}}'_{z(L)} + \widehat{\boldsymbol{\Theta}}_{z(L)},$$

the estimated residual covariance structure of $\mathbf{y}_{z(L)}$. For factor models the term $\widehat{\boldsymbol{\Psi}}_{(L)}\widehat{\boldsymbol{\Lambda}}'_{z(L)}\widehat{\boldsymbol{\Sigma}}_{z(L)}^{-1}$ in (7.3) is often called the *factor scoring matrix* for the regression method.

The unconditional expectation (over $\mathbf{y}_{z(L)}$) becomes

$$\text{E}_{\mathbf{y}}\left(\widetilde{\boldsymbol{\zeta}}_{z(L)}^{\text{EB}} | \mathbf{X}_{z(L)}; \widehat{\boldsymbol{\theta}}\right) = 0,$$

because the expectation of the term in brackets in (7.3) is zero. If we condition on the true realized latent variables $\boldsymbol{\zeta}_{z(L)}$, this is no longer the case with $\text{E}_{\mathbf{y}}(\mathbf{y}_{z(L)} - \mathbf{X}_{z(L)}\widehat{\boldsymbol{\beta}} | \boldsymbol{\zeta}_{z(L)}, \mathbf{X}_{z(L)}; \widehat{\boldsymbol{\theta}}) = \widehat{\boldsymbol{\Lambda}}_{z(L)}\boldsymbol{\zeta}_{z(L)}$. The conditional expectation given $\boldsymbol{\zeta}_{z(L)}$ therefore is

$$\text{E}_{\mathbf{y}}\left(\widetilde{\boldsymbol{\zeta}}_{z(L)}^{\text{EB}} | \boldsymbol{\zeta}_{z(L)}, \mathbf{X}_{z(L)}; \widehat{\boldsymbol{\theta}}\right) = \left(\widehat{\boldsymbol{\Psi}}_{(L)}\widehat{\boldsymbol{\Lambda}}'_{z(L)}\widehat{\boldsymbol{\Sigma}}_{z(L)}^{-1}\widehat{\boldsymbol{\Lambda}}_{z(L)}\right)\boldsymbol{\zeta}_{z(L)}$$

$$= \boldsymbol{\zeta}_{z(L)} - \left(\mathbf{I} + \widehat{\boldsymbol{\Psi}}_{(L)}\widehat{\boldsymbol{\Omega}}_{z(L)}\right)^{-1}\boldsymbol{\zeta}_{z(L)}, \quad (7.4)$$

where

$$\widehat{\boldsymbol{\Omega}}_{z(L)} \equiv \widehat{\boldsymbol{\Lambda}}'_{z(L)}\widehat{\boldsymbol{\Theta}}_{z(L)}^{-1}\widehat{\boldsymbol{\Lambda}}_{z(L)}.$$

Hence, the empirical Bayes predictor is unconditionally unbiased but conditionally biased since the last term in (7.4) does not in general equal a zero vector. In the 'linear case', the empirical Bayes predictor is the 'Best Linear Unbiased Predictor' BLUP (Goldberger, 1962; Robinson, 1991) since it is linear in $\mathbf{y}_{(L)}$, unconditionally unbiased and best in the sense that it minimizes the marginal sampling variance of the prediction error, if the parameters are treated as known (see also Lawley and Maxwell, 1971). Note that the BLUP concept is more general than parametric empirical Bayes in the sense that it does not rely on distributional assumptions (e.g. McCulloch and Searle, 2001, p.256).

It turns out that many results from the statistical and psychometric literature can be derived as special cases of the above formulae. For the conventional common factor model (see Sections 3.3.2 and 3.3.3), the empirical Bayes predictor is just the *regression method* for factor scoring discussed by Thomson (1938) and Thurstone (1935) in their seminal treatments of factor analysis. Interestingly, the predictor proposed by Spearman (1927) for his one-factor model is a special case of the above, and consequently an early application of empirical Bayes methodology.

For random effects models we obtain the results reported by for instance Rao (1975) and Strenio *et al.* (1983). For example, in a two-level random intercept model with homoscedastic level-1 variances $\theta_{ii} = \theta$, the empirical Bayes predictor reduces to

$$
\begin{aligned}
\widetilde{\zeta}_{j(2)}^{\mathrm{EB}} &= \frac{\widehat{\psi}}{\widehat{\psi} + \widehat{\theta}/n_j} \left(\frac{1}{n_j} \sum_{i=1}^{n_j} (y_{ij} - \mathbf{x}_{ij}' \widehat{\boldsymbol{\beta}}) \right) \\
&= \widehat{R}_{j(2)} \left(\frac{1}{n_j} \sum_{i=1}^{n_j} (y_{ij} - \mathbf{x}_{ij}' \widehat{\boldsymbol{\beta}}) \right).
\end{aligned}
\tag{7.5}
$$

The term in parentheses is the mean 'raw' or total residual for cluster j and $\widehat{R}_{j(2)}$ is a *shrinkage* factor which pulls the empirical Bayes prediction towards zero, the mean of the prior distribution. The shrinkage factor can be interpreted as the estimated reliability of the mean raw residual as a 'measurement' of $\zeta_{j(2)}$ (the variance of the 'true score' over the total variance). The reliability is smallest when n_j is small and when θ is large compared with ψ; the conditional density of the responses $\prod g^{(1)}(y_{ij}|\zeta_{j(2)}, \mathbf{x}_{ij}; \widehat{\boldsymbol{\theta}})$ then becomes flat and uninformative compared with the prior density $h(\zeta_{j(2)}; \widehat{\boldsymbol{\theta}})$.

In empirical Bayes prediction the effect of the prior for small clusters is to pull the predictions $\mathbf{x}_{ij}' \widehat{\boldsymbol{\beta}} + \widetilde{\zeta}_{j(2)}$ toward $\mathbf{x}_{ij}' \widehat{\boldsymbol{\beta}}$ (where all clusters contribute to the estimation of $\boldsymbol{\beta}$), often referred to as 'borrowing strength' from the other clusters. We will show in Section 7.6 how the concept of shrinkage also applies to empirical Bayes predictions for the general 'linear case'.

A concrete example of a three-level linear random intercept model is helpful, and we will repeatedly return to this example in the chapter.

Example: A level-3 unit contains two level-2 units with two level-1 units within each. As shown in Display 3.2 on page 59, the structure matrix $\widehat{\boldsymbol{\Lambda}}_{1(3)}$ (referred to as $\mathbf{Z}_{1(3)}$ in the display) and latent variable vector $\boldsymbol{\zeta}_{1(3)}$ are

$$
\widehat{\boldsymbol{\Lambda}}_{1(3)} = \begin{bmatrix} 1 & 0 & 1 \\ 1 & 0 & 1 \\ 0 & 1 & 1 \\ 0 & 1 & 1 \end{bmatrix} \quad \text{and} \quad \boldsymbol{\zeta}_{1(3)} = \begin{bmatrix} \zeta_{11}^{(2)} \\ \zeta_{21}^{(2)} \\ \zeta_{11}^{(3)} \end{bmatrix}.
$$

Assuming that the random effects covariance matrix was estimated as

$$
\widehat{\boldsymbol{\Psi}}_{(3)} = \begin{bmatrix} 1 & 0 & 0 \\ 0 & 1 & 0 \\ 0 & 0 & 2 \end{bmatrix},
$$

and the level-1 variance as $\widehat{\theta} = 1$, the empirical Bayes predictions become

$$
\widetilde{\boldsymbol{\zeta}}_{z(L)}^{\text{EB}} =
\begin{bmatrix}
0.21 & 0.21 & -0.12 & -0.12 \\
-0.12 & -0.12 & 0.21 & 0.21 \\
0.18 & 0.18 & 0.18 & 0.18
\end{bmatrix}
\begin{bmatrix}
y_{11} - \mathbf{x}'_{11}\widehat{\boldsymbol{\beta}} \\
y_{21} - \mathbf{x}'_{21}\widehat{\boldsymbol{\beta}} \\
y_{31} - \mathbf{x}'_{31}\widehat{\boldsymbol{\beta}} \\
y_{41} - \mathbf{x}'_{41}\widehat{\boldsymbol{\beta}}
\end{bmatrix},
$$

where the 'weights' have been rounded to two decimal places. Note that all responses for the same top-level unit provide information on all latent variables for that unit. The conditional expectations given the true realized latent variables are

$$
\mathbf{E}_{\mathbf{y}}\left(\widetilde{\boldsymbol{\zeta}}_{z(L)}^{\text{EB}} | \boldsymbol{\zeta}_{z(L)}\right) =
\begin{bmatrix}
0.42 & -0.24 & 0.18 \\
-0.24 & 0.42 & 0.18 \\
0.36 & 0.36 & 0.73
\end{bmatrix}
\begin{bmatrix}
\zeta_{11}^{(2)} \\
\zeta_{21}^{(2)} \\
\zeta_{11}^{(3)}
\end{bmatrix}.
$$

Here the true realization of each latent variable affects the mean predictions of *all* latent variables in the same cluster. For example, a large true level-3 random intercept will lead, on average, to larger predictions for the level-2 random intercepts.

It is important to note that latent scores generally cannot simply be plugged into non linear functions of latent variables to obtain predictions of the function. For instance, in a dichotomous random intercept logit model, the probability of a positive response takes the form $[1 + \exp(-\mathbf{x}'_{ij}\boldsymbol{\beta} - \zeta_j)]^{-1}$. To obtain the empirical Bayes predictor of the probability we must integrate the nonlinear function with respect to the posterior distribution of the latent variable instead of plugging in $\widetilde{\zeta}_j$.

7.3.2 Empirical Bayes variances and covariances

We consider four types of variances and covariances of latent scores (we will use the term 'covariances' to stand for both):

- Posterior covariances: $\text{Cov}(\boldsymbol{\zeta}_{(L)} | \mathbf{y}_{(L)}, \mathbf{X}_{(L)}, \widehat{\boldsymbol{\theta}})$

- Marginal sampling covariances: $\text{Cov}_{\mathbf{y}}(\widetilde{\boldsymbol{\zeta}}_{(L)} | \mathbf{X}_{(L)}, \widehat{\boldsymbol{\theta}})$

- Conditional sampling covariances: $\text{Cov}_{\mathbf{y}}(\widetilde{\boldsymbol{\zeta}}_{(L)} | \boldsymbol{\zeta}_{(L)}, \mathbf{X}_{(L)}, \widehat{\boldsymbol{\theta}})$

- Prediction error covariances (marginal): $\text{Cov}_{\mathbf{y}}(\widetilde{\boldsymbol{\zeta}}_{(L)} - \boldsymbol{\zeta}_{(L)} | \mathbf{X}_{(L)}, \widehat{\boldsymbol{\theta}})$

These covariances are relevant to all scoring methods discussed in this chapter. We have again substituted estimates $\widehat{\boldsymbol{\theta}}$ for the structural parameters $\boldsymbol{\theta}$. Note that the posterior covariance is not fully Bayesian since it is not marginal with respect to a random parameter vector $\boldsymbol{\theta}$ and the sampling covariances are not fully frequentist since the sampling variability of $\widehat{\boldsymbol{\theta}}$ is ignored.

Posterior variances and covariances

The *posterior* covariance matrix is the covariance matrix of the latent variables over the posterior distribution, given the observed responses and covariates. The *empirical posterior* covariance matrix, which we will rely on, is the posterior covariance matrix with parameter estimates plugged in. Confidence intervals based on the posterior mean and posterior standard deviation are analogous to Bayesian credible intervals based on approximate normality of the posterior.

The empirical posterior covariance matrix is given by

$$\text{Cov}(\boldsymbol{\zeta}_{(L)}|\mathbf{y}_{(L)}, \mathbf{X}_{(L)}; \widehat{\boldsymbol{\theta}})$$

$$= \int \left(\boldsymbol{\zeta}_{(L)} - \widetilde{\boldsymbol{\zeta}}_{(L)}^{\text{EB}}\right)\left(\boldsymbol{\zeta}_{(L)} - \widetilde{\boldsymbol{\zeta}}_{(L)}^{\text{EB}}\right)' \omega(\boldsymbol{\zeta}_{(L)}|\mathbf{y}_{(L)}, \mathbf{X}_{(L)}; \widehat{\boldsymbol{\theta}}) \, d\boldsymbol{\zeta}_{(L)},$$

which can be obtained by numerical integration as shown for adaptive quadrature in Section 6.3.2.

The empirical posterior variances are *biased downwards* compared with the fully Bayesian posterior variances since the structural parameters $\boldsymbol{\theta}$ are treated as known. This can be seen by writing the fully Bayesian posterior variance matrix as

$$\begin{aligned}\text{Cov}(\boldsymbol{\zeta}_{(L)}|\mathbf{y}_{(L)}, \mathbf{X}_{(L)}) &= \text{E}_{\boldsymbol{\theta}}[\text{Cov}(\boldsymbol{\zeta}_{(L)}|\mathbf{y}_{(L)}, \mathbf{X}_{(L)}; \boldsymbol{\theta})] \\ &+ \text{Cov}_{\boldsymbol{\theta}}[\text{E}(\boldsymbol{\zeta}_{(L)}|\mathbf{y}_{(L)}, \mathbf{X}_{(L)}; \boldsymbol{\theta})],\end{aligned} \qquad (7.6)$$

where the first term is approximated by the empirical Bayesian posterior covariance matrix. Importantly, the first term of (7.6) will dominate when the number of clusters is large and the number of units per cluster is small (Kass and Steffey, 1989). Kass and Steffey also suggest approximations for the second term. Assuming uniform priors for all parameters, their first-order approximation (in terms of the estimated fundamental parameters $\boldsymbol{\vartheta}$) is simply

$$\text{Cov}_{\boldsymbol{\vartheta}}\left[\text{E}(\boldsymbol{\zeta}_{(L)}|\mathbf{y}_{(L)}, \mathbf{X}_{(L)}; \widehat{\boldsymbol{\vartheta}})\right]$$

$$\approx -\mathbf{H}^{-1}\left(\frac{\partial \text{E}(\boldsymbol{\zeta}_{(L)}|\mathbf{y}_{(L)}, \mathbf{X}_{(L)}; \widehat{\boldsymbol{\vartheta}})}{\partial \boldsymbol{\vartheta}}\right)\left(\frac{\partial \text{E}(\boldsymbol{\zeta}_{(L)}|\mathbf{y}_{(L)}, \mathbf{X}_{(L)}; \widehat{\boldsymbol{\vartheta}})}{\partial \boldsymbol{\vartheta}}\right)',$$

where $\mathbf{H}$ is the Hessian of the log-likelihood at $\widehat{\boldsymbol{\vartheta}}$ and the terms in brackets are partial derivatives of the empirical Bayes predictions with respect to $\boldsymbol{\vartheta}$, evaluated at the estimates $\widehat{\boldsymbol{\vartheta}}$. In linear mixed models, this approximation has a closed form and is discussed by Goldstein (2003, Appendix 2.2). Ten Have and Localio (1999) use numerical integration to evaluate this approximation for mixed effects logistic regression. See Section 9.5 for a comparison of fully Bayesian and empirical Bayesian credible intervals for study-specific treatment effects in meta-analysis.

The posterior standard deviation is often used by frequentists as a standard error of the empirical Bayes prediction. This can be justified in the 'linear case' where the posterior standard deviation equals the sampling standard deviation of the prediction error (see page 234). However, the posterior standard deviation is also commonly used in IRT (e.g. Embretson and Reise, 2000) and generalized linear mixed models (e.g. Ten Have and Localio, 1999), apparently without any frequentist justification.

The 'linear case': In this case the empirical posterior covariance matrix becomes

$$\text{Cov}(\boldsymbol{\zeta}_{z(L)}|\mathbf{y}_{z(L)}, \mathbf{X}_{z(L)}; \widehat{\boldsymbol{\theta}}) = \widehat{\boldsymbol{\Psi}}_{(L)} - \widehat{\boldsymbol{\Psi}}'_{(L)}\widehat{\boldsymbol{\Lambda}}'_{z(L)}\widehat{\boldsymbol{\Sigma}}^{-1}_{z(L)}\widehat{\boldsymbol{\Lambda}}_{z(L)}\widehat{\boldsymbol{\Psi}}_{(L)}, \quad (7.7)$$

which for the special case of the random intercept model reduces to

$$\text{Var}(\zeta_{j(2)}|\mathbf{y}_{j(2)}, \mathbf{X}_{j(2)}; \widehat{\boldsymbol{\theta}}) = \frac{\widehat{\theta}/n_j}{\widehat{\psi} + \widehat{\theta}/n_j} \, \widehat{\psi} = (1 - \widehat{R}_{j(2)}) \, \widehat{\psi}.$$

As expected, the posterior variance is smaller than the prior variance due to the information gained regarding the random intercept by knowing the responses $\mathbf{y}_{j(2)}$.

Example: Returning to the three-level example, the posterior covariance matrix becomes

$$\text{Cov}(\boldsymbol{\zeta}_{1(3)}|\mathbf{y}_{1(3)}, \mathbf{X}_{1(3)}; \widehat{\boldsymbol{\theta}}) = \begin{bmatrix} 0.58 & & \\ 0.24 & 0.58 & \\ -0.36 & -0.36 & 0.55 \end{bmatrix},$$

where there are two important things to note. First, even though the random intercepts of the two level-2 units are uncorrelated under the prior distribution, the posterior covariance is nonzero (equal to 0.24). Second, even though there are no cross-level covariances under the prior, the level-2 random intercepts have nonzero covariances with the level-3 random intercept (equal to -0.36).

Marginal sampling variances and covariances

The *marginal sampling* covariances are the covariances of the predictions under repeated sampling of clusters and units within clusters, keeping both the covariates and the parameter estimates fixed.

In contrast to the conditional sampling covariances discussed in the next section, the marginal sampling covariances also reflect variability due to sampling of the latent variables from their prior distribution. The marginal sampling standard deviation can therefore be used for detecting clusters that appear inconsistent with the model. For this reason, Goldstein (2003) refers to this quantity as the 'diagnostic standard error'. See Section 8.6.2 for further discussion.

The marginal sampling covariance matrix of the empirical Bayes predictor is

$$\text{Cov}_{\mathbf{y}}\left(\widetilde{\boldsymbol{\zeta}}^{\text{EB}}_{(L)}|\mathbf{X}_{(L)}; \widehat{\boldsymbol{\theta}}\right) = \text{Cov}_{\mathbf{y}}\left[\text{E}(\boldsymbol{\zeta}_{(L)}|\mathbf{y}_{(L)}, \mathbf{X}_{(L)}; \widehat{\boldsymbol{\theta}})\right]$$

$$= \int \widetilde{\boldsymbol{\zeta}}^{\text{EB}}_{(L)}\widetilde{\boldsymbol{\zeta}}^{\text{EB}\prime}_{(L)} \, g^{(L)}(\mathbf{y}_{(L)}|\mathbf{X}_{(L)}; \widehat{\boldsymbol{\theta}}) \, d\mathbf{y}_{(L)},$$

where $g^{(L)}(\mathbf{y}_{(L)}|\mathbf{X}_{(L)}; \widehat{\boldsymbol{\theta}})$ is the joint marginal distribution of the responses for the top-level cluster. Note that we have dropped the z subscript to simplify notation and will continue to do so in the remainder of this chapter.

There is unfortunately no closed form expression for the general model. However, Skrondal (1996) suggested using the relation

$$\mathrm{Cov}(\boldsymbol{\zeta}_{(L)}|\mathbf{X}_{(L)};\widehat{\boldsymbol{\theta}}) = \mathrm{E}_{\mathbf{y}}\Big[\mathrm{Cov}(\boldsymbol{\zeta}_{(L)}|\mathbf{y}_{(L)},\mathbf{X}_{(L)};\widehat{\boldsymbol{\theta}})\Big]$$
$$+ \mathrm{Cov}_{\mathbf{y}}\Big[\mathrm{E}(\boldsymbol{\zeta}_{(L)}|\mathbf{y}_{(L)},\mathbf{X}_{(L)};\widehat{\boldsymbol{\theta}})\Big],$$

to derive an approximate expression. Recognizing that the last term is the required covariance matrix of the empirical Bayes predictor, this can be rewritten as

$$\mathrm{Cov}_{\mathbf{y}}\left(\widetilde{\boldsymbol{\zeta}}_{(L)}^{\mathrm{EB}}|\mathbf{X}_{(L)};\widehat{\boldsymbol{\theta}}\right) = \mathrm{Cov}(\boldsymbol{\zeta}_{(L)}|\mathbf{X}_{(L)};\widehat{\boldsymbol{\theta}}) - \mathrm{E}_{\mathbf{y}}\Big[\mathrm{Cov}(\boldsymbol{\zeta}_{(L)}|\mathbf{y}_{(L)},\mathbf{X}_{(L)};\widehat{\boldsymbol{\theta}})\Big].$$

The first term on the right-hand-side is the estimated prior covariance matrix $\widehat{\boldsymbol{\Psi}}$ of the vector of latent variables $\boldsymbol{\zeta}_{(L)}$ and the second term is the expectation of the posterior covariance matrix which can be approximated by the posterior covariance matrix. Therefore we obtain

$$\mathrm{Cov}_{\mathbf{y}}\left(\widetilde{\boldsymbol{\zeta}}_{(L)}^{\mathrm{EB}}|\mathbf{X}_{(L)};\widehat{\boldsymbol{\theta}}\right) \approx \widehat{\boldsymbol{\Psi}}_{(L)} - \mathrm{Cov}(\boldsymbol{\zeta}_{(L)}|\mathbf{y}_{(L)},\mathbf{X}_{(L)};\widehat{\boldsymbol{\theta}}). \qquad (7.8)$$

An alternative approximation would be via simulation, first sampling the latent variables from the prior distribution and then the responses from their conditional distribution given the latent variables and covariates. An advantage of this approach is that uncertainty in the parameter estimates $\widehat{\boldsymbol{\theta}}$ is easily accommodated by drawing new samples from their sampling distribution before sampling the latent variables as suggested by Longford (2001) in a related context.

The 'linear case': In this case, (7.8) holds perfectly and the marginal sampling variance becomes

$$\mathrm{Cov}_{\mathbf{y}}\left(\widetilde{\boldsymbol{\zeta}}_{(L)}^{\mathrm{EB}}|\mathbf{X}_{(L)};\widehat{\boldsymbol{\theta}}\right) = \widehat{\boldsymbol{\Psi}}_{(L)} - \mathrm{Cov}(\boldsymbol{\zeta}_{(L)}|\mathbf{y}_{(L)},\mathbf{X}_{(L)};\widehat{\boldsymbol{\theta}})$$
$$= \widehat{\boldsymbol{\Psi}}_{(L)}\widehat{\boldsymbol{\Lambda}}'_{(L)}\widehat{\boldsymbol{\Sigma}}^{-1}_{(L)}\widehat{\boldsymbol{\Lambda}}_{(L)}\widehat{\boldsymbol{\Psi}}_{(L)}. \qquad (7.9)$$

The diagonal elements of this covariance matrix are clearly smaller than those of the prior covariance matrix since the posterior variances are positive. This is due to shrinkage.

Shrinkage has led some researchers (e.g. Louis, 1984) to suggest *adjusted* empirical Bayes predictors with the same covariances as the prior distribution. This predictor minimizes the posterior expectation of the summed quadratic loss function (for given parameter estimates) in (7.2) subject to the side condition that the predictions satisfy the estimated first and second order moments of the prior distribution. In the factor analysis literature, the idea of obtaining factor scores with the same covariance matrix as the prior distribution dates back to Anderson and Rubin (1956), who considered models with orthonormal factors. This 'covariance preserving' approach has been extended to general prior covariance matrices (e.g. Ten Berge, 1983; Ten Berge *et al.*, 1999).

For a simple random intercept model, the marginal sampling variance reduces to

$$\text{Var}_{\mathbf{y}} \left(\widetilde{\zeta}_{j(2)}^{\text{EB}} | \mathbf{X}_{j(2)}; \widehat{\boldsymbol{\theta}} \right) = \frac{\widehat{\psi}}{\widehat{\psi} + \widehat{\theta}/n_j} \, \widehat{\psi} = \widehat{R}_{j(2)} \, \widehat{\psi}.$$

Example: Returning to the example of a three-level random intercept model, the marginal sampling covariance matrix of the random effects of the level-3 unit becomes

$$\text{Cov}_{\mathbf{y}} \left(\widetilde{\zeta}_{1(3)}^{\text{EB}} | \mathbf{X}_{1(3)}; \widehat{\boldsymbol{\theta}} \right) = \begin{bmatrix} 0.42 & & \\ -0.24 & 0.42 & \\ 0.36 & 0.36 & 1.45 \end{bmatrix}.$$

Note that there are nonzero covariances between the level-2 random intercepts for different level-2 units as for the posterior covariance matrix. Also observe that there are again nonzero covariances across levels.

Nonzero sampling covariances across levels complicate diagnostics in multilevel models considerably. Unfortunately, this problem has often been overlooked in the multilevel literature, for example by Langford and Lewis (1998) and Goldstein (2003).

Conditional sampling variances and covariances

The *conditional sampling* covariances are the covariances of the predictions under repeated sampling of units from the same cluster with fixed 'true' latent variables $\boldsymbol{\zeta}_{(L)}$ (in addition to fixed covariates and parameter estimates).

Note that the conditional sampling standard deviation should not be confused with the 'comparative standard error' used by Goldstein (2003, p.23). Goldstein describes this as being conditional on the true latent variables but it is actually the *marginal* (over $\boldsymbol{\zeta}$) sampling variance of the prediction errors, which we will show to be equal to the posterior variance in the 'linear case' on page 235.

The conditional sampling covariance matrix of the empirical Bayes predictors $\widetilde{\boldsymbol{\zeta}}_{(L)}^{\text{EB}}$, given the latent variables, is

$$\begin{aligned} \text{Cov}_{\mathbf{y}} \left(\widetilde{\boldsymbol{\zeta}}_{(L)}^{\text{EB}} | \boldsymbol{\zeta}_{(L)}, \mathbf{X}_{(L)}; \widehat{\boldsymbol{\theta}} \right) &= \text{Cov}_{\mathbf{y}} \left[\text{E}(\boldsymbol{\zeta}_{(L)} | \mathbf{y}_{(L)}, \mathbf{X}_{(L)}; \widehat{\boldsymbol{\theta}}) | \boldsymbol{\zeta}_{(L)} \right] \\ &= \int \widetilde{\boldsymbol{\zeta}}_{(L)}^{\text{EB}} \widetilde{\boldsymbol{\zeta}}_{(L)}^{\text{EB}\prime} \prod g^{(1)}(y_{ij...z} | \boldsymbol{\zeta}_{(L)}, \mathbf{X}_{(L)}; \widehat{\boldsymbol{\theta}}) \, \mathrm{d}\mathbf{y}_{(L)}, \end{aligned}$$

where the product represents the joint conditional distribution of the responses to all level-1 units within the level-L unit.

The 'linear case': Here, the conditional sampling covariance matrix becomes

$$\text{Cov}_{\mathbf{y}} \left(\widetilde{\boldsymbol{\zeta}}_{(L)}^{\text{EB}} | \boldsymbol{\zeta}_{(L)}, \mathbf{X}_{(L)}; \widehat{\boldsymbol{\theta}} \right) = \widehat{\boldsymbol{\Psi}}_{(L)}' \widehat{\boldsymbol{\Lambda}}_{(L)}' \widehat{\boldsymbol{\Sigma}}_{(L)}^{-1} \widehat{\boldsymbol{\Theta}}_{(L)} \widehat{\boldsymbol{\Sigma}}_{(L)}^{-1} \widehat{\boldsymbol{\Lambda}}_{(L)} \widehat{\boldsymbol{\Psi}}_{(L)}, (7.10)$$

which for the special case of the random intercept model can be expressed as

$$\text{Var}_{\mathbf{y}} \left(\widetilde{\zeta}_{j(2)}^{\text{EB}} | \zeta_{j(2)}, \mathbf{X}_{j(2)}; \widehat{\boldsymbol{\theta}} \right) = \widehat{R}_{j(2)} (1 - \widehat{R}_{j(2)}) \, \widehat{\psi}.$$

Example: In the three-level random intercept model, the conditional sampling covariance matrix for the random effects of the level-3 unit becomes

$$\text{Cov}_{\mathbf{y}}\left(\tilde{\boldsymbol{\zeta}}_{1(3)}^{\text{EB}}|\boldsymbol{\zeta}_{1(3)}, \mathbf{X}_{1(3)}; \widehat{\boldsymbol{\theta}}\right) = \begin{bmatrix} 0.12 & & \\ -0.10 & 0.12 & \\ 0.03 & 0.03 & 0.13 \end{bmatrix}.$$

As expected, the variances are considerably lower than the marginal sampling variances. Note also the difference between the conditional sampling covariance matrix and the corresponding posterior covariance matrix in this example as these variances are often confused.

Prediction error variances and covariances

The *prediction error* covariances are the covariances of the prediction errors $\tilde{\boldsymbol{\zeta}}_{(L)}^{\text{EB}} - \boldsymbol{\zeta}_{(L)}$ under repeated sampling of the responses from their marginal distribution,

$$\text{Cov}_{\mathbf{y}}\left(\tilde{\boldsymbol{\zeta}}_{(L)}^{\text{EB}} - \boldsymbol{\zeta}_{(L)}|\mathbf{X}_{(L)}; \widehat{\boldsymbol{\theta}}\right)$$
$$= \int \left(\tilde{\boldsymbol{\zeta}}_{(L)}^{\text{EB}} - \boldsymbol{\zeta}_{(L)}\right)\left(\tilde{\boldsymbol{\zeta}}_{(L)}^{\text{EB}} - \boldsymbol{\zeta}_{(L)}\right)' g^{(L)}(\mathbf{y}_{(L)}|\mathbf{X}_{(L)}; \widehat{\boldsymbol{\theta}}) \, d\mathbf{y}_{(L)}.$$

The covariances are marginal since they reflect variability due to sampling of the latent variables as well as the sampling of responses given the latent variables. The standard deviation of the prediction errors is perhaps the most obvious measure of prediction uncertainty.

Using a multivariate generalization of a derivation in Waclawiw and Liang (1994), the prediction error covariance matrix can be expressed as (omitting the conditioning on $\mathbf{X}_{(L)}$ and $\widehat{\boldsymbol{\theta}}$ for brevity)

$$\text{Cov}_{\mathbf{y}}\left(\tilde{\boldsymbol{\zeta}}_{(L)}^{\text{EB}} - \boldsymbol{\zeta}_{(L)}\right)$$
$$= \text{E}_{\mathbf{y}}\left[\left(\tilde{\boldsymbol{\zeta}}_{(L)}^{\text{EB}} - \boldsymbol{\zeta}_{(L)}\right)\left(\tilde{\boldsymbol{\zeta}}_{(L)}^{\text{EB}} - \boldsymbol{\zeta}_{(L)}\right)'\right] - \text{E}_{\mathbf{y}}\left(\tilde{\boldsymbol{\zeta}}_{(L)}^{\text{EB}} - \boldsymbol{\zeta}_{(L)}\right)\text{E}_{\mathbf{y}}\left(\tilde{\boldsymbol{\zeta}}_{(L)}^{\text{EB}} - \boldsymbol{\zeta}_{(L)}\right)'$$
$$= \text{E}_{\mathbf{y}}\left[\left(\tilde{\boldsymbol{\zeta}}_{(L)}^{\text{EB}} - \boldsymbol{\zeta}_{(L)}\right)\left(\tilde{\boldsymbol{\zeta}}_{(L)}^{\text{EB}} - \boldsymbol{\zeta}_{(L)}\right)'\right]$$
$$= \text{E}_{\mathbf{y}}\left[\left(\text{E}(\boldsymbol{\zeta}_{(L)}|\mathbf{y}_{(L)}) - \boldsymbol{\zeta}_{(L)}\right)\left(\text{E}(\boldsymbol{\zeta}_{(L)}|\mathbf{y}_{(L)}) - \boldsymbol{\zeta}_{(L)}\right)'\right]$$
$$= \text{E}_{\mathbf{y}}\left\{\text{E}_{\mathbf{y}}\left[\left(\text{E}(\boldsymbol{\zeta}_{(L)}|\mathbf{y}_{(L)}) - \boldsymbol{\zeta}_{(L)}\right)\left(\text{E}(\boldsymbol{\zeta}_{(L)}|\mathbf{y}_{(L)}) - \boldsymbol{\zeta}_{(L)}\right)'\Big|\mathbf{y}_{(L)}\right]\right\}$$
$$= \text{E}_{\mathbf{y}}\left[\text{Cov}(\boldsymbol{\zeta}_{(L)}|\mathbf{y}_{(L)})\right],$$

the expectation of the posterior covariance matrix over the marginal sampling distribution. In the above derivation, the second equality relies on unconditional unbiasedness of the empirical Bayes predictor, the fourth equality exploits the double expectation rule and the final equality simply recognizes that the term in square brackets represents the posterior covariance matrix.

The above expression suggests the following approximation:

$$\text{Cov}_\mathbf{y}\left(\widetilde{\boldsymbol{\zeta}}^{\text{EB}}_{(L)} - \boldsymbol{\zeta}_{(L)}|\mathbf{X}_{(L)}; \widehat{\boldsymbol{\theta}}\right) \approx \text{Cov}\left(\boldsymbol{\zeta}_{(L)}|\mathbf{y}_{(L)}, \mathbf{X}_{(L)}; \widehat{\boldsymbol{\theta}}\right). \qquad (7.11)$$

An alternative approximation would be to use simulation as for the marginal sampling covariances: Draw latent variables from their prior and subsequently responses from their conditional distribution given the latent variables. The 'true' latent variable realizations are then just the simulated ones and subtracting these from the empirical Bayes predictions, we can estimate the prediction error variances. To reflect the imprecision of the parameter estimates, the above could be preceded by sampling the parameters from their estimated sampling distribution.

The 'linear case': The posterior covariance matrix for the 'linear case' in (7.7) does not involve the responses $\mathbf{y}_{(L)}$ so that the approximation in (7.11) becomes exact in this case,

$$\begin{aligned} \text{Cov}_\mathbf{y}\left(\widetilde{\boldsymbol{\zeta}}^{\text{EB}}_{(L)} - \boldsymbol{\zeta}_{(L)}|\mathbf{X}_{(L)}; \widehat{\boldsymbol{\theta}}\right) &= \widehat{\boldsymbol{\Psi}}_{(L)} - \widehat{\boldsymbol{\Psi}}'_{(L)}\widehat{\boldsymbol{\Lambda}}'_{z(L)}\boldsymbol{\Sigma}^{-1}_{z(L)}\widehat{\boldsymbol{\Lambda}}_{z(L)}\widehat{\boldsymbol{\Psi}}_{(L)} \\ &= \text{Cov}(\boldsymbol{\zeta}_{(L)}|\mathbf{y}_{(L)}, \mathbf{X}_{(L)}, \widehat{\boldsymbol{\theta}}), \end{aligned}$$

i.e., the prediction error covariances are just the posterior covariances given in equation (7.7) on page 231.

7.4 Empirical Bayes modal (EBM)

7.4.1 Empirical Bayes modal prediction

Instead of using the posterior mean as in empirical Bayes prediction, we could use the posterior mode. It can be shown that the posterior mode minimizes the posterior expectation of the zero-one loss function:

$$L^{\text{BM}}(\boldsymbol{\zeta}_{(L)}, \widetilde{\boldsymbol{\zeta}}_{(L)}) = \begin{cases} 0 & \text{if } |\boldsymbol{\zeta}_{(L)} - \widetilde{\boldsymbol{\zeta}}_{(L)}| \le \boldsymbol{\epsilon} \\ 1 & \text{if } |\boldsymbol{\zeta}_{(L)} - \widetilde{\boldsymbol{\zeta}}_{(L)}| > \boldsymbol{\epsilon}, \end{cases}$$

where $\boldsymbol{\epsilon}$ is a vector of minute numbers such that $L^{\text{BM}}(\boldsymbol{\zeta}_{(L)}, \widetilde{\boldsymbol{\zeta}}_{(L)})$ is zero when $\widetilde{\boldsymbol{\zeta}}_{(L)}$ is in the close vicinity of $\boldsymbol{\zeta}_{(L)}$ and one otherwise. Unlike the mean squared error loss function underlying empirical Bayes, the above loss function is also meaningful for latent classes since the loss will be 1 whenever the predicted latent class is not the true latent class and zero otherwise. The loss function is in this case simply the number of misclassifications.

Plugging in estimates $\widehat{\boldsymbol{\theta}}$ for $\boldsymbol{\theta}$, we obtain

$$\widetilde{\boldsymbol{\zeta}}^{\text{EBM}}_{(L)} = \underset{\boldsymbol{\zeta}_{(L)}}{\max \arg}\ \omega(\boldsymbol{\zeta}_{(L)}|\mathbf{y}_{(L)}, \mathbf{X}_{(L)}; \widehat{\boldsymbol{\theta}}).$$

We suggest denoting this predictor the _empirical Bayes modal_ (EBM). Some authors use the terms Bayes modal (e.g. Samejima, 1969) or 'modal a posteriori (MAP) estimators' (e.g. Bock and Aitkin, 1981; Bock and Mislevy, 1982;

Bock, 1983, 1985), which do not explicitly acknowledge that estimates have
been plugged in.

Empirical Bayes modal is the standard classification method in latent class
modeling. An alternative loss-function assigns different weights to different
misclassifications reflecting the different costs incurred. The Bayes risk criterion then uses the classification that minimizes the expected cost. Proportional
prediction, on the other hand, randomly assigns classes to clusters according
to their posterior probabilities. Clogg *et al.* (1991) suggest using this method
for imputing latent classes in multiple imputation.

Generally, there is no analytical expression for the empirical Bayes modal
predictor and we must resort to numerical methods. The posterior mode is
the solution of

$$\frac{\partial}{\partial \boldsymbol{\zeta}_{(L)}} \ln \omega(\boldsymbol{\zeta}_{(L)} | \mathbf{y}_{(L)}, \mathbf{X}_{(L)}; \widehat{\boldsymbol{\theta}}) = \mathbf{0},$$

(provided second order conditions are fulfilled) which can be obtained using
gradient methods (see Section 6.4.2), for instance the Newton-Raphson algorithm.

Since the denominator of the posterior distribution does not depend on $\boldsymbol{\zeta}_{(L)}$,
as seen in equation (7.1), we can use the numerator in place of ω in the above
expressions,

$$\frac{\partial}{\partial \boldsymbol{\zeta}_{(L)}} \ln h(\boldsymbol{\zeta}_{(L)}; \widehat{\boldsymbol{\theta}}) + \frac{\partial}{\partial \boldsymbol{\zeta}_{(L)}} \ln \prod g^{(1)}(y_{ij...z} | \mathbf{x}_{ij...z}, \boldsymbol{\zeta}_{(L)}; \widehat{\boldsymbol{\theta}}) = \mathbf{0}. \qquad (7.12)$$

In contrast to empirical Bayes, this method does not require numerical integration. For this reason, empirical Bayes modal is often used as an approximation
to empirical Bayes when the posterior density is approximately multivariate
normal. As pointed out in the previous chapter, Lindley and Smith (1972)
suggest using Bayes modal as an approximation to the Bayes predictor in the
truly Bayesian setting. Using this method corresponds to maximizing the *h*-
likelihood with respect to the latent variables for given parameter values (see
page 164).

Since integration is avoided, empirical Bayes modal is more computationally efficient than empirical Bayes for latent variable models with non-normal
responses. Samejima (1969) therefore introduced empirical Bayes modal in the
context of her 'graded response model', an ordinal probit or logit one-factor
model with normally distributed factor, and provided a rigorous derivation
of its properties. Interestingly, Samejima also reported the somewhat surprising result that EB and EBM predictions were virtually indistinguishable for
a model with just six dichotomous responses. Muthén (1977) subsequently
extended Samejima's approach to probit multiple-factor models with dichotomous responses. Generalizing the results of Samejima and Muthén, Skrondal
(1996, Chapter 7) discussed empirical Bayes modal prediction for a general
class of multidimensional latent variable models with multivariate normal latent responses.

The 'linear case': Here, the posterior is multivariate normal so that the expectation equals the mode. Hence, the empirical Bayes and empirical Bayes modal predictors coincide in this case, see equation (7.3) on page 227.

7.4.2 Empirical Bayes modal covariances and classification error

Empirical Bayes modal variances and covariances

For general latent variable models we are not aware of methods for deriving the different types of sampling (co)variances for empirical Bayes modal, apart from simulation. However, for large clusters where the posterior approaches normality and the posterior mean is close to the mode, the empirical Bayes sampling covariances should be good approximations. In the 'linear case' the empirical Bayes covariances equal the empirical Bayes modal covariances; see equation (7.7) on page 231 for the prediction error covariances, equation (7.9) on page 232 for the marginal sampling covariances, and equation (7.10) on page 233 for the conditional sampling covariances.

Having used gradient methods to find the posterior mode, it is natural to use the negative inverse of the Hessian at the mode as an approximation to the posterior covariance matrix,

$$\text{Cov}(\boldsymbol{\zeta}_{z(L)}|\mathbf{y}_{z(L)}, \mathbf{X}_{z(L)}; \widehat{\boldsymbol{\theta}}) \approx -\left(\frac{\partial^2}{\partial \boldsymbol{\zeta}^2} \ln \omega(\boldsymbol{\zeta}_{(L)}|\mathbf{y}_{(L)}, \mathbf{X}_{(L)}; \widehat{\boldsymbol{\theta}})\right)^{-1}.$$

The approximation becomes exact as the posterior approaches a multivariate normal, i.e. as the cluster size increases.

Classification error

For discrete latent variables, the quality of the classification of a given cluster can be assessed using the estimated conditional probability of misclassification given $\mathbf{y}_j$ and $\mathbf{X}_j$,

$$f_j = 1 - \max_c \omega(e_c|\mathbf{y}_j, \mathbf{X}_j; \widehat{\boldsymbol{\theta}}). \tag{7.13}$$

The overall misclassification rate can be estimated by the sample mean of f_j over clusters. This is the basis of the proportional reduction of classification error criterion which compares this misclassification rate with the rate when $\mathbf{y}_j$ and $\mathbf{X}_j$ are not available (see Sections 9.3 and 13.5 for applications).

7.5 Maximum likelihood

7.5.1 Latent score estimation

Latent variables are sometimes taken to be nonrandom or *fixed*. In this situation it is natural to interpret the latent scores as unknown *parameters* to be estimated. In Section 6.10.1 we considered *joint* estimation of model parameters and latent variables. In contrast, we now assume that the model parameters have been estimated (using one of the estimation methods discussed in the previous chapter) and consider the estimation of the latent variables ζ for given $\widehat{\boldsymbol{\theta}}$.

The estimation approach to scoring or classification is based on the conditional distribution of the responses, given the latent variables, with the estimates $\widehat{\boldsymbol{\theta}}$ of the model parameters plugged in,

$$\prod g^{(1)}(y_{ij...z}|\mathbf{x}_{ij...z}, \zeta_{z(L)}; \widehat{\boldsymbol{\theta}}).$$

This conditional distribution is interpreted as a 'likelihood' with the values of the latent variables for the cluster as unknown parameters.

Analogously to maximum likelihood estimation of model parameters, the conditional distribution is maximized with respect to the unknown latent variables (parameters) by solving the likelihood equations

$$\frac{\partial}{\partial \zeta_{z(L)}} \ln \prod g^{(1)}(y_{ij...z}|\mathbf{x}_{ij...z}, \zeta_{z(L)}; \widehat{\boldsymbol{\theta}}) = \mathbf{0}.$$

It is interesting to note that this corresponds to maximizing the second term of (7.12) for empirical Bayes modal. In empirical Bayes modal the log prior density may hence be regarded as a penalty term for deviations from the prior mode. The empirical Bayes modal predictions are therefore shrunken towards the prior mode relative to the maximum likelihood estimates.

As would be expected, the estimates for a cluster are asymptotically unbiased as the number of units in the cluster tends to infinity. However, this result may not be useful since the number of units in the cluster is often small, for instance in longitudinal or family studies. For the Rasch model Hoijtink and Boomsma (1995) give an overview of the literature on the finite sample properties of maximum likelihood, empirical Bayes modal and the weighted maximum likelihood estimator suggested by Warm (1989).

Special problems arise for clusters with sparse information because the prior distribution of the latent variables is not utilized. For example, consider a unidimensional factor model with dichotomous responses. If all the responses for a given cluster (typically subject) are zero, the likelihood contribution for that cluster does not have a maximum and the factor score would have to be $-\infty$ to satisfy the likelihood equation (Samejima, 1969). Another example is a growth curve model with a random intercept and slope where the slope for a cluster cannot be estimated if only one response is observed for the cluster. Neither example would pose any problems for empirical Bayes prediction which utilizes the prior distribution. The first example benefits from shrinkage, pulling the prediction away from $-\infty$, whereas the second benefits from the information in the random intercept through its posterior covariance with the random slope.

Furthermore, maximum likelihood estimation of the latent variables requires that the latent variables are considered fixed parameters. This is inconsistent with our model framework and the marginal maximum likelihood method of parameter estimation. We would in general not recommend the maximum likelihood scoring method for these reasons. However, the maximum likelihood method may be useful for assessing the normality assumption for the latent variables (see Section 8.6.2).

The 'linear case': Solving the likelihood equations for $\boldsymbol{\zeta}_{(L)}$ gives

$$\widetilde{\boldsymbol{\zeta}}_{(L)}^{\mathrm{ML}} = \left(\widehat{\boldsymbol{\Lambda}}'_{(L)}\widehat{\boldsymbol{\Theta}}_{(L)}^{-1}\widehat{\boldsymbol{\Lambda}}_{(L)}\right)^{-1}\widehat{\boldsymbol{\Lambda}}'_{(L)}\widehat{\boldsymbol{\Theta}}_{(L)}^{-1}\left(\mathbf{y}_{(L)} - \mathbf{X}_{(L)}\widehat{\boldsymbol{\beta}}\right). \tag{7.14}$$

It follows that

$$\mathrm{E_y}\left(\widetilde{\boldsymbol{\zeta}}_{(L)}^{\mathrm{ML}}|\boldsymbol{\zeta}_{(L)}, \mathbf{X}_{(L)}; \widehat{\boldsymbol{\theta}}\right) = \boldsymbol{\zeta}_{(L)},$$

so unlike the empirical Bayes and Bayes modal predictors, the maximum likelihood estimator is *conditionally unbiased*, given the values of the latent variables $\boldsymbol{\zeta}_{(L)}$.

For common factor models (see Sections 3.3.2 and 3.3.3) maximum likelihood corresponds to the *Bartlett* factor scoring method (Bartlett, 1937; 1938). That this method can be interpreted as a maximum likelihood estimator for the 'linear case' is not transparent in the conventional treatments, where the Bartlett method is derived as either the minimizer of the sum of squares of the standardized residuals (Anderson and Rubin, 1956; Lawley and Maxwell, 1971) or as the minimizer of summed quadratic loss among *conditionally* unbiased estimators (Lawley and Maxwell, 1971). On the other hand, these derivations demonstrate that the Bartlett method can be motivated without making distributional assumptions.

In random effects modeling, this estimator is also known as the ordinary least squares (OLS) estimator of the random effects. In the special case of a two-level random intercept model, the estimator is just the mean raw residual

$$\widetilde{\boldsymbol{\zeta}}_{j(2)}^{\mathrm{ML}} = \frac{1}{n_j}\sum_{i=1}^{n_j}(y_{ij} - \mathbf{x}'_{ij}\widehat{\boldsymbol{\beta}}).$$

Note that multilevel models with fixed intercepts at several levels are not identified unless constraints are imposed, for instance that the sum of the level-two intercepts within the same level-three cluster add to zero. This implies that maximum likelihood estimates do not exist for the running example in this chapter since the matrix $(\widehat{\boldsymbol{\Lambda}}'_{(L)}\widehat{\boldsymbol{\Theta}}_{(L)}^{-1}\widehat{\boldsymbol{\Lambda}}_{(L)})^{-1}$ in equation (7.14) is singular.

7.5.2 Variances and covariances of ML estimator

Using likelihood theory, the asymptotic covariance matrix of the ML estimator $\widetilde{\boldsymbol{\zeta}}_{(L)}^{\mathrm{ML}}$ becomes

$$\mathrm{Cov_y}\left(\widetilde{\boldsymbol{\zeta}}_{(L)}^{\mathrm{ML}}|\boldsymbol{\zeta}_{(L)}, \mathbf{X}_{(L)}; \widehat{\boldsymbol{\theta}}\right) \approx -\left(\frac{\partial^2}{\partial\boldsymbol{\zeta}_{(L)}^2}\ln\prod g^{(1)}(y_{ij...z}|\mathbf{x}_{ij...z}, \boldsymbol{\zeta}_{(L)}; \widehat{\boldsymbol{\theta}})\right)^{-1}.$$

Interestingly, $\mathrm{Cov_y}(\widetilde{\boldsymbol{\zeta}}_{(L)}^{\mathrm{ML}}|\boldsymbol{\zeta}_{(L)}, \mathbf{X}_{(L)}; \widehat{\boldsymbol{\theta}})$ can be interpreted as the *conditional sampling* covariance matrix of the scores, given the true latent variable $\boldsymbol{\zeta}_{z(L)}$.

The asymptotics require that the number of units (or items) in a cluster tends to infinity, and not the number of clusters. Thus, the utility of this result may be questionable in practice, where there are often few units in a cluster.

For a given cluster, the inverse of the above covariance matrix, the observed information, is the sum of contributions $-\partial^2 \ln g^{(1)}(y_{ij...z}|\mathbf{x}_{ij...z}, \boldsymbol{\zeta}_{(L)}; \widehat{\boldsymbol{\theta}})/\partial \boldsymbol{\zeta}_{(L)}^2$ from the individual units. In unidimensional IRT ($\boldsymbol{\zeta}_{(L)} \equiv \zeta_j$), these contributions from individual items, plotted against ζ_j, are known as 'item information curves', and the sum of the contributions over the items is called the 'test information' (e.g. Birnbaum, 1968). Item information functions can be used to assess how much 'information' is gained about an individual's ability from knowing her response to an item as a function of her true ability. Information functions play a central role in IRT and are helpful in test construction, item selection, assessment of precision of measurement, comparison of tests, comparison of scoring methods, and tailored or adaptive testing (e.g. Hambleton and Swaminathan, 1985). For instance, in adaptive testing the information function can be useful for choosing the next item to present to an examinee given the current estimate of his ability. The item is noninformative if it is too simple or too difficult for the examinee, making the response too predictable.

Observe that the covariance matrix of the maximum likelihood estimator tends to the posterior covariance matrix as the cluster size tends to infinity – the likelihood swamping the prior (e.g. DeGroot, 1970).

The marginal sampling covariance matrix is

$$\mathrm{Cov}_\mathbf{y}\left(\widetilde{\boldsymbol{\zeta}}_{(L)}^{\mathrm{ML}}|\mathbf{X}_{(L)}; \widehat{\boldsymbol{\theta}}\right) = \mathrm{Cov}_\zeta\left[\mathrm{E}_\mathbf{y}\left(\widetilde{\boldsymbol{\zeta}}_{(L)}^{\mathrm{ML}}|\boldsymbol{\zeta}_{(L)}, \mathbf{X}_{(L)}; \widehat{\boldsymbol{\theta}}\right)\right]$$
$$+ \mathrm{E}_\zeta\left[\mathrm{Cov}_\mathbf{y}\left(\widetilde{\boldsymbol{\zeta}}_{(L)}^{\mathrm{ML}}|\boldsymbol{\zeta}_{(L)}, \mathbf{X}_{(L)}; \widehat{\boldsymbol{\theta}}\right)\right].$$

The 'linear case':   Here, the marginal sampling covariance matrix of the maximum likelihood estimator becomes

$$\mathrm{Cov}_\mathbf{y}\left(\widetilde{\boldsymbol{\zeta}}_{(L)}^{\mathrm{ML}}|\mathbf{X}_{(L)}; \widehat{\boldsymbol{\theta}}\right) = \widehat{\boldsymbol{\Psi}} + \left(\widehat{\boldsymbol{\Lambda}}'_{(L)}\widehat{\boldsymbol{\Theta}}_{(L)}^{-1}\widehat{\boldsymbol{\Lambda}}_{(L)}\right)^{-1}, \qquad (7.15)$$

and the conditional sampling covariance matrix is

$$\mathrm{Cov}_\mathbf{y}\left(\widetilde{\boldsymbol{\zeta}}_{(L)}^{\mathrm{ML}}|\boldsymbol{\zeta}_{(L)}, \mathbf{X}_{(L)}; \widehat{\boldsymbol{\theta}}\right) = \left(\widehat{\boldsymbol{\Lambda}}'_{(L)}\widehat{\boldsymbol{\Theta}}_{(L)}^{-1}\widehat{\boldsymbol{\Lambda}}_{(L)}\right)^{-1} = \mathrm{Cov}_\mathbf{y}\left(\widetilde{\boldsymbol{\zeta}}_{(L)}^{\mathrm{ML}} - \boldsymbol{\zeta}_{(L)}|\mathbf{X}_{(L)}; \widehat{\boldsymbol{\theta}}\right),$$

the same as the unconditional prediction error variance.

For a linear random intercept model, the marginal sampling variance is

$$\mathrm{Var}_\mathbf{y}\left(\widetilde{\zeta}_{(2)}^{\mathrm{ML}}|\mathbf{X}_{(2)}; \widehat{\boldsymbol{\theta}}\right) = \widehat{\psi} + \widehat{\theta}/n_j,$$

and the conditional variance and prediction error variance are simply

$$\mathrm{Var}_\mathbf{y}\left(\widetilde{\zeta}_{(2)}^{\mathrm{ML}}|\zeta_{(2)}, \mathbf{X}_{(2)}; \widehat{\boldsymbol{\theta}}\right) = \mathrm{Var}_\mathbf{y}\left(\widetilde{\zeta}_{(2)}^{\mathrm{ML}} - \zeta_{(2)}|\mathbf{X}_{(2)}; \widehat{\boldsymbol{\theta}}\right) = \widehat{\theta}/n_j.$$

7.6 Relating the scoring methods in the 'linear case'

We are now in a position to present a very instructive expression which relates the empirical Bayes, empirical Bayes modal and maximum likelihood methods in the 'linear case'.

The marginal sampling covariance matrix of the maximum likelihood predictor in (7.15) can be written as

$$\text{Cov}_\mathbf{y}\left(\tilde{\boldsymbol{\zeta}}_{(L)}^{\text{ML}} \mid \mathbf{X}_{(L)}; \boldsymbol{\theta}\right) = \widehat{\boldsymbol{\Psi}}_{(L)} + \widehat{\mathbf{A}}_{(L)},$$

where

$$\widehat{\mathbf{A}}_{(L)} \equiv \left(\widehat{\boldsymbol{\Lambda}}'_{(L)} \widehat{\boldsymbol{\Theta}}_{(L)}^{-1} \widehat{\boldsymbol{\Lambda}}_{(L)}\right)^{-1},$$

and $\widehat{\boldsymbol{\Psi}}_{(L)}$ represent the *intra-cluster* and *inter-cluster* contributions to the covariance matrix, respectively. The *multivariate reliability* of the maximum likelihood estimator can then be defined as

$$\widehat{\mathbf{R}}_{(L)} \equiv \text{Cov}_\mathbf{y}(\boldsymbol{\zeta}_{(L)}) \left[\text{Cov}_\mathbf{y}\left(\tilde{\boldsymbol{\zeta}}_{(L)}^{\text{ML}}\right)\right]^{-1} = \widehat{\boldsymbol{\Psi}}_{(L)} \left(\widehat{\boldsymbol{\Psi}}_{(L)} + \widehat{\mathbf{A}}_{(L)}\right)^{-1}.$$

Using the same line of proof as in Bock (1983), the following identity can be demonstrated:

$$\tilde{\boldsymbol{\zeta}}_{(L)}^{\text{EB}} = \tilde{\boldsymbol{\zeta}}_{(L)}^{\text{EBM}} = \widehat{\mathbf{R}}_{(L)} \tilde{\boldsymbol{\zeta}}_{(L)}^{\text{ML}}.$$

This identity is a multivariate representation of the phenomenon dubbed *shrinkage* in the statistical literature (e.g. James and Stein, 1961). We note that the empirical Bayes predictor is pulled toward the prior expectation $\mathbf{0}$ of the latent variables whenever the estimated level-1 variation $\widehat{\boldsymbol{\Theta}}$ is large relative to the estimated inter-cluster variation $\widehat{\boldsymbol{\Psi}}_{(L)}$. On the other hand, the empirical Bayes predictor is pulled toward the maximum likelihood estimator when the inter-cluster variation is large compared to the intra-cluster variation (for instance due to large cluster sizes). In the limit, where $\widehat{\mathbf{R}}_{(L)} = \mathbf{I}$, we obtain $\tilde{\boldsymbol{\zeta}}_{(L)}^{\text{EB}} = \tilde{\boldsymbol{\zeta}}_{(L)}^{\text{EBM}} = \tilde{\boldsymbol{\zeta}}_{(L)}^{\text{ML}}$; all three methodologies coincide. Of course, all these results are in perfect accordance with our intuition regarding a sensible latent scoring methodology.

For a random intercept model, we encountered this reliability in (7.5) with

$$\widehat{R}_{(2)} = \frac{\widehat{\psi}}{\widehat{\psi} + \widehat{\theta}/n_j}.$$

Note that $\widehat{\mathbf{R}}_{1(3)}$ is not defined for the three-level numerical example since $\mathbf{A}_{1(3)}$ is singular for any higher-level model.

7.7 Ad hoc scoring methods

For hypothetical constructs common in the social and behavioral sciences (see Section 1.3), scores are often assigned by ad hoc methods such as simply summing the values of responses from a number of indicators or items. An important assumption underlying these methods is that the items contributing to the score measure a unidimensional construct. For multidimensional constructs, the methods are sometimes applied to subsets of items, producing 'subscales'.

7.7.1 Raw Sumscores

The most common ad hoc approach for continuous, ordinal and dichotomous responses is undoubtedly the use of raw or unweighted sumscores, defined as

$$y_{\cdot j} = \sum_{i=1}^{n_j} y_{ij}.$$

The resultant score is denoted a Likert scale (Likert, 1932) when the responses are ordinal and given successive integer codes.

In psychology, psychiatry and related fields, measurement scales based on questionnaires or structured interviews are usually defined as raw sum scores of dichotomous or ordinal responses. Researchers using such 'validated instruments' are expected to adhere to the scoring method described in the manual accompanying the questionnaire. This practice effectively discourages serious measurement modeling.

In some cases the sumscore method can be given a theoretical justification. For continuous responses, raw sumscores are proportional to factor scores produced by either the Bartlett or regression methods for unidimensional parallel measurement models (see Section 3.3.2), where all factor loadings are equal and all unique factor variances are equal (see also Jöreskog, 1971b and Maxwell, 1971). For dichotomous responses the raw sumscore forms a sufficient statistic for estimating the 'ability' in the Rasch model discussed in Section 3.3.4. However, both the unidimensional parallel measurement and Rasch models are very restrictive models and unlikely to hold in practice.

In general, use of raw sumscores as a measurement strategy cannot be given a theoretical motivation and usually implies a rejection of measurement *modeling*. Torgerson (1958) therefore describes the sumscore strategy as an example of 'measurement by fiat', in contrast to the more respectable 'fundamental measurement' obtained from measurement modeling. In practice, some form of modeling is often employed to justify the use of sum scores. For example, unidimensionality is typically investigated through the use of factor modeling. It appears somewhat inconsistent to use modeling arguments as a justification for the use of an ad hoc scoring method.

A standard argument in favor of the raw sumscore methodology is that it has been demonstrated repeatedly that the Pearson correlation between the sumscore and scores from more sophisticated methodologies often approaches one, especially when the number of variables is relatively large. This holds for continuous items (see Wilks, 1938; Gulliksen, 1950; Wang and Stanley, 1970; Wainer, 1976) as well as for dichotomous items (see Muthén, 1977; Kim and Rabjohn, 1978). The bright side of these results is that an extremely simple approach appears to work as well as much more cumbersome methodologies. However, the use of the high Pearson correlations as evidence has been criticized. The point is that differences on specific parts of the latent scale of special interest, for instance a cut-off point for admission in an ability test, can be masked by the use of a summary statistic such as the correlation (e.g. Hambleton and Swaminathan, 1985).

A major limitation of the sumscore approach is that it cannot be directly applied to clusters with missing items. An ad hoc strategy in this case is to impute the responses for the missing items using the cluster mean of the nonmissing units. Finally, the sumscore methodology cannot incorporate covariate information or relationships (regressions or covariances) between latent variables in contrast to the model-based approach.

7.7.2 Other methods

One version of the raw sumscore strategy is to discard 'nonsalient' items from the sumscore (see e.g. Thurstone, 1947; Gorsuch, 1983). A basic problem with this variant is that the employed definitions of salience are arbitrary.

The *representative item* strategy discards all but one particular item, presumably easily measured and valid, and takes the response for that item as the latent score (e.g. Rummel, 1967). This approach is often used in quality of life questionnaires where one item asks directly about quality of life. Here the answer to that single question is often used as the gold standard with which scores derived from the remaining items are validated. Obviously, the representative item strategy amounts to wasting information and presupposes data of extremely high quality (e.g. Adams, 1975).

In the *factor loadings as weights* strategy the scores are obtained by using a weighted sumscore using the factor loadings as weights (e.g. Fruchter, 1954; Blalock, 1960). One problem with this strategy is that it lacks a theoretical rationale. Another problem is that the items are individually credited with an influence that they share with other items, yielding redundant solutions (see also Glass and Maguire, 1966; Harris, 1967; Halperin, 1976).

Finally, 'linear case' factor scoring methods using the factor scoring matrix are sometimes applied to noncontinuous responses, an ad hoc strategy that should be avoided.

7.8 Some uses of latent scoring and classification

7.8.1 Introduction

There are many applications of latent scoring and classification including measurement, ability scoring, disease mapping, small area estimation, medical diagnosis, image analysis and model diagnostics. For continuous latent variables we can distinguish between two kinds of latent scores. *Factor scores* for measurement are discussed in Section 7.8.2 and *random effects scores* in Section 7.8.3. In Section 7.8.4 we briefly discuss classification and in Section 7.8.5

we point out that latent scoring is useful for model diagnostics, a topic we return to in more detail in Section 8.6.2 of the next chapter.

7.8.2 Factor scoring as measurement proper

A conventional definition of measurement is due to Stevens (1951), who defined measurement as merely the assignment of a number to an attribute according to a rule. This definition is extremely broad, and in our opinion not particularly fruitful. Many phenomena of interest are best construed as latent variables or factors and it is natural to focus on *factor scoring*.

Clogg (1988) defines the *measurement process* culminating in measurement as embodying the following steps:

1. Selection of items

2. Specification of tentative latent trait models

3. Choice of retained latent trait model

4. Estimation and interpretation of model parameters

5. Measurement of latent traits.

Much has been written in the psychometric and statistical literature regarding the first four steps of the measurement process. Somewhat surprisingly, and not without irony, most treatments of measurement modeling stop short of measurement. Consider for instance the chapter denoted 'Measurement' by Bohrnstedt (1983). Although giving a nice introduction to measurement models, it fails to address precisely what is expected from the title, namely measurement per se.

At the other extreme, the purpose of latent variable modeling sometimes is to derive *scoring procedures* (e.g. Gorsuch, 1983). The scoring procedures or keys are subsequently employed in other samples, making latent variable modeling superfluous in future research. Unfortunately, the entire strategy of generalizing scoring procedures across populations is highly problematic. In particular, results from the theory of factorial invariance (e.g. Skrondal and Laake, 1999; Meredith, 1964, 1993) suggest that scoring weights are not expected to be invariant across populations. Thus, we recommend that the model parameters are estimated or 'calibrated' on the same population for which latent scores are desired, possibly by using multi-group modeling.

Apart from the need for measurement, there are several other motivations for obtaining factor scores. Factor scores can help in model interpretation. In particular, plots of latent scores often prove useful in discussing issues of dimensionality in factor models (e.g. McDonald, 1967; Etezadi-Amoli and McDonald, 1983), see Figures 10.5 and 10.7 for applications. Factor scores can also be useful in *classification* of units. Examples include the admission of students according to ability tests and treatment of patients according to mental health tests (e.g. Duncan-Jones *et al.*, 1986; Muthén, 1989a). However, for this purpose it appears more natural to use a discrete latent variable and classify units using the posterior probabilities.

Factor scores are also useful in *tailored or adaptive testing*. Here the scores are updated sequentially as new item responses are obtained. The 'current' score determines the choice of subsequent items in order to maximize the obtained 'information' (e.g. Bock and Mislevy, 1982).

Conventionally, factor scores have served as *vehicles for further analysis*. Specifically, factor scores are often used to impute latent variables in structural equation models. This use has been hailed as one of the major motivations for obtaining factor scores by a number of authors (e.g. Kim and Mueller, 1978; Gorsuch, 1983; Johnson and Wichern, 1983). Kim and Mueller (1978, p.60) stated that

> "In fact, with the exception of the psychometric literature, factor analysis seems to have been used more often as a means of creating factor scales for other studies than as a means of studying factor structures per se".

This statement is also apt for the present situation as is apparent in some software packages where factor analysis is one of the options in the 'data reduction' menu, another option being principal component analysis. In view of the frequent use of factor scores as vehicles for further analysis, it is important to point out that this approach can be problematic. Biased estimates of model parameters will result unless care is exercised in obtaining the scores and standard errors are underestimated (Skrondal and Laake, 2001).

The modern approach to estimating latent variable models is to estimate the model parameters directly, without resorting to imputed latent variables. This fact may explain the remarkable paucity of research on scoring for latent variable models.

7.8.3 Random effects scores as cluster-specific effects

In random effects models, individual clusters are construed as having their own regression 'parameters', sampled from some distribution. Consequently, random effects scores represent an assessment of the *cluster specific effects* of explanatory variables. Obviously, such effects are often of considerable substantive interest. Random effects scores are extremely useful in growth curve or development modeling. In this case the scores form the basis for plotting *growth trajectories* for the individual clusters or groups of clusters (e.g. Strenio *et al.*, 1983). An application for epileptic seizures is given in Figure 11.1.

Another application of random effects scores is *classification* and *ranking* of clusters. Examples include the classification of organizations as more or less effective according to their random effects scores (Aitkin and Longford, 1986), ranking of different industries in terms of the gender gap in earnings (Kreft and de Leeuw, 1994) and ranking of schools in terms of exam performance (Goldstein and Spiegelhalter, 1996). An appropriate standard error for comparing the random effects of two units is the standard deviation of the prediction error. Rankings can however be extremely variable and their precision is not easily expressed in terms of standard errors (Goldstein and Spiegelhalter, 1996).

Random effects models are often used to combine effect size estimates from different studies in a *meta-analysis*. In this case, it is natural to replace the original effect size estimates with the posterior means of the random effects model to 'borrow strength' from other studies. Posterior standard deviations are in this case typically used to represent the confidence intervals which also represent the prediction error variances in the continuous case. In Section 9.5 we use this approach in a meta-analysis of nicotine gum for smoking cessation.

Random effects scoring is popular for *disease mapping* and *small area estimation*. In regions with small populations, the raw incidence estimates can be very imprecise and the resulting map 'noisy'. Borrowing strength from other regions can result in more reliable and smoother maps. Ideally, the models should in this case exploit spatial information; see Clayton and Kaldor (1987), Langford *et al.* (1999) and Section 11.4.

7.8.4 Classification

Sometimes a latent variable is inherently discrete, the canonical example being medical diagnosis where a patient either has a particular illness or not. In this case classification is crucial for prescribing the correct treatment. In marketing, customers are sometimes classified as belonging to one of several 'market segments', characterized by specific sets of preferences, for the purpose of targeted advertising (e.g. Wedel and Kamakura, 2000).

Often the latent variable may be best perceived as continuous, but classification is required to make a decision. In this case it might be preferable to specify a discrete latent variable model (such as a latent class model). We can then use empirical Bayes modal to classify the units instead of applying arbitrary thresholds to a continuous score. In education, this approach has been used when mastery of a subject is of interest rather than ability on a continuous scale (e.g. Bergan, 1988; MacReady and Dayton, 1992).

A common problem in image analysis is image segmentation or restoration where pixels (picture elements on a square grid) or voxels (volume elements on a cubic grid) are classified as belonging to one of several regions. An example in brain imaging is delineating a brain tumor. Here, spatial models such as Markov random fields (e.g. Besag, 1986) are sometimes specified for the latent region labels or latent classes, and conditionally on the latent classes the responses are normally distributed representing 'noise'. Segmentation is then achieved by finding the modal a posteriori region labels using for instance 'simulated annealing' (e.g. Geman and Geman, 1984).

In this book we mostly consider latent class models for classification when the true classification is not known for any of the units. If the true classification is known for a subsample of units, sometimes called the 'training set' or 'validation sample', latent class models can be extended as shown in Section 14.3 in the context of covariate misclassification. Wedel (2002a) discusses the problem where 'core variables' (the responses) and 'concomitant variables' (explanatory variables) are observed for the training set, but only concomitant variables are observed for new units to be classified. A different problem is prediction of a single categorical response variable y, where the discrete latent variables serve as a 'hidden layer' of intervening variables as in neural networks (see Vermunt and Magidson (2003b) for an overview of such models).

As in latent scoring, latent classifications are sometimes used as 'observed' variables in subsequent analyses. A related approach mentioned by Wedel (2002b) is to regress estimated posterior probabilities on explanatory variables. Both these ad hoc approaches are problematic and should be avoided (e.g. Croon, 2002). As for continuous latent variables, it is preferable to model the relationship between latent classes and observed responses and/or covariates directly. In Section 14.3 we estimate models with latent classes as covariates and in Section 13.5 latent classes are regressed on covariates. Note that prediction of latent class membership in this case uses information from all variables included in the model and not only the items 'measuring' the latent classes.

7.8.5 Model diagnostics

An important application of latent scoring is model *diagnostics*. It might be tempting to use latent scores to study the assumptions underlying statistical models in much the same way as if observed variables were investigated. However, it should be remembered that the distribution of the predicted scores is not the same as the theoretical distribution of the latent variables, making this approach problematic, particularly for nonnormal responses. See also Section 8.6.2.

Latent scores can be treated as estimated residuals for outlier detection, for instance by comparing the scores with their approximate sampling standard deviation. The use of latent scoring in diagnostics for general latent variable models will be discussed in Section 8.6.2; see Section 11.3.3 for a concrete application.

7.9 Summary and further reading

For all the model based scoring and classification methods discussed in this chapter, missing responses do not pose any problems as long as they can be assumed to be missing at random (MAR).

By far the most common approach for assigning values to continuous latent variables is empirical Bayes prediction. The advantage of this approach is that the predictions minimize a mean squared error loss function (if the model parameters are assumed known) and are unconditionally unbiased. The conditional bias or shrinkage associated with the method can be seen as an advantage when sparse information is available on some units.

The maximum likelihood method is sometimes used since, in contrast to empirical Bayes, the scores are conditionally unbiased. However, this approach is not consistent with the modeling assumptions and will not yield predictions for clusters with insufficient information. Furthermore, the method cannot be applied to truly multilevel models.

There are various ways of defining standard errors. The most commonly used are the posterior standard deviation, equal to the standard deviation of the prediction error in the continuous case, and the marginal sampling standard deviation. Whichever scoring method is used, it is important to use the standard error appropriate for the particular application.

For discrete latent variables, the most common classification method is Empirical Bayes modal since it minimizes the expected misclassification rate.

In contrast to estimation of latent variable models, the literature on latent scoring and classification is relatively scant. However, useful books on empirical Bayes are Maritz and Lwin (1989) and Carlin and Louis (2000). Empirical Bayes prediction in linear random coefficient models is reviewed by Strenio *et al.* (1983). A nice overview of the Bartlett and regression methods for factor analysis is given in Lawley and Maxwell (1971, Chapter 8).

Appendix: Some software

We now provide an incomplete list of software implementing the different methods of assigning values to latent variables. We do not provide addresses or links to the software since such information is quickly out of date and can easily be found on the internet.

- Empirical Bayes

  - aML uses quadrature for multilevel and multiprocess models (Lillard and Panis, 2000),
  - BILOG-MG uses quadrature for binary logistic item-response models (Zimowski *et al.*, 1996; Du Toit, 2003)
  - `gllapred`, the prediction command for `gllamm`, uses adaptive quadrature for generalized linear latent and mixed models (Rabe-Hesketh *et al.*, 2004b, 2005b)
  - MIXNO, MIXOR, MIXREG, MIXPREG and MIXSUR use quadrature for two-level generalized linear mixed models (Hedeker and Gibbons, 1996ab; Hedeker, 1999),
  - TESTFACT uses adaptive quadrature for multidimensional probit factor models (Bock *et al.*, 1999)

- Empirical Bayes modal

  - Scoring

    * BILOG-MG for binary logistic item-response models (Zimowski *et al.*, 1996; Du Toit, 2003)
    * HLM uses PQL or LaPlace6 for multilevel generalized linear mixed models (Raudenbush *et al.*, 2004)
    * MLwiN uses PQL for multilevel generalized linear mixed models (Rasbash *et al.*, 2004)
    * Mplus for structural equation models with continuous, dichotomous, ordinal and censored responses (Muthén and Muthén, 1998, 2003)
    * SAS NLMIXED for two-level generalized linear mixed models (Wolfinger, 1999)

  - Classification:

    * Mplus for latent class models with continuous, dichotomous, ordinal and censored responses (Muthén and Muthén, 1998, 2003)
    * Latent GOLD for most response types (Vermunt and Magidson, 2000, 2003a)
    * `gllamm` (posterior probabilities) for generalized linear latent and mixed models in **Stata** (Rabe-Hesketh *et al.*, 2004b, 2005b)

- Maximum likelihood

  - BILOG-MG for binary logistic item-response models (e.g. Du Toit, 2003)

CHAPTER 8

Model specification and inference

8.1 Introduction

In Chapter 6 we discussed the problem of estimating the parameters of a given statistical model without considering why that particular model was specified. In this chapter we consider the perhaps more difficult task of finding an appropriate model.

Before considering how to proceed with model specification we reflect upon statistical modeling in Section 8.2. Specifically, we discuss the roles and purposes of different kinds of statistical models as well as modeling strategies. In Section 8.3 we review maximum likelihood inference which is useful for assessing the uncertainty of estimates and forms the basis of many 'relative fit' criteria for model selection discussed in Section 8.4. It could be argued that relying solely on relative fit, all we can say is that a model appears to be better than the competitors, but little is known concerning how good or bad the better model is in an absolute sense. Furthermore, misspecification of the candidate models could invalidate model selection. In Section 8.5 we therefore discuss methods of ascertaining how good the best model is using 'global absolute fit' criteria. In contrast to global absolute fit criteria, 'local absolute fit' criteria can be used not only to discover that a model is inadequate but also to diagnose where a model is misspecified. Such diagnostics are discussed in Section 8.6.

The organization of this chapter may seem to imply that model building proceeds in the following sequence: (1) select the 'best' model among a set of models, (2) assess the adequacy of the selected model (if feasible), (3) use diagnostics to pinpoint misspecification, which may suggest different ways of elaborating the model taking us back to (1). However, this sequence has no particular theoretical justification and other sequences may be just as useful.

8.2 Statistical modeling

8.2.1 Types of statistical model

Broadly speaking, in applications of *mathematical models* in empirical research the relationship between variables is formalized in terms of one or more deterministic equations. Such models are appropriate in situations where there are known deterministic laws governing the relationships, as is sometimes the case in the natural sciences, a typical example being Newton's laws of motion. In contrast, *statistical models* are mathematical models which also include a random or stochastic component in addition to the deterministic component. The random component may represent measurement error, making statistical models useful also in studying deterministic laws. More importantly, it could represent 'natural variation' or stochastic causal laws, for instance those of Mendelian inheritance. Finally, the random component may reflect our incomplete knowledge regarding a deterministic 'law' governing the empirical phenomena under consideration. Note that some statistical tools do not involve statistical models since they have no random component. An example is principal component analysis which is merely an orthogonal transformation of the data.

Two types of statistical models have typically been delineated in the literature. We adopt the terms *substantive* and *empirical* models used by Cox (1990). Lehmann (1990) describes similar distinctions put forth by Neyman (e.g. Neyman, 1939) who contrasted 'explanatory models' versus 'interpolatory formulae' and Box and colleagues (e.g. Box *et al.*, 1978) who contrast 'theoretical' or 'mechanistic' models versus 'empirical models'.

Substantive models

The most appealing statistical models are substantive models which connect directly with subject matter considerations and background information, constituting an effort to achieve understanding and *explanations*, i.e., answers to 'why questions' in the terminology of philosophers. Typically the researcher believes that there is a single 'true' model that has generated the data.

'Directly substantive models' explain what is observed in terms of explicit mechanisms, usually via quantities that are not directly observed and some theoretical notions as to how the system under study 'works'. In the natural sciences there may be one or at most a few stringent competing theories purporting to explain the observations in terms of lawlike relationships with a specified functional form. Neyman's favorite example was Mendelian inheritance. To 'test' a theory the researcher can sometimes vary a 'treatment' of interest under controlled conditions using randomization.

A weaker type of substantive model merely posits substantive hypotheses about dependencies, for instance in graphical or structural equation modeling where some variables may be specified as conditionally independent given other variables. An example is the structural equation model positing 'complete mediation' mentioned in Section 1.3, where a variable only has an indirect effect on the outcome via an intermediate variable and no direct effect. Such models are useful in the social sciences and medicine where there are often several rather loose and sometimes conflicting ideas about how more or less fuzzy phenomena are associated. Here, information often stems from an existing body of *empirical research* on the problem under investigation and related problems. However, studies are typically based on observational designs where the researcher is merely a passive observer of the empirical process, making it daunting to investigate causality. A somewhat stronger design is the quasi-experiment (e.g. Cook and Campbell, 1979) where the researcher can vary the treatment but randomization is for some reason unfeasible.

Since a substantive statistical model is a simplified representation of the data generating process, it should be possible to simulate data directly from the model. A good substantive model should be parsimonious and yet capture the main features of the data generating process; it should neither be too complex nor too simplistic. An overly complex model is of little use since it merely mirrors the realized data instead of the underlying process and is likely to be a poor representation of other realizations of the process. On the other hand, an overly simplistic model may fail to capture important aspects of the data generating process and lead to incorrect inferences.

Cox and Wermuth (1996) suggest that a satisfactory substantive statistical model should:

1. establish a link with background knowledge

2. set up a connection with previous work

3. give some pointer toward a data generating process

4. have primary parameters with clear subject-specific substantive interpretations

5. specify haphazard aspects well enough to provide meaningful assessment of precision

6. have adequate fit

Empirical models

Empirical models are the more common type of model in many applications where background information is relatively scarce. According to Box, empirical models are used as a guide to action, with emphasis on *prediction*. The models are intended to provide guidance for the particular situation at hand, using all special circumstances, which means that good approximations can only be expected over the area of interest. Empirical models may be obtained from a family of models selected largely for convenience, on the basis solely of the data without much input from the underlying situation. Instead of believing that there is one true model, the researcher is looking for one among several potentially useful models. A modern version is the *algorithmic models* advocated by Breiman (e.g. Breiman, 2001), a black-box strategy focusing on effective and flexible algorithms for prediction of output from input. No effort whatsoever is made to explicate the black-box whose contents are treated as unknown. Typical examples would be neural networks or 'regression forests'.

Cox (1990) considers a less extreme kind of empirical model which is not based on specific substantive considerations but rather aims to represent in idealized form dependencies, often 'smooth' dependencies, thought to be present. According to Cox, the first and most common role for empirical models is to estimate effects and their precision. The widespread use of regression models is a canonical example, for instance the estimation of associations from logistic regression in epidemiology. It is important to note that these kinds of empirical models are not void of substantive considerations, for instance in the choice of confounders in epidemiology. Another role of empirical models is 'correction of deficiencies in data' such as measurement error, missing data and complex sampling. This less extreme notion of empirical model is similar to the idea of a weak substantive model.

8.2.2 Modeling strategies

The *problem of specification*, the task of specifying a low-dimensional parametric statistical model, was the first kind of problem of statistics mentioned by Fisher (1922). Interestingly, his discussion of specification was confined to a single paragraph dominated by the first sentence:

> "As regards problems of specification, these are entirely a matter for the practical statistician,..."

Lehmann (1990) interprets Fisher's statement to imply that there can be no theory of modeling and no modeling strategies, but that instead each problem must be considered entirely on its own merits.

However, in practice one of four kinds of modeling strategies is typically adopted (see Jöreskog (1993) for a similar classification). For substantive models a natural modeling strategy is the *strictly confirmatory* approach involving one or perhaps two models. For instance, in measurement modeling one might wish to determine whether a particular independent clusters structure (see Section 3.3.3) holds by either retaining or rejecting this model based on absolute fit criteria; see Section 8.5. Clinical trials often involve two models where the null model is typically that there is no effect of a drug and the alternative model is that there is an effect. Model selection then proceeds by hypothesis testing. It is important that the models are specified in advance (giving rise to the term 'planned comparisons' in ANOVA) and not suggested by the same data on which they are tested. Unfortunately, models or 'theories' suggested by the data are often presented as if they had been formulated in advance to make conclusions more credible. To prevent such malpractice in drug development (where it could have lethal consequences), it is becoming common to prepare a detailed analysis plan before a clinical trial is conducted. The analysis plan specifies the 'primary hypotheses' and the exact manner in which they are to be tested.

Another modeling strategy for substantive models is the *competing models* approach where a moderate number of alternative models are specified from which one is selected. This is appropriate if there are a few competing theories purporting to explain a phenomenon and one desires to dispel faulty ones.

For empirical or weak substantive models a natural modeling strategy is the *model generating* approach where an initial tentative model is specified based on the available background information. If this model is deemed to be unacceptable according to diagnostic and/or fit criteria, it is modified either according to background theory or to achieve a better fit to the data. This iterative process of specification, estimation, confrontation with data and re-specification proceeds until the model is found acceptable.

For empirical models the typical modeling approach is *strictly exploratory* where models are 'derived' from the data. In practice, this corresponds to starting with a very large number of competing models and using purely statistical criteria for choosing amongst them. A common example is best-subset linear regression where any subset of a large number of covariates is considered. This approach may yield useful models for prediction if combined with some form of cross-validation (see page 271 for further discussion) to avoid overfitting.

Importantly, the exploratory approach should not be used to derive substantive models. It is well known that results from such 'data-dredging' are at best suggestive and should be assessed on independent data using a confirmatory approach. Freedman (1983) shows that theories can easily be derived from pure noise. He simulated 51 independent standard normal variables for 100 units, treating the last as response variable and the remaining as potential covariates in a linear regression. Selecting only covariates with coefficients significant at the 25% level produced 'convincing' results with many significant coefficient at the 5% level. A similar criticism in the context of 'automatic interaction detection', regression analysis, factor analysis and nonmetric multidimensional scaling was presented by Einhorn (1969). It is well worth citing from the conclusion in his paper 'Alchemy in the Behavioral Sciences':

> "It should be clear that proceeding *without* a theory and *with* powerful data analytic techniques can lead to large numbers of Type I errors. Just as the ancient alchemists were not successful in turning base metal into gold, the modern researcher cannot rely on the "computer" to turn his data into meaningful and valuable scientific information."

Although the distinction between confirmatory and exploratory approaches and its implications apply to any kind of modeling, we now discuss it in the context of factor analysis.

Example: Confirmatory versus exploratory factor analysis

Confirmatory factor analysis (CFA) is a hypotheticist procedure designed to test hypotheses about the relationships between items and factors, where the number and interpretation of the factors are given in advance. Hence, in the confirmatory mode, particular parameters are set to prescribed values.

Exploratory factor analysis (EFA), on the other hand, can be construed as an inductivist method designed to *discover* an optimal set of factors that accounts for the covariation among the items (see Mulaik, 1988b; Holzinger and Harman, 1941). Mulaik (1988ab) gave three reasons why EFA cannot deliver what is promised in this inductivist programme: First, there are no rationally optimal ways to extract knowledge from experience without making prior assumptions. Second, the interpretation of an EFA is not unique, due to the factor indeterminacy problem (Guttmann, 1955). Third, it is difficult to justify the results of a model which, in principle, can never be falsified. An ambiguity in EFA was pointed out by McDonald (1985, p.102):

> "In the exploratory approach, it might be claimed, we do not behave consistently. We first fit the model with many parameters and no constraint due to simple structure. We then transform the result to an equally fitting approximation to simple structure that may be very poor and speak as though we now have fewer parameters. But either the low numbers in the simple structure are consistent with exact zeros in the population or they are not. If they are, we should estimate only the nonzeros. If they are not, we do not in fact have simple structure at all."

It has be argued that the results from EFA have heuristic and suggestive value (e.g. Anderson, 1963) and may uncover hypotheses which are capable of more objective testing by other methods of multivariate analysis (Hotelling, 1957) and in new datasets. However, the prospects of obtaining sensible hypotheses from EFA are bleak, as was forcefully demonstrated by Armstrong (1967) (see also Mukherjee, 1973). Armstrong argues that meaningful EFA is only possible when considerable prior information is available, in which case CFA should be used in the first place (see also Section 1.3).

The use of confirmatory models in scale development can also be criticized. Here, researchers sometimes have 'pet' models such as the unidimensional Rasch model discussed in Section 3.3.4 and discard items contradicting the model to ensure a good fit. Goldstein (1994) points out that this approach is dubious because good fit is seen as supporting unidimensionality of the latent variable.

From now on we assume that the aim is to select a weak substantive model using modeling strategies ranging from the strictly confirmatory to model generating. It is important to note that a number of crucial decisions are usually made on a more or less heuristic basis before formal statistical modeling is undertaken, including:

- Selection of a model class (e.g. multilevel models)
- 'Causal' ordering of the variables:
  - Regression models: Typically classification into a set of explanatory variables and another set of response variables (sometimes just one).
  - Structural equation models: More elaborate ordering with explanatory variables, intermediate variables and response variables
  - Latent variable models: Selection of variables 'measuring' the latent variables
- Specification of probability distributions

Given these decisions, the role of statistical modeling is usually to help make decisions regarding *model form*; for instance which explanatory variables to include, which interactions to include, which 'paths' to include in structural equations models and which items measure which latent variables.

The adopted modeling strategy typically depends on the model class as well as the subject matter area. For instance, for linear mixed modeling in a biostatistical context, Verbeke and Molenberghs (2000, Chapter 9) suggest the following sequence for model building. First, find a preliminary model for the fixed part, selecting covariates and determining their functional relationships with the response. Second, find a preliminary model for the random part, deciding which effects to treat as random. Third, find a reasonable model for the level-1 error, for instance whether heteroscedasticity and/or autocorrelation should be specified or not.

8.3 Inference (likelihood based)

8.3.1 Properties of the fundamental and structural parameter estimates

Having estimated the parameters by maximum likelihood, the next question concerns the properties of the parameter estimates.

Consider first the estimated fundamental parameters $\widehat{\boldsymbol{\vartheta}}$. Since $\widehat{\boldsymbol{\vartheta}}$ is an ML estimator, it follows from e.g. Cox and Hinkley (1974) that it has a number of nice theoretical properties under suitable regularity conditions. Specifically, $\widehat{\boldsymbol{\vartheta}}$ is consistent, asymptotically normal, and asymptotically efficient. Consider then the estimators $\widehat{\boldsymbol{\theta}}$ of the structural parameters $\boldsymbol{\theta} = \mathbf{h}(\boldsymbol{\vartheta})$. First, since it is well known that ML-estimators are invariant under transformations (e.g. Cox and Hinkley, 1974, p.287), the ML estimator of $\boldsymbol{\theta}$, $\widehat{\boldsymbol{\theta}}$, is given by

$$\widehat{\boldsymbol{\theta}} = \mathbf{h}(\widehat{\boldsymbol{\vartheta}}). \tag{8.1}$$

It also follows that $\widehat{\boldsymbol{\theta}}$ inherits the asymptotic optimality properties of $\widehat{\boldsymbol{\vartheta}}$.

Rubin (1976) shows that consistency is retained for maximum likelihood estimators if responses are missing at random (MAR). This requires that the probability that a response is missing does not depend on the value of the response had it been observed, although it may depend on covariates included in the model and other responses. Importantly, responses are not required to be missing completely at random (MCAR) where missingness does not depend on either covariates, observed responses or missing responses. Little (1995) points out that methods such as GEE which are often said to require MCAR really remain valid when missingness is covariate dependent.

In the context of random effects models for longitudinal data, Little (1995) distinguishes between *covariate dependent* dropout, *missing at random* dropout, *nonignorable outcome-based* dropout and *nonignorable random-coefficient-based* dropout. Let $\mathbf{y}_j$ denote a vector of both observed and unobserved (missing) responses $\mathbf{y}_j' = [\mathbf{y}_{\mathrm{obs},j}, \mathbf{y}_{\mathrm{mis},j}]$ and let $\mathbf{r}_j$ be a vector of missingness indicators for a unit j. The different types of dropout can be defined as:

- Covariate dependent dropout

$$\Pr(\mathbf{r}_j | \mathbf{y}_j, \mathbf{X}_j, \boldsymbol{\zeta}_j) = \Pr(\mathbf{r}_j | \mathbf{X}_j),$$

- Missing at random dropout

$$\Pr(\mathbf{r}_j|\mathbf{y}_j,\mathbf{X}_j,\boldsymbol{\zeta}_j) \;=\; \Pr(\mathbf{r}_j|\mathbf{y}_{\mathrm{obs},j},\mathbf{X}_j),$$

- Nonignorable outcome-based dropout

$$\Pr(\mathbf{r}_j|\mathbf{y}_j,\mathbf{X}_j,\boldsymbol{\zeta}_j) \;=\; \Pr(\mathbf{r}_j|\mathbf{y}_{\mathrm{obs},j},\mathbf{y}_{\mathrm{mis},j},\mathbf{X}_j),$$

- Nonignorable random-coefficient-based dropout

$$\Pr(\mathbf{r}_j|\mathbf{y}_j,\mathbf{X}_j,\boldsymbol{\zeta}_j) \;=\; \Pr(\mathbf{r}_j|\mathbf{y}_{\mathrm{obs},j},\mathbf{X}_j,\boldsymbol{\zeta}_j).$$

We believe that this classification is also useful for general latent variable models with nonmonotone missing data patterns (intermittent missingness in the longitudinal setting).

If missingness is either covariate dependent or at random (and the parameters of the substantive process and the missingness process are distinct) inference can be based solely on the likelihood for the observed responses (the substantive process). This is because the joint likelihood of the substantive and missingness processes decomposes into separate components. Unfortunately, this useful result does not hold for the two nonignorable missingness processes where both substantive and missingness processes must be modeled jointly (e.g. Heckman, 1979; Hausman and Wise, 1979; Wu and Carroll, 1988; Diggle and Kenward, 1994). We refer to Little and Rubin (2002) for an extensive discussion.

8.3.2 Model-based standard errors

The asymptotic covariance matrix of $\widehat{\boldsymbol{\vartheta}}$ is

$$\mathrm{Cov}(\widehat{\boldsymbol{\vartheta}}) \;=\; -\mathrm{E}(\mathbf{H}(\widehat{\boldsymbol{\vartheta}}))^{-1} \;\cong\; -\mathbf{H}(\widehat{\boldsymbol{\vartheta}})^{-1}, \tag{8.2}$$

where $-\mathrm{E}(\mathbf{H}(\widehat{\boldsymbol{\vartheta}}))$ is the *Fisher information* or expected information and $-\mathbf{H}(\widehat{\boldsymbol{\vartheta}})$, minus the Hessian of the log-likelihood, is the *observed information*. The observed information approximates the expected information due to the strong law of large numbers.

There are three motivations for using the observed information in place of the expected information. The first reason is purely practical. The observed information is a by-product of the Newton-Raphson algorithm and therefore a natural choice when this algorithm is used for parameter estimation. Second, as was pointed out by Laird (1988) and Schluchter (1988) among others, use of the expected information is problematic in the context of missing data satisfying MAR but not MCAR. This is because one would have to integrate over the missing data mechanism to obtain the correct expected information. Third, Efron and Hinkley (1978, p.459) argue that the observed information is 'closer to the data' than the expected information (see also Kendall and Stuart, 1979), and that it tends to agree more closely with Bayesian analyses.

We can derive the covariance matrix of the estimated structural parameters from that of the estimated fundamental parameters using the multivariate delta-method (e.g. Serfling, 1980) applied to the function in (8.1):

$$\mathrm{Cov}(\widehat{\boldsymbol{\theta}}) \cong \left(\frac{\partial \mathbf{h}(\boldsymbol{\vartheta})}{\partial \widehat{\boldsymbol{\vartheta}}}\right) \mathrm{Cov}(\widehat{\boldsymbol{\vartheta}}) \left(\frac{\partial \mathbf{h}(\boldsymbol{\vartheta})}{\partial \widehat{\boldsymbol{\vartheta}}}\right)'.$$

8.3.3 Robust standard errors

The total log-likelihood is the sum of the top-level cluster contributions

$$\ell(\boldsymbol{\vartheta}) = \sum_{z=1}^{n^L} \ell_z(\boldsymbol{\vartheta}),$$

where

$$\ell_z(\boldsymbol{\vartheta}) = \ln g^{(L)}(\mathbf{y}_{z(L)}|\mathbf{X}_{z(L)}; \boldsymbol{\vartheta}).$$

Therefore maximum likelihood estimators satisfy the likelihood equations

$$\mathbf{g}(\widehat{\boldsymbol{\vartheta}}) = \sum_{z=1}^{n^{(L)}} \mathbf{g}^z = \mathbf{0}, \tag{8.3}$$

where $\mathbf{g}^z$ is the score vector for the zth level-L unit,

$$\mathbf{g}^z = \frac{\partial \ell_z(\boldsymbol{\vartheta})}{\partial \boldsymbol{\vartheta}}.$$

Using the delta-method, we can write the covariance matrix of $\mathbf{g}(\widehat{\boldsymbol{\vartheta}})$ as

$$\mathrm{Cov}[\mathbf{g}(\widehat{\boldsymbol{\vartheta}})] = \left(\frac{\partial \mathbf{g}(\widehat{\boldsymbol{\vartheta}})}{\partial \boldsymbol{\vartheta}}\right) \mathrm{Cov}(\widehat{\boldsymbol{\vartheta}}) \left(\frac{\partial \mathbf{g}(\widehat{\boldsymbol{\vartheta}})}{\partial \boldsymbol{\vartheta}}\right)'.$$

Solving for $\mathrm{Cov}(\widehat{\boldsymbol{\vartheta}})$ gives

$$\begin{aligned}
\mathrm{Cov}(\widehat{\boldsymbol{\vartheta}}) &= \left\{\frac{\partial \mathbf{g}(\widehat{\boldsymbol{\vartheta}})}{\partial \boldsymbol{\vartheta}}\right\}^{-1} \mathrm{Cov}[\mathbf{g}(\widehat{\boldsymbol{\vartheta}})] \left\{\left(\frac{\partial \mathbf{g}(\widehat{\boldsymbol{\vartheta}})}{\partial \boldsymbol{\vartheta}}\right)'\right\}^{-1} \\
&= \mathbf{H}^{-1} \mathrm{Cov}[\mathbf{g}(\widehat{\boldsymbol{\vartheta}})] \, \mathbf{H}^{-1}, \tag{8.4}
\end{aligned}$$

where $\mathbf{H} \equiv \mathbf{H}(\widehat{\boldsymbol{\vartheta}})$ is the Hessian of the log-likelihood at the parameter estimates. If the model is correct, $\mathrm{Cov}[\mathbf{g}(\widehat{\boldsymbol{\vartheta}})] = -\mathrm{E}(\mathbf{H}) \cong -\mathbf{H}$ and therefore $\mathrm{Cov}(\widehat{\boldsymbol{\vartheta}}) = -\mathrm{E}(\mathbf{H})^{-1} \cong -\mathbf{H}^{-1}$ as in equation (8.2).

Instead of relying on the model being correctly specified, we can utilize that $\mathbf{g}(\widehat{\boldsymbol{\vartheta}})$ in (8.3) is a sum of independent score vectors $\mathbf{g}^z$ with mean $\mathbf{0}$, so that the empirical covariance matrix becomes

$$\mathrm{Cov}[\mathbf{g}(\widehat{\boldsymbol{\vartheta}})] = \frac{n^{(L)}}{n^{(L)} - 1} \sum_{z=1}^{n^{(L)}} \mathbf{g}^z (\mathbf{g}^z)'. \tag{8.5}$$

Substituting (8.5) into (8.4) gives the so-called sandwich estimator; see for instance Huber (1967) and White (1982). This can be seen as an estimator of the covariance matrix of the design-based sampling distribution for the estimates defined as (implicit) functions of the data values (Binder, 1983). The sandwich estimator is popular in generalized estimating equations (see Section 6.9) and for complex survey data with sample weights, where the log-likelihood contributions are weighted by the inverse selection probabilities, giving a 'pseudo-likelihood'.

If the highest level units of the multilevel model are not mutually independent but clustered in n_c mutually exclusive clusters with index sets C_m, $m = 1, \ldots, n_c$, then

$$
\text{Cov}[\mathbf{g}(\widehat{\boldsymbol{\vartheta}})] = \frac{n_c}{n_c - 1} \sum_{m=1}^{n_c} \left(\sum_{z \in C_m} \mathbf{g}^z \right) \left(\sum_{z \in C_m} \mathbf{g}^z \right)', \tag{8.6}
$$

see Wooldridge (2002, Section 13.8.2) and Williams (2000) for proofs. Muthén and Satorra (1995) suggest using this approach for linear structural equation models for clustered data with inverse probability weighting.

Obvious alternatives to the sandwich estimator are resampling methods such as the bootstrap and the jackknife. Meijer *et al.* (1995) and Busing *et al.* (1994) discuss parametric and nonparametric bootstrapping for linear two-level models. There are two types of nonparametric bootstrapping, one based on resampling cases and the other on resampling residuals or errors. Neither are straightforward for multilevel data. For the 'cases bootstrap', it is not entirely clear whether to resample clusters and units within clusters, only clusters or only units within clusters. For the 'error bootstrap' it is not clear how to estimate the higher-level residuals because of shrinkage (see also Carpenter *et al.*, 1999). Patterson *et al.* (2002) use the jackknife for a latent class model with sample weights and Busing *et al.* (1994) for two-level linear models.

8.3.4 Likelihood ratio, Wald and score tests

Let $\mathcal{M}_1$ and $\mathcal{M}_2$ denote two contending models with v_1 and v_2 fundamental parameters $\boldsymbol{\vartheta}_{\mathcal{M}_1}$ and $\boldsymbol{\vartheta}_{\mathcal{M}_2}$, respectively. Assume that $\mathcal{M}_2$ is nested in $\mathcal{M}_1$, in the sense that restrictions are imposed on the structural parameters of $\mathcal{M}_1$ to yield a model $\mathcal{M}_2$ with $v_1 - v_2$ fewer fundamental parameters. Let the maximized log-likelihoods for the two models be denoted $\ell(\widehat{\boldsymbol{\vartheta}}_{\mathcal{M}_1} | \mathbf{y}, \mathbf{X})$ and $\ell(\widehat{\boldsymbol{\vartheta}}_{\mathcal{M}_2} | \mathbf{y}, \mathbf{X})$.

Conventional likelihood-ratio testing can then be performed using the statistic

$$
D_{\mathcal{M}_1 \mathcal{M}_2} = 2 \left[\ell(\widehat{\boldsymbol{\vartheta}}_{\mathcal{M}_1} | \mathbf{y}, \mathbf{X}) - \ell(\widehat{\boldsymbol{\vartheta}}_{\mathcal{M}_2} | \mathbf{y}, \mathbf{X}) \right], \tag{8.7}
$$

which under regularity conditions is asymptotically χ^2-distributed with $v_1 - v_2$ degrees of freedom under the restricted model $\mathcal{M}_2$ (e.g. Cox and Hinkley, 1974).

Wald-tests can be derived from $\text{Cov}(\widehat{\boldsymbol{\vartheta}}_{\mathcal{M}_1})$ and Lagrange multiplier or score tests from $\mathrm{Cov}(\widehat{\boldsymbol{\vartheta}}_{\mathcal{M}_2})$. These test statistics, which only necessitate the estimation of *one* model, $\mathcal{M}_1$ or $\mathcal{M}_2$ respectively, can be regarded as quadratic approximations to the likelihood ratio test statistic. It is well known that Wald and Lagrange multiplier tests are asymptotically equivalent to the likelihood ratio test (e.g. Cox and Hinkley, 1974; Buse, 1982; Engle, 1984).

However, in the finite sample situation the choice of test statistic can be important. The Wald test performs poorly if the log-likelihood is not well approximated by a quadratic function in the neighborhood of the parameter estimates. Hauck and Donner (1977) show how this can happen in logistic regression. Note that, unlike the likelihood ratio test, the Wald test is not invariant to nonlinear transformations of the parameter, some transformations being preferable to others. If the Wald and likelihood ratio tests yield different results, the likelihood ratio test is preferable.

The score test can be justified using a central limit theorem argument, not just as an approximation to the likelihood ratio test (e.g. Pawitan, 2001, p.235). In some situations the score test performs better than the likelihood ratio test. An advantage of the score test over the Wald test is that it is invariant to nonlinear transformations of the parameters. We refer to Pawitan (2001) for an excellent and accessible account of likelihood theory.

Unfortunately, standard asymptotic results for the likelihood ratio, Wald and score test statistics do not hold if the null hypothesis is on the boundary of the parameter space which would violate regularity conditions (e.g. Moran, 1971; Self and Liang, 1987). A common example is testing null hypotheses regarding random effects (e.g. Miller, 1977; Stram and Lee, 1994, 1995; Berkhof and Snijders, 2001). Consider two two-level models, $\mathcal{M}_2$ with $M-1$ random effects and $\mathcal{M}_1$ containing an additional random effect with variance ψ_{MM} and covariances ψ_{Mj}, $j = 1, \ldots, M-1$. Since ψ_{MM} is nonnegative, it lies on the boundary of the parameter space, $\psi_{MM} = 0$, under $\mathcal{M}_2$. Importantly, the correct asymptotic distribution of the likelihood ratio statistic under the null depends on software implementation. Imposing $\widehat{\psi}_{MM} \geq 0$ and letting $\widehat{\psi}_{Mj} = 0$ if $\widehat{\psi}_{MM} = 0$, the distribution becomes a 50:50 mixture of a point mass at zero and a χ^2_M. The correct p-value is then simply obtained by halving the 'naive' p-value based on the χ^2_M. Verbeke and Molenberghs (2000) give a nontechnical discussion and Crainiceanu and Ruppert (2004) a critical appraisal of likelihood ratio testing of random effects. Verbeke and Molenberghs (2003) derive general one-sided score tests for random effects.

Another violation of regularity conditions occurs in latent class models where a $K-1$ class model cannot be obtained by imposing a simple restriction on the K class model. For instance, fixing the probability of one class to zero renders the corresponding location nonidentified. Alternatively, setting the locations of two classes equal implies that only the sum of the corresponding probabilities becomes identified. Therefore, likelihood ratio statistics do not have a 'naive' χ^2 distribution in this setting (e.g. Aitkin and Rubin, 1985; Titterington *et al.*, 1985; McLachlan and Basford, 1988; Everitt, 1988). A possible solution is parametric bootstrapping where data are simulated from the

$K-1$ class model followed by estimation of both the $K-1$ class and K class model to compute the likelihood ratio statistic. The empirical distribution of these statistics over bootstrap replications is then used to obtain approximate significance probabilities. We refer to Böhning (2000, Chapter 4) for a detailed discussion and simulations.

8.3.5 Confidence intervals

Confidence intervals can be constructed by inverting the likelihood ratio, Wald or score (Lagrange multiplier) tests. To construct a $100(1 - \alpha)\%$ confidence interval for a parameter β, we need to find a lower confidence limit β_l so that the one-sided test of the null hypothesis $\beta = \beta_l$ with alternative hypothesis $\beta > \beta_l$ has a p-value equal to $\alpha/2$. The upper confidence limit is obtained analogously.

The Wald-based confidence interval simply becomes

$$\widehat{\beta} \pm z_{1-\alpha/2}\mathrm{SE}(\widehat{\beta}),$$

where $z_{1-\alpha/2}$ is the $1 - \alpha/2$ fractile of the standard normal distribution and $\mathrm{SE}(\widehat{\beta})$ is the estimated standard error of $\widehat{\beta}$.

However, for score and likelihood based intervals, a search is required to find the confidence limits. For the likelihood based interval, this requires evaluating the log-likelihood at the maximum likelihood estimates of the other parameters for each fixed value of the parameter β of interest, giving the *profile log-likelihood*. The confidence limits are those values of β where the profile log-likelihood is $\chi^2_{1-\alpha}(1)$ lower than the log-likelihood maximized with respect to all parameters, where $\chi^2_{1-\alpha}(1)$ is the $1-\alpha$ fractile of the chi-squared distribution with one degree of freedom (3.84 for a 95% confidence interval).

In the application chapters we present estimated standard errors for all parameter estimates, allowing Wald-based confidence intervals to be constructed if desired. The profile log-likelihood method (based on the deviance) is used in Section 9.7 for deriving a confidence interval for the population size of snowshoe hares. In Section 9.4 we use profile log-likelihood based confidence intervals for the guessing parameter in a three-parameter logistic item response model.

Construction of confidence intervals for variance components is problematic, particularly if the estimates are close to zero. Bottai (2003) recommends that confidence regions should be based on a score test using the expected information.

8.4 Model selection: Relative fit criteria

Competing models are usually compared using relative fit criteria (e.g. Jöreskog, 1974; Tanaka *et al.*, 1990; Tanaka, 1993).

As pointed out in Section 5.3, equivalent models cannot be distinguished empirically although they may represent different or even contradictory substantive processes. The same applies to nonequivalent models which happen to yield similar fit for the particular data set being analyzed. For these cases model selection must proceed by substantive or other nonstatistical arguments.

8.4.1 Significance testing for nested models

Likelihood ratio, Wald and Score tests can be used to compare nested models. This approach is, however, not well suited to model selection for at least five reasons.

One major problem is that we have seldom decided which models to compare a priori. The tests are, on the contrary, suggested by the same data which are to be employed for model assessment. In other words, we are in a 'model generating' situation. Clearly, this situation does not fit the conventional test paradigm and the sampling properties of the overall model selection strategy are unknown (e.g. Freedman, 1983). A second objection to the conventional strategy is that conditioning on a single selected model ignores model uncertainty and leads to underestimation of standard errors (e.g. Miller, 1984). A third problem with the conventional testing approach is that the power of hypothesis tests depends on sample size. While acknowledging that more observations imply more information, it nevertheless appears unreasonable to base model selection on the test criterion. This point can be made clear by considering a situation with a very large number of observations. Here, we expect all but extremely complicated models to be rejected and we are left with 'models' which merely mirror the particular data set at hand. On the other hand, if few observations are available, we expect that oversimplifications tend to be retained. Fourth, it should be recognized that models of interest may be *nonnested*. Hence, in this case, investigation of fit cannot be based on the traditional test criterion. However, tests for nonnested models such as those suggested by Cox (1961, 1962) and related tests surveyed by Davidson and MacKinnon (1993, Chapter 11) can be used. Fifth, it has been argued that significance probabilities and evidence are often in conflict, even for the unrealistic case of solely two nested models (Berger and Sellke, 1987; Berger and Delampady, 1987).

However, it should be noted that significance probabilities need not be interpreted in a strict sense, but merely as representing less formal indices of fit (e.g. Jöreskog, 1969, 1978).

8.4.2 Bayesian model selection

Posterior odds, Bayes factors and the Bayesian information criterion (BIC)

Suppose that we want to use the data $\mathbf{D}$ to compare a set of possibly nonnested competing models. A Bayesian approach is to compare the posterior probabilities of the models given the data.

By Bayes theorem, the required posterior probability for a given model $\mathcal{M}_k$

is

$$\Pr(\mathcal{M}_k|\mathbf{D}) = \Pr(\mathbf{D}|\mathcal{M}_k)\Pr(\mathcal{M}_k)/\Pr(\mathbf{D}),$$

where $\Pr(\mathcal{M}_k)$ is the prior probability of the model, $\Pr(\mathbf{D}|\mathcal{M}_k)$ the marginal probability of the data given the model and $\Pr(\mathbf{D})$ the marginal probability of the data. The extent to which the data support $\mathcal{M}_k$ over a competing model $\mathcal{M}_l$ can be assessed by the *posterior odds* of $\mathcal{M}_k$ against $\mathcal{M}_l$,

$$\frac{\Pr(\mathcal{M}_k|\mathbf{D})}{\Pr(\mathcal{M}_l|\mathbf{D})} = \left[\frac{\Pr(\mathcal{M}_k)}{\Pr(\mathcal{M}_l)}\right]\left[\frac{\Pr(\mathbf{D}|\mathcal{M}_k)}{\Pr(\mathbf{D}|\mathcal{M}_l)}\right],$$

where the first term in square brackets is the prior odds and the second term is the *Bayes factor*. Often, we have little reason to favor one model over another a priori and therefore assign equal prior probabilities to the models so that the posterior odds reduce to the Bayes factor.

Note that many Bayesians would not select a single model but rather base inference regarding the parameter(s) of interest on the posterior distribution of the parameter(s), averaging over models. Such *model averaging* is often accomplished by using approximate posterior probabilities of the models given the data as weights (see e.g. Wasserman, 2000).

The marginal probability of the data given the model, often called the integrated likelihood, is given by

$$\Pr(\mathbf{D}|\mathcal{M}_k) = \int \Pr(\mathbf{D}|\boldsymbol{\vartheta}_k, \mathcal{M}_k)\Pr(\boldsymbol{\vartheta}_k|\mathcal{M}_k)\,d\boldsymbol{\vartheta}_k, \tag{8.8}$$

where $\boldsymbol{\vartheta}_k$ are the parameters for model $\mathcal{M}_k$. It has been pointed out that the integrated likelihoods (and therefore the Bayes factor) depend heavily on the prior distributions, even if the priors are vague (e.g. Kass and Raftery, 1995).

The integral can rarely be evaluated analytically and various approximations have therefore been suggested. The simplest and most commonly used approximation for twice the Bayes factor is the *Bayesian Information Criterion* (BIC) (e.g. Schwarz, 1978),

$$\text{BIC} = 2\left[\ell(\widehat{\boldsymbol{\vartheta}}_{\mathcal{M}_k}|\mathbf{y}, \mathbf{X}) - \ell(\widehat{\boldsymbol{\vartheta}}_{\mathcal{M}_l}|\mathbf{y}, \mathbf{X})\right] - (v_k - v_l)\ln N. \tag{8.9}$$

The BIC can be derived using the *Laplace approximation* introduced in Section 6.3.1 for the integral in (8.8). A further approximation is to replace the posterior mode of the parameters by the maximum likelihood estimates and the Hessian of the log of the integrand by the Hessian of the log-likelihood, i.e. the likelihood is assumed to dominate the prior distribution. It can be shown that the BIC is a good approximation to the Bayes factor if a 'unit information prior' is used, a multivariate normal with covariance matrix equal to the inverse of N^{-1} times the Fisher information (e.g. Kass and Wasserman, 1995). For relatively nontechnical derivations of the BIC; see Kass and Raftery (1995) and Raftery (1995).

Although only a crude approximation, the BIC is popular among some Bayesians as well as frequentists because it is easily obtained from standard output of statistical software. The BIC for a given model $\mathcal{M}_k$, here denoted

BIC_k, is usually defined as

$$\text{BIC}_k = -2\ell(\widehat{\vartheta}_{\mathcal{M}_k}|\mathbf{y}, \mathbf{X}) + v_k \ln N, \qquad (8.10)$$

and the model with the lowest BIC_k is selected. Frequentists find the BIC attractive since it handles nonnested models in contrast to the likelihood ratio criterion and because it does not require specification of a prior distribution for the model parameters.

Unfortunately, BIC is difficult to apply to models with latent variables because it is not clear what 'N' should be in the second term of (8.10). For instance, in two-level models, including factor, latent class, and structural equation models where the level-1 units are items, either the number of clusters J or the total number of level-1 units N have been used. In latent class modeling, the predominant approach is to use J (e.g. Vermunt and Magidson, 2000; Clogg, 1995; McCulloch et al., 2002), but the latent class program GLIMMIX (Wedel, 2002b) uses N. In structural equation modeling, Bollen (1989) and Raftery (1993) use N, but Raftery (1995) recommends using J. The BIC is little used in multilevel regression modeling. Hoijtink (2001) uses the Bayes factor for latent class models, avoiding the BIC approximation and the choice between J and N.

Another issue with latent variable models is determining the degrees of freedom (effective number of parameters). Hodges and Sargent (2001), Burnham and Anderson (2002) and Vaida and Blanchard (2004) argue that the degrees of freedom lie somewhere between the number of model parameters (in a frequentist sense, excluding the latent variables) and the sum of the number of model parameter and the number of realizations of the latent variables.

The deviance information criterion (DIC)

Spiegelhalter et al. (2002) base a measure of model complexity on the concept of the 'excess of the true over the estimated residual information', defined as

$$d\{\vartheta, \widetilde{\vartheta}, \mathbf{y}\} = -2\ell(\vartheta|\mathbf{y}, \mathbf{X}) + 2\ell(\widetilde{\vartheta}|\mathbf{y}, \mathbf{X}),$$

where ϑ is the true parameter vector and $\widetilde{\vartheta}$ is the estimated parameter vector. This can be thought of as the degree of overfitting since it represents how much less the data deviate from the model with estimated parameters than they do from the model with true parameters. Spiegelhalter et al. (2002) propose using the posterior expectation of this measure,

$$p_D \equiv \text{E}_{\vartheta}(d\{\vartheta, \widetilde{\vartheta}, \mathbf{y}\}|\mathbf{y}) = \text{E}_{\vartheta}(-2\ell(\vartheta|\mathbf{y}, \mathbf{X})|\mathbf{y}) + 2\ell(\widetilde{\vartheta}|\mathbf{y}, \mathbf{X}),$$

as a Bayesian measure of model complexity or 'effective number of parameters'. They also suggest using the posterior expectation of minus twice the log-likelihood (the first term above) as a Bayesian measure of fit, the deviance information criterion (DIC),

$$\text{DIC} \equiv \text{E}_{\vartheta}(-2\ell(\vartheta|\mathbf{y}, \mathbf{X})|\mathbf{y}) = -2\ell(\widetilde{\vartheta}|\mathbf{y}, \mathbf{X}) + p_D. \qquad (8.11)$$

In hierarchical Bayesian models, Spiegelhalter et al. (2002) point out that we cannot define model complexity without specifying the level of the hierarchy that is the focus of the modeling exercise (conditional or marginal focus in the terminology of Vaida and Blanchard, 2004). The focus determines how the full probability model (see also Section 6.11)

$$\Pr(\mathbf{D}, \boldsymbol{\zeta}, \boldsymbol{\vartheta}_1, \boldsymbol{\vartheta}_2) = \Pr(\mathbf{D}|\boldsymbol{\zeta}, \boldsymbol{\vartheta}_1)\Pr(\boldsymbol{\zeta}|\boldsymbol{\vartheta}_2)\Pr(\boldsymbol{\vartheta}_2)\Pr(\boldsymbol{\vartheta}_1),$$

is factorized into the likelihood and prior components. Here $\mathbf{D}$ are the data, $\boldsymbol{\zeta}$ are latent variables, $\boldsymbol{\vartheta}_1$ are 'fixed' parameters and $\boldsymbol{\vartheta}_2$ are hyperparameters.

A conditional (cluster-specific) focus corresponds to defining the likelihood as conditional on the latent variables and 'fixed' parameters, $\Pr(\mathbf{D}|\boldsymbol{\zeta}, \boldsymbol{\vartheta}_1)$, and the prior as marginal with respect to the hyperparameters

$$\Pr(\boldsymbol{\vartheta}_1)\Pr(\boldsymbol{\zeta}) = \Pr(\boldsymbol{\vartheta}_1) \int \Pr(\boldsymbol{\zeta}|\boldsymbol{\vartheta}_2)\Pr(\boldsymbol{\vartheta}_2)\mathrm{d}\boldsymbol{\vartheta}_2.$$

A marginal (population) focus corresponds to defining the likelihood as marginal with respect to the latent variables

$$\Pr(\mathbf{D}|\boldsymbol{\vartheta}_1, \boldsymbol{\vartheta}_2) = \int \Pr(\mathbf{D}|\boldsymbol{\zeta}, \boldsymbol{\vartheta}_1)\Pr(\boldsymbol{\zeta}|\boldsymbol{\vartheta}_2)\mathrm{d}\boldsymbol{\zeta}$$

and the prior as $\Pr(\boldsymbol{\vartheta}_1)\Pr(\boldsymbol{\vartheta}_2)$. The effective number of parameters will obviously be larger for the conditional focus than the marginal one.

8.4.3 The Akaike information criterion (AIC)

The Akaike Information Criterion (AIC) (e.g. Akaike, 1987) or its variants (e.g. Bozdogan, 1987) are often used for model selection.

Let $f(\mathbf{y}|\mathbf{X}; \boldsymbol{\vartheta})$ denote the distribution of the responses given the parameters for a specified model (i.e. the likelihood) and $f^*(\mathbf{y}|\mathbf{X})$ the distribution for the true model. The 'information lost' when $f(\mathbf{y}|\mathbf{X}; \boldsymbol{\vartheta})$ is used to approximate $f^*(\mathbf{y}|\mathbf{X})$ can be defined as

$$I(f, f^*, \boldsymbol{\vartheta}) = \int f^*(\mathbf{y}|\mathbf{X})[\ln f^*(\mathbf{y}|\mathbf{X}) - \ln f(\mathbf{y}|\mathbf{X}; \boldsymbol{\vartheta})]\,\mathrm{d}\mathbf{y}, \qquad (8.12)$$

known as the Kullback-Leibler information between the two models. This expectation (over the true distribution of $\mathbf{y}|\mathbf{X}$) of the difference in true and approximate log-likelihoods is large if data from the true model tend to be unlikely under the specified model. The measure is zero if $f(\mathbf{y}|\mathbf{X}; \boldsymbol{\vartheta}) = f^*(\mathbf{y}|\mathbf{X})$ and positive otherwise.

It would be natural to plug in parameter estimates $\widehat{\boldsymbol{\vartheta}}(\mathbf{y}^\bullet)$, obtained from some data $\mathbf{y}^\bullet$, into the Kullback-Leibler information. The expectation in (8.12) is then over samples $\mathbf{y}$ that are independent of $\mathbf{y}^\bullet$ but from the same distribution and is therefore sometimes interpreted as a cross-validation measure. The expectation of the Kullback-Leibler information over repeated samples $\mathbf{y}^\bullet|\mathbf{X}$,

$$\int f^*(\mathbf{y}^\bullet)I(f, f^*, \widehat{\boldsymbol{\vartheta}}(\mathbf{y}^\bullet|\mathbf{X}))\,\mathrm{d}\mathbf{y}^\bullet,$$

(a double-expectation with respect to $\mathbf{y}|\mathbf{X}$ and $\mathbf{y}^\bullet|\mathbf{X}$) forms the basis of the Akaike information criterion. The first term of $I(f, f^*, \widehat{\boldsymbol{\vartheta}})$ is constant across models and can therefore be ignored in model comparison. Akaike (1973) showed that twice the expectation of the second term can be approximated as

$$-2 \int f^*(\mathbf{y}^\bullet|\mathbf{X})\ln f(\mathbf{y}|\mathbf{X}; \widehat{\boldsymbol{\vartheta}}(\mathbf{y}^\bullet|\mathbf{X}))\, \mathrm{d}\mathbf{y}^\bullet \approx -2\ell(\widehat{\boldsymbol{\vartheta}}|\mathbf{y}, \mathbf{X}) + 2v \equiv \mathrm{AIC},$$

where $\widehat{\boldsymbol{\vartheta}}$ are the parameter estimates for the observed data and v is the number of model parameters. The term $2v$ serves to correct the bias in using the maximized log-likelihood as an estimator of its double expectation.

Note that AIC is identical to Mallows C_p for conventional linear regression models. The AIC, BIC and DIC can all be viewed as deviances with a penalty for model complexity. For the BIC this penalty is greater than for the AIC so that more parsimonious models tend to be selected. In latent variable models, it is not clear what the number of parameters v should be, for the same reason discussed for the BIC in Section 8.4.2, see e.g. Vaida and Blanchard (2004).

See Zucchini (2000) and Wasserman (2000) for discussions of the AIC and BIC, the latter from a Bayesian perspective. Recently, a focussed information criterion (FIC) has been proposed by Claeskens and Hjort (2003) to select the 'best' model for inference regarding a given parameter of interest.

We use the AIC and BIC to compare nonnested models for overdispersed count data in Section 11.2. There we somewhat arbitrarily use the number of fixed parameters (in the fixed and random parts of the models) for v and the number of clusters for N which in this case equals the number of units.

8.5 Model adequacy: Global absolute fit criteria

Misspecifications can occur in one or more of the model components of the general model framework. For the conditional response model all misspecifications conceivable in generalized linear models may happen, including omitted variable problems, inappropriate link functions, inappropriate variance functions and inappropriate distributional assumptions. In addition, the assumption of conditional independence may be violated, random regression coefficients mistakenly specified as fixed and inappropriate constraints imposed on factor loadings and measurement error variances. The structural equations for latent variables may be misspecified by omitting relevant observed or latent covariates, mistakenly specifying the relations as linear and misspecifying the distribution of the disturbances at the different levels of the multilevel model.

Misspecifications manifest themselves through a lack of fit of the specified model to the available dataset. The major challenge is to distinguish between lack of fit due to sampling variability, which is not a problem, and lack of fit due to using an inappropriate model, which is a problem. Due to the multiplicity of potential misspecification problems, it is clear that model diagnostics is a daunting task for complex models.

A natural approach would seem to be to first assess whether there is evidence for any type of misspecification before proceeding to identify the source. Two common approaches are global tests for misspecification and 'fit indices'. While some of the tests and indices might be sensitive to specific departures from assumptions, there are generally several possible sources of discrepancy between model and data.

8.5.1 Tests for misspecification

Tests of absolute fit

Tests of absolute fit presuppose the existence of a benchmark. The benchmark is typically the saturated model for a given set of variables. When both responses and covariates are categorical, the saturated model is the unrestricted multinomial model with expected counts equal to observed counts for all cells in the full contingency table (e.g. Bock and Lieberman, 1970; Bock and Aitkin, 1981). For multivariate normal response variables (and no covariates) the saturated model is a multivariate normal density with unrestricted means, variances and covariances. Importantly, tests of absolute fit cannot detect omitted variables because the benchmark is relative to the specific variables included. Also note that there is no absolute standard available when more general models are considered.

Any of the relative fit criteria discussed above can be used to compare the model of interest to the saturated model. The likelihood ratio test is the most common because it has a known asymptotic distribution under the null hypothesis that the restricted model is true. Twice the difference in log-likelihood between a model and the saturated model is called the deviance. For categorical data an alternative statistic is the Pearson X^2. The deviance is used to assess absolute fit of a latent class model in Section 13.5. Both the deviance and Pearson X^2 are used for item response models in Section 9.4.

These tests are problematic for sparse contingency tables since asymptotic results cannot be relied on. In the context of latent trait and latent class models, Glas (1988), Reiser (1996), Reiser and Lin (1999) and Bartholomew and Leung (2002) suggest tests based on a collection of marginal tables, such as tables for all pairs of variables.

The logic of hypothesis testing is undermined when absolute fit is tested in log-linear or covariance structure modeling (e.g. Bishop *et al.*, 1975; Fornell, 1983). In covariance structure modeling the null hypothesis corresponds to a restricted model and the alternative to the empirical covariance matrix. The important thing to note is that the researcher in this case desires to *retain* the null hypothesis in favor of the alternative. Consequently, the null hypothesis is maintained when it cannot be rejected. It is clear that the status of null and alternative hypothesis is reversed in this case, compared to the standard framework for statistical testing. Fornell (1983) points out the associated problem that models are often retained due to small sample size and resulting lack of power. Furthermore, and perhaps more surprisingly, weak observed relationships among variables (small correlations) increase the chances of retaining any model considered.

The Hausman and White tests

The Hausman (1978) misspecification test considers two estimators $\widehat{\boldsymbol{\beta}}$ and $\widetilde{\boldsymbol{\beta}}$ which are both consistent if the model is correctly specified but converge to different limits when the model is misspecified.

Consider for example estimation of the fixed regression coefficients in a random intercept model. Both the usual maximum likelihood estimator $\widehat{\boldsymbol{\beta}}$ for the random intercept model and the ordinary least squares estimator $\widehat{\boldsymbol{\beta}}_{\mathrm{FE}}$ for the fixed intercepts model (see Section 3.6.1) are consistent if the model is correctly specified. However, if the random intercept correlates with one of the covariates (see Section 3.2.1 on page 52), the maximum likelihood estimator of the random intercept model becomes inconsistent whereas the ordinary least squares estimator remains consistent. Therefore a difference between the estimates suggests that the random intercept model is misspecified.

Formalizing this idea, Hausman suggests the following test statistic:

$$w_h = (\widehat{\boldsymbol{\beta}} - \widehat{\boldsymbol{\beta}}_{\mathrm{FE}})' \left[\mathrm{Cov}(\widehat{\boldsymbol{\beta}} - \widehat{\boldsymbol{\beta}}_{\mathrm{FE}}) \right]^{-1} (\widehat{\boldsymbol{\beta}} - \widehat{\boldsymbol{\beta}}_{\mathrm{FE}}),$$

where $\mathrm{Cov}(\widehat{\boldsymbol{\beta}} - \widehat{\boldsymbol{\beta}}_{\mathrm{FE}})$ is the covariance matrix of the difference if the model is correctly specified. The test statistic is asymptotically χ^2 distributed with degrees of freedom equal to the rank of $\mathrm{Cov}(\widehat{\boldsymbol{\beta}} - \widehat{\boldsymbol{\beta}}_{\mathrm{FE}})$. Hausman shows that, asymptotically,

$$\mathrm{Cov}(\widehat{\boldsymbol{\beta}} - \widehat{\boldsymbol{\beta}}_{\mathrm{FE}}) = \mathrm{Cov}(\widehat{\boldsymbol{\beta}}) - \mathrm{Cov}(\widehat{\boldsymbol{\beta}}_{\mathrm{FE}}),$$

making the test easy to implement since it requires only the estimated covariance matrices of the two estimators.

Although easy to implement and potentially useful, there are some limitations of the approach. First, in common with other tests of fit, the test is sensitive to different kinds of misspecification making it hard to pinpoint the problem. Second, simulation studies indicate that the power of the test may be low for typical sample sizes (e.g. Long and Trivedi, 1993). Finally, the sampling distribution of the test statistic in finite samples may not be well approximated by a χ^2 distribution.

It is known from maximum likelihood theory that the estimated covariance matrix of the parameter estimates is given by the sandwich estimator in (8.4). If the model is correctly specified, the sandwich estimator reduces to the inverse of the information matrix $-\mathbf{H}^{-1}$. As for the Hausman test, a difference between the two estimators suggests that the model is misspecified. White's (1982) information matrix test therefore compares the two covariance matrices using the test statistic

$$\mathbf{d}' \widehat{\mathbf{C}}^{-1} \mathbf{d},$$

where $\mathbf{d}$ is a vector of differences of a subset of the elements of the estimated covariance matrices with associated covariance matrix $\widehat{\mathbf{C}}$. This test shares with the Hausman test the problem of being sensitive to many types of mis-specification. In addition the test is difficult to implement since it requires an estimator of $\mathbf{C}$.

8.5.2 Goodness of Fit Indices

As noted earlier the roles of null and alternative hypotheses are reversed in tests of absolute fit, where the researcher is hoping to 'fail' to reject the null hypothesis that the model is true. Curiously, high power then becomes a problem.

One reaction to this problem is to use so-called goodness of fit indices (GFI), a typical example being the incremental fit index Δ proposed for linear struc-tural equation models by Bentler and Bonett (1980)

$$\Delta = \frac{F_b - F_m}{F_b}.$$

Here, F_m and F_b are the values of the fitting function used to estimate the parameters (see Section 6.2) for the model of interest and a baseline model, respectively. This index can be interpreted as the proportional reduction in fit-ting function between the baseline model and model of interest. In covariance structure modeling, a common choice of baseline model is a model imposing independence among the response variables. Note that the squared multiple correlation coefficient R^2 in linear regression can be defined the same way where the fitting function is the sum of squared residuals and the baseline model has no covariates.

An attraction of the GFIs is that they are generally normed between 0 and 1, with values in the 0.90s typically said to indicate 'good fit'. A major problem of this approach is that model choice gets an arbitrary flair. This situation is not helped by the large number of fit indices proposed, see e.g. Bollen (1989), Marsh et al. (1988) and Mulaik et al. (1989) for surveys, and produced by standard software.

The choice of baseline or null model is important (Sobel and Bohrnstedt, 1985). It could be argued that it does not make sense to define goodness of fit of a given model relative to a baseline model known to be inadequate. For instance in measurement and longitudinal models a baseline model specify-ing independent responses would a priori be expected to be very wrong. It therefore comes as no surprise that models generally obtain high GFIs in this case.

Furthermore, how badly the baseline model fits the data depends greatly on the magnitude of the parameters of the true model. For instance, con-sider estimating a simple parallel measurement model. If the true model is a congeneric measurement model (with considerable variation in factor loadings and measurement error variances between items), the fit index could be high simply because the null model fits very poorly, i.e. because the reliabilities of the items are high. However, if the true model is a parallel measurement model with low reliabilities the fit index could be low although we are estimating the correct model. Similarly, estimating a simple linear regression model can yield a high R^2 if the relationship is actually quadratic with a considerable linear trend and a low R^2 when the model is true but with a small slope (relative to the overall variance).

Goldberger (1991, p.177) puts it the following way:

"Nothing in the CR (Classical Regression) model requires that R^2 be high. Hence, a high R^2 is not evidence in favor of the model, and a low R^2 is not evidence against it."

Perhaps the fit indices should therefore be better described as 'coefficients of determination', a description often used for the R^2.

8.5.3 Error of approximation

In the context of covariance structure modeling Cudeck and Henly (1991) consider discrepancies among four different covariance matrices. When estimating a model the fitting function $\widehat{F} \equiv F(\widehat{\Sigma}, \mathbf{S})$ compares the $n \times n$ sample covariance matrix $\mathbf{S}$ with the estimated model implied covariance matrix $\widehat{\Sigma}$ based on v parameters. If the model were estimated in the population, the analogous matrices would be the true covariance matrix Σ_0 and the covariance matrix implied by the approximating model $\widetilde{\Sigma}$. The *discrepancy due to approximation*, defined as $F_0 \equiv F(\widetilde{\Sigma}, \Sigma_0)$, is unknown since the population matrices are unknown.

It can be shown that the *sample discrepancy function* $\widehat{F}$ is a biased estimator of the discrepancy due to approximation F_0. A less biased estimator is $\widehat{F}_0 = \widehat{F} - J^{-1}d$ (see McDonald, 1989), where d is the degrees of freedom ($d = \frac{1}{2}n(n+1) - v$). If $\widehat{F}_0$ is negative, Browne and Cudeck (1993) suggest setting it to zero.

To penalize for model complexity, Steiger (1990) proposes the *root mean square error of approximation* (RMSEA), estimated as

$$\widehat{\text{RMSEA}} = \sqrt{\frac{\widehat{F}_0}{d}}.$$

Browne and Cudeck (1993, p.144) state: "We are of the opinion that a value of about 0.08 or less for the RMSEA would indicate a reasonable error of approximation and would not want to employ a model with a RMSEA greater than 0.1." They also show how confidence intervals and tests for the RMSEA can be constructed. We refer to Browne and Arminger (1995) for further discussion.

8.5.4 Cross-validation

Validating a model on the same data for which the model was built using for instance goodness of fit indices leads to overoptimistic assessments. The same problem applies to model selection using relative fit criteria. A remedy for these problems is the use of cross-validation, which can be implemented in several ways.

An obvious approach is to split the sample randomly into a calibration sample used to estimate candidate models and a confirmation sample for testing the models (see Section 10.3.3, page 334 for an example). However, this approach is wasteful since a major portion of the data is discarded both for calibration and validation.

Another approach is to repeatedly estimate the model leaving out one unit at a time. The estimates produced when unit i is omitted are used to obtain the contribution to the discrepancy measure for that unit (e.g. Stone, 1974; Geisser, 1975). Obviously this method can be used only if the discrepancy measure is a function of the contributions from the individual units. For instance, it does not work for discrepancy measures based on covariance matrices commonly used in structural equation modeling (see Section 6.2).

A general approach to cross-validation is the use of *resampling* techniques (e.g. Efron and Tibshirani, 1993, Chapter 17) such as bootstrapping. A simple version is to estimate the parameters in each bootstrap sample and obtain the goodness of fit index for the original data. The mean index over the bootstrap samples is then used as a measure of cross-validation.

In some situations it is possible to obtain an estimate of the expectation of a cross-validation index (over repeated validation and confirmation samples) based on data from a single sample. Examples include the adjusted R^2 in multiple regression and the expected cross-validation index (ECVI) suggested by Browne and Cudeck (1989) for covariance structure models; see also Browne and Cudeck (1993) and Browne (2000).

8.6 Model diagnostics: Local absolute fit criteria

After finding an indication that something is wrong with the model, the next logical step is to diagnose the problem. Model diagnostics are procedures designed to suggest violations of more or less specific model assumptions.

A first step is usually to derive some statistics reflecting specific model departures such as residuals. The next step is 'inspect' the statistics and devise more or less formal criteria for detecting problems. This step covers a wide range of approaches including graphics, ad hoc application of thresholds and formal tests based on theoretical distributions or on simulations such as posterior predictive checks. Finally, we have to decide whether to take action and if so, which action.

Another approach is to investigate specific forms of misspecification directly by elaborating or extending the model. The problem of diagnostics is then transformed into a problem of model selection.

Different types of residuals can be defined according to the kind of model used and the type of responses. In some cases residuals may be defined as differences between model implied and 'observed' summary statistics (see Section 8.6.1). More generally, residuals are often defined for individual units. However, in latent variable models there are 'units' at different levels and residuals can therefore be defined at different levels. In analogy with residuals in linear regression, natural higher-level residuals are predictions of the latent disturbances or residuals in the model. However, any discrepancy functions could in principle be used at any of the levels depending on the type of model violation investigated. When there are residuals at different levels, it is not obvious in which order these should be considered. Snijders and Berkhof (2004) propose an upward approach, starting at level 1, whereas Langford and Lewis (1998) suggest a downward approach. An argument in favor of the upward approach in linear multilevel models is that it is in this case possible to define level-1 residuals that are unconfounded by level-2 residuals but not vice versa (Hilden-Minton, 1995).

8.6.1 Residuals for summary statistics

In covariance structure analysis residuals are typically defined as the differences between the model-implied and empirical covariances or correlations (e.g. Costner and Schoenberg, 1973), which can suggest how the model might be elaborated. In a contingency table the obvious residual is based on the difference between model-implied and observed cell counts, often standardized (see e.g. Agresti, 2002). Instead of considering the full contingency table, it is easier to investigate residuals or goodness of fit in all pairwise (marginal) tables. For instance, in latent variable models such pairwise tables may suggest that conditional independence is violated for particular pairs of variables (e.g. Glas, 1988; Vermunt and Magidson, 2000, Appendix).

8.6.2 Higher-level residuals

The theoretical residuals for the clusters at the different levels l of a multi-level dataset are the corresponding disturbances $\zeta^{(l)}$. The scoring methods discussed in Chapter 7 can be used to predict or estimate these residuals, empirical Bayes being the most common. For linear models, the maximum likelihood estimator (called OLS in linear mixed models) may be preferable because, as pointed out by Snijders and Berkhof (2004), they are less dependent on model assumptions. However, if the assumptions of the level-1 model have been checked, Waternaux et al. (1989) recommend using empirical Bayes. Another argument in favor of empirical Bayes is that the approach can also be used for categorical and discrete responses, where the maximum likelihood method can be problematic. For instance, in the case of dichotomous responses, estimates for clusters with all responses equal to 1 or all responses equal to 0 (precisely the outlier candidates) cannot be obtained.

The appropriate standard error for empirical Bayes predictions in diagnostics is the unconditional sampling standard deviation since this reflects the variability in the estimated residuals under repeated sampling from the model. This 'diagnostic standard error' was used by Lange and Ryan (1989), Goldstein (2003), Langford and Lewis (1998) and Lewis and Langford (2001) for linear mixed models. In Section 11.3.3 we use the approximate sampling standard deviation (see equation (7.8) on page 232) of empirical Bayes predictions for count data to define standardized residuals.

As pointed out in Section 7.3.2, page 233, the predicted residuals are mutually correlated not just within the same level but also across levels, although this is commonly not recognized. Furthermore, the predictions depend on the true values of other latent variables in the same cluster, see Section 7.3.1, page 229. For these reasons it is difficult to pinpoint the source of any problem. For example, in a two-level model with a random intercept and slope, a cluster with a large true slope but moderate true intercept could have a large predicted intercept due to the correlation between predicted intercept and true slope. Similarly, a level-2 unit could have a large predicted residual because the true residual of another level-2 unit in the same level-3 unit is large.

Instead of attempting to assess the different residuals for a top-level unit individually, it may therefore be more advisable to use a discrepancy measure based on all the residuals for the top-level unit. For two-level linear mixed models, Snijders and Berkhof (2004) define the standardized level-2 residual as

$$\widetilde{\boldsymbol{\eta}}_j^{(2)\prime} \left[\mathrm{Cov}_{\mathbf{y}}(\widetilde{\boldsymbol{\eta}}_j^{(2)} | \mathbf{X}_j; \widehat{\boldsymbol{\vartheta}}) \right]^{-1} \widetilde{\boldsymbol{\eta}}_j^{(2)},$$

where the covariance matrix is the marginal sampling covariance matrix discussed in Section 7.3.2, page 231. They show that this residual is identical to the maximum likelihood (OLS) counterpart for linear mixed models. Treating the estimated covariance matrix as known, Snijders and Berkhof (2004) point out that this residual has an approximate chi-squared distribution with M degrees of freedom (where M is the number of latent variables at level 2).

Another possible residual for a top-level cluster in a multilevel model is the change in log-likelihood when the cluster is specifically accommodated, for instance by including a separate fixed intercept for the cluster (Longford, 2001). Alternatively, we could use the log-likelihood contribution of a top-level cluster, a similar idea to Snijders and Berkhof's (2004) multivariate residual for two-level linear mixed models.

8.6.3 Assessing latent variable distributions

In models with normally distributed responses and latent variables, the empirical distribution of the empirical Bayes predictions of the residuals is often used to assess normality of the latent variables. Lange and Ryan (1989) use this idea to produce weighted normal quantile plots of standardized linear combinations of latent variable predictions in linear mixed models. Note that empirical Bayes predictions should not be used to assess the normality assumption in models with nonnormal responses because the sampling distribution of the empirical Bayes predictions is in this case unknown. Goldstein (2003, p. 100) nevertheless uses this diagnostic in models with dichotomous responses.

Even for linear mixed models this approach is problematic since, to some extent, what you put in is what you get out. Specifically, if the multivariate normal prior distribution $h(\boldsymbol{\zeta})$ of the latent variables 'dominates' (i.e. has a much sharper peak than) the conditional response distribution given the latent variables $g(\mathbf{y}|\boldsymbol{\zeta}, \mathbf{X})$, the posterior distribution of the latent variables $\omega(\boldsymbol{\zeta}|\mathbf{y}, \mathbf{X})$ will appear normal regardless of the true distribution $h(\boldsymbol{\zeta})$ leading to a normal sampling distribution of the empirical Bayes predictions. This problem was demonstrated by Verbeke and Lasaffre (1996) using simulations. Although the true latent variable distribution was a mixture of two well-separated normal densities, the posterior distribution of the latent variable (wrongly assuming a normal latent variable distribution) appeared normal due to shrinkage. For these reasons it will often be difficult to detect departures from normality using empirical Bayes predictions.

A solution to this problem could be to use maximum likelihood estimates of the residuals since these only depend on $g(\mathbf{y}|\boldsymbol{\zeta}, \mathbf{X})$ and are not affected by the assumed latent variable distribution. An alternative is to relax the normality assumption for the latent variables. Verbeke and Lesaffre (1996) suggest specifying a mixture of normal densities with a known number of components. We prefer using nonparametric maximum likelihood estimation (NPMLE) since this semiparametric approach does not require any distributional assumption for the latent variables. Rabe-Hesketh et al. (2003a) showed that empirical Bayes predictions based on NPMLE are virtually indistinguishable from those assuming normality if normality holds, but not as affected by shrinkage when the true latent variable distribution is skewed. This approach is used in Section 11.3.3 for longitudinal count data.

Atikin et al. (2004) suggest a likelihood ratio test to compare the NPMLE (with C masses) with the conventional model assuming normality. For the conventional model, C-point Gauss-Hermite quadrature is used so that the model can be viewed as nested within the semiparametric model (with locations and probabilities constrained equal to the quadrature locations and weights). A potential problem with this approach is that the quadrature approximation may be poor if C is small.

8.6.4 Level-1 residuals

For the linear regression model, the *standardized residual* is defined as

$$\frac{y_i - \widehat{\mu}_i}{\widehat{\sigma}},$$

where $\widehat{\sigma}$ is the estimated residual standard deviation.

The *deletion residual* for a unit is the residual using parameter estimates derived from the sample when the unit is omitted. The idea is that an outlier could lead to an overestimate of the residual standard deviation making the standardized residual too small. Langford and Lewis (1998) give expressions for deletion residuals for linear mixed models.

For generalized linear models, the most common residuals are the Pearson, deviance and Anscombe residuals, shown for the Bernoulli and Poisson distributions in Table 8.1.

Table 8.1 *Pearson, deviance and Anscombe residuals*

	Bernoulli
Pearson	$\dfrac{y_i - \widehat{\mu}_i}{\sqrt{\widehat{\mu}_i(1 - \widehat{\mu}_i)}}$
Deviance	$\begin{aligned}&\text{sign}(y_i - \widehat{\mu}_i)\sqrt{-2\ln(1 - \widehat{\mu}_i)} && \text{if } y_i = 0 \\ &\text{sign}(y_i - \widehat{\mu}_i)\sqrt{-2\ln(\widehat{\mu}_i)} && \text{if } y_i = 1\end{aligned}$
Anscombe	$\dfrac{B(y_i, 2/3, 2/3) - B(\widehat{\mu}_i, 2/3, 2/3)}{(\widehat{\mu}_i(1 - \widehat{\mu}_i))^{1/6}}, \quad B(z, a, b) = \int_0^z t^{a-1}(1-t)^{b-1}\mathrm{d}t$

	Poisson
Pearson	$\dfrac{y_i - \widehat{\mu}_i}{\sqrt{\widehat{\mu}_i}}$
Deviance	$\begin{aligned}&\text{sign}(y_i - \widehat{\mu}_i)\sqrt{2\widehat{\mu}_i} && \text{if } y_i = 0 \\ &\text{sign}(y_i - \widehat{\mu}_i)\sqrt{2(y_i\ln(y_i/\widehat{\mu}_i) - (y_i - \widehat{\mu}_i))} && \text{if } y_i \neq 0\end{aligned}$
Anscombe	$\dfrac{1.5(y_i^{2/3} - \widehat{\mu}_i^{2/3})}{\widehat{\mu}_i^{1/6}}$

For models with a latent response formulation such as the probit and logit models, Albert and Chib (1995, 1997) and Gelman *et al.* (2000) use 'latent data residuals' $y^* - \mu$ within a fully Bayesian framework. Similarly, in the frequentist setting 'generalized residuals' have been defined as the conditional expectation of the latent data residual given the observed response y (e.g. Gourieroux *et al.*, 1987a; Chesher and Irish, 1987) and 'simulated residuals' as draws from the posterior distribution of the latent residual given the observed response (Gourieroux *et al.*, 1987b).

If the latent variables were known, we could simply substitute their values into the linear predictor and use $\widehat{\mu} = g^{-1}(\widehat{\nu})$ in expressions for conventional residuals to obtain level-1 residuals. However, since the values of the latent variables are not known, it is not clear how to define the residuals. In linear mixed models, empirical Bayes predictions for the latent variables are often substituted (e.g. Langford and Lewis, 1998), yielding the posterior mean of the residual. For linear factor and structural equation models with latent variables, Bollen and Arminger (1991) use either the empirical Bayes predictor (regression method) or the maximum likelihood estimator (Bartlett method) to define residuals for the items. They also standardize these residuals using the appropriate sampling standard deviation to identify outlying items for units.

In models with nonlinear link functions, substituting empirical Bayes predictions for latent variables does not produce the posterior mean of the residual. In a Bayesian setting, Dey *et al.* (1998) and Albert and Chib (1995) use the posterior distribution of the raw (unstandardized) residual $y - \mu$ given the observed responses, whereas Albert and Chib also use the posterior distribution of the latent data residual $y^* - \mu$ given the observed responses. There is surprisingly little work on level-1 residuals for latent variable models with nonnormal responses in the frequentist setting.

For linear mixed models Hilden-Minton (1995) suggests estimating a separate model for each cluster to define level-1 residuals that are not 'confounded' with level 2 residuals; see also Snijders and Bosker (1999).

8.6.5 Identifying outliers

We let an outlier be a unit or cluster which appears to be inconsistent with the specified model. This presupposes that most units or clusters appear consistent with the model. Since we do not know the true residuals, outlier detection must be based on estimated residuals and their sampling distribution. If the sampling distribution of the residuals is normal, as in the 'linear case' with normal latent variables and responses, a residual can be defined as an outlier if it exceeds a certain normal fractile. For other response-types, the sampling distribution is generally unknown and simulations can be used to obtain a reference distribution.

We will use T_j to denote a residual or discrepancy statistic for cluster (or unit) j. A natural approach would be to flag the cluster (or unit) with largest statistic, $T_{\max}$, as a potential outlier. Testing often proceeds as if the corresponding cluster j^* had been selected a priori, that is, by comparing $T_{\max}$ with the sampling distribution of T_{j^*}. However, the correct reference distribution, taking the post-hoc selection into account, is the sampling distribution of the largest statistic $T_{\max}$. This can easily be accomplished by simulation or parametric bootstrapping (e.g. Longford, 2001). In each replication k, responses are first simulated from the model, parameters are then estimated and the statistics T_j^k computed. The empirical distribution of the largest statistics $T_{\max}^k$ is then used to obtain a p-value. If we use parameter estimates based on the original data for simulating responses from the model, we are unrealistically treating the parameters as known. To take estimation uncertainty into account, Longford (2001) suggests sampling the parameters from their estimated sampling distribution (multivariate normal with covariance matrix from the information matrix).

Somewhat ironically, significance testing for diagnostics has recently become popular among Bayesians (e.g. Gelman *et al.*, 2003; Marshall and Spiegelhalter, 2003). The most common approach is posterior predictive checking (e.g. Rubin, 1984), where the 'predictive distribution' of a discrepancy statistic T

is defined as

$$\Pr(T|\mathbf{y}^{\mathrm{obs}}) = \int \Pr(T(\mathbf{y})|\mathbf{L})\Pr(\mathbf{L}|\mathbf{y}^{\mathrm{obs}})\,\mathrm{d}\mathbf{L}.$$

Here $\Pr(T(\mathbf{y})|\mathbf{L})$ is the sampling distribution of $T(\mathbf{y})$ for given parameters $\mathbf{L}$ and $\Pr(\mathbf{L}|\mathbf{y}^{\mathrm{obs}})$ is the posterior distribution of the parameters, so that $\Pr(T)$ can be loosely interpreted as the sampling distribution of T averaged over the posterior of the parameters $\mathbf{L}$. Posterior predictive checking is straightforward using Markov chain Monte Carlo (MCMC); see Section 6.11.5. For each draw of $\mathbf{L}$ from its posterior, $\mathbf{y}$ is sampled from its conditional distribution given $\mathbf{L}$ (the normed likelihood). The empirical distribution of $T(\mathbf{y})$ is then the required reference distribution.

In latent variable models, or hierarchical Bayesian models, the parameter vector $\mathbf{L}$ includes latent variables, parameters and hyperparameters, $\mathbf{L} = (\boldsymbol{\vartheta}, \boldsymbol{\zeta})$. In this case posterior predictive checking has been criticized for being too lenient or conservative (e.g. Dey et al., 1998; Bayarri and Berger, 2000; Marshall and Spiegelhalter, 2003). This is because the latent variables $\boldsymbol{\zeta}_j$ are sampled from their posterior distribution given the responses for cluster j, $\mathbf{y}_j^{\mathrm{obs}}$. New responses simulated for these $\boldsymbol{\zeta}_j$ will then resemble the observed responses too closely. This problem can be avoided by sampling the latent variables from their prior distribution to reflect sampling of clusters as well as sampling of units within clusters. The reference distribution for a discrepancy statistic T_j for cluster j then becomes

$$\Pr(T_j) = \int \Pr(T_j(\mathbf{y}_j)|\boldsymbol{\vartheta}, \boldsymbol{\zeta}_j)\Pr(\boldsymbol{\zeta}_j|\boldsymbol{\vartheta})\Pr(\boldsymbol{\vartheta}|\mathbf{y}_j^{\mathrm{obs}})\,\mathrm{d}\boldsymbol{\vartheta}.$$

Marshall and Spiegelhalter (2003) view this 'full-data mixed replication' approach as a computationally convenient approximation to the ideal method of cross validation. The reference distribution for cross validation equals the above, with the difference that it uses the posterior $p(\boldsymbol{\vartheta}|\mathbf{y}_{(-j)}^{\mathrm{obs}})$, based on all responses excluding those for cluster j. These ideas are easily adapted to the frequentist setting.

8.6.6 Influence

We have defined outliers as units (or clusters) that appear to be inconsistent with the rest of the data. Another type of extreme unit is one with great influence on the parameter estimates, in the sense that omitting the unit will cause substantial changes.

The influence of individual top-level clusters on the parameter estimates can be assessed using Cook's distance for the jth cluster defined as

$$C_j = -2\mathbf{g}^{j\prime}\mathbf{H}^{-1}\mathbf{g}^j,$$

where $\mathbf{g}^j$ is the score vector (first derivatives of log-likelihood contribution) for cluster j and $\mathbf{H}$ is the Hessian of the total log-likelihood.

Another measure of influence is the change in parameter estimates when a cluster is deleted. Let $\widehat{\boldsymbol{\vartheta}}$ be the parameter estimates using the full sample and $\widehat{\boldsymbol{\vartheta}}_{(-j)}$ the estimates when cluster j is deleted. DFBETAS$_{s(-j)}$ for a parameter ϑ_s is then defined as

$$\text{DFBETAS}_{s(-j)} = \frac{\widehat{\vartheta}_s - \widehat{\vartheta}_{s(-j)}}{\text{SE}(\widehat{\vartheta}_s)}.$$

It can be computationally very heavy to re-estimate the model with each cluster deleted in turn. In the context of generalized linear models, Pregibon (1981) suggests using one step of the Newton-Raphson algorithm to obtain an approximation for $\widehat{\boldsymbol{\vartheta}}_{(-j)}$ using $\widehat{\boldsymbol{\vartheta}}$ as starting values,

$$\widehat{\boldsymbol{\vartheta}}^1_{(-j)} = \widehat{\boldsymbol{\vartheta}} - \mathbf{H}^{-1}_{(-j)}\mathbf{g}_{(-j)}, \tag{8.13}$$

where $\mathbf{H}_{(-j)}$ is the Hessian without cluster j and $\mathbf{g}_{(-j)}$ is the gradient vector without cluster j, given by

$$\mathbf{g}_{(-j)} = \sum_{k \neq j}\mathbf{g}^k = -\mathbf{g}^j, \tag{8.14}$$

since the total gradient vector is $\mathbf{0}$ at the maximum likelihood estimates.

There is a simple relationship between Cook's distance and DFBETAS obtained using the one-step approximation in (8.13). To show this, we first use (8.13) and (8.14) to write the score vector as

$$\mathbf{g}^j = \mathbf{H}_{(-j)}(\widehat{\boldsymbol{\vartheta}} - \widehat{\boldsymbol{\vartheta}}^1_{(-j)}).$$

Cook's distance can then be approximated as

$$\begin{aligned} C_j &= -2(\widehat{\boldsymbol{\vartheta}} - \widehat{\boldsymbol{\vartheta}}^1_{(-j)})'\mathbf{H}'_{(-j)}\mathbf{H}^{-1}\mathbf{H}_{(-j)}(\widehat{\boldsymbol{\vartheta}} - \widehat{\boldsymbol{\vartheta}}^1_{(-j)}), \\ &\approx 2(\widehat{\boldsymbol{\vartheta}} - \widehat{\boldsymbol{\vartheta}}^1_{(-j)})' \left[\text{Cov}(\widehat{\boldsymbol{\vartheta}})\right]^{-1} (\widehat{\boldsymbol{\vartheta}} - \widehat{\boldsymbol{\vartheta}}^1_{(-j)}), \end{aligned} \tag{8.15}$$

since $\mathbf{H}_{(-j)} \approx \mathbf{H}$ in large samples (where a given cluster makes a small contribution to the Hessian) and $\mathbf{H} = -[\text{Cov}(\widehat{\boldsymbol{\vartheta}})]^{-1}$. The expression on the right-hand-side of (8.15) is twice Pregibon's (1981) one-step influence diagnostic.

A typical cut-point for Cook's distance is four times the number of parameters divided by the number of observations (clusters in this case). For DFBETAS, two divided by the square root of the number of observations is often used.

Cook's distances were applied to linear mixed models by Lesaffre and Verbeke (1998) and to generalized linear mixed models by Ouwens et al. (2001) and Xiang et al. (2002). Ouwens et al. (2001) also developed methods to assess the influence of level-1 units. We use influence diagnostics in Section 11.3.3 to identify influential subjects in a longitudinal dataset.

8.7 Summary and further reading

We have discussed some approaches to model specification and diagnostics without providing much guidance on how to proceed. One reason for this is that there may not be one optimal strategy. Another reason is that there has been relatively little research in this area, particularly for latent variable models. Finally, bringing together suggestions from disparate literatures has been a daunting task.

Given that statistical modeling pervades much if not most empirical research, it is somewhat surprising that the literature on statistical modeling per se appears to be scarce. Two useful papers are Cox (1990) and Lehmann (1990).

Leamer (1978) gives an interesting treatment of 'specification searches' from a Bayesian viewpoint whereas Harrell (2001) provides an extensive treatment of model building for 'empirical models'. Strategies for model building and diagnostics in linear mixed models are discussed by Langford and Lewis (1998), Snijders and Bosker (1999), Verbeke and Molenberghs (2000, Chapter 9) and Snijders and Berkhof (2004). Different strategies for model building in structural equation models are discussed in Bollen and Long (1993).

Useful books on statistical inference include Cox and Hinkley (1974), Lindsey (1996) and Pawitan (2001).

Research on diagnostics for latent variable models still appears to be in its infancy, especially for models with noncontinuous responses. A recent book on diagnostics for linear mixed models (particularly growth curve models) is Pan and Fang (2002). For ordinary linear regression models there are several books on diagnostics that may also be useful for more general models, including Barnett and Lewis (1984), Belsley *et al.* (1980), Cook and Weisberg (1982) and Chatterjee and Hadi (1988).

It is important to remember that the 'final model' may be misspecified, however carefully diagnosed and checked. When inference is required for particular parameters such as treatment effects, it may therefore be advisable to investigate the sensitivity to model assumptions in some way. Bayesians sometimes use model averaging, yielding credible intervals for the parameters that attempt to take model uncertainty into account. A less ambitious approach is sensitivity analysis where model assumptions are modified to investigate the 'robustness' of the inferences.

It should also be remembered that standard errors tend to underestimate uncertainty because model building is usually performed on the same data used to estimate parameters. In an extremely exploratory analysis, standard errors are therefore not presented at all, whereas they are generally taken at face value in somewhat exploratory analyses. We agree with Cox (1990, p.173) that standard errors should instead be interpreted as lower bounds in this case:

"Most applications are in any case somewhat in between the confirmatory-exploratory extremes and some notion, however approximate, of precision seems highly desirable in the exploratory portions of the analysis, if extremes of overinterpretation are to be avoided. The attachment of standard errors, etc. to the main features of an exploratory analysis, e.g. an exploratory multiple regression, seems often enlightening as indicating a minimum uncertainty."

PART II

Applications

CHAPTER 9

Dichotomous responses

9.1 Introduction

Since this is the first application chapter, we begin by discussing some of the classical models described in Chapter 3; a random intercept model in Section 9.2, a latent class model in Section 9.3, item response and MIMIC models in Section 9.4 and a random coefficient model in Section 9.5. The random coefficient model is used in the somewhat unusual context of a meta-analysis. A more common application for longitudinal data is described in Section 11.3 of the chapter on counts.

The subsequent sections consider a wide variety of less conventional models. In Section 9.6 we model longitudinal data using models incorporating both state dependence and unobserved heterogeneity. In Section 9.7 we use capture-recapture models with unobserved heterogeneity to estimate population sizes. Finally, we consider multilevel item response models in Section 9.8.

The applications discussed in this chapter come from a wide range of different disciplines, namely education, clinical medicine, epidemiology, biology, economics and sociology or social psychology.

9.2 Respiratory infection in children: A random intercept model

Sommer *et al.* (1983) describe a cohort study of Indonesian preschool children examined up to six consecutive quarters for the presence of respiratory infection.

Zeger and Karim (1991), Diggle *et al.* (2002) and others estimate a logistic-normal random intercept model for a subset of 275 of the children[1]. The model for the ith quarter and the jth child can be written as

$$\text{logit}[\Pr(y_{ij} = 1|\mathbf{x}_{ij}, \zeta_j)] = \mathbf{x}'_{ij}\boldsymbol{\beta} + \zeta_j,$$

where

$$\zeta_j \sim \text{N}(0, \psi).$$

Zeger and Karim used the following covariates:

- [Age] age in months (centered around 36)

- [Xero] a dummy variable for presence of xeropthalmia, an ocular manifestation of chronic vitamin A deficiency

---

[1] The data can be downloaded from `gllamm.org/books` or Patrick Hegearty's webpage `http://faculty.washington.edu/heagerty/Books/AnalysisLongitudinal/xerop.data`.

- [Cosine] cosine term of the annual cycle to capture seasonality
- [Sine] sine term of the annual cycle to capture seasonality
- [Female] a dummy variable for female gender
- [Height] height for age as percent of the National Center for Health Statistics (NCHS) standard (centered at 90%), which indicates lower nutritional status
- [Stunted] a dummy variable for stunting, defined as being below 85% in height for age.

Maximum likelihood estimates for the random intercept model using 12-point adaptive quadrature are given in the left part of Table 9.1. Respiratory

Table 9.1 *Estimates for random intercept logistic model and GEE*

	Random intercept model			Marginal 'model' (GEE)		
	Est	(SE)	OR	Est	(SE)	OR
β_1 [Age]	−0.034	(0.007)	0.967	−0.032	(0.006)	0.969
β_2 [Xero]	0.624	(0.480)	1.867	0.621	(0.441)	1.861
β_3 [Female]	−0.436	(0.257)	0.646	−0.418	(0.236)	0.658
β_4 [Cosine]	−0.594	(0.174)	0.552	−0.568	(0.170)	0.567
β_5 [Sine]	−0.165	(0.174)	0.848	−0.162	(0.146)	0.850
β_6 [Height]	−0.048	(0.027)	0.953	−0.048	(0.031)	0.954
β_7 [Stunted]	0.202	(0.441)	1.224	0.149	(0.411)	1.161
β_0 [Const]	−2.673	(0.224)		−2.421	(0.178)	
$\sqrt{\psi}$	0.806	(0.217)				
Log-likelihood		−334.65				

infection is related to [Age], season ([Cosine] and [Sine]) and [Height]. The odds ratio for [Age] is estimated as 0.967, corresponding to a decrease of the odds by 3.3% every month. This is the conditional effect of [Age] given the random intercept ζ_j (and the other covariates). Estimates of marginal effects using generalized estimating equations (GEE) are given in the second set of columns of Table 9.1. Here the structure of the working correlation matrix was specified as exchangeable (see Display 3.3A on page 82) and the standard errors are based on the sandwich estimator described in Section 8.3.3. The marginal and conditional effect estimates are very similar here since the random effects variance is estimated as only 0.65, so that the attenuation factor is about 0.90 (see Section 4.8.1).

Figure 9.1 displays the conditional and population averaged relationships between age and respiratory infection in the first quarter for girls who are not stunted, do not have xeropthalmia and whose height equals the average 1.83 at the first quarter. The three dashed curves show the conditional relationships for $\zeta_j = -0.8$, 0 and 0.8 (from bottom to top). The circles show the population averaged relationship as estimated by GEE, whereas the solid curve coinciding with these circles represents the population averaged curve for the random intercept model, obtained by integrating over the random intercept,

$$E_\zeta \left[\left. \frac{\exp(\mathbf{x}'_{ij}\boldsymbol{\beta} + \zeta_j)}{1 + \exp(\mathbf{x}'_{ij}\boldsymbol{\beta} + \zeta_j)} \right| \mathbf{x}_{ij} \right].$$

Here the random intercept model and GEE imply nearly the same marginal relationship between respiratory infection and [Age], although this need not always be so.

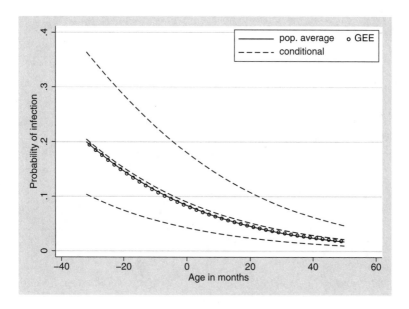

Figure 9.1 *Conditional and population averaged effects of* [Age]. *Circles: population averaged curve from GEE; solid curve: population averaged curve from random intercept model; dashed curves: conditional relationships for* $\zeta_j = -0.8, 0, 0.8$

The random intercept variance is estimated as $\widehat{\psi} = 0.650$. Using the latent response formulation of the logit model (see Section 2.4), the correlation between the latent responses, conditional on covariates, is therefore $\widehat{\psi}/(\widehat{\psi} + \pi^2/3) = 0.165$. The conditional correlations between the observed responses depend on the covariate values. In contrast, GEE assumes a constant Pearson correlation between observed responses (conditional on the covariates), estimated as 0.045.

Figure 9.2 shows a boxplot of the empirical Bayes predictions (see Section 7.3.1) of the children's random intercepts (first boxplot). The distribution is skewed and there are some extreme values. However, with the important exception of linear mixed models, the distribution of the empirical Bayes predictions is generally not normal in generalized linear mixed models. Therefore it is difficult to judge whether the extreme values are a cause for concern (see Section 8.6.2). To assess this informally, we simulated the responses from the estimated model (keeping the covariate values from the data), estimated the parameters and predicted the random intercepts. We repeated this three times, giving the second to fourth boxplots in Figure 9.2. The boxplots for the simulated data resemble that for the real data, so there appears to be no cause for concern.

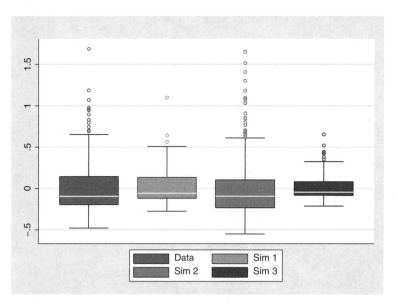

Figure 9.2 *Boxplots of empirical Bayes predictions of random intercept for data (first boxplot) and for three simulated datasets (boxplots 2 to 4)*

Figure 9.3 shows a graph of the predicted random intercepts versus their rank, for every fifth rank, with error bars representing ± one posterior standard deviation.

9.3 Diagnosis of myocardial infarction: A latent class model

Rindskopf and Rindskopf (1986) analyze data[2] from a coronary care unit in New York City where patients were admitted to rule out myocardial infarction (MI) or 'heart attack'.

Each of 94 patients was assessed on four diagnostic criteria:

- [Q-wave] a dummy variable for presence of a Q-wave in the ECG

---

[2] The data can be downloaded from `gllamm.org/books`

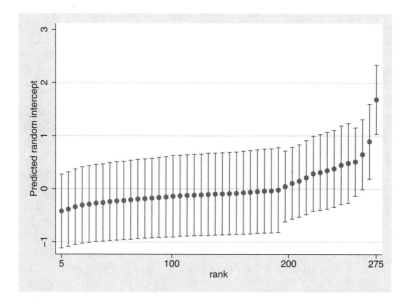

Figure 9.3 *Empirical Bayes predictions of random intercept versus rank with error bars representing ± one posterior standard deviation*

- [History] a dummy for presence of a classical clinical history
- [LDH] a dummy for having a flipped LDH
- [CPK] a dummy for presence or absence of a CPK-MB

The data are shown in Table 9.2. Since the patients have either had MI or not, it seems reasonable to specify two latent classes. Let π_1 be the probability of being in the first latent class,

$$\text{logit}(\pi_1) = \varrho_0.$$

If the second latent class corresponds to MI, the prevalence of MI is $\pi_2 = 1 - \pi_1$. The conditional response probabilities can be specified as

$$\text{logit}[\Pr(y_{ij} = 1|c)] = e_{ic}.$$

The probabilities $\Pr(y_{ij} = 1|c = 2)$ represent the sensitivities of the diagnostic tests (the probabilities of a correct diagnosis for people with the illness), whereas $1 - \Pr(y_{ij} = 1|c = 1)$ represent the specificities (the probabilities of a correct diagnosis for people without the illness). Note that this model is equivalent to a two-class one-factor model since we can replace e_{ic} by $\lambda_i e_c$, where $\lambda_1 = 1$.

Parameter estimates are given in Table 9.3. The estimates $\widehat{e}_{11}$ for [Q-wave] and $\widehat{e}_{42}$ for [CPK] take on very large negative and positive values corresponding to conditional response probabilities very close to 0 and 1, respectively, and therefore represent a so-called *boundary solution*. The corresponding standard errors are extremely large. This is because the likelihood changes very

Table 9.2 *Diagnosis of myocardial infarction data*

[Q-wave] ($i=1$)	[History] ($i=2$)	[LDH] ($i=3$)	[CPK] ($i=4$)	Obs. count	Exp. count	Prob. of MI ($c=2$)
1	1	1	1	24	21.62	**1.000**
0	1	1	1	5	6.63	**0.992**
1	0	1	1	4	5.70	**1.000**
0	0	1	1	3	1.95	**0.889**
1	1	0	1	3	4.50	**1.000**
0	1	0	1	5	3.26	0.420
1	0	0	1	2	1.19	**1.000**
0	0	0	1	7	8.16	0.044
1	1	1	0	0	0.00	0.017
0	1	1	0	0	0.22	0.000
1	0	1	0	0	0.00	0.001
0	0	1	0	1	0.89	0.000
1	1	0	0	0	0.00	0.000
0	1	0	0	7	7.78	0.000
1	0	0	0	0	0.00	0.000
0	0	0	0	33	32.11	0.000

Source: Rindskopf and Rindskopf (1986)

Table 9.3 *Estimates for diagnosis of MI*

Parameter	Class 1 ('No MI')			Class 2 ('MI')		
	Est	(SE)	Prob.	Est	(SE)	Prob.
			1-Spec.			Sens.
e_{1c} [Q-wave]	−17.58	(953.49)	0.00	1.19	(0.42)	0.77
e_{2c} [History]	−1.42	(0.39)	0.20	1.33	(0.39)	0.79
e_{3c} [LDH]	−3.59	(1.01)	0.03	1.57	(0.47)	0.83
e_{4c} [CPK]	−1.41	(0.41)	0.20	16.86	(706.04)	1.00
			1-Prev.			Prev.
ϱ_0 [Cons]	0.17	0.22	0.54			0.46

little when these extreme parameters change, as the predicted probabilities remain close to 0 and 1 for a large range of values. For example, logits of 5 and 20 correspond to probabilities of 0.993 and 1.000, respectively.

Following Rindskopf and Rindskopf (1986), we will nevertheless interpret these parameter estimates. For each set of test results in Table 9.2, the expected count was obtained by multiplying the likelihood contribution for a patient with these test results,

$$g(\mathbf{y}_j) = \pi_1 \prod_i \Pr(y_{ij}|c = 1) + \pi_2 \prod_i \Pr(y_{ij}|c = 2),$$

by 94, the number of patients. Comparing the expected counts with the observed counts in Table 9.2, the model appears to fit well.

From Table 9.3, the prevalence of MI is estimated as 0.46. The specificity of [Q-wave] is estimated as 1, implying that all patients without MI will have a negative result on that test. [History] has the lowest specificity of 0.70. The estimated sensitivities range from 0.77 for [Q-wave] to 1.00 for [CPK], so that 77% of MI cases test positively on [Q-wave] and 100% on [CPK].

We can obtain the posterior probabilities (similar to 'positive predictive values') of MI given the four test results using Bayes theorem (see Section 7.2),

$$\omega(e_2|\mathbf{y}_j;\widehat{\boldsymbol{\theta}}) \equiv \Pr(c = 2|\mathbf{y}_j) = \frac{\pi_2 \prod_i \Pr(y_{ij}|c = 2)}{\pi_1 \prod_i \Pr(y_{ij}|c = 1) + \pi_2 \prod_i \Pr(y_{ij}|c = 2)}.$$

These probabilities are presented in the last column of Table 9.2, where bold-face means that a patient with these test results has a higher posterior probability of being in class 2 than class 1 and is therefore diagnosed as MI using the empirical Bayes modal classification rule (see Section 7.4). For most patients the diagnosis (classification) is very clear with posterior probabilities close to 0 and 1. For each patient we can work out the probability of misclassification (using the empirical Bayes modal classification rule) as

$$f_j = 1 - \max_c \omega(e_c|\mathbf{y}_j;\widehat{\boldsymbol{\theta}}),$$

see (7.13) on page 237. For instance, for $\mathbf{y}_j = (0,1,0,1)$, the patient would be classified as 'no myocardial infarction' and the probability of misclassification would be 0.42. This is the only test result with a large probability of misclassification and is fortunately expected to occur for only $3.26/94 = 3.5\%$ of patients.

We can estimate the proportion of classification errors in the population using the sample average of f_j, giving 0.030. If we had no test results, we would have to diagnose patients according to the prior probabilities. Everyone would be diagnosed as 'no myocardial infarction' since this is more likely ($\pi_1 = 0.54$) than myocardial infarction ($\pi_2 = 0.46$). The estimated probability of misclassification would be 0.46. The estimated proportional reduction of classification error due to knowing the test results is therefore $(0.4579 - 0.0296)/0.4579 = 0.94$. If we use the expectation of f_j instead of the sample average (using model-based expected frequencies instead of the observed frequencies), the proportional reduction of classification error becomes 0.95. Use of covariate information such as age and sex would be likely to improve diagnostic accuracy even further; see Section 13.5 for an example of a latent class model with covariates.

9.4 Arithmetic reasoning: Item response models

We will analyze data[3] from the Profile of American Youth (U.S. Department of Defense, 1982), a survey of the aptitudes of a national probability sample of Americans aged 16 through 23. The data for four items of the arithmetic reasoning test of the Armed Services Vocational Aptitude Battery (Form 8A) are shown in Table 9.4 for samples of white males and females and black males and females. These data were previously analyzed by Mislevy (1985).

Table 9.4 *Arithmetic reasoning data*

Item Response				White	White	Black	Black
1	2	3	4	Males	Females	Males	Females
0	0	0	0	23	20	27	29
0	0	0	1	5	8	5	8
0	0	1	0	12	14	15	7
0	0	1	1	2	2	3	3
0	1	0	0	16	20	16	14
0	1	0	1	3	5	5	5
0	1	1	0	6	11	4	6
0	1	1	1	1	7	3	0
1	0	0	0	22	23	15	14
1	0	0	1	6	8	10	10
1	0	1	0	7	9	8	11
1	0	1	1	19	6	1	2
1	1	0	0	21	18	7	19
1	1	0	1	11	15	9	5
1	1	1	0	23	20	10	8
1	1	1	1	86	42	2	4
		Total:		263	228	140	145

Source: Mislevy (1985)

We first estimate a one-parameter logistic item response model (see Section 3.3.4) for item i and subject j,

$$\text{logit}[\Pr(y_{ij} = 1|\eta_j)] = \beta_i + \eta_j.$$

[3] The data can be downloaded from gllamm.org/books

The parameter estimates are given in Table 9.5 where we note that the estimated item difficulties $\widehat{\beta}_i$ increase from item 1 to item 4. This model assumes that the effect of increasing ability is the same for all items (on the logit scale), an assumption that can be relaxed using the two-parameter logistic item response model

$$\text{logit}[\Pr(y_{ij} = 1|\eta_j)] \;=\; \beta_i + \lambda_i \eta_j,$$

where we set $\lambda_1 = 1$ for identification. The model can be written in GRC formulation as

$$\text{logit}[\Pr(y_{ij} = 1|\eta_j)] \;=\; \mathbf{d}_i'\boldsymbol{\beta} + \eta_j \mathbf{d}_i'\boldsymbol{\lambda}, \qquad (9.1)$$

where $\mathbf{d}_i$ is a four-dimensional vector with ith element equal to 1 and all other elements equal to 0. The parameter estimates are also given in Table 9.5 where we see that the estimated discrimination parameters or factor loadings $\widehat{\lambda}_i$ for items 2 and 3 are lower than for the other two items. However, the two-parameter model does not fit much better than the one-parameter model.

Table 9.5 *Estimates for one, two and three-parameter item response models using 20-point adaptive quadrature*

Parameter	One-parameter Est	(SE)	Two-parameter Est	(SE)	Three-parameter Est	(SE)
Intercepts						
β_1 [Item1]	0.58	(0.10)	0.64	(0.12)	-0.05	(0.18)
β_2 [Item2]	0.24	(0.10)	0.22	(0.09)	-0.54	(0.16)
β_3 [Item3]	-0.22	(0.09)	-0.22	(0.09)	-1.64	(0.40)
β_4 [Item4]	-0.59	(0.10)	-0.63	(0.11)	-5.07	(3.73)
Factor loadings						
λ_1 [Item1]	1		1		1	
λ_2 [Item2]	1		0.67	(0.16)	0.66	(0.22)
λ_3 [Item3]	1		0.73	(0.18)	0.97	(0.35)
λ_4 [Item4]	1		0.93	(0.23)	2.42	(1.92)
Guessing parameter						
c					0.22	
Variance						
ψ	1.63	(0.21)	2.47	(0.84)	6.66	(3.09)
Log-likelihood	-2004.94		-2002.76		-1992.78	

The estimated item characteristic curves for the one and two-parameter item response models are given by

$$\Pr(y_{ij} = 1|\eta_j) \;=\; \frac{\exp(\nu_{ij})}{1 + \exp(\nu_{ij})},$$

where the linear predictor ν_{ij} is $\beta_i + \eta_j$ for the one-parameter model and $\beta_i + \lambda_i \eta_j$ for the two-parameter model. These curves tend to 0 as ability tends to $-\infty$.

If it is possible to guess the right answer, as in multiple choice questions, a more realistic model is the three-parameter logistic item response model which can be written as

$$\Pr(y_{ij} = 1|\eta_j) \;=\; c_i + (1 - c_i)\frac{\exp(\nu_{ij})}{1 + \exp(\nu_{ij})}.$$

Here, c_i is often called a 'guessing parameter' which can be interpreted as the probability of a correct answer from a subject with ability minus infinity. This model does not fit into the general model framework described in Chapter 4 since the response model is not a generalized linear model (conditional on the latent variable). However, if we fix the guessing parameter to some constant, for example 0.1, the response model can be expressed as a generalized linear model with a composite link (see Section 2.3.5, equation (2.14)),

$$\Pr(y_{ij} = 1|\eta_j) \;=\; 0.1g_1^{-1}(1) + 0.9g_2^{-1}(\nu_{ij}),$$

where g_1 is the identity link and g_2 is the logit link. Assuming that the guessing parameter is the same for all items, we tried different values of c (from 0 to 0.4 in steps of 0.02) and maximized the likelihood with respect to the other parameters, giving the profile log-likelihood. This profile log-likelihood is plotted against c in Figure 9.4 and has a maximum at $c=0.22$. Approximate 95% confidence limits are those values of c where the profile log-likelihood is 3.84/2 lower than the maximum as indicated by the horizontal dotted line in the figure. The approximate 95% confidence interval for c therefore is from 0.14 to 0.28.

The parameter estimates for the three-parameter logistic item response model with $c = 0.22$ are given in Table 9.5. Note that the standard errors are underestimated because c is treated as known. The model fits substantially better than the two-parameter model. Unfortunately, the parameter estimates are not very reliable because the likelihood appears to be somewhat flat. In particular, the correlation between the estimates $\widehat{\beta}_4$ and $\widehat{\lambda}_4$ is estimated as -0.95. Furthermore, different starting values lead to quite different estimates but very similar log-likelihood values. Empirical identification (see Section 5.2.5) thus appears to be fragile.

The item characteristic curves for all three models are shown in Figure 9.5. Unlike the one-parameter model, the curves for the two-parameter model intersect with items 1 and 4 having higher slopes than items 2 and 3, clearly violating double monotonicity. It is clear that the curves for the three-parameter model approach an asymptote of 0.22 as ability tends to $-\infty$.

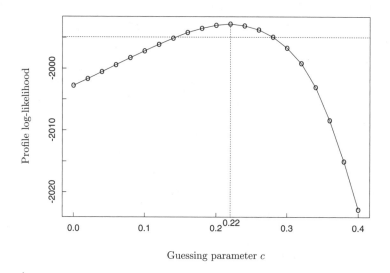

Figure 9.4 *Profile log-likelihood for guessing parameter in three-parameter logistic item response model*

Returning to a two-parameter item response model, we now consider the following covariates:

- [Female] a dummy variable for subject being female
- [Black] a dummy variable for subject being black

We can specify a structural model for ability η_j, allowing the mean abilities to differ between groups,

$$\eta_j = \gamma_0 + \gamma_1 F_j + \gamma_2 B_j + \gamma_3 F_j B_j + \zeta_j,$$

where F_j represents [Female] and B_j [Black]. Since we have included a constant in the structural model, we have to fix one of the constants in the response model for identification and set $\beta_1 = 0$. This is a MIMIC model of the kind discussed in Section 3.5 where the covariates affect the response via a latent variable only.

Table 9.6 gives parameter estimates for this model ($\mathcal{M}_2$) and the model without covariates, $\gamma_1 = \gamma_2 = \gamma_3 = 0$ ($\mathcal{M}_1$), which is equivalent (see Section 5.3) to the simple two-parameter item response model in Table 9.5. Deviance and Pearson X^2 statistics are also reported in the table, from which we see that $\mathcal{M}_2$ fits better than $\mathcal{M}_1$. The variance estimate of the disturbance decreases from 2.47 for $\mathcal{M}_1$ to 1.88 for $\mathcal{M}_2$ because some of the variability in ability is 'explained' by the covariates. There is some evidence for a [Female] by [Black] interaction. While being female is associated with lower ability among white

One-parameter item response model

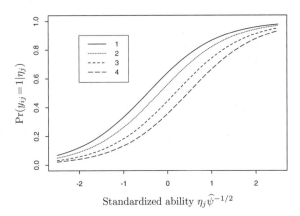

Two-parameter item response model

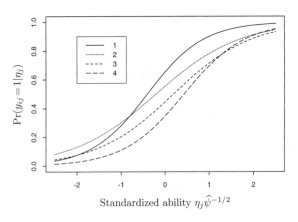

Three-parameter item response model

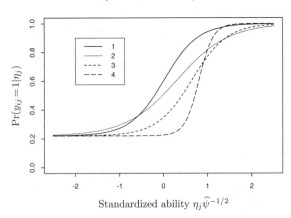

Figure 9.5 *Item characteristic curves for one-parameter (top), two-parameter (middle) and three-parameter (bottom) logistic item response models*

Table 9.6 *Estimates for MIMIC models*

Parameter	$\mathcal{M}_1$ Est	(SE)	$\mathcal{M}_2$ Est	(SE)	$\mathcal{M}_3$ Est	(SE)
Intercepts						
β_1 [Item1]	0		0		0	
β_2 [Item2]	−0.21	(0.12)	−0.22	(0.12)	−0.13	(0.13)
β_3 [Item3]	−0.68	(0.14)	−0.73	(0.14)	−0.57	(0.15)
β_4 [Item4]	−1.22	(0.19)	−1.16	(0.16)	−1.10	(0.18)
β_5 [Item1]× [Black]×[Female]	0		0		−1.07	(0.69)
Factor loadings						
λ_1 [Item1]	1		1		1	
λ_2 [Item2]	0.67	(0.16)	0.69	(0.15)	0.64	(0.17)
λ_3 [Item3]	0.73	(0.18)	0.80	(0.18)	0.65	(0.14)
λ_4 [Item4]	0.93	(0.23)	0.88	(0.18)	0.81	(0.17)
Structural model						
γ_0 [Cons]	0.64	(0.12)	1.41	(0.21)	1.46	(0.23)
γ_1 [Female]	0		−0.61	(0.20)	−0.67	(0.22)
γ_2 [Black]	0		−1.65	(0.31)	−1.80	(0.34)
γ_3 [Black]×[Female]	0		0.66	(0.32)	2.09	(0.86)
ψ	2.47	(0.84)	1.88	(0.59)	2.27	(0.74)
Log-likelihood	−2002.76		−1956.25		−1954.89	
Deviance	204.69		111.68		108.96	
Pearson X^2	190.15		102.69		100.00	

people, this is not the case among black people where males and females have similar abilities. Black people have lower mean abilities than both white men and white women.

We can also investigate if there are direct effects of the covariates on the responses, in addition to the indirect effects via the latent variable. This could be interpreted as 'item bias' or 'differential item functioning' (DIF), i.e., where the probability of responding correctly to an item differs for instance between men and women with the same ability. Such item bias would be a problem since it suggests that candidates cannot be fairly assessed by the test. Bartholomew (1987, 1991) found that the black women performed worse on the first item. We will investigate whether this is the case after allowing for group differences in mean ability by adding the term $\beta_5 F_j B_j d_{i1}$ to (9.1). The parameter estimates are given under $\mathcal{M}_3$ in Table 9.6. Here there is no evidence that item 1 functions differently for black females. See also Section 10.3.3 for an

investigation of item bias in two-dimensional ordinal item response models for
political efficacy.

Note that none of the models appear to fit well according to absolute fit criteria. For example, for $\mathcal{M}_2$, the deviance is 111.68 with 53 degrees of freedom,
although the table is perhaps too sparse to rely on the χ^2 distribution.

We can predict people's abilities on the basis of their responses to the four
questions using empirical Bayes as described in Section 7.3.1. The empirical
Bayes predictions for all possible response patterns are given in Table 9.7 for
$\mathcal{M}_1$ and $\mathcal{M}_2$. The abilities can be interpreted as logits of the probability of
a correct response to item 1 (since $\beta_1 = 0$ and $\lambda_1 = 1$), with 0 corresponding
to a probability of 50% and ± 1 to probabilities 73% and 27%. For $\mathcal{M}_2$, the
predicted abilities depend on group, with for instance black males given lower
'scores' for the same performance than white males since black males have a
lower mean ability than white males. Statistically, these predictions may be
better than those ignoring the covariates, but the scoring method certainly
does not appear to be fair!

Table 9.7 *Empirical Bayes predictions of ability*

Item				$\mathcal{M}_1$	$\mathcal{M}_2$			
Response				All	White	White	Black	Black
1	2	3	4	Groups	Males	Females	Males	Females
0	0	0	0	−1.18	−0.53	−0.85	−1.43	−1.41
0	0	0	1	−0.13	0.27	−0.02	−0.53	−0.51
0	0	1	0	−0.34	0.20	−0.09	−0.60	−0.58
0	0	1	1	0.62	0.98	0.70	0.21	0.23
0	1	0	0	−0.40	0.11	−0.18	−0.71	−0.68
0	1	0	1	0.57	0.88	0.60	0.11	0.13
0	1	1	0	0.36	0.82	0.53	0.04	0.07
0	1	1	1	1.31	1.62	1.32	0.82	0.84
1	0	0	0	−0.05	0.38	0.09	−0.42	−0.39
1	0	0	1	0.90	1.16	0.87	0.38	0.40
1	0	1	0	0.69	1.09	0.80	0.31	0.34
1	0	1	1	1.66	1.91	1.60	1.09	1.12
1	1	0	0	0.64	0.99	0.71	0.22	0.24
1	1	0	1	1.61	1.80	1.50	0.99	1.02
1	1	1	0	1.39	1.73	1.43	0.93	0.95
1	1	1	1	2.45	2.64	2.29	1.73	1.76

9.5 Nicotine gum and smoking cessation: A meta-analysis

9.5.1 Introduction

Systematic reviews of available evidence regarding the efficacy of medical treatments are of obvious importance for informing clinical practice. Such reviews have become increasingly popular, forming a vital part of the gospel of 'evidence based medicine'. Sackett *et al.* (1991) give the following advice to clinicians:

> "If a rigorous scientific overview has been conducted on the clinical question you are attempting to answer, your time is better spent studying it rather than a grab (and perhaps distorted) sample of its citations."

The importance of systematic reviews is reflected in the formation in 1993 of the 'Cochrane Collaboration', which produces and updates vast numbers of reviews of clinical trials for most areas of medical research as well as setting up guidelines, offering courses, etc.

Meta-analysis is the statistical approach to combining evidence from different studies to obtain an overall estimate of treatment effect. Although modern meta-analysis originates in education and psychology (e.g. Glass, 1976; Hunt, 1997), its recent proliferation in medical research has led to an upsurge of interest within biostatistics.

Here we discuss meta-analysis of clinical trials of nicotine replacement therapy for smoking cessation using data[4] from Silagy *et al.* (2003). Following Silagy *et al.*, we carry out a separate analysis of studies using nicotine gum (rather than for instance nicotine patches) combined with a high level of support, including formal therapy or 'assessment and reinforcement' visits.

In each trial, patients were randomized to a treatment group given nicotine gum or a control group. In most studies the control group received placebo gum which had the same appearance as the nicotine gum but lacked the active ingredient nicotine. In some studies, the control group had no gum. Smoking cessation at least 6 months after treatment was the outcome considered. The most rigorous definition of abstinence for each trial was used. The results for the trials can be summarized by two-by-two tables (treatment arm by outcome), which can be derived from the rows of Table 9.8.

We will consider estimation of the overall odds ratio, the odds of quitting smoking if treated divided by the odds of quitting if not treated. For an individual study j, this odds ratio can be estimated as

$$o_j = \frac{d_{1j}/(n_{1j} - d_{1j})}{d_{0j}/(n_{0j} - d_{0j})}, \tag{9.2}$$

where d_{1j} and d_{0j} are the numbers of quitters in the treatment and control groups, respectively, whereas n_{1j} and n_{0j} are the total numbers of subjects in these groups. Other measures of treatment effect include the risk ratio and the risk difference.

---

[4] The data can be downloaded from `gllamm.org/books`

Table 9.8 *Randomized studies of nicotine gum and smoking cessation*

	Treated		Control	
	Quitters	Total	Quitters	Total
Study	d_1	n_1	d_0	n_0
Blondal 1989	37	92	24	90
Campbell 1991	21	107	21	105
Fagerstrom 1982	30	50	23	50
Fee 1982	23	180	15	172
Garcia 1989	21	68	5	38
Garvey 2000	75	405	17	203
Gross 1995	37	131	6	46
Hall 1985	18	41	10	36
Hall 1987	30	71	14	68
Hall 1996	24	98	28	103
Hjalmarson 1984	31	106	16	100
Huber 1988	31	54	11	60
Jarvis 1982	22	58	9	58
Jensen 1991	90	211	28	82
Killen 1984	16	44	6	20
Killen 1990	129	600	112	617
Malcolm 1980	6	73	3	121
McGovern 1992	51	146	40	127
Nakamura 1990	13	30	5	30
Niaura 1994	5	84	4	89
Niaura 1999	1	31	2	31
Pirie 1992	75	206	50	211
Puska 1979	29	116	21	113
Schneider 1985	9	30	6	30
Tonnesen 1988	23	60	12	53
Villa 1999	11	21	10	26
Zelman 1992	23	58	18	58

Source: Silagy *et al.* (2003)

9.5.2 Approaches to meta-analysis

There are essentially two different approaches to meta-analysis: *fixed effects* and *random effects*. Fixed effects meta-analysis assumes that there is a single true treatment effect and that any variability between the studies' estimated treatment effects is completely due to within-study sampling variability. Note that this use of the term 'fixed effects' is somewhat misleading since the term usually implies that there is a fixed effect for each cluster (here study); see Section 3.6.1. The assumption of a common treatment effect is often tested using Cochran's Q-test of homogeneity (e.g. Cochran, 1950; DerSimonian and Laird, 1986).

In contrast, random effects meta-analysis assumes that the true treatment effect varies between studies. This variation could be due to differences in populations and trial protocols including drug dosage, duration of treatment, definition and measurement of outcomes and length of follow-up. The aim of the meta-analysis then becomes to estimate the mean treatment effect for an imagined population of studies.

Fleiss (1993) and Bailey (1987) discuss two considerations for choosing between the two competing approaches. First, the random effects approach attempts to generalize conclusions to the population of studies including future studies, whereas the fixed effects approach restricts conclusions to the studies contributing to the analysis. Second, the random effects explicitly allow for study-to-study variation in contrast to the fixed effects approach. In many cases, the studies differ from one another so fundamentally that it might be easier to argue that it is nonsensical to pool the effect sizes at all than that there is a single true treatment effect.

Here we adopt the random effects approach because we consider it unlikely a priori that the treatment effects do not differ between the studies. For instance, the 'Blondal 1989' study used gum containing 4mg of nicotine for one month whereas 'Barcia 1989' used gum containing 2mg of nicotine for three to four months. The former study considered 12 months sustained abstinence whereas the latter considered 6 months sustained abstinence. The studies also differed in the nature and intensity of additional support and in the types of smokers considered. For instance, 'Campbell 1991' treated only patients with smoking-related illnesses whereas most other studies treated any smokers interested in quitting. Furthermore, studies were conducted in a wide range of countries including Iceland, Sweden, Spain and the USA.

The predominant approach to meta-analysis, whether fixed effects or random effects, is to analyze the estimated study-level treatment effects instead of the original patient-level data. When the effect size of interest is an odds ratio as here, log-odds ratios are often analyzed instead of odds ratios since their sampling distribution is likely to be better approximated by a normal distribution. Random effects meta-analysis then consists of estimating the following linear random intercept model,

$$\ln(o_j) = \beta_0 + \zeta_{0j} + \epsilon_j, \quad \epsilon_j \sim N(0, \theta_j) \tag{9.3}$$

where o_j is the estimated odds ratio for study j as defined in (9.2), β_0 is the mean log odds ratio of interest and ζ_{0j}, the random intercept, represents the deviation of the study's true log odds-ratio from the mean log-odds ratio. The within-study standard deviations $\sqrt{\theta_j}$ are simply set equal to the standard errors of the log-odds ratios estimated using Woolf's method (Woolf, 1955),

$$\sqrt{\theta_j} = \sqrt{\frac{1}{d_{0j}} + \frac{1}{n_{0j} - d_{0j}} + \frac{1}{d_{1j}} + \frac{1}{n_{1j} - d_{1j}}}. \tag{9.4}$$

Recently, 'meta-regression' where the heterogeneity between studies is 'explained' by including study-specific covariates such as drug dosage in (9.3) has attracted considerable attention (e.g. Berkey *et al.*, 1995; van Houwelingen *et al.*, 2002).

Unfortunately, analyzing study-level estimates of effect size is problematic because the normality assumption will often be violated. In the case of log odds ratios, this will be the case when the studies are small and/or the outcome of interest is rare. If d_0 and/or d_1 are zero this approach requires ad-hoc practices such as adding 0.5 to the counts (a practice also recommended by Gart and Zweifel (1967) for small counts to reduce bias). It is therefore preferable to model the observed patient-level dichotomous responses directly. Surprisingly, analysis of study-level estimates is common not just in applied papers, but also in methodological work (e.g. Normand, 1999), including Bayesian treatments (see e.g. Carlin, 1992; DuMouchel *et al.*, 1996; Gelman *et al.*, 2003) where 'proper' modeling using Markov chain Monte Carlo methods is straightforward. Analyzing study-level estimates of effect sizes is perhaps justified only if patient-level data is not available (see also Chalmers, 1993).

9.5.3 Random effects modeling of patient-level data

For patient-level data, where we let i index patients and j index studies, Agresti and Hartzel (2000) consider the random coefficient model

$$\text{logit}(\Pr(y_{ij} = 1 | x_{ij}, \zeta_{0j}, \zeta_{1j}) = \beta_0 + \beta_1 x_{ij} + \zeta_{0j} + \zeta_{1j} x_{ij},$$

where

$$x_{ij} = \begin{cases} 0.5 & \text{for treated patients} \\ -0.5 & \text{for control patients} \end{cases},$$

and

$$(\zeta_{0j}, \zeta_{1j})' \sim N_2(\mathbf{0}, \boldsymbol{\Psi}).$$

Here β_0 and ζ_{0j} are fixed and random intercepts, respectively, and β_1 and ζ_{1j} fixed and random slopes of x_{ij}. β_1 represents the log odds ratio of interest whereas $\beta_1 + \zeta_{1j}$ represents the 'true' log odds ratio of study j. The study-specific intercepts are sometimes treated as fixed; see for instance Turner *et al.* (2000) and Thompson *et al.* (2001).

Agresti and Hartzel assume that the random intercept and slope are uncorrelated with $\psi_{10} = 0$. Note that the coding of x_{ij} becomes important in this case; a model with for instance $x_{ij} = 0, 1$ would not be equivalent to a model with $x_{ij} = -0.5, 0.5$ (see page 54). Agresti and Hartzel argue that an advantage of the latter coding or 'centering' is that with $\psi_{10} = 0$ the total variance $\text{Var}(\zeta_{0j} + \zeta_{1j}x_{ij})$ of the log odds is the same for both groups.

We will nevertheless investigate the validity of the assumption of zero correlation for the nicotine gum data. The models were estimated by maximum likelihood using 20-point adaptive quadrature. The correlation between random intercept and slope was estimated as 0.17 and the difference in log-likelihoods between the models allowing for a correlation and the model with zero corre-

Table 9.9 *Empirical Bayes, full Bayes and NPML estimates*

Parameter	Empirical Bayes		Full Bayes		NPMLE	
	Est	(SE)	Est	(SE)	Est	(SE)
Fixed part						
β_0 [Cons]	-1.16	(0.14)	-1.17	(0.15)	-1.20	(0.15)
β_1 [Treat]	0.57	(0.09)	0.59	(0.09)	0.59	(0.10)
Random part						
$\sqrt{\psi_{00}}$ [Cons]	0.70	(0.11)	0.73	(0.12)		
$\sqrt{\psi_{11}}$ [Treat]	0.22	(0.10)	0.20	(0.10)		

lation was only 0.05. Thus, we present estimates for the model with $\psi_{10} = 0$ under 'Empirical Bayes' (maximum likelihood) in Table 9.9. There appears to be clear evidence that nicotine gum increases the odds of quitting, with an estimated odds ratio of $\exp(0.57) = 1.77$. There is some heterogeneity in the overall prevalence of quitting, reflected in the estimate $\widehat{\psi}_{00}^{1/2} = 0.70$ and small variability in the treatment effects estimated as $\widehat{\psi}_{11}^{1/2} = 0.22$

We also considered a fully Bayesian approach using noninformative priors as described in Section 6.11. The prior distributions of β_0 and β_1 were specified as $N(0, 10^6)$ and the priors of ζ_{0j} and ζ_{1j} as $N(0, \psi_{00})$ and $N(0, \psi_{11})$, respectively. The hyperpriors of ψ_{00} and ψ_{11} were specified as IG(0.001,0.001), where IG is the inverse gamma density given on page 210. Note that this specification is similar to that used for a different meta-analysis in Chapter 10 of Volume 1 of the BUGS Examples Manual (Spiegelhalter *et al.*, 1996b). The difference is that we treat the study-specific intercept as a random effect with a prior and hyperprior for the variance whereas the manual treats it as a 'fixed effect' with no hyperprior. Gibbs sampling, as implemented in BUGS, was used to estimate the parameters (see Section 6.11.5). A burn-in of 10000 iterations was used and the means and standard deviations of the parameters were obtained from a further 10000 iterations. The results are given in Table 9.9 under 'full Bayes' and agree quite closely with the empirical Bayes (maximum likelihood) results.

We could use (9.2) and (9.4) to estimate the individual log odds ratios and standard errors for each study. However, if we believe in the Bayesian random effects model, all inferences regarding the individual log odds ratios $\beta_1 + \zeta_{1j}$ should be based on their marginal posterior distribution, integrating over all other model parameters, here β_0, ψ_{00} and ψ_{11}. Within an MCMC algorithm (see Section 6.11.5), this amounts to using the empirical distributions of the sampled log odds ratios $\beta_1^{(r)} + \zeta_{1j}^{(r)}$. In empirical Bayes, the conditional posterior distribution of the ζ_{1j} is used, given that the other parameters are equal to the maximum likelihood estimates (see Section 7.3.1). The predicted log odds ratios then become $\hat{\beta}_1 + \tilde{\zeta}_{1j}$, where $\tilde{\zeta}_{1j}$ is the empirical Bayes prediction (mean of conditional posterior).

Bayesians use credible intervals instead of confidence intervals. For a 95% credible interval, the posterior probability that the parameter lies in the interval is 95%. Figure 9.6 shows approximate Bayes 95% credible intervals (using estimated percentiles of the posterior distribution based on 10000 draws) of the true effect $\beta_1 + \zeta_{1j}$ of each study, as well as the empirical Bayes counterparts. Unlike the fully Bayesian intervals, the empirical Bayes intervals are derived by treating the model parameters as known and assuming that the posterior distribution of the random slopes is normal. Here the intervals are constructed as posterior mean ± 1.96 times the posterior standard deviation. We would expect the fully Bayes intervals to be wider since they attempt to account for parameter uncertainty. However, the differences are generally very small with the possible exception of the 'Huber 1988' study. The raw log odds ratios $\ln(o_j)$, shown as 'x's, tend to be further from the average log odds, shown as a solid vertical line, than the empirical and full Bayes predictions. This is due to shrinkage as discussed in Section 7.3.1.

Instead of assuming bivariate normality for the random intercept and slope, Aitkin (1999b) leaves their joint distribution unspecified by using nonparametric maximum likelihood estimation (NPMLE) (see Section 4.4.2). Here a discrete distribution is used with locations $\zeta_{0j} = e_{0c}$, $\zeta_{1j} = e_{1c}$ and masses or probabilities π_c, $c = 1, \ldots, C$, giving a mixture regression model. The number of masses C is determined to maximize the likelihood. Note that the intercepts and slopes are now no longer independent. Using the Gateaux derivative method described in Section 6.5, where the number of masses is increased one at a time until the derivative is negative, we found that $C = 10$.

The estimates of the mean intercept and slope, given in Table 9.9 under NPMLE, are remarkably similar to the estimates assuming bivariate normality and $\psi_{10} = 0$. The log-likelihood for NPMLE was -3061.5 compared with -3074.2 assuming bivariate normality and $\psi_{10} = 0$. In NPMLE, the variances and covariances for the random effects are not model parameters but can be derived from the discrete distribution. The standard deviation of the random intercepts was 0.76, the standard deviation of the random slopes 0.30 and the correlation 0.08.

Figure 9.7 shows the NPMLE masses. In the top panel, the locations of the circles are $\hat{e}_{0c}, \hat{e}_{1c}$ whereas the areas are proportional to the probabilities $\hat{\pi}_c$. In the bottom panel, the probabilities are instead shown as the heights of spikes. Figure 9.8 shows the log odds for the control and treatment (gum) groups for each of the locations. The thickness of the lines reflects the probabilities which are also shown in the figure as percentages.

In summary, all three approaches considered produce practically identical estimates. The overall conclusion is that nicotine gum increases the odds of smoking cessation by about 80%. The 95% confidence and credible intervals for the odds-ratio (derived from the estimated log odds ratio and its standard

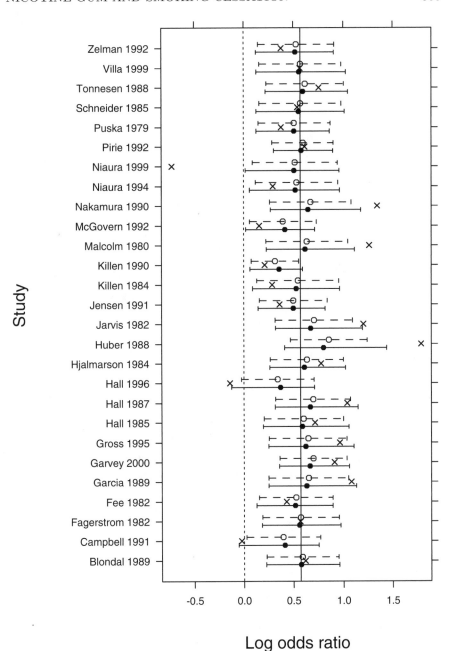

Figure 9.6 *Empirical Bayes (dashed lines) and full Bayes (solid lines) intervals for the study's true effects $\beta_1 + \zeta_{1j}$. The 'x's are the raw log odds ratios $\ln(o_j)$. The solid vertical line is the maximum likelihood estimate of the mean log odds ratio $\widehat{\beta}_1$, whereas the dotted vertical line at zero corresponds to no treatment effect*

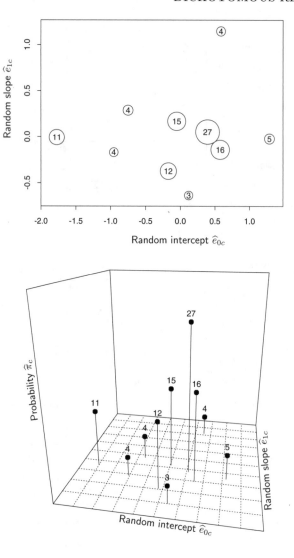

Figure 9.7 *NPML masses. Numbers shown are probabilities in percent*

error) are nearly the same for all three methods, the interval for empirical Bayes was from 1.5 to 2.1. There does not appear to be much heterogeneity in the treatment effect between studies.

It is important to note that meta-analysis is not without its critics, see for instance Thompson and Pocock (1991) and Oakes (1993). One problem that is generally acknowledged is publication bias. This is due to small studies being difficult to publish if the findings are not significant, leading to overestimated treatment effects in the meta-analysis (e.g. Sterlin, 1959; Sutton *et al.*, 2000).

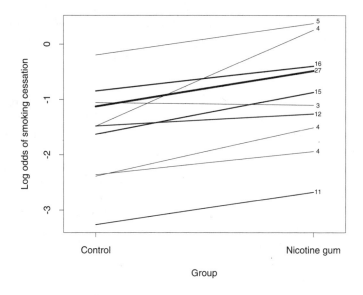

Figure 9.8 *Predicted log odds by treatment group for each 'class'*

9.6 Wives' employment transitions: Markov models with unobserved heterogeneity

We will now analyze data[5,6] on wives' employment from a panel survey or repeated measurement study. The 'Social Change and Economic Life Initiative', described in Davies *et al.* (1992) and Davies (1993), followed the employment status of wives from their marriage to the end of the survey in 1987. Here, we consider a subsample from Rochdale, one of the six localities studied.

The response is whether a wife is in paid employment (state 1) or not (state 0). The explanatory variables are all time-varying:

- [HUnemp] a dummy variable taking the value 1 if the wife's husband is unemployed and 0 otherwise
- [Time] the year of observation − 1975
- [Child1] a dummy for the wife having children under the age of 1
- [Child5] a dummy for the wife having children under the age of 5
- [Age] the wife's age in years − 35
- [Agesq] the square of [Age]

There are two competing explanations of the empirical regularity that people having experienced an event in the past are more likely to experience the event in the future than others. One explanation is 'causal'; employment

---

[5] We thank Dave Stott for providing us with these data.
[6] The data can be downloaded from gllamm.org/books

changes people, having been employed in itself changes a wife's future probability of employment, inducing dependence over time. This is called *true state
dependence* by Heckman (1978, 1981ac). This causal explanation is very common but is often naive because an alternative explanation may be just as plausible. Here, the observed dependence over time is interpreted as having been
induced by permanent components that represent unobserved heterogeneity.
Thus, the higher probability of being employed could be due to unobserved
characteristics of the wives, some having a high probability of future employment regardless of their previous employment history. This is called *spurious
state dependence* by Heckman who argues that distinguishing between true
and spurious state dependence is crucial in observational studies.

Consider first a conventional first-order discrete-time Markov model, where
employment status y_{tj} at time t is conditionally independent of employment
history given the previous employment status $y_{t-1,j}$, and current covariates.
Using a logit link, the model can be written as

$$\Pr(y_{tj}=1|\mathbf{x}_{tj}, y_{t-1,j}) \;=\; \frac{\exp(\mathbf{x}_{tj}'\boldsymbol{\beta} + \gamma y_{t-1,j})}{1 + \exp(\mathbf{x}_{tj}'\boldsymbol{\beta} + \gamma y_{t-1,j})}.$$

Unobserved heterogeneity can be included in the Markov model using a
random intercept ζ_j (e.g. Heckman, 1981a)

$$\Pr(y_{tj}=1|\mathbf{x}_{tj}, y_{t-1,j}, \zeta_j) \;=\; \frac{\exp(\mathbf{x}_{tj}'\boldsymbol{\beta} + \gamma y_{t-1,j} + \zeta_j)}{1 + \exp(\mathbf{x}_{tj}'\boldsymbol{\beta} + \gamma y_{t-1,j} + \zeta_j)}.$$

According to this model there is true state dependence if $\gamma \neq 0$ and spurious
state dependence if $\gamma = 0$.

We can also allow the random intercept variance to depend on the previous
state by specifying a factor model,

$$\Pr(y_{tj}=1|\mathbf{x}_{it}, y_{t-1,j}=0, \zeta_j) \;=\; \frac{\exp(\mathbf{x}_{tj}'\boldsymbol{\beta} + \zeta_j)}{1 + \exp(\mathbf{x}_{tj}'\boldsymbol{\beta} + \zeta_j)},$$

and

$$\Pr(y_{tj}=1|\mathbf{x}_{tj}, y_{t-1,j}=1, \zeta_j) \;=\; \frac{\exp(\mathbf{x}_{tj}'\boldsymbol{\beta} + \gamma + \lambda\zeta_j)}{1 + \exp(\mathbf{x}_{tj}'\boldsymbol{\beta} + \lambda\zeta_j)}.$$

The first probability is conditional on the subject previously being in state 0,
whereas the second is conditional on being in state 1. The other two transition
probabilities are $\Pr(y_{tj}=0|y_{t-1,j}=0) = 1 - \Pr(y_{tj}=1|y_{t-1,j}=0)$ and $\Pr(y_{tj}=0|y_{t-1,j}=1) = 1 - \Pr(y_{tj}=1|y_{t-1,j}=1)$. The factor ζ_j represents subject-
specific unobserved heterogeneity for the transition process, with variance
$\lambda^2\psi$ when $y_{t-1,j}=1$ and variance ψ when $y_{t-1,j}=0$.

The parameter estimates for these three models, denoted $\mathcal{M}_1$, $\mathcal{M}_2$ and
$\mathcal{M}_3$, respectively, are given in Table 9.10. The largest effect estimates are for
[Child1] and [HUnemp], both variables decreasing the odds of wives' employment. A negative effect of having children under the age of one on employment
is hardly surprising. The negative effect of [HUnemp] might for instance be due
to a considerable reduction in the husband's unemployment benefits if his wife

Table 9.10 *Estimates for simple Markov models with and without unobserved heterogeneity*

Parameter	$\mathcal{M}_1$		$\mathcal{M}_2$		$\mathcal{M}_3$	
	Est	(SE)	Est	(SE)	Est	(SE)
Fixed part						
β_0 [Cons]	-1.277	(0.240)	-1.114	(0.299)	-1.222	(0.339)
β_1 [HUnemp]	-1.406	(0.373)	-1.512	(0.412)	-1.524	(0.388)
β_2 [Time]	-0.012	(0.025)	-0.011	(0.026)	0.006	(0.032)
β_3 [Child1]	-3.008	(0.392)	-2.953	(0.408)	-2.767	(0.409)
β_4 [Child5]	-0.165	(0.253)	-0.241	(0.272)	-0.383	(0.294)
β_5 [Age]	-0.005	(0.014)	0.000	(0.016)	-0.009	(0.018)
β_6 [Agesq]	-0.001	(0.001)	-0.001	(0.001)	-0.002	(0.001)
γ [Lag]	4.391	(0.209)	4.226	(0.264)	4.380	(0.326)
Random part						
ψ			0.308	(0.326)	2.177	(0.932)
λ					-0.119	(0.164)
Log-likelihood	-411.50		-410.89		-401.25	

is employed or damaged male self-esteem. Comparing models 1 and 3, there is clearly evidence for both state dependence and unobserved heterogeneity, the random effect variance being considerably larger when the previous state is unemployment.

Model $\mathcal{M}_4$ suggested by Francis *et al.* (1996) allows both the regression parameters and the effect of unobserved heterogeneity to depend on previous state,

$$\Pr(y_{tj}=1|\mathbf{x}_{tj}, y_{t-1,j}=0, \zeta_j) = \frac{\exp(\mathbf{x}'_{tj}\boldsymbol{\beta}^0 + \zeta_j)}{1 + \exp(\mathbf{x}'_{tj}\boldsymbol{\beta}^0 + \zeta_j)},$$

and

$$\Pr(y_{tj}=1|\mathbf{x}_{tj}, y_{t-1,j}=1, \zeta_j) = \frac{\exp(\mathbf{x}'_{tj}\boldsymbol{\beta}^1 + \lambda\zeta_j)}{1 + \exp(\mathbf{x}'_{tj}\boldsymbol{\beta}^1 + \lambda\zeta_j)}.$$

The fixed effects of the covariates $\mathbf{x}_{tj}$ are $\boldsymbol{\beta}^0$ if the previous state is 0 and $\boldsymbol{\beta}^1$ if the previous state is 1. There is true state dependence if $\boldsymbol{\beta}^0 \neq \boldsymbol{\beta}^1$ after accounting for unobserved heterogeneity. In this model, spurious state dependence arises if $\boldsymbol{\beta}^0 \neq \boldsymbol{\beta}^1$ before introducing unobserved heterogeneity but $\boldsymbol{\beta}^0 = \boldsymbol{\beta}^1$ after taking the heterogeneity into account. The model can be written in the GRC formulation described in Section 4.2.3 as

$$\text{logit}[\Pr(y_{tj}=1|\mathbf{x}_{tj}, y_{t-1,j}, \zeta_j)] = \nu_{tj},$$

where the linear predictor is

$$\nu_{tj} = (1 - y_{t-1,j})\mathbf{x}'_{tj}\boldsymbol{\beta}^0 + y_{t-1,j}\mathbf{x}'_{tj}\boldsymbol{\beta}^1 + \zeta_j[(1 - y_{t-1,j}) + \lambda y_{t-1,j}].$$

The parameter estimates for this general Markov model are given in Table 9.11. The greatest difference in coefficients is for [Child1]. For employed

Table 9.11 *Estimates for general Markov model $\mathcal{M}_4$*

Parameter	Getting job $y_{t-1,j}=0,\ a=0$ Est	(SE)	Keeping job $y_{t-1,j}=1,\ a=1$ Est	(SE)
Fixed part				
β_0^a [Cons]	−1.521	(0.429)	3.391	(0.353)
β_1^a [HUnemp]	−1.916	(0.735)	−1.361	(0.493)
β_2^a [Time]	0.013	(0.042)	−0.013	(0.040)
β_3^a [Child1]	−1.439	(0.695)	−2.776	(0.490)
β_4^a [Child5]	−0.253	(0.396)	−0.554	(0.413)
β_5^a [Age]	−0.031	(0.027)	0.031	(0.019)
β_6^a [Agesq]	0.009	(0.002)	−0.002	(0.002)
Random part				
λ			−0.212	(0.223)
ψ		1.411 (0.717)		
Log-likelihood		−396.99		

wives, [Child1] substantially increases the odds of leaving employment (odds ratio $\exp(2.776) = 16.1$), whereas for unemployed wives, this variable increases the odds of staying out of employment to a lesser extent (odds ratio $\exp(1.439) = 4.2$).

In order to interpret the dependence structure, we will formulate the model as a latent response model,

$$y_{tj}^* = \mathbf{x}'_{tj}\boldsymbol{\beta}^0 + \zeta_j + \epsilon_{tj} \quad \text{if} \quad y_{t-1,j}=0$$

and

$$y_{tj}^* = \mathbf{x}'_{tj}\boldsymbol{\beta}^1 + \lambda\zeta_j + \epsilon_{tj} \quad \text{if} \quad y_{t-1,j}=1,$$

where

$$y_{tj} = \begin{cases} 1 & \text{if } y_{tj}^* > 0 \\ 0 & \text{otherwise.} \end{cases}$$

It follows that the latent response correlations between two nonadjacent occasions s and u, $|s - u| > 1$, are

$$\rho_{00} = \text{Cor}(y_{sj}^*, y_{uj}^* | y_{s-1,j}=0, y_{u-1,j}=0, \mathbf{x}_{sj}, \mathbf{x}_{uj}) = \frac{\psi}{\psi + \pi^2/3},$$

$$\rho_{11} = \mathrm{Cor}(y_{sj}^*, y_{uj}^* | y_{s-1,j} = 1, y_{u-1,j} = 1, \mathbf{x}_{sj}, \mathbf{x}_{uj}) = \frac{\lambda^2 \psi}{\lambda^2 \psi + \pi^2/3},$$

and

$$\rho_{01} = \mathrm{Cor}(y_{sj}^*, y_{uj}^* | y_{s-1,j} = 0, y_{u-1,j} = 1, \mathbf{x}_{sj}, \mathbf{x}_{uj}) = \frac{\lambda \psi}{\sqrt{\lambda^2 \psi + \pi^2/3}\sqrt{\psi + \pi^2/3}}.$$

Here, ρ_{00} represents the within-wife residual correlation in the propensity to become employed when unemployed, estimated as $\hat{\rho}_{00} = 0.30$. ρ_{11} is the within-wife correlation in the propensity to remain employed, estimated as $\hat{\rho}_{11} = 0.02$. ρ_{01} is the correlation between the propensity to become employed when not employed and the propensity to remain employed when employed, estimated as $\hat{\rho}_{01} = -0.08$.

Initial conditions must be addressed in dynamic models. Conditions often invoked include that initial states are exogenous (the approach taken here for simplicity) or that the process is in equilibrium. However, a common problem is that the process under investigation is not studied from its beginning, implying that the first state observed cannot be exogenous. Heckman (1981b) suggests an ad hoc approach to approximate the initial conditions for dichotomous dynamic models. See Hsiao (2002) for a discussion of this and other approaches.

9.7 Counting snowshoe hares: Capture-recapture models with heterogeneity

Capture-recapture studies are often used to ascertain the size of a population, for example the number of heroin users in a city or the number of animals of a given species in some geographical area. The idea is to 'capture' individuals from the population on different occasions or using different methods and record the identity of the captured individuals (for animal populations this requires marking the animals). The capture histories of those individuals captured at least once can then be used to estimate the number of individuals never captured and hence the total population size. A basic assumption here is that the population is constant or 'closed' throughout the study.

Consider the simple example of two captures. If all individuals have the same chance of being captured and the probabilities of being captured on the two occasions are independent, then a large proportion of individuals captured on both occasion indicates a small population. However, an alternative explanation for a large number of recaptures is that some individuals are much easier to catch than others and it is these individuals who were captured twice. With only two captures, we cannot distinguish between these two explanations empirically and independence is usually assumed. With more than two captures, we can use the observed capture histories to estimate the degree of unobserved heterogeneity in catchability. Failing to account for unobserved heterogeneity can lead to biased estimates of population size (e.g. Otis *et al.*, 1978), although bias can to some extent be mitigated by design or by including observed covariates in the models.

Burnham and Cuschwa (see Otis *et al.*, 1978) laid out a livetrapping grid in a black spruce forest in Alaska to estimate the (closed) population of snowshoe hares. The basic grid was 10×10 with traps spaced 200 feet apart. Trapping was carried out for 9 consecutive days in early winter but traps were not baited for the first three days. Data were obtained on 68 captures and recaptures from the last 6 days of trapping. Table 9.12 shows the number of hares experiencing each of the possible capture histories, represented by indicators for capture ($1 =$ capture, $0 =$ no capture) at each of the six occasions[7]. The count for the cell corresponding to no captures is unknown and our aim is to estimate this number.

Table 9.12 *Results of capture-recapture of snowshoe hares*

Captures 6,5,4	Captures 3,2,1							
	000	001	010	011	100	101	110	111
000	?	3	6	0	5	1	0	0
001	3	2	3	0	0	1	0	0
010	4	2	3	1	0	1	0	0
011	1	0	0	0	0	0	0	0
100	4	1	1	1	2	0	2	0
101	4	0	3	0	1	0	2	0
110	2	0	1	0	1	0	1	0
111	1	1	1	0	0	0	1	2

Source: Agresti (1994)

A common approach to estimating the population size is Sanathanan's (1972) conditional method. First a model is specified for the probabilities π_y of the capture histories **y** where **y** denotes a sequence of indicators (0 or 1) for capture on each of the occasions. The parameters of this model are estimated by maximizing the conditional likelihood of the observable capture histories given that the individuals were captured at least once. The conditional probability of capture history **y** given that the individual was captured at least once is

$$\pi_{cy} = \pi_y/(1 - \pi_{0...0}),$$

where $\pi_{0...0}$ is the probability of never getting caught. The conditional log-likelihood therefore is

$$L_c = \sum_y n_y \ln \pi_{cy} = \sum_y n_y \ln \pi_y - \sum_y n_y \ln(1 - \pi_{0...0}),$$

[7] The data can be downloaded from gllamm.org/books

where the sum is over the observable capture histories, i.e. all possible histories excluding $0 \ldots 0$, and $n_{\mathbf{y}}$ is the number of individuals with history $\mathbf{y}$. The total population size is then estimated by maximizing the binomial probability of capturing $n = \sum_{\mathbf{y}} n_{\mathbf{y}}$ individuals at least once where the probability of success is $(1 - \pi_{0\ldots0})$, giving

$$\widehat{N} = \frac{n}{1 - \pi_{0\ldots0}}.$$

Unobserved heterogeneity in catchability is likely in the capture-recapture study of snowshoe hares for several reasons. We would generally expect heterogeneity due to behavioral differences such as trap-attraction or trap-avoidance. Moreover, hares with larger foraging areas are also exposed to more traps than those with smaller areas and hares near grid boundaries are less prone to capture.

Coull and Agresti (1999) use a random intercept logistic model to account for unobserved heterogeneity. The conditional probability of capture of individual j on occasion i is modeled as

$$\text{logit}[\Pr(y_{ij} = 1 | \zeta_j)] = \beta_i + \zeta_j,$$

where ζ_j, representing the 'catchability' of animal j, is normally distributed with mean 0 and variance ψ. This is the one-parameter logistic item response model discussed in Section 3.3.4. The probability of a given history $\pi_{\mathbf{y}}$ with $\mathbf{y} = (y_1, \ldots, y_I)'$ then is

$$\pi_{\mathbf{y}} = \int \prod_{i=1}^{I} \frac{\exp(y_i(\beta_i + \sqrt{\psi}z))}{1 + \exp(\beta_i + \sqrt{\psi}z)} \, \phi(z) \, \mathrm{d}z.$$

Instead of assuming a normal distribution for the random intercept, we can assume that the population consists of latent classes with different constant levels of catchability, i.e. we can allow ζ_j to be discrete so that

$$\pi_{\mathbf{y}} = \sum_c \prod_{i=1}^{I} \frac{\exp(y_i(\beta_i + e_c))}{1 + \exp(\beta_i + e_c)} \pi_c,$$

where e_c is the catchability of latent class c and π_c is the probability of belonging to class c. For identification, we restrict the mean of ζ_j to be zero,

$$\mathrm{E}(\zeta_j) = \sum_c \pi_c e_c = 0,$$

so that the variance of η_j is

$$\text{Var}(\zeta_j) = \sum_c \pi_c e_c^2.$$

The estimates for the homogenous population model, the model with a normal random intercept and the two-class model are shown in Table 9.13. The two-class solution has $\widehat{e}_1 = 4.14$, $\widehat{e}_2 = -0.13$, $\widehat{\pi}_1 = 0.03$ and $\widehat{\pi}_2 = 0.97$. Increasing the number of classes to three only results in a small increase of 0.30 in the conditional log-likelihood. Note that the estimated population sizes are

Table 9.13 *Estimates for capture-recapture of snowshoe hares*

Parameter	Homogen. population Est	(SE)	Random intercept Est	(SE)	Two classes Est	(SE)	Random int. & prev. hist. Est	(SE)
β_1 [Cap1]	−1.30	(0.29)	−1.83	(0.47)	−1.36	(0.33)	−1.56	(0.38)
β_2 [Cap2]	−0.52	(0.25)	−0.98	(0.43)	−0.51	(0.28)	−0.39	(0.40)
β_3 [Cap3]	−1.01	(0.27)	−1.51	(0.45)	−1.03	(0.30)	−0.59	(0.53)
β_4 [Cap4]	−0.64	(0.25)	−1.11	(0.44)	−0.64	(0.29)	−0.03	(0.60)
β_5 [Cap5]	−0.82	(0.26)	−1.30	(0.44)	−0.83	(0.29)	−0.10	(0.67)
β_6 [Cap6]	−0.29	(0.25)	−0.74	(0.43)	−0.29	(0.28)	0.59	(0.74)
γ [Prev]							−1.10	(0.60)
$\mathrm{Var}(\zeta_j)$			0.93	(0.63)	0.56	(−)	1.01	(0.56)
L_c	−254.5		−250.74		−249.16		−249.41	
$\widehat{N}$	75.1		92.1		77.1		75.0	
95% CI	(69,84)		(74,153)		(70,88)		(68,116)	

quite different for the two approaches to including unobserved heterogeneity (92 and 77), the latent class estimate being close to the conventional estimate, although the fit of the heterogeneity models is similar.

The models assume that the dependence among the responses is purely due to unobserved heterogeneity. However, it is also possible that capture on one occasion directly affects the probability of capture on subsequent occasions (state dependence), particularly if the same method of trapping is used. Huggins (1989) therefore included a time-varying indicator of previous capture x_{ij} ([Prev]) in the random intercept model, where $x_{ij} = 1$ if the animal has been captured before and $x_{ij} = 0$ otherwise,

$$\mathrm{logit}[\mathrm{Pr}(y_{ij} = 1|\zeta_j, x_{ij})] = \beta_i + \zeta_j + \gamma x_{ij}.$$

For the snowshoe hare data and a normally distributed random intercept ζ_j, the estimates are shown in the last column of Table 9.13. Here $\widehat{\gamma} = -1.10$, indicating that animals are less likely to be caught again if they have previously been caught. The effect is not quite significant at the 5% level ($p = 0.07$). The estimated population size is now 75, equal to the conventional estimate.

Cormack (1992) suggests constructing confidence intervals for the true population size N using the (unconditional) profile likelihood for N. The approach is to substitute different values for $n_{0...0}$, estimate the model parameters by maximizing the unconditional likelihood and evaluate the deviance of the model. The confidence limits are then the values of $N = n + n_{0...0}$ that yield a deviance differing from the minimum by a prespecified value, 3.84 in the case of a 95% confidence interval. It is important to use the deviance rather than the log-likelihood itself since the log-likelihood of the saturated model changes with $n_{0...0}$. Cormack shows that the parameter estimates and deviance for the

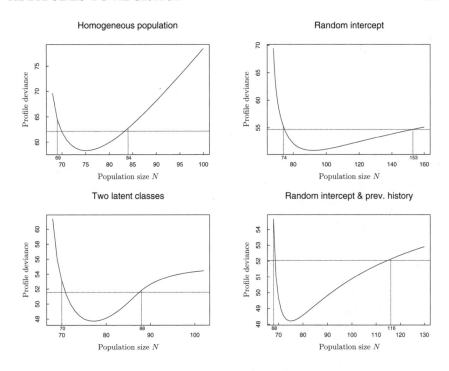

Figure 9.9 *Profile deviances for population size for four capture-recapture models*

conditional likelihood are identical to those for the unconditional likelihood when $n_{0...0}$ is equal to the conditional estimate.

The profile deviances for the four models are shown in Figure 9.9. The horizontal lines represent the minimum deviance plus 3.84. The vertical lines indicate approximate 95% confidence limits for the population size - integer values of N with deviances as close as possible, and no less than the value indicated by the horizontal line. The confidence intervals, given in Table 9.13, are quite wide, particularly for the random intercept model.

Summarizing the findings, we can make the conservative statement that the population size lies somewhere between 68 and 153.

9.8 Attitudes to abortion: A multilevel item response model

In the British Social Attitudes Survey Panel 1983-1986 (Social and Community Planning Research, 1987)[8] respondents were asked whether or not abortion should be allowed by law under the following circumstances:

1. [Woman] the woman decides on her own she does not wish to have the child

---

[8] Data were supplied by the UK Data Archive. Neither the original data collectors nor the archive are responsible for the present analyses.

2. [Couple] the couple agree that they do not wish to have the child

3. [Marriage] the woman is not married and does not wish to marry the man

4. [Financial] the couple cannot afford any more children

5. [Defect] there is a strong chance of a defect in the baby

6. [Risk] the woman's health is seriously endangered by the pregnancy

7. [Rape] the woman became pregnant as a result of rape

The data have a three-level structure with occasions or 'panel waves' nested in individuals nested in polling districts. There were 14143 responses to the 7 items over the four panel waves from 734 individuals in 57 polling districts. The multilevel design is highly unbalanced with 49% of subjects responding to at least one item in all four panel waves, 12% in three waves, 13% in two waves and 25% in one wave. Unit nonresponse was therefore common, but if an interview took place, item nonresponse occurred in only 7% of cases. We will not explicitly model unit or item nonresponse and therefore assume that the data are missing at random (MAR).

Previous multilevel analyses of these data have used raw sumscores or scores constructed from item response models as response variable (Knott *et al.*, 1990; Wiggins et al., 1990). However, using such constructed scores as proxies for latent variables has been demonstrated to be highly problematic, leading to biased standard errors and often to inconsistent parameter estimates (Skrondal and Laake, 2001). Hence, we use multilevel factor models with a logit link for the dichotomous items (Rabe-Hesketh *et al.*, 2004a). The change in deviance is used to choose between competing models. Each model is fitted a number of times using adaptive quadrature comparing solutions with different numbers of quadrature points per dimension to ensure reliable results.

Initially, we focus on between-subject heterogeneity and subsequently also include heterogeneity between polling districts. It is plausible that in addition to a 'general attitude' factor $\eta_{Gjk}^{(3)}$ measured by all items there may be an independent 'extreme circumstance' factor $\eta_{Ejk}^{(3)}$ representing people's additional inclination to be in favor of abortion when there is a strong chance of a defect in the baby, a high risk to the woman, or where the pregnancy was a result of rape (items 5, 6 and 7). Using indices i for item or circumstance (level 1), t for occasion (level 2), j for subject (level 3) and k for polling district (level 4), the two-factor model can be written in GRC notation as

$$\nu_{itjk} = \mathbf{d}_i'\boldsymbol{\beta} + \eta_{Gjk}^{(3)}\mathbf{d}_i'\boldsymbol{\lambda}_G + \eta_{Ejk}^{(3)}\boldsymbol{\delta}_{Ei}'\boldsymbol{\lambda}_E, \qquad (9.5)$$

where $\mathbf{d}_i$ is a 7-dimensional vector with ith element equal to 1 and all other elements equal to 0 and $\boldsymbol{\delta}_{Ei}$ is a 3-dimensional vector of indicators for items 5, 6 and 7, equal to the last three elements of $\mathbf{d}_i$, for example $\boldsymbol{\delta}_{E6} = (0, 1, 0)'$. A unidimensional factor model appears to be inadequate since removing the extreme circumstance factor increases the deviance by 207.7 with 3 degrees of freedom.

Since there are repeated responses for each subject and item, item specific unique factors can be included at the subject level:

$$\nu_{itjk} = \mathbf{d}_i'\boldsymbol{\beta} + \eta_{Gjk}^{(3)}\mathbf{d}_i'\boldsymbol{\lambda}_G + \eta_{Ejk}^{(3)}\boldsymbol{\delta}_{Ei}'\boldsymbol{\lambda}_E + \sum_{m=1}^{7} \eta_{Umjk}^{(3)}d_{im}, \qquad (9.6)$$

where the latent variables are mutually independent. The unique factors $\eta_{Umjk}^{(3)}$ in the last term can be interpreted as heterogeneity between subjects in their attitudes to specific items which induces additional dependence between responses over time not accounted for by the common factors. Evaluation of the log-likelihood for this model requires integration over 9 dimensions at level 3. To reduce the dimensionality, the items i can be treated as level-2 units so that time becomes level 1 and the model is reparameterized as

$$\nu_{tijk} = \mathbf{d}_i'\boldsymbol{\beta} + \eta_{Uijk}^{(2)}\mathbf{d}_i'\boldsymbol{\lambda}_U + \eta_{Gjk}^{(3)}\mathbf{d}_i'\boldsymbol{\lambda}_G + \eta_{Ejk}^{(3)}\boldsymbol{\delta}_{Ei}'\boldsymbol{\lambda}_E. \qquad (9.7)$$

Here the last term in (9.6) which evaluates to $\eta_{Uijk}^{(3)}$ for item i has been replaced by $\eta_{Uijk}^{(2)}\lambda_{Ui}$. Whereas the $\eta_{Uijk}^{(3)}$ are treated as separate latent variables for the items, $i = 1, \ldots, 7$, $\eta_{Uijk}^{(2)}$ is a single latent variable with different realizations for different items i. The purpose of λ_U is to allow the unique factor variances to differ between the items. The models are equivalent since both $\eta_{Uijk}^{(3)}$ and $\eta_{Uijk}^{(2)}\lambda_{Ui}$ vary between items, are uncorrelated across items and have item-specific variances. The advantage of (9.7) is that a nine-dimensional integral at level 3 has been replaced by a one-dimensional integral at level 2 and a two-dimensional integral at level 3. It is often possible to reduce the dimensionality of integration by reparameterization to an equivalent model (see also Section 5.3.2). Adding unique factors at the subject level to the two-factor model decreases the deviance by 12.6, a small change for seven additional parameters.

Introducing district-level latent variables in addition to subject level latent variables, the common factors can be allowed to vary between polling districts giving two-dimensional variance components factor models (see Section 4.3). Allowing the general attitude factor to vary between districts decreases the deviance by 8.2 with one extra parameter whereas the deviance decreases by only 3.2 for the extreme circumstance factor. The retained model is therefore the response model in (9.5) plus the structural model

$$\eta_{Gjk}^{(3)} = \eta_{Gk}^{(4)} + \zeta_{Gij}^{(3)}.$$

A path diagram for the retained model is given in Figure 9.10. It should be noted than the paths to the responses do not represent linear effects on the responses; the paths from $\eta_G^{(3)}$ and $\eta_E^{(3)}$ represent linear effects on the log odds, whereas the short arrows represent random (Bernoulli) variability of the responses given the model-implied probabilities.

Including unique factors at the district level increases the dimension of integration at level 4 from 1 to 8. The dimensionality cannot be reduced by

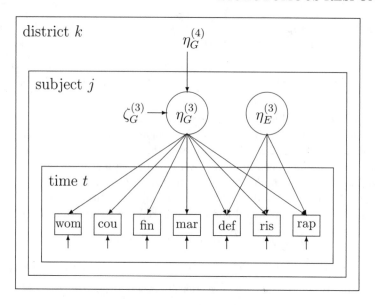

Figure 9.10 *Path diagram for the multilevel variance components factor model*

reparameterization in this case. We therefore included each unique factor separately, but the changes in deviance were small. Estimates for the retained multilevel variance components logit factor model are given in Table 9.14. These were obtained using adaptive quadrature with 10 points per dimension which gave very similar results to 8 and 5 points per dimension. As expected, the intercepts for the extreme circumstance items were much larger than for the others due the larger prevalence of endorsing these items. A general attitude and an extreme circumstance factor were required at the subject level. Only the general attitude factor appeared to vary at the polling district level, but with a relatively small standard deviation.

9.9 Summary and further reading

We first considered longitudinal data on respiratory infection. We used a random intercept logistic regression model and compared the results to GEE. Useful reviews of analysis of clustered binary data include Neuhaus (1992) and Pendergast *et al.* (1996). Discussions of the pros and cons of conditional versus marginal approaches are provided in Lindsey and Lambert (1998), Lindsey (1999) and Crouchley and Davies (1999). A two-level random intercept model was used to model change in condom use after HIV diagnosis by Skrondal *et al.* (2000) and a three-level random intercept model used for repeated neuropsychological measures in schizophrenics, their healthy relatives and unrelated controls by Rabe-Hesketh *et al.* (2001c). A two-level random coefficient model for longitudinal data on thought disorder has been considered by Dig-

Table 9.14 *Estimates for the multilevel variance components logit factor model*

Fixed part		
Intercepts:		
β_1 [Woman]	−0.83 (0.14)	
β_2 [Couple]	−0.17 (0.15)	
β_3 [Marriage]	−0.28 (0.16)	
β_4 [Financial]	−0.01 (0.14)	
β_5 [Defect]	3.79 (0.27)	
β_6 [Risk]	5.90 (0.56)	
β_7 [Rape]	4.82 (0.39)	
Random part: Subject level		
Factor loadings	General	Extreme
λ_{G1} & λ_{E1} [Woman]	1	0
λ_{G2} & λ_{E2} [Couple]	1.13 (0.08)	0
λ_{G3} & λ_{E3} [Marriage]	1.21 (0.09)	0
λ_{G4} & λ_{E4} [Financial]	1.01 (0.08)	0
λ_{G5} & λ_{E5} [Defect]	0.78 (0.09)	1
λ_{G6} & λ_{E6} [Risk]	0.73 (0.13)	1.53 (0.26)
λ_{G7} & λ_{E7} [Rape]	0.72 (0.11)	1.23 (0.21)
Factor variances		
$\psi_G^{(2)}$ & $\psi_E^{(2)}$	5.22 (0.67)	3.30 (0.80)
Random part: District level		
Factor variances		
$\psi_G^{(3)}$ & $\psi_E^{(3)}$	0.36 (0.17)	0
Log-likelihood	−5160.9	

Source: Rabe-Hesketh *et al.* (2004a)

gle *et al.* (2002), Skrondal and Rabe-Hesketh (2003c) and Rabe-Hesketh and Everitt (2003), among others.

The next application was a latent class model for the diagnosis of myocardial infarction. Other medical applications of such models are given in Formann and Kohlmann (1996). See also Section 13.5 for latent class models for rankings, Section 13.6 for first choices, Section 12.4.4 for durations and Sections 14.3 and 14.4 for multiple processes. A good overview of latent class modeling is given by Clogg (1995).

We also considered one, two and three-parameter item response models with covariates for ability testing. Multidimensional versions of the two-parameter model for ordinal responses will be discussed in the next chapter and a multilevel version was discussed in Section 9.8.

A meta-analysis of the effectiveness of nicotine gum for smoking cessation was conducted, using random effects models for patient-level data. Results from Bayesian and likelihood methods were compared. Although extremely popular in medicine, it should be noted that meta-analysis is also gaining popularity in other disciplines such as economics (e.g. Granger, 2002) and sociology (e.g. DiPrete, 2002). Useful books on meta-analysis include Hedges and Olkin (1985) and Whitehead (2002), and useful reviews are given by Fleiss (1993) and Normand (1999).

Wives' employment transition data were then analyzed using different types of Markov models with random effects to explore the issue of true versus spurious state dependence. In this chapter the response has been treated as dichotomous, but could alternatively be viewed as a discrete time duration; see Chapter 12.

Another application concerned the estimation of the population size of snowshoe hares using capture-recapture models with unobserved heterogeneity. A useful review of these methods is given by Chao *et al.* (2001).

Finally, we described a multilevel item response model for attitudes to abortion. Fox (2001) explores multilevel structural equation models for dichotomous responses in an educational setting. Ansari and Jedidi (2000) and Fox and Glas (2001) discuss Bayesian estimation of multilevel item response models.

All models considered in this chapter have used the logit link, but could also have been formulated in terms of probit or complementary log-log links.

CHAPTER 10

Ordinal responses

10.1 Introduction

The first theme of this chapter is 'growth curve' models for analyzing the effect of a cluster randomized intervention on ordinal responses measured at several occasions. Initially, we discuss multilevel growth models for repeated measures of a particular ordinal observed response. These models are subsequently extended to growth models for a latent variable that is repeatedly measured by several ordinal items at each occasion.

The other theme of the chapter is 'psychometric validation' of measurement instruments with ordinal items. In particular, we demonstrate how properties such as factor dimensionality, item reliability and item bias can be investigated.

10.2 Cluster randomized trial of sex education: Latent growth curve model

10.2.1 Introduction

A cluster randomized trial is one where clusters of units rather than the units themselves are randomized to treatment groups. A typical application is the evaluation of nontherapeutic interventions, for instance the effect of different modes of sex education on use of contraceptives.

Cluster randomized trials have several merits: First, some treatments are most naturally applied at the cluster level. This is obviously the case with sex education which takes place in school classes, making randomization of individual students impractical. Second, cluster randomized trials reduce experimental contamination. In the sex education example, contamination would occur if students receiving the intervention would share their knowledge with students not receiving the intervention. Such contamination can be minimized by randomizing at the school level, assuming that there is little communication among students from different schools.

There are, however, some disadvantages of cluster randomized trials: First, units in a cluster are often more similar than units in different clusters. This implies that units cannot be treated as independent in statistical modeling; the dependence among units within clusters must be accounted for. Second, and related to the first issue, cluster randomized trials are usually less efficient than classical randomized trials.

In this section we will analyze data from a cluster randomized trial of sex

education for 15 and 16 year olds in Norwegian schools (Træen, 2003)[1]. A
school book and curriculum for sex education was developed. This included
dramas created for students to perform, debates of specific questions and
practical tasks such as finding out how to get hold of contraception. The
intervention was designed to 'make adolescents competent actors in sexual
contexts in the sense that they dared handle contraception and put up limits'.

The outcome of interest, whether contraception was being used, was only
available on a minority of the adolescents who were sexually active. Instead
of actual behavior, the hypothetical construct 'contraceptive self-efficacy' was
hence studied. This construct has previously been shown to be a good predictor of contraceptive use (e.g. Kvalem and Træen, 2000). In this section we
focus on the constituent construct 'situational contraceptive communication',
measured by three questionnaire items:

"If my partner and I were about to have intercourse without either of us
having mentioned contraception ...

- [Tell] I could easily tell him/her that I didn't have any contraception"

- [Ask] I could easily ask him/her if he/she had any contraception"

- [Get] I could easily get out a condom (if I had one with me)"

The questions were answered in terms of five ordinal response categories:

1. Not at all true of me

2. Slightly true of me

3. Somewhat true of me

4. Mostly true of me

5. Completely true of me

Schools were randomized to receive the intervention or not. Questionnaires
were completed prerandomization and 6 months and 18 months postrandom-
ization. The data therefore have a three-level structure with occasions t nested
in students j nested in schools k. 46 schools and 1184 students contributed
to the analysis. Only 570 students always responded, 400 responded on some
occasions and 114 never responded. The two predictors we will use here are

- [Treat] dummy variable for student being in school receiving treatment x_{1jk}
  (yes=1, no=0)

- [Time] time since randomization in 6-month periods x_{2tjk} (0, 1, 3)

10.2.2 Growth curve modeling

We will initially estimate a multilevel proportional odds model for one of the
items, [Get]. We will allow the mean of the latent response to depend on [Time]
(a linear trend), [Treat] and [Time]×[Treat] and include random intercepts

---

[1] We thank Bente Træen for providing us with these data.

for students $\zeta_{jk}^{(2)}$ and schools $\zeta_k^{(3)}$. The latent response y_{tjk}^* at occasion t for student j and school k is therefore modeled as

$$
\begin{aligned}
y_{tjk}^* &= \beta_1 x_{1jk} + \beta_2 x_{2tjk} + \beta_3 x_{1jk} x_{2tjk} + \zeta_{jk}^{(2)} + \zeta_k^{(3)} + \epsilon_{tjk} \\
&= \mathbf{x}_{tjk}' \boldsymbol{\beta} + \zeta_{jk}^{(2)} + \zeta_k^{(3)} + \epsilon_{tjk},
\end{aligned}
$$

where ϵ_{tjk} has a logistic distribution. The observed responses are generated by the threshold model

$$
y_{tjk} = \begin{cases}
1 & \text{if} & y_{tjk}^* \leq \kappa_1 \\
2 & \text{if} & \kappa_1 < y_{tjk}^* \leq \kappa_2 \\
3 & \text{if} & \kappa_2 < y_{tjk}^* \leq \kappa_3 \\
4 & \text{if} & \kappa_3 < y_{tjk}^* \leq \kappa_4 \\
5 & \text{if} & \kappa_4 < y_{tjk}^*.
\end{cases}
$$

Note that the constant has been omitted in the latent response model so that all four thresholds κ_s, $s = 1, \ldots, 4$, are identified.

The estimates are shown in Table 10.1. Surprisingly, the variance of the

Table 10.1 *Estimates for multilevel proportional odds model*

Parameter	Single-level Est	(SE)	Two-level Est	(SE)	Three-level Est	(SE)
Regression coefficients						
β_1 [Time]	-0.12	(0.06)	-0.13	(0.06)	-0.13	(0.06)
β_2 [Treat]	-0.05	(0.14)	-0.02	(0.19)	-0.02	(0.19)
β_3 [Time]×[Treat]	0.17	(0.08)	0.17	(0.09)	0.17	(0.09)
Variances						
student-level						
$\psi^{(2)}$			2.03	(0.31)	2.03	(0.31)
school-level						
$\psi^{(3)}$					0.00	(0.00)
Thresholds						
κ_1	-3.54	(0.17)	-4.41	(0.23)	-4.41	(0.23)
κ_2	-2.43	(0.13)	-3.15	(0.19)	-3.15	(0.19)
κ_3	-1.18	(0.12)	-1.58	(0.16)	-1.58	(0.16)
κ_4	0.16	(0.12)	0.25	(0.15)	0.25	(0.15)
Log-likelihood	-2531.21		-2470.88		-2470.88	

random intercept for school is estimated as nearly zero. However, there is evidence for unobserved heterogeneity between students with a variance estimated as 2.03, corresponding to a residual intraclass correlation between the

latent responses of 0.37. There is a small treatment effect [Time]×[Treat], the
estimate 0.17 corresponding to an odds ratio of 1.19. Therefore the percentage
increase in the odds of high versus low responses per six month period is 19%
higher in the treatment group than the control group. (Here high versus low
response can either mean response 5 versus 1 to 4, or responses 4 or 5 versus
1 to 3 or responses 3 to 5 versus 1 or 2 or responses 2 to 5 versus 1.)

To visualize this treatment effect, Figure 10.1 shows the population averaged
probabilities $\Pr(y_{tjk} \geq s|\mathbf{x}_{tjk}; \widehat{\boldsymbol{\theta}})$ of response s or above, $s = 2, 3, 4, 5$, by time
and treatment group. These probabilities were obtained by integrating the
conditional response probabilities, given the random effects, over the random
effects distribution. The corresponding observed proportions are also shown.
It is worth pointing out that linear relationships on the logit scale do not
generally look this linear on the probability scale.

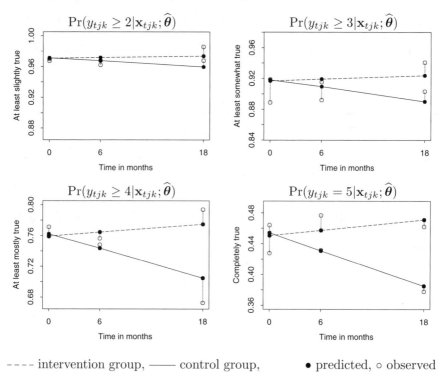

---- intervention group, —— control group,   • predicted, ∘ observed

Figure 10.1 *Predicted and observed marginal response probabilities*

10.2.3 Latent growth curve modeling

Since there is no evidence for variability between schools, we will henceforth
omit the k subscript. We first develop a measurement model for contraceptive
self-efficacy η_{tj}, measured by the three ordinal items y_{itj}; $i = 1$ [Tell], $i = 2$

[Ask] and $i=3$ [Get]. One-factor models with three different specifications of thresholds κ_{is} and intercepts δ_i were considered:

- Different thresholds for each item i and no intercepts (12 parameters)

$$
y_{itj} = \begin{cases}
1 & \text{if} & y^*_{itj} \leq \kappa_{i1} \\
2 & \text{if} & \kappa_{i1} < y^*_{itj} \leq \kappa_{i2} \\
3 & \text{if} & \kappa_{i2} < y^*_{itj} \leq \kappa_{i3} \\
4 & \text{if} & \kappa_{i3} < y^*_{itj} \leq \kappa_{i4} \\
5 & \text{if} & \kappa_{i4} < y^*_{itj}.
\end{cases}
$$

- One set of thresholds for all three items and no intercepts (4 parameters)

$$
y_{itj} = \begin{cases}
1 & \text{if} & y^*_{itj} \leq \kappa_{1} \\
2 & \text{if} & \kappa_{1} < y^*_{itj} \leq \kappa_{2} \\
3 & \text{if} & \kappa_{2} < y^*_{itj} \leq \kappa_{3} \\
4 & \text{if} & \kappa_{3} < y^*_{itj} \leq \kappa_{4} \\
5 & \text{if} & \kappa_{4} < y^*_{itj}.
\end{cases}
$$

- One set of thresholds for all three items and intercepts δ_2 and δ_3 for items 2 and 3 (6 parameters).

The log-likelihoods are -6946, -6990 and -6950, respectively, so that the last model is retained. The latent response y^*_{itj} for the ith item at occasion t for student j is therefore modeled as

$$
y^*_{itj} = \delta_i + \lambda_i \eta^{(2)}_{tj} + \epsilon_{itj}, \quad \lambda_1 = 1, \quad \delta_1 = 0,
$$

with constant thresholds κ_s, $s = 1, 2, 3, 4$.

We then combine the measurement model with a structural model for contraceptive self-efficacy

$$
\eta^{(2)}_{tj} = \gamma_1 x_{1j} + \gamma_2 x_{2tj} + \gamma_3 x_{1j} x_{2tj} + \eta^{(3)}_{j} + \zeta^{(2)}_{tj}, \quad \eta^{(3)}_{j} = \zeta^{(3)}_{j},
$$

where $\eta^{(3)}_j$ is a random intercept at the student level and $\zeta^{(2)}_{tj}$ an occasion specific random intercept. A path diagram of this latent growth curve model is shown in Figure 10.2 where the three latent variables $\eta^{(2)}_{1j}$, $\eta^{(2)}_{2j}$ and $\eta^{(2)}_{3j}$ represent $\eta^{(2)}_{tj}$, $t = 1, 2, 3$, and analogously for $\zeta^{(2)}_{tj}$ and $\mathbf{x}_{tj}$. We can alternatively place $\eta^{(2)}_{tj}$ and $\mathbf{x}_{tj}$ into a 'level-2' box and present the model as in Figure 10.3.

The parameter estimates for the latent growth curve model are given in Table 10.2. The treatment effect is of a similar magnitude as before. Surprisingly, there is a decline in contraceptive self-efficacy in the control group, but efficacy increases in the treatment group as expected. There are large variances both between students and between occasions within students. [Ask] and [Get] appear to be 'easier' than [Tell] since the estimates of δ_2 and δ_3 are positive. Although the factor loadings are quite close to 1, the low standard errors suggest that they should not be constrained to 1.

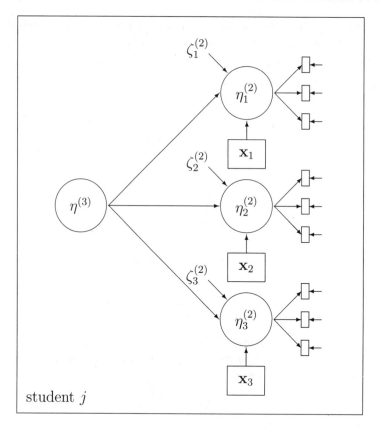

Figure 10.2 *Path diagram for latent growth curve model*

10.3 Political efficacy: Factor dimensionality and item-bias

10.3.1 Introduction

Political efficacy is a central construct in theories of political attitudes, behavior and participation. This hypothetical construct was originally introduced and operationalized by the Survey Research Center of the University of Michigan in 1952 (Campbell *et al.*, 1954). Campbell *et al.* (p.187) defined political efficacy as

"...the feeling that individual political action does have, or can have, an impact upon the political process...".

"The feeling that political and social change is possible, and that the individual citizen can play a part in bringing about this change".

Originally, the construct was operationalized in terms of four dichotomous items. In 1968, the instrument was supplemented by two additional items, giving the following six attitude statements, $i = 1, \ldots, 6$:

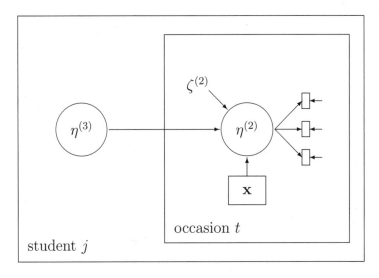

Figure 10.3 *Alternative path diagram for latent growth curve model*

- [Nosay] people like me have no say in what the government does
- [Voting] voting is the only way that people like me can have any say about how the government runs things
- [Complex] sometimes politics and government seem so complicated that a person like me cannot really understand what is going on
- [Nocare] I don't think that public officials care much about what people like me think
- [Touch] generally speaking, those we elect to Congress in Washington lose touch with the people pretty quickly
- [Interest] parties are only interested in people's votes, but not in their opinions

In addition, the dichotomous response categories were replaced by four ordered categories:

1. Disagree strongly
2. Disagree
3. Agree
4. Agree strongly

The efficacy items have been used in a number of empirical investigations. One example is the eight nation survey where the so-called 'political action data'[2] were collected (Wieken-Mayser *et al.*, 1979; see also Barnes *et al.*, 1979).

---

[2] The data used in this section were compiled by S.H.Barnes, R.Inglehart, M.K.Jennings and B.Farah and made available by the Norwegian Social Science Data Services (NSD). Neither the original collectors nor NSD are responsible for the analysis reported here.

Table 10.2 *Estimates for latent growth curve model*

Parameter	Est	(SE)
Structural model		
Regression coefficients		
γ_1 [Time]	−0.06	(0.09)
γ_2 [Treat]	−0.28	(0.24)
γ_3 [Time]×[Treat]	0.20	(0.11)
Variances		
occasion-level		
$\psi^{(2)}$	4.57	(0.41)
student-level		
$\psi^{(3)}$	3.72	(0.43)
Measurement model		
Intercepts		
δ_1 [Tell]	0	
δ_2 [Ask]	0.42	(0.08)
δ_3 [Get]	0.52	(0.08)
Factor loadings		
λ_1 [Tell]	1	
λ_2 [Ask]	1.09	(0.04)
λ_3 [Get]	0.91	(0.04)
Thresholds		
κ_1	−5.46	(0.24)
κ_2	−3.74	(0.23)
κ_3	−1.82	(0.22)
κ_4	0.87	(0.21)
Log-likelihood	−6860.96	

Following Skrondal (1996), we will consider the 1719 respondents of the 1974 cross-section of the American subsample. For the present purposes, we have included 'Don't know' responses as missing values (see also Rubin *et al.*, 1995). The univariate frequency distributions of the efficacy items are presented in Table 10.3 and the frequency distribution of the number of items with missing values is reported in Table 10.4. The 1710 respondents who responded to at least one efficacy item are analyzed here under the missing at random (MAR) assumption.

Table 10.3 *Univariate frequency distributions of the political efficacy items*

	4	3	2	1	Missing
[Nosay]	175	518	857	130	39
[Voting]	283	710	609	80	37
[Complex]	343	969	323	63	21
[Nocare]	250	701	674	57	37
[Touch]	273	881	462	26	77
[Interest]	264	762	581	31	81

Table 10.4 *Frequency distribution of number of items with missing values*

Number of missing items	0	1	2	3	4	5	6
Frequency	1554	106	26	18	4	2	9

10.3.2 Factor dimensionality and reliability

Let us first consider the factor dimensionality of the political efficacy items. For this purpose we will use an ordinal probit factor model of the form

$$\mathbf{y}_j^* = \mathbf{\Lambda}\boldsymbol{\eta}_j + \boldsymbol{\epsilon}_j.$$

We have omitted the constants so that we can identify all four thresholds $\kappa_{1i}, \ldots, \kappa_{4i}$ for each item i. In the case of a unidimensional ordinal probit factor model, we then obtain what is referred to as the *graded response model* (Samejima, 1969) in item response theory (IRT).

The unidimensional factor model provides a formalization of the concept of unidimensionality which appears to concur with the ideas of applied scientists (McDonald, 1981). When it comes to multidimensionality the picture is less clear. For instance, it is possible to formulate a number of factor models which are consistent with different notions of *bidimensionality*. Here, we will explicate four kinds of factor bidimensionality, ordered in degree from strict to weak:

1. Strict factor bidimensionality is formalized in terms of an *independent clusters* factor model where the items only measure the dimensions they are purported to measure. Thus, $\mathbf{\Lambda}$ is specified as block-diagonal. The factor dimensions are moreover a priori specified as *orthogonal*; the covariance matrix of the factors $\mathbf{\Psi}$ is diagonal. If this model is retained, the dimensional validity of the items is maximal.

2. Strong factor bidimensionality is also formalized in terms of an independent clusters factor model, but the factors are permitted to be correlated (see

top panel of Figure 10.4 for an example). Hence, this concept of strong bidimensionality is somewhat weaker than strict factor dimensionality.

3. Intermediate factor bidimensionality applies if one or more items, but not all, measure *both* dimensions. Such composite items are problematic, since they measure different phenomena.

4. Weak factor bidimensionality corresponds to an *unrestricted* or exploratory factor model (Jöreskog, 1969; Lawley and Maxwell, 1971). All items are permitted to reflect both dimensions of political efficacy (subject to iden- tification restrictions). Hence, this is the weakest possible formalization of bidimensionality in factor models. If this model is retained, the dimensional validity of the items is minimal. How to specify exploratory factor models as equivalent confirmatory factor models was discussed in Section 3.3.3; see the lower panel of Figure 10.4 for a parameterization of the exploratory two-factor model.

There is a voluminous literature on the measurement properties of the po- litical efficacy items, and a number of alternative measurement models have been proposed. Some authors have argued for the unidimensionality of efficacy, whereas the predominant position clearly favors bidimensionality. Controversy reigns, however, when it comes to which items measure what dimension.

Here, we will confine the discussion to the so-called NES (National Elec- tion Studies)-model (Miller *et al.*, 1980). Miller *et al.* (1980, p. 253) present the following interpretation of the two dimensions of political efficacy: one dimension is interpreted as "individuals' self-perceptions that they are capa- ble of understanding politics and competent enough to participate in political acts such as voting", and the other as "individuals' beliefs about political institutions rather than perceptions about their own abilities". The first di- mension is dubbed 'internal efficacy' (or personal political competence) and the second dimension is called 'external efficacy' (or political system respon- siveness). Miller *et al.* suggest that [Nosay], [Voting] and [Complex] measure internal efficacy, whereas [Nocare], [Touch] and [Interest] measure external ef- ficacy. Hence, it appears to be reasonable to interpret the NES-model as a model with strong bidimensionality (see top panel of Figure 10.4).

At this point, it is important to point out that determination of the dimen- sionality of factor models is often treated in a somewhat superficial manner in the literature. Caution should be exercised for a number of reasons. First, a problem of examining absolute fit is the multiple sources of discrepancy between model and data. Lack of fit may not exclusively be due to misspecifi- cation of the dimensionality, but may reflect any misspecifaction of the factor model. This problem is compounded with categorical data, where scalar spec- ifications of thresholds will influence the absolute fit. Second, as was pointed out in Section 8.5.2, there is a multiplicity of possibly contradictory goodness of fit criteria. Third, the equivalence problem must be faced. This can be il- lustrated from the literature on political efficacy. Craig and Maggiotto (1982) specified a two-factor model where [Nosay], [Nocare], [Touch] and [Interest] measure one dimension, whereas [Voting] and [Complex] measure the other dimension. In contrast, Mason *et al.* (1985) specify a model with three factors where the second factor of Craig and Maggiotto is split into two factors, measured by [Voting] and [Complex] respectively. Both models are identified, notwithstanding a somewhat contrived identification of the second model, and equivalent. It follows that no empirical arguments can be used in choosing between these competing two and three-dimensional conceptualizations of political efficacy (see also Section 5.1).

Keeping the above problems in mind, we now proceed to compare the NES-model $\mathcal{M}2$ with the weak dimensionality model $\mathcal{M}1$. These models are depicted in Figure 10.4. We note that $\mathcal{M}2$ is nested in $\mathcal{M}1$ since it results from setting the factor loadings λ_{41}, λ_{51}, λ_{22} and λ_{32} of the latter model to zero. Consequently, likelihood-ratio tests can be used to compare the fit of the two competing models. The maximum likelihood estimators for the different models are now obtained under the specification of bivariate normal latent traits, utilizing all available data. The estimated factor loadings and factor variances and covariances are displayed in Table 10.5 and the thresholds in Table 10.6.

Table 10.5 *Estimated factor loadings and (co)variances*

	$\mathcal{M}2$		$\mathcal{M}1$	
	Internal	External	Internal	External
Factor loadings				
λ_{1k} [Nosay]	1	0	1	0
λ_{2k} [Voting]	0.52 (0.05)	0	0.69 (0.10)	−0.18 (0.10)
λ_{3k} [Complex]	0.77 (0.07)	0	0.56 (0.08)	0.15 (0.08)
λ_{4k} [Nocare]	0	1	0.72 (0.12)	1
λ_{5k} [Touch]	0	0.74 (0.05)	−0.09 (0.17)	1.41 (0.26)
λ_{6k} [Interest]	0	0.86 (0.06)	0	1.53 (0.16)
Factor (co)variances				
ψ_{kk}	0.81 (0.09)	2.67 (0.31)	0.91 (0.13)	1.02 (0.21)
ψ_{12}	1.24 (0.10)		0.73 (0.10)	
Log-likelihood	−9950.43		−9924.04	

The two dimensions of political efficacy appear to be rather highly correlated (0.84 for $\mathcal{M}2$ and 0.76 for $\mathcal{M}1$), which is also reflected in the empirical Bayes or factor score plot in Figure 10.5. Comparing $\mathcal{M}2$ with $\mathcal{M}1$, the likelihood ratio statistic is 52.79 with 4 degrees of freedom. We see that strong bidimensionality is clearly implausible for the political efficacy items, and weak bidimensionality must be retained. Thus, the dimensional validity of the efficacy items is low. We note in passing that different models with intermediate bidimensionality have been proposed (e.g. Aish and Jöreskog, 1989), but such models will not be pursued here.

Consider next the reliabilities of the different items under the retained

Strong bidimensionality model $\mathcal{M}2$

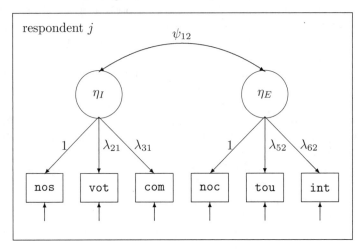

Weak bidimensionality model $\mathcal{M}1$

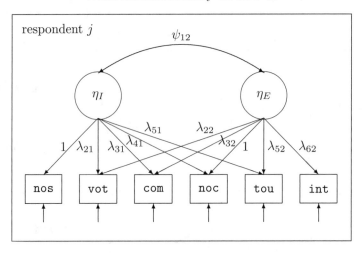

Figure 10.4 *Graphical representations of models $\mathcal{M}2$ and $\mathcal{M}1$*

model. As a development of Jöreskog (1971b) and Werts *et al.* (1974), we define ρ_i as the 'reliabilities' of the latent responses y_i^* as:

$$\rho_i = \frac{\mathbf{\Lambda}_i \mathbf{\Psi} \mathbf{\Lambda}_i'}{\mathbf{\Lambda}_i \mathbf{\Psi} \mathbf{\Lambda}_i' + \theta_{ii}}, \tag{10.1}$$

where $\mathbf{\Lambda}_i$ is the ith row of the factor loading matrix and $\theta_{ii} = 1$ for the probit models estimated in this section. As was pointed out in Section 3.3, ρ_i can be regarded as *lower bounds* of the true reliabilities, unless the specific variance

Table 10.6 *Estimated thresholds for models $\mathcal{M}2$ and $\mathcal{M}1$*

	κ_{i1}	κ_{i2}	κ_{i3}
$\mathcal{M}2$			
[Nosay]	$-1.70\ (0.06)$	$-0.28\ (0.04)$	$1.89\ (0.07)$
[Voting]	$-1.06\ (0.04)$	$0.25\ (0.04)$	$1.84\ (0.06)$
[Complex]	$-1.01\ (0.05)$	$0.91\ (0.04)$	$2.13\ (0.07)$
[Nocare]	$-2.00\ (0.10)$	$0.35\ (0.06)$	$3.36\ (0.15)$
[Touch]	$-1.53\ (0.07)$	$0.86\ (0.06)$	$3.27\ (0.12)$
[Interest]	$-1.69\ (0.07)$	$0.58\ (0.06)$	$3.50\ (0.15)$
$\mathcal{M}1$			
[Nosay]	$-1.74\ (0.08)$	$-0.29\ (0.05)$	$1.94\ (0.08)$
[Voting]	$-1.08\ (0.04)$	$0.26\ (0.04)$	$1.90\ (0.07)$
[Complex]	$-0.99\ (0.05)$	$0.90\ (0.04)$	$2.09\ (0.07)$
[Nocare]	$-1.96\ (0.10)$	$0.35\ (0.06)$	$3.30\ (0.15)$
[Touch]	$-1.63\ (0.09)$	$0.91\ (0.07)$	$3.48\ (0.16)$
[Interest]	$-1.80\ (0.10)$	$0.62\ (0.06)$	$3.72\ (0.20)$

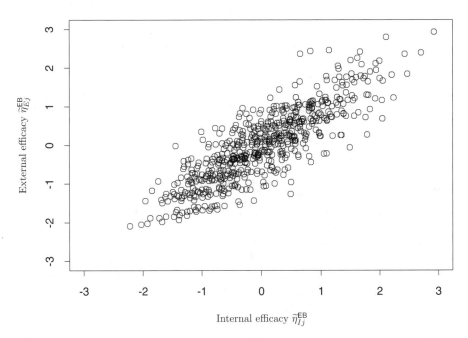

Figure 10.5 *Empirical Bayes factor scores of political efficacy ($\mathcal{M}1$)*

of item i is assumed to be zero. The estimated lower bounds of the reliabilities for $\mathcal{M}1$ are 0.48, 0.23 and 0.30 for the internal efficacy items [Nosay], [Voting] and [Complex], respectively, and 0.72, 0.65 and 0.71 for the external efficacy items [Nocare], [Touch] and [Interest]. The lower bounds are rather small for the internal efficacy items.

10.3.3 Item bias

Let us now consider the *item bias* or *differential item functioning* (DIF) of the political efficacy items. In item response theory (IRT) an item is 'biased' if the response to the item is dependent on extraneous information apart from the factors. (See also Section 9.4, page 297 for an investigation of item-bias in a dichotomous item response model.) Since item-bias is closely related to the 'fairness' of tests, it comes as no surprise that claims of racial and ethnic bias have led to a heated public debate and even lawsuits. Our analysis of item-bias is based on the validation approach introduced by Muthén (1985, 1988b, 1989d). We find this methodology more direct and elegant than the standard approaches in IRT, surveyed by e.g. Hambleton and Swaminathan (1985) and Hambleton *et al.* (1991).

We specify the model

$$\mathbf{y}_j^* = \mathbf{X}_j\boldsymbol{\beta} + \boldsymbol{\Lambda}\boldsymbol{\eta}_j + \boldsymbol{\epsilon}_j,$$
$$\boldsymbol{\eta}_j = \boldsymbol{\Gamma}\mathbf{w}_j + \boldsymbol{\zeta}_j.$$

First consider a model without item-bias, $\boldsymbol{\beta} = \mathbf{0}$, a MIMIC model where the efficacy factors are regressed on covariates. It follows from this specification that the expectations of the factors become heterogenous, and the response probabilities are no longer homogeneous. The items are nevertheless not biased, since all heterogeneity is transmitted through the factors. From our previous results on factor dimensionality, an unstructured factor model is specified. Following the discussion in Abramson (1983) and Listhaug (1989), we have selected two covariates:

- [Educ] standardized education in years
- [Black] dummy variable for being black

The resultant MIMIC model is denoted $\mathcal{M}4$, where we have specified $\gamma_{22} = 0$ ([Black] has no effect on external efficacy) based on a preliminary analysis. Discarding three respondents with missing values on either [Black] or [Educ] yields a sample size of 1707.

A model incorporating item bias is now specified as a generalized MIMIC model where the covariates have direct regression effects on some items, in addition to the indirect effects via the factors. In terms of the model parameters, this means that all elements of $\boldsymbol{\beta}$ are no longer zero. This model is called $\mathcal{M}3$. Note that $\mathcal{M}3$ is not given a priori, but on the contrary to be suggested by exploratory analysis of our data. Performing a simple cross-validation (see Section 8.5.4), our sample has been randomly divided into an exploration sample of size 840 and a confirmation sample of size 867. We are then free to delve in our exploration sample, and can subsequently test the competing models on the confirmation sample. Exploration suggests that [Educ] and [Black] have direct effects on [Voting] (β_1 and β_2) and [Educ] has a direct effect on [Complex] (β_3). Path diagrams of models $\mathcal{M}3$ and $\mathcal{M}4$ are shown in Figure 10.6.

Table 10.7 *Estimates for MIMIC models*

	$\mathcal{M}4$		$\mathcal{M}3$	
	Internal	External	Internal	External
Structural model				
Factor regressions				
γ_{k1} [Educ]	0.34 (0.04)	0.24 (0.08)	0.38 (0.07)	0.28 (0.06)
γ_{k2} [Black]	−0.25 (0.08)	0	−0.37 (9.16)	0
Factor (co)variances				
Ψ_{kk}	0.67 (0.05)	0.92 (0.23)	0.97 (0.17)	1.05 (0.15)
Ψ_{12}		0.73 (0.09)		0.69 (0.07)
Measurement model				
Factor loadings				
λ_{1k} [Nosay]	1	0	1	0
λ_{2k} [Voting]	1.67 (0.45)	−0.85 (0.31)	0.33 (0.10)	0.04 (0.10)
λ_{3k} [Complex]	1.64 (0.41)	−0.60 (0.26)	0.20 (0.09)	0.32 (0.10)
λ_{4k} [Nocare]	0.82 (0.31)	1	0.63 (0.18)	1
λ_{5k} [Touch]	−0.01 (0.30)	1.34 (0.44)	−0.11 (0.18)	1.31 (0.30)
λ_{6k} [Interest]	0	1.79 (0.43)	0	1.59 (0.25)
Item regression				
β_1 [Voting]				
× [Educ]	0		0.20 (0.05)	
β_2 [Voting]				
× [Black]	0		−0.34 (0.14)	
β_3 [Complex]				
× [Educ]	0		0.28 (0.04)	
Log-likelihood	−4978.08		−4973.36	

The parameter estimates for the confirmation sample are reported in Table 10.7, except for the thresholds which are given in Table 10.8. We note that [Black] is negatively related to internal political efficacy, whereas [Educ] is positively related to both kinds of political efficacy, as might be expected. Since model $\mathcal{M}4$ is clearly nested in $\mathcal{M}3$, likelihood-ratio tests can be performed on the confirmation sample. From Table 10.7, the likelihood ratio statistic is 9.43 with 3 degrees of freedom indicating that the model with item-bias $\mathcal{M}3$ should be retained, although the evidence is not overwhelming. It can be seen from Table 10.7 that there is substantial item bias for two of the internal effi-

MIMIC model $\mathcal{M}4$

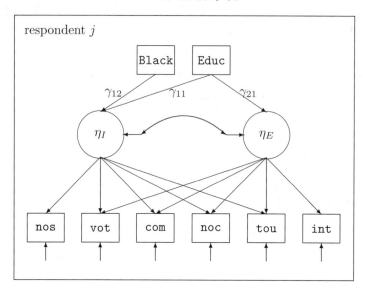

MIMIC model with item bias $\mathcal{M}3$

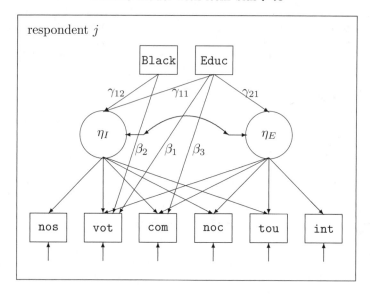

Figure 10.6 *Graphical representations of models $\mathcal{M}4$ and $\mathcal{M}3$*

Table 10.8 *Estimated thresholds*

	κ_{1i}	κ_{2i}	κ_{3i}
$\mathcal{M}4$			
[Nosay]	−1.51 (0.08)	−0.32 (0.06)	1.74 (0.08)
[Voting]	−1.15 (0.07)	0.26 (0.05)	1.92 (0.10)
[Complex]	−1.04 (0.07)	0.92 (0.06)	2.09 (0.11)
[Nocare]	−1.85 (0.12)	0.30 (0.08)	3.28 (0.17)
[Touch]	−1.52 (0.10)	0.85 (0.08)	3.19 (0.16)
[Interest]	−1.77 (0.13)	0.66 (0.09)	3.88 (0.27)
$\mathcal{M}3$			
[Nosay]	−1.75 (0.17)	−0.38 (0.08)	2.02 (0.19)
[Voting]	−1.10 (0.06)	0.24 (0.05)	1.82 (0.09)
[Complex]	−0.95 (0.06)	0.90 (0.06)	2.01 (0.09)
[Nocare]	−1.92 (0.14)	0.31 (0.09)	3.39 (0.20)
[Touch]	−1.57 (0.12)	0.89 (0.09)	3.32 (0.20)
[Interest]	−1.80 (0.15)	0.66 (0.09)	3.90 (0.29)

cacy items. [Black] has negative direct effects on the response for [Voting] and [Educ] has positive direct effects on the responses for [Voting] and [Complex].

10.3.4 Conclusion

In summary, the psychometric validity of the political efficacy items appears to be dubious. Only a weak kind of bidimensionality is retained, the reliabilities appear to be low and substantial item-bias is found. We note that the two latter problems are compounded for the internal efficacy items. The problems unmasked here may be due to the conceptual gap between measures with quasi-theoretic status on the one hand and theories subsequently developed on the other hand (Mason *et al.*, 1985). We conclude that the NES measurement instrument for political efficacy investigated here might best be abandoned. It is interesting to note in this connection that Converse (1972, p.334), one of the elder statesmen of this area, stated that:

> "The political efficacy scale with which we have worked since 1952 involves a considerable blend... ".

Finally, we observe that a new instrument for political efficacy was implemented in NES 1988 (e.g. Niemi *et al.*, 1991).

10.4 Life satisfaction: Ordinal scaled probit factor models

10.4.1 Introduction

Satisfaction of life is a phenomenon which has attracted both normative and empirical interest. The most common approach in empirical research on life-satisfaction appears to be based on people's *reported perceptions* of satisfaction (e.g. Campbell *et al.*, 1976; Andrews and Withey, 1976).

In this section, we present an empirical analysis of reported perceptions of life satisfaction among Americans, previously reported in Skrondal (1996). This methodology enables us to investigate the dimensionality of life-satisfaction and the quality of the individual items. Having obtained a retained model for life-satisfaction, the properties of the model are presented by means of a graphical procedure advocated by Lazarsfeld (1950).

The data employed here are based on the 1989 version of the General Social Survey[3] (GSS). The GSS is a cross-sectional survey of the noninstitutionalized residential population of the continental USA aged 18 and over (NORC, 1989). It has been conducted almost annually by the National Opinion Research Center (NORC) at the University of Chicago since 1972. The purpose of the GSS is to monitor social trends in attitudes and behavior.

The question wording of the life satisfaction items is:

"For each area of life I am going to name, tell me the number that shows how much satisfaction you get from that area".

Five different areas are evaluated by the respondents:

- [City] the city or place you live in
- [Hobby] your nonworking activities – hobbies and so on
- [Family] your family life
- [Friend] your friendships
- [Health] your health and physical condition

The respondents' numerical answers to the items correspond to a *rating form* labeled as

1. A very great deal

2. A great deal

3. Quite a bit

4. A fair amount

5. Some

6. A little

7. None

---

[3] The data used in this section were compiled by the National Opinion Research Center (NORC) and made available by the Norwegian Social Science Data Services (NSD). Neither NORC nor NSD are responsible for the analysis presented here.

Such a rating form is often denoted a *Likert form* in attitude measurement after Likert (1932).

The GSS has employed a split-ballot design (Smith, 1988) since 1988. Specifically, some items are permanent in all surveys, whereas others, including the life satisfaction items, are rotated among three ballots. We confine the analysis to the 1035 respondents of ballots B and C who were presented the life satisfaction items in 1989. The univariate frequency distributions of the items are presented in Table 10.9. and the frequency distribution of the number of

Table 10.9 *Univariate frequency distributions of the life satisfaction items.*

	7	6	5	4	3	2	1	Missing
[City]	178	283	217	196	63	69	22	2
[Hobby]	243	368	178	109	40	53	34	5
[Family]	433	344	102	67	22	44	15	3
[Friend]	343	397	156	78	14	32	8	2
[Health]	260	324	174	170	30	48	22	2

items with missing values is reported in Table 10.10. The 1030 respondents

Table 10.10 *Frequency distribution of the number of items with missing values*

Number of missing items	0	1	2	3	4	5
Frequency	1016	14	0	0	0	5

who responded to at least one efficacy item are analyzed here under the missing at random (MAR) assumption. We note that there are remarkably few missing values for the life satisfaction items of GSS 1989.

Specification of a graded response model would lead to the following unrestricted threshold model for each item i:

$$
y_{ij} = \begin{cases}
1 & \text{if} & y^*_{ij} \leq \kappa_{i1} \\
2 & \text{if} & \kappa_{i1} < y^*_{ij} \leq \kappa_{i2} \\
3 & \text{if} & \kappa_{i2} < y^*_{ij} \leq \kappa_{i3} \\
4 & \text{if} & \kappa_{i3} < y^*_{ij} \leq \kappa_{i4} \\
5 & \text{if} & \kappa_{i4} < y^*_{ij} \leq \kappa_{i5} \\
6 & \text{if} & \kappa_{i5} < y^*_{ij} \leq \kappa_{i6} \\
7 & \text{if} & \kappa_{i6} < y^*_{ij},
\end{cases}
$$

all in all 30 threshold parameters.

Due to the large number of response categories, this model has an excessive number of parameters. Clogg (1979) and Masters (1985) therefore collapsed response categories. Their particular collapsing was criticized by Thissen and

Steinberg (1988) who suggested a more sensible collapsing. However, a better solution is feasible here since the items are *homogeneous* in the sense that the same Likert rating form is used for all items. In this case it makes sense to constrain the thresholds to be equal across items and introduce an intercept for each item as in Section 10.2.2,

$$y_{ij}^* = \beta_i + \lambda_i \eta_j + \epsilon_{ij}, \quad \kappa_{is} = \kappa_s, \quad \kappa_1 = 0.$$

Here the intercepts for all items are identified because κ_1 is set to 0. As shown on page 147 in Section 5.2.3, setting the thresholds equal across items does not only identify the intercepts but also the relative scales of the latent responses, giving a scaled ordinal probit model (see also Section 2.3.4),

$$\epsilon_{ij} \sim N(0, \theta_{ii}), \quad \theta_{11} = 1.$$

Hence the locations and scales of the latent responses differ between the items.

10.4.2 Factor dimensionality

We consider the analysis of all five life-satisfaction items from GSS. This is in line with Muraki (1990), but in contrast to Clogg (1979), Masters (1985) and Thissen and Steinberg (1988) who, for undisclosed reasons, confined the analysis to three of the items. Clogg (1988), on the other hand, considered four of the items.

A *unidimensional* model $\mathcal{M}1$ for the life-satisfaction items is first specified, with identification restrictions

$$\kappa_1 = 0, \quad \lambda_1 = 1, \quad \theta_{11} = 1.$$

Estimated parameters and standard errors are reported in the second column of Table 10.11. Inspection of the estimated parameters strongly suggests that all diagonal elements of Θ are close to unity, apart from θ^{44}. Hence, we next specify model $\mathcal{M}2$ incorporating the restrictions

$$\theta_{22} = \theta_{33} = \theta_{55} = 1 \tag{10.2}$$

in $\mathcal{M}1$. The estimated parameters and standard errors of this model are given in the third column of Table 10.11. The log-likelihoods for models $\mathcal{M}2$ and $\mathcal{M}1$ are given in the same table. The likelihood ratio statistic is 0.96 with 3 degrees of freedom so the restrictions appear to be innocuous, although we are guilty of 'data snooping' here.

Consider now $\mathcal{M}3$, which is obtained from $\mathcal{M}1$ by imposing the restriction

$$\Theta = I,$$

or alternatively from the specification of $\theta_{44} = 1$ in $\mathcal{M}2$. The likelihood ratio statistic for comparing models $\mathcal{M}3$ and $\mathcal{M}2$ is 58.52 with 1 degree of freedom from which it follows that $\mathcal{M}3$ is clearly rejected. Thus, the residual variance seems to be considerably lower for the [Friend] item than the other items, which have the same residual variance.

Table 10.11 *Estimated parameters of models $\mathcal{M}1$ and $\mathcal{M}2$*

	$\mathcal{M}1$	$\mathcal{M}2$
Fixed part		
Thresholds		
κ_1	0	0
κ_2	1.08 (0.04)	1.10 (0.02)
κ_3	1.64 (0.05)	1.67 (0.03)
κ_4	2.25 (0.07)	2.29 (0.03)
κ_5	2.64 (0.08)	2.69 (0.04)
κ_6	3.13 (0.09)	3.19 (0.05)
Intercepts		
β_1 [City]	1.13 (0.05)	1.15 (0.04)
β_2 [Hobby]	0.82 (0.05)	0.84 (0.04)
β_3 [Family]	0.22 (0.05)	0.22 (0.05)
β_4 [Friend]	0.44 (0.04)	0.45 (0.04)
β_5 [Health]	0.81 (0.05)	0.83 (0.04)
Random part		
Factor loadings		
λ_1 [City]	1	1
λ_2 [Hobby]	1.44 (0.18)	1.43 (0.17)
λ_3 [Family]	1.98 (0.25)	1.96 (0.23)
λ_4 [Friend]	1.81 (0.22)	1.82 (0.19)
λ_5 [Health]	1.44 (0.18)	1.43 (0.17)
Factor variance		
ψ_{11}	0.48 (0.04)	0.49 (0.04)
Residual variances		
θ_{11} [City]	1	1
θ_{22} [Hobby]	0.97 (0.09)	1
θ_{33} [Family]	0.93 (0.09)	1
θ_{44} [Friend]	0.45 (0.05)	0.47 (0.05)
θ_{55} [Health]	0.93 (0.08)	1
Log-likelihood	-7669.26	-7669.75

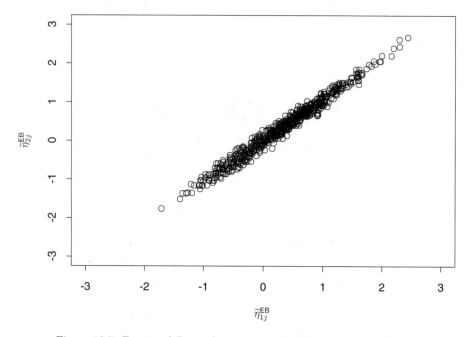

Figure 10.7 *Empirical Bayes factor scores for life-satisfaction* ($\mathcal{M}4$)

Let us now consider an alternative *bidimensional* model for the life-satisfaction items. Since we have no preconceptions regarding the patterning of the factor loadings for these items, we specify an exploratory factor model reflecting weak bidimensionality. In order to identify the model, we impose

$$\lambda_{21} = 1 = \lambda_{42},$$

and

$$\lambda_{11} = 0 = \lambda_{52}.$$

The restrictions in (10.2) are furthermore still imposed. The resulting model, incorporating the identification restrictions $\kappa_1 = 0$ and $\theta_{11} = 1$, is denoted $\mathcal{M}0$. Note that the preferred unidimensional model $\mathcal{M}2$ is nested in the two-dimensional contender $\mathcal{M}0$. The likelihoods for $\mathcal{M}2$ and $\mathcal{M}0$ are fairly similar and we conclude that the restrictions leading to the unidimensional model for life-satisfaction are acceptable and retain this model. This argument is corroborated by the empirical Bayes plot of the factor scores for the two dimensions of life-satisfaction presented in Figure 10.7. Note that the scores from $\mathcal{M}0$ are nearly linearly related, which is also reflected in the estimated correlation $\dfrac{\widehat{\psi_{21}}}{\sqrt{(\widehat{\psi_{11}\psi_{22}})}} = 0.93$. This result is even more compelling when we remember that an exploratory factor model is used.

We see that $\mathcal{M}2$ outperforms all contenders considered above and is thus chosen as our retained model. Following Lazarsfeld (1950), we now present the item characteristic curves of $\mathcal{M}2$ in Figures 10.8 and 10.9. An item characteristic curve represents a plot of the conditional response distribution (see also Section 3.3.4), here given by

$$\Pr(y_{ij}=s|\eta_j) \;=\; \Phi\left(\frac{\kappa_s - \nu_{ij}}{\sqrt{\theta_{ii}}}\right) - \Phi\left(\frac{\kappa_{s-1} - \nu_{ij}}{\sqrt{\theta_{ii}}}\right), \quad s = 1,\ldots,7,$$

where $\kappa_0 = -\infty$ and $\kappa_7 = \infty$. The curves represent the probability of responding in a particular category s for a given item i as a function of degree of life satisfaction (we have reversed the life-satisfaction scale in the figures so that more satisfaction is associated with lower response categories). It is evident from Figure 10.8 that the [Friend] item functions better than the [City] item in the sense that the response categories discriminate well between different degrees of life satisfaction. We see from Figure 10.9 that the remaining items occupy an intermediate position in this regard.

10.4.3 Reliabilities

Consider next the reliabilities of the different life satisfaction items under the unidimensional model. Analogously to the definition introduced on page 332, we define the lower bounds of the reliabilities ρ_i in the present model as:

$$\rho_i = \frac{\Lambda_i \psi_{11} \Lambda_i'}{\Lambda_i \psi_{11} \Lambda_i' + \theta_{ii}}.$$

The estimated lower bounds of the estimated reliabilities for models $\mathcal{M}1$ and $\mathcal{M}2$ are reported in the second and third columns of Table 10.12 respectively.

Table 10.12 *Estimated lower bounds of reliabilities for models $\mathcal{M}1$, $\mathcal{M}2$ and $\mathcal{M}4$*

	$\mathcal{M}1$	$\mathcal{M}2$	$\mathcal{M}4$
[City]	0.19	0.19	0.21
[Hobby]	0.33	0.33	0.35
[Family]	0.49	0.48	$0.51^=$
[Friend]	0.62	0.63	$0.51^=$
[Health]	0.34	0.33	0.35

We observe that the lower bound of the reliability of the [City] item is very low, the [Hobby] and [Health] items occupy an intermediate position, whereas the [Family] and [Friend] items stand out as most promising. The utility of the [City] item is questionable, since measurement error and/or an item specific component are of considerable magnitude.

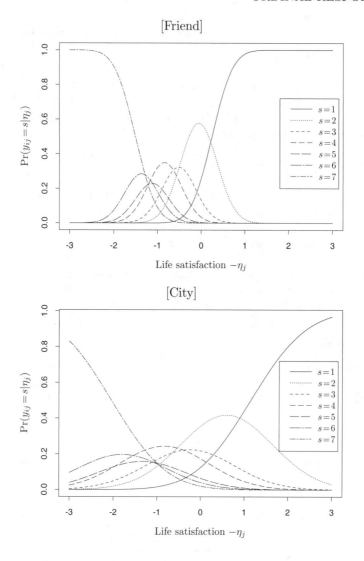

Figure 10.8 *Conditional response distribution for* [Friend] *and* [City] *items (M2)*

Keeping in mind that we are inspecting *estimated* lower bounds of reliabilities, the next question concerns whether the lower bounds of the two most promising items [Family] and [Friend] are equal. Investigating this question also presents an opportunity to demonstrate the power of the scalar parameter restriction strategy discussed in Section 4.5. We wish to impose the restriction

$$\rho_3 = \rho_4.$$

Substituting from the definition of ρ_i, and remembering that $\theta_{33} = 1$ in $\mathcal{M}2$,

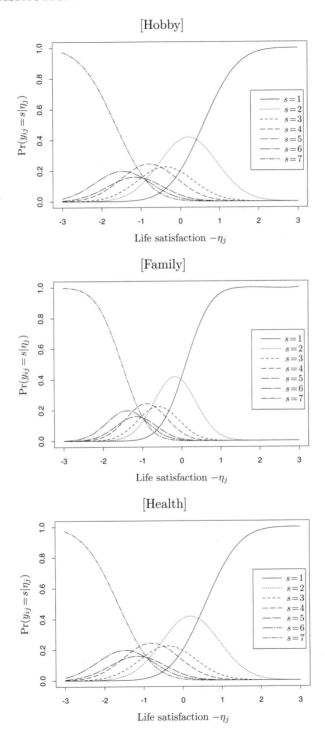

Figure 10.9 *Conditional response distribution for* [Hobby], [Family] *and* [Health] *items (*$\mathcal{M}2$*)*

we obtain

$$\frac{(\lambda_3)^2\psi_{11}}{(\lambda_3)^2\psi_{11}+1} = \frac{(\lambda_4)^2\psi_{11}}{(\lambda_4)^2\psi_{11}+\theta_{44}}.$$

Solving the above equation for θ_{44} yields

$$\theta_{44} = \left(\frac{\lambda_4}{\lambda_3}\right)^2,$$

which is clearly a *nonlinear* parameter restriction. Observe that θ_{44} is definitely not represented by a fundamental parameter in this case, since it is expressed as a function of the structural parameters λ_3 and λ_4. The resultant model, including the nonlinear parameter restriction, is denoted $\mathcal{M}4$. The translation table between structural and fundamental parameters for this model is presented in Table 10.13. Note in particular the expression for θ_{44}.

Table 10.13 *Translation table - Fundamental and structural parameters ($\mathcal{M}4$).*

Structural Parameters	Fundamental Parameters
$\kappa_{12},\kappa_{22},\kappa_{32},\kappa_{42},\kappa_{52}$	ϑ_1
$\kappa_{13},\kappa_{23},\kappa_{33},\kappa_{43},\kappa_{53}$	ϑ_2
$\kappa_{14},\kappa_{24},\kappa_{34},\kappa_{44},\kappa_{54}$	ϑ_3
$\kappa_{15},\kappa_{25},\kappa_{35},\kappa_{45},\kappa_{55}$	ϑ_4
$\kappa_{16},\kappa_{26},\kappa_{36},\kappa_{46},\kappa_{56}$	ϑ_5
μ_1	ϑ_6
μ_2	ϑ_7
μ_3	ϑ_8
μ_4	ϑ_9
μ_5	ϑ_{10}
λ_{21}	ϑ_{11}
λ_{31}	ϑ_{12}
λ_{41}	ϑ_{13}
λ_{51}	ϑ_{14}
ψ_{11}	ϑ_{15}
θ_{44}	$\vartheta_{13}/\vartheta_{12}$

Regarding the estimated parameters, we report $\widehat{\lambda}_3 = 1.95$ and $\widehat{\lambda}_4 = 1.67$, from which it follows that $\widehat{\theta}_{44} = 0.73$. The estimated standard error of $\widehat{\theta}_{44}$ can, if desired, be obtained via the delta method as described in Section 8.3.1. In fact, this was the motivation for insisting that the function h_k presented in Section 4.5 was one time differentiable. The estimated lower bounds of the reliabilities are reported in the third column of Table 10.7, where the superscript '$=$' indicates that equality restrictions are imposed.

Comparing the restricted model $\mathcal{M}4$ with $\mathcal{M}2$, the likelihood ratio statistic is 18.89 with one degree of freedom so that there is considerable evidence against the assertion of equal lower bounds for the reliabilities of the [Family] and [Friend] items.

10.4.4 Special cases

Muraki (1990) presented an IRT model which is a special case of the model discussed here. We note in passing that many special cases of the categorical factor model are equivalent to models formulated in the IRT literature (e.g. Takane and de Leeuw, 1987). Estimation of this model using the EM algorithm is implemented in the accompanying PARSCALE software written by Muraki and Bock (1993).

Regarding the Muraki-Bock model, we first observe that only the unidimensional model is accommodated. Moreover, in addition to our identification restrictions they impose $\Theta = I$ and, if the factor loadings are not all equal, $\kappa_{S-2} = a$, where a is a fixed value. Muraki and Bock consider both restrictions necessary for identification. It is evident, however, from our analysis in Section 5.2.3 that these restrictions are not required. The failure to recognize this is apparently due to an exclusive focus on the threshold structure of individual items (see Muraki, 1990, p. 64), so that the information contained in the correlation structure and the simultaneous threshold structure of the items is not used. It follows that some of their identification restrictions are empirically falsifiable. For instance, the restriction $\Theta = I$ was clearly rejected when we compared models $\mathcal{M}2$ and $\mathcal{M}3$. Similarly, we retained the unidimensional model $\mathcal{M}2$ in favor of the bidimensional $\mathcal{M}2$. A final limitation of the Muraki-Bock methodology is that covariates cannot be included. An important virtue of the analysis of identification outlined in Section 5.2 is that it can reveal that models conventionally regarded as not identified turn out to be so.

10.4.5 Conclusion

We conclude that life-satisfaction, as measured in the GSS, appears to be a unidimensional phenomenon. Regarding the quality of the individual items, the [Friend] item appears to be the best item, both in terms of discrimination and reliability. The [City] item, on the other hand, stands out in a negative direction, since measurement error and/or an item specific component are of a considerable magnitude. Hence, one should seriously consider discarding this item.

10.5 Summary and further reading

The sex education example involved a complex research design with a cluster randomized intervention coupled with multiple ordinal measurements of a latent variable repeated over time. The first model considered was a random intercept proportional odds model. Other models for clustered ordered categorical data are reviewed in Agresti and Natarajan (2001). Instead of specifying a random intercept, Wolfe and Firth (2002) allow the thresholds to vary randomly. King et al. (2003) describe a method of making ordered responses comparable across different cultures by anchoring the thresholds using vignettes.

A latent growth curve model was then used to model a latent outcome, 'contraceptive efficacy'. Such models can also be used when the latent outcome has been measured by continuous response (e.g. Rabe-Hesketh *et al.*, 2001d) or responses of mixed types (e.g. Gueorguieva and Sanacora, 2003). Skrondal *et al.* (2002) extended the latent growth curve model for the sex education data to accommodate nonignorable dropout.

The political efficacy and life-satisfaction examples served to illustrate the use of latent variable models for psychometric validation of measurement instruments with ordinal items. We investigated in what sense the instruments could be called uni- or bidimensional. Item-bias was investigated using generalized MIMIC models. The items in the life-satisfaction example have seven categories. Since the number of threshold parameter proliferates as the number of ordinal categories increases, we set thresholds equal across items but allowed the intercepts and residual variances of the latent responses to differ yielding a scaled ordinal probit model. Ordinal IRT or factor models are discussed by Johnson and Albert (1999, Chapters 6 and 7) and Moustaki (2000). Bivariate multilevel ordinal response models are developed in Grilli and Rampichini (2003).

In this chapter we have used cumulative models for ordinal responses (the proportional odds and ordinal probit model). Other possibilities include the adjacent category logit or continuation ratio logit. The latter is used to model discrete time duration data in Section 12.3. Furthermore, we did not consider latent class models for ordinal responses; see Vermunt (2001).

CHAPTER 11

Counts

11.1 Introduction

In this chapter we discuss both Poisson and binomial models for counts. Since counts are aggregated data, obtained by summing dichotomous variables representing the occurrence of an event or presence of a feature, even apparently simple data structures with one count per unit can be considered as two-level datasets. An important consequence is that there might be unobserved heterogeneity leading to overdispersion.

In the first example we consider different approaches to handling overdispersion, including the zero-inflated Poisson and zero-inflated binomial models. In a second example we estimate random coefficient models for longitudinal count data and use various model diagnostics discussed in Section 8.6. Finally, we discuss disease mapping and small area estimation using models with a spatial dependence structure for the random effects.

11.2 Prevention of faulty teeth in children: Modeling overdispersion

11.2.1 Introduction

We consider dental data on 797 Brazilian children who participated in a dental health trial (Mendonça and Böhning, 1994)[1]. Each of six schools was assigned to one of six treatments aiming to prevent tooth decay:

- [Control] no treatment
- [Educ] oral health education
- [Enrich] school diet enriched with ricebran
- [Rinse] mouthrinse with 0.2% NaF solution
- [Hygiene] oral hygiene
- [All] all four treatments above

The outcome was the number of decayed, missing or filled teeth (DMFT). In addition to school or treatment arm, there were two other covariates:

- [Male] dummy variable for child a male
- Ethnic group: (reference group 'brown')

---

[1] These data can be downloaded from `gllamm.org/books` or the Royal Statistical Society Datasets Website at `http://www.blackwellpublishing.com/rss/Volumes/av162p2.htm`.

- [White] dummy variable for white
- [Black] dummy variable for black

The observed distribution of DMFT counts (marginal over the covariates) is shown in the first two columns of Table 11.1. An obvious model to consider is the Poisson model presented in Section 2.2 with mean μ_i for child i, modeled as

$$\log(\mu_i) = \mathbf{x}_i'\boldsymbol{\beta},$$

where the covariates $\mathbf{x}_i$ are dummies for the treatment arms, sex and ethnic groups. The third column of Table 11.1 shows the predicted frequencies for this model, again marginal over the covariates. The predicted frequencies were obtained by computing the probability for each possible count from 0 to 20 for each of the observed covariate pattern, multiplying the probability by the total number of observations and then aggregating over the covariates. Although all available covariates have been included in the model, there are still large discrepancies between observed and expected counts. The largest discrepancy is for zero DMFT (231 observed compared with 134 expected).

11.2.2 Modeling overdispersion

Because of the large number of observed zeros, Böhning *et al.* (1999) estimated a *zero-inflated Poisson* (ZIP) model (Lambert, 1992) for these data. A ZIP model is a mixture of two Poisson distributions, one having zero mean and the other having a mean that depends on covariates, giving the response probability

$$\Pr(y_i|\mathbf{x}_i) = \pi_1 g(y_i; \mu_i=0) + \pi_2 g(y_i; \mu_i=\exp(\mathbf{x}_i'\boldsymbol{\beta})), \qquad (11.1)$$

where π_1 and $\pi_2 = 1-\pi_1$ are the component weights or latent class probabilities and $g(y_i; \mu_i)$ is the Poisson probability for count y_i with mean μ_i,

$$g(y_i; \mu_i) = \frac{\mu_i^{y_i} \exp(-\mu_i)}{y_i!}.$$

When $y_i > 0$, the first term in (11.1) is zero, so units with counts greater than zero belong to the second class. However, when $y_i = 0$, both terms are greater than zero because a zero count can result from a Poisson model with mean zero (class 1) or mean greater than zero (class 2). Therefore units with counts equal to zero could belong to either class. For instance, if the count is the number of drinks consumed in the last two weeks, a zero response could be from a teetotaller (who never drinks) or from a drinker who happened not to drink during the period. In ZIP models the latent class probability π_1 determines the number of excess zeros compared with an ordinary Poisson model.

It may however not be necessary to allow explicitly for an excess number of zeros in this way since other forms of overdispersion are also consistent with larger numbers of zeros. We therefore also consider random intercept models

$$\ln(\mu_i) = \mathbf{x}_i'\boldsymbol{\beta} + \zeta_i,$$

where ζ_i is either $N(0, \psi)$ or discrete,

$$\zeta_i = e_c, \quad c = 1, \ldots, C$$

with probabilities π_c. In this example, $C = 2$ gave the largest likelihood and therefore corresponds to the nonparametric maximum likelihood estimator (NPMLE) (see Section 4.4.2). Note that the intercept ζ_i varies at level 1 here, unlike other applications in the book where it varies at higher levels.

Table 11.1 *Observed and expected frequencies (marginal w.r.t. covariates and random effects) for Poisson models with different kinds of overdispersion*

DMFT Count	Observed Frequency	Expected Frequencies			
		Poisson	Normal Intercept	Two-class Intercept	ZIP
0	231	133.96	186.35	227.19	227.04
1	163	229.64	221.64	168.24	148.44
2	140	205.75	165.50	147.66	165.05
3	116	128.14	101.72	115.94	125.70
4	70	62.21	56.95	73.12	73.63
5	55	25.03	30.51	38.06	35.31
6	22	8.66	16.06	16.92	14.42
7	0	2.64	8.44	6.59	5.15
8	0	0.72	4.46	2.29	1.64
9	0	0.18	2.39	0.72	0.47
10	0	0.04	1.29	0.21	0.12
11	0	0.01	0.71	0.06	0.03
12	0	0.00	0.40	0.01	0.01
13	0	0.00	0.23	0.00	0.00
14	0	0.00	0.13	0.00	0.00
15	0	0.00	0.08	0.00	0.00
16	0	0.00	0.05	0.00	0.00
17	0	0.00	0.03	0.00	0.00
18	0	0.00	0.02	0.00	0.00
19	0	0.00	0.01	0.00	0.00
20	0	0.00	0.01	0.00	0.00

Estimates for the conventional Poisson model, the two-class (NPMLE) and normal random intercept Poisson models and the ZIP model are shown in Table 11.2 with corresponding expected frequencies given in Table 11.1. The BIC and AIC defined in Sections 8.4.2 and 8.4.3, respectively, are also reported in

Table 11.2, where we have used the number of children (797) for N and the number of estimated parameters for v. (As discussed in Section 8.4.2, determining N and v for AIC and BIC is not obvious in latent variable models.) The smallest AIC and BIC are shown in bold in the table.

According to the AIC, the two-class random intercept model provides the best fit, but according to the BIC, the ZIP model provides a better fit. The discrepancy between observed and expected frequencies is particularly large for five DMFTs, even for the best-fitting models.

Table 11.2 *Estimates for various Poisson models*

Parameter	Poisson Est (SE)	Normal Intercept Est (SE)	Two-class Intercept Est (SE)	ZIP Est (SE)
Regression coefficients				
β_0 [Cons]	0.76 (0.07)	0.63 (0.09)		0.94 (0.08)
Treatment:				
β_1 [Educ]	−0.23 (0.09)	−0.23 (0.11)	−0.24 (0.11)	−0.22 (0.09)
β_2 [Enrich]	−0.09 (0.09)	−0.09 (0.11)	−0.08 (0.09)	−0.06 (0.09)
β_3 [Rinse]	−0.35 (0.08)	−0.37 (0.11)	−0.26 (0.10)	−0.22 (0.09)
β_4 [Hygiene]	−0.30 (0.09)	−0.32 (0.11)	−0.22 (0.11)	−0.23 (0.10)
β_5 [All]	−0.59 (0.10)	−0.61 (0.12)	−0.49 (0.11)	−0.47 (0.11)
Sex:				
β_6 [Male]	0.13 (0.05)	0.13 (0.07)	0.10 (0.06)	0.10 (0.06)
Ethnicity:				
β_7 [White]	0.09 (0.06)	0.10 (0.07)	0.09 (0.07)	0.08 (0.06)
β_8 [Black]	−0.14 (0.09)	−0.16 (0.11)	−0.12 (0.10)	−0.12 (0.10)
Variance				
ψ		0.29 (0.05)		
Log odds parameter				
ϱ_0^1			−0.86 (0.21)	−1.39 (0.12)
Location parameters				
e_1			−1.11 (0.37)	$-\infty$
e_2			1.04 (0.09)	
Log-likelihood	−1469.05	−1432.53	−1406.03	−1410.27
AIC	2956.04	2885.07	**2834.06**	2840.54
BIC	3058.35	2998.68	2959.04	**2954.16**

The parameter estimates in Table 11.2 represent estimated adjusted log ratios of the expected numbers of DMFTs. For instance, for the ZIP model, the sex and ethnicity adjusted ratio of the expected count in the group receiving all treatments [All] divided by the expected count in the control group is estimated as $\exp(-0.47) = 0.63$. The other treatments also reduce the expected number of DMFTs, but [All] is the most effective and [Enrich] has a negligible effect. Comparing the ordinary Poisson model with the normal random intercept Poisson model, we see that the effects of the covariates are almost identical. This is because for a random intercept model with a log-link, the conditional effects equal the marginal effects; see Section 4.8.1.

Use of a Poisson distribution is questionable here where the count represents the number of decayed, missing or filled teeth ('successes') out of a total of eight deciduous molars ('trials'). We will therefore instead consider models assuming a binomial distribution with denominator 8 for the counts. Table 11.3 shows that a simple binomial logistic regression model does not produce a sufficiently large expected frequency of zero counts.

To handle this problem we introduce a *zero-inflated binomial* (ZIB) model analogous to the ZIP model discussed above. The ZIB model is a mixture of two binomial distributions, one having probability parameter equal to zero and the other having probability parameter depending on covariates via a logit link, giving the response probability

$$\Pr(y_i|\mathbf{x}_i) \;=\; \pi_1 g(y_i; \mu_i{=}0) + \pi_2 g(y_i; \mathrm{logit}(\mu_i){=}\mathbf{x}_i'\boldsymbol{\beta}), \qquad (11.2)$$

where $g(y_i; \mu_i)$ is now a binomial probability with parameter μ_i and denominator 8,

$$g(y_i; \mu_i) \;=\; \frac{y!}{y!(8-y)!} \mu_i^{y_i}(1-\mu_i)^{8-y_i}.$$

Similarly, we can estimate binomial logistic regression models with normal or nonparametric random intercepts. The estimates for these models are given in Table 11.4 and the expected frequencies in Table 11.3.

Table 11.3 *Observed and predicted frequencies (marginal w.r.t. covariates and random effects) for binomial logistic regression models with different kinds of overdispersion*

DMFT Count	Observed Frequency	Predicted Frequencies			
		Binomial	Normal Intercept	Three-class Intercept	ZIB
0	231	107.21	207.22	226.73	227.88
1	163	230.82	202.14	170.45	120.50
2	140	233.84	149.66	143.67	174.21
3	116	145.04	101.13	108.65	150.21
4	70	59.90	64.65	78.02	84.21
5	55	16.76	38.81	45.52	31.33
6	22	3.08	21.09	18.70	7.53
7	0	0.34	9.51	4.71	1.06
8	0	0.02	2.78	0.55	0.07

Table 11.4 *Estimates for various binomial logistic regression models*

Parameter	Binomial Est (SE)	Normal Intercept Est (SE)	Three-class Intercept Est (SE)	ZIB Est (SE)
Regression coefficients				
β_0 [Cons]	−1.00 (0.08)	−1.29 (0.14)		−0.69 (0.09)
Treatment:				
β_1 [Educ]	−0.32 (0.10)	−0.35 (0.17)	−0.41 (0.17)	−0.32 (0.11)
β_2 [Enrich]	−0.12 (0.10)	−0.15 (0.17)	−0.16 (0.16)	−0.09 (0.11)
β_3 [Rinse]	−0.47 (0.10)	−0.57 (0.16)	−0.49 (0.16)	−0.26 (0.11)
β_4 [Hygiene]	−0.40 (0.10)	−0.50 (0.18)	−0.36 (0.17)	−0.29 (0.11)
β_5 [All]	−0.76 (0.11)	−0.91 (0.17)	−0.80 (0.17)	−0.59 (0.13)
Sex:				
β_6 [Male]	0.17 (0.06)	0.20 (0.10)	0.16 (0.09)	0.13 (0.07)
Ethnicity:				
β_7 [White]	0.13 (0.07)	0.14 (0.11)	0.13 (0.10)	0.11 (0.08)
β_8 [Black]	−0.18 (0.10)	−0.23 (0.16)	−0.17 (0.15)	−0.14 (0.12)
Variance				
ψ		1.05 (0.12)		
Log odds parameters				
ϱ_0^1			−1.01 (0.40)	−1.16 (0.10)
ϱ_0^2			0.27 (0.30)	
Location parameters				
e_1			−32.09*	−∞
e_2			−1.53 (0.26)	
e_3			−0.06 (0.18)	
Log-likelihood	−1546.78	−1409.39	−1397.53	−1431.09
AIC	3111.56	2838.78	2821.07	2882.19
BIC	2113.81	2952.40	2968.77	2995.80

*Boundary solution

It is interesting to note that the binomial logistic normal random intercept model fits considerably better than the Poisson normal random intercept
model (log-likelihood −1409.39 versus −1432.53 with the same number of parameters), mainly because the latter produces quite large expected frequencies
for large counts whereas the former cannot generate any counts exceeding 8.
The binomial logistic NPML solution has three classes (or masses) and this
model fits best among all models considered according to the AIC, but not according to the BIC, according to which the normal random intercept binomial
logistic model would be chosen.

The parameter estimates for the binomial logistic models represent estimated log odds ratios. For instance, for the three-class model, the sex and ethnicity adjusted ratio of the odds of having a DMFT in the [All] group divided by the odds of having a DMFT in the control group is estimated as $\exp(-0.80)=0.45$. Again, all treatments appear to be beneficial, although the effect of [Enrich] appears to be negligible.

Comparing the estimates for random intercept binomial models, which represent conditional effects given the random intercept, with the estimates for the ordinary binomial model, which represent marginal effects, the attenuation discussed in Section 4.8.1 is evident. Note that the estimates for the binomial NPML model represent a boundary solution since the first class has an estimated location of -32.09, corresponding to a binomial probability parameter of virtually zero. The log-odds of belonging to this class is estimated as -1.01, similar to the log-odds of -1.16 of belonging to the zero-probability class in the ZIB model (the corresponding probabilities are 0.27 and 0.24).

11.3 Treatment of epilepsy: A random coefficient model

11.3.1 Introduction

The longitudinal epilepsy data from Leppik *et al.* (1987), have previously been analyzed by Thall and Vail (1990), Breslow and Clayton (1993), Lindsey (1999), Diggle *et al.* (2002) and many others. The data[2] come from a randomized controlled trial comparing an anti-epileptic drug with placebo. For each patient the number of epileptic seizures was recorded during a baseline period of eight weeks. Patients were then randomized to treatment with progabide or to placebo (in addition to standard chemotherapy). The outcomes are counts of epileptic seizures during the two weeks before each of four consecutive clinic visits. Breslow and Clayton considered the following covariates:

- [Lbas] logarithm of a quarter of the number of seizures in the eight weeks preceding entry into the trial
- [Treat] dummy variable for treatment group
- [LbasTrt] interaction between two variables above
- [Lage] logarithm of age
- [V4] dummy for visit 4
- [Visit] time at visit, coded as -0.3, -0.1, 0.1 and 0.3

11.3.2 Modelling repeated counts

Model II in Breslow and Clayton is a log-linear (Poisson regression) model including all the covariates listed above except [Visit] as well as a random intercept for subjects. The seizure count y_{ij} for subject j at visit i is assumed to be conditionally Poisson distributed with mean μ_{ij} modeled as

$$\nu_{ij} = \log(\mu_{ij}) = \mathbf{x}'_{ij}\boldsymbol{\beta} + \zeta_{1j}.$$

[2] The data can be downloaded from `gllamm.org/books`

The subject-specific random intercept ζ_{1j} is assumed to have a normal distribution with zero mean and variance ψ_{11}.

Model IV in Breslow and Clayton includes the predictor [Visit] instead of [V4] and has a random slope ζ_{2j} of [Visit] in addition to the random intercept,

$$\nu_{ij} = \log(\mu_{ij}) = \mathbf{x}'_{ij}\boldsymbol{\beta} + \zeta_{1j} + \zeta_{2j}z_{ij}.$$

The intercept and slope are assumed to have a bivariate normal distribution with variances ψ_{11} and ψ_{22}, respectively, and covariance ψ_{21}.

Table 11.5 *Parameter estimates and standard errors for Models II and IV using PQL-1 (Breslow and Clayton, 1993) and maximum likelihood using adaptive Gaussian quadrature*

	Model II		Model IV	
	PQL-1	AGQ	PQL-1	AGQ
	Est (SE)	Est (SE, SE_R)[†]	Est (SE)	Est (SE, SE_R)[†]
Fixed effects				
β_0 [Cons]	−1.25 (1.2)	2.11 (0.22, 0.21)	−1.27 (1.2)	2.10 (0.22, 0.21)
β_1 [Lbas]	0.87 (0.14)	0.88 (0.13, 0.11)	0.87 (0.14)	0.89 (0.13, 0.11)
β_2 [Treat]	−0.91 (0.41)	−0.93 (0.40, 0.40)	−0.91 (0.41)	−0.93 (0.39, 0.40)
β_3 [LbasTrt]	0.33 (0.21)	0.34 (0.20, 0.20)	0.33 (0.21)	0.34 (0.20, 0.20)
β_4 [Lage]	0.47 (0.35)	0.48 (0.35, 0.30)	0.46 (0.36)	0.48 (0.35, 0.33)
β_5 [V4]	−0.16 (0.05)	−0.16 (0.05, 0.07)		
β_6 [Visit]			−0.26 (0.16)	−0.27 (0.16, 0.17)
Random effects				
$\sqrt{\psi_{11}}$	0.53 (0.06)	0.50 (0.06, 0.06)	0.52 (0.06)	0.50 (0.06, 0.06)
$\sqrt{\psi_{22}}$			0.74 (0.16)	0.73 (0.16, 0.16)
ψ_{21}			−0.01 (0.03)	0.00 (0.09, 0.11)
Log-likelihood		−665.29		−655.68

[†] SE_R denotes 'robust' standard errors based on the sandwich estimator

Rabe-Hesketh *et al.* (2002) showed that the parameters of these models can be reliably estimated using adaptive quadrature whereas ordinary quadrature is extremely unstable (see Section 6.3.2 for a description of quadrature methods). We therefore use adaptive quadrature with 15 points for Model II and 8 points per dimension for Model IV. Maximum likelihood estimates and standard errors are shown in Table 11.5 together with PQL-1 estimates (see Section 6.3.1) reported by Breslow and Clayton. The estimates produced by adaptive quadrature and PQL-1 were very similar (the constants are not comparable since we have centered the predictors around their means), which is reassuring since PQL-1 is expected to work well in this particular case.

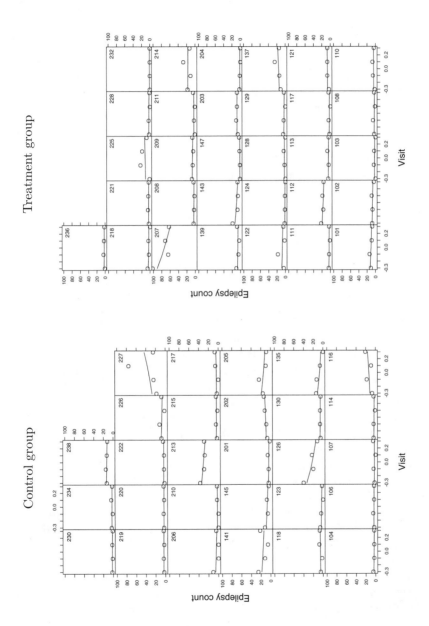

Figure 11.1 *Observed and predicted epilepsy counts versus* [Visit] *with subject numbers*

Note that the parameter estimates for Model II reported in Yau and Kuk (2002) using 16-point ordinary quadrature are considerably different from the PQL-1 and adaptive quadrature estimates (for example the treatment effect estimate is -0.52), suggesting that their solution is not reliable. We also report 'robust' standard errors SE_R based on the sandwich estimator (see Section 8.3.3) for the maximum likelihood estimates.

Figure 11.1 shows growth curves of the predicted epilepsy counts over visits for each subject by treatment group for Model IV together with observed counts shown as circles. Here, the predicted counts are the posterior means of the exponential of the linear predictor,

$$E_\zeta[\exp(\mathbf{x}'_{ij}\boldsymbol{\beta} + \zeta_{1j} + \zeta_{2j}z_{ij})|\mathbf{y}_j, \mathbf{X}_j, \mathbf{z}_j].$$

As was pointed out in Section 7.8, we must integrate the above exponential function with respect to the posterior distribution to obtain the expectation and cannot simply plug in the empirical Bayes predictions $\tilde{\zeta}_{1j}$ and $\tilde{\zeta}_{2j}$ in the exponential.

The variability in slopes is most apparent for subjects with larger counts. The observed counts of subject 227 deviate substantially from the predicted counts, particularly at visit 3.

11.3.3 Model diagnostics

We now consider model diagnostics of the kind discussed in Section 8.6 for Model II which only included a random intercept.

Normality of the random effects can be assessed by estimating the model with a nonparametric random intercept distribution. The NPML solution (not mean-centered) has six masses at -30, 1.00, 1.75, 2.03, 2.37 and 2.90 with probabilities 0.02, 0.15, 0.45, 0.22, 0.15 and 0.09, respectively (note that Yau and Kuk (2002) only found four masses). The five-mass solution was a boundary solution with a very large negative location for one of the classes, giving an expected count of zero for that class. To avoid a very flat log-likelihood and obtain the NPMLE, this location was fixed at -30. Figure 11.2 shows the predicted counts (exponentials of the locations) when all the mean-centered covariates are zero. The estimated distribution is highly asymmetric (also on the log scale), suggesting that the normality assumption may be dubious. However, the log-likelihood based on normality, -665.29, is not much lower than that from NPMLE, -655.06, when taking into account that 9 extra parameters are estimated.

The existence of masses at predicted counts of 0 and 18 suggests that there may be outlying subjects who could be influential. To assess influence we therefore computed Cook's distances (see Section 8.6.6) for the model assuming a normal random intercept. We also obtained standardized predictions of the random intercepts, so-called standardized empirical Bayes residuals, using the approximate sampling standard deviation in equation (7.8) on page 232 for the standardization. The standardized residuals were computed using both

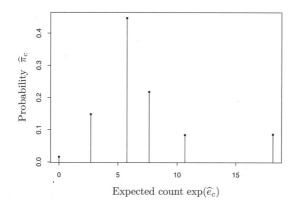

Figure 11.2 *Estimated probabilities* $\widehat{\pi}_c$ *and expected counts* $\exp(\widehat{e}_c)$ *for components* $c = 1, \ldots, 5$ *when* $\mathbf{x}_{ij} = \mathbf{0}$

the parameter estimates $\widehat{\theta}_{(-j)}$ when subject j is deleted and the estimates based on the full sample, $\widehat{\theta}$.

These diagnostics are reported in Table 11.6 for the six subjects whose Cook's distances exceed 1 as well as two subjects with lower distances but with standardized residuals exceeding 2 in absolute value. For these subjects we also give DFBETAS for [Treat], [V4] and the random intercept standard deviation $\sqrt{\widehat{\psi}_{11}}$. These were obtained by actually deleting the subject and re-estimating the parameters instead of using the approximate one-step method described in Section 8.6.6. We also show the corresponding responses $\mathbf{y}_j$, pre-treatment seizure count divided by 4 [Base] and treatment group [Treat].

Since there were 59 subjects, the expected number of standardized residuals exceeding an absolute value of 2.39 is about one if the residuals are standard normally distributed. Therefore, subjects with standardized empirical Bayes predictions exceeding this value may be considered outliers (shown in bold for subjects 225 and 232). As expected, subjects 225 and 232 therefore also have large DFBETAS for the random intercept standard deviation and [Treat] (since [Treat] is a between-subject covariate). Subjects 135, 227, 206 and 112 are also possible outliers in the random intercept distribution and have considerable influence on the estimate of the random intercept standard deviation. Subject 207 has a relatively small residual and small influence on the standard deviation although the responses are extremely high. This is probably because the baseline count (the log of which is a covariate) was also considerably higher than for subjects 225 and 232. Nevertheless, removing subject 207, who was in the treatment group, would substantially reduce the estimate of the treatment effect as reflected in the large DFBETAS for [Treat].

Table 11.6 *Influence statistics (Cook's D and DFBETAS) and various residuals*

Subj.	[Base]	y_j				[Treat]	Cook's D	DFBETAS			Normal		NPML	
								[Treat]	[V4]	$\sqrt{\psi_{11}}$	$\frac{\tilde{\zeta}_{j(-j)}}{\sigma_{j(-j)}}$	$\frac{\tilde{\zeta}_j}{\sigma_j}$	ω_1	ω_6
126	13.0	40	20	23	12	0	1.10	−0.02	**0.52**	0.02	1.04	0.89	0.00	0.00
135	2.5	14	13	6	0	0	1.52	0.39	**0.40**	**−0.33**	2.23	1.97	0.00	**0.99**
227	13.8	18	24	76	25	0	1.46	−0.14	**0.39**	**−0.33**	2.19	1.93	0.00	**1.00**
207	37.8	102	65	72	63	1	1.68	**0.58**	0.24	−0.16	1.97	1.37	0.00	**1.00**
225	5.5	1	23	19	8	1	1.05	**−0.23**	0.18	**−0.44**	**2.47**	2.26	0.00	**1.00**
232	3.3	0	0	0	0	1	1.57	**0.34**	0.00	**−0.44**	**−2.92**	−2.77	**0.94**	0.00
206	12	11	0	0	5	0	0.52	0.13	−0.08	**−0.32**	−2.10	−1.91	0.00	0.00
112	7.75	22	17	19	16	1	0.72	−0.03	0.02	**−0.32**	2.26	2.07	0.00	**1.00**

Furthermore, removing subject 207 virtually eliminates the interaction effect of [LbasTrt]. Subjects 126, 135 and 227 all have large influence on the estimate of [V4]. This is because these subjects had a marked drop in epilepsy count at the fourth visit.

For the NPMLE, we now consider the posterior probabilities of the smallest and largest locations, denoted ω_1 and ω_6. All subjects in the table except 126 and 206 have posterior probabilities close to 1 of belonging to one of these extreme classes; subjects 126 and 206 have posterior probabilities of 0.99 and 1.00 of belonging to classes 5 and 2, respectively. For subjects not shown in the table, the largest value of ω_6 was 0.03 and p_1 was zero for everyone. Simulations could be used to attach p-values to the various measures of 'outlyingness'.

11.4 Lip cancer in Scotland: Disease mapping

11.4.1 Introduction

We now consider models for disease mapping or small area estimation. Clayton and Kaldor (1987) presented and analyzed data on lip cancer for each of the 56 (pre-reorganization) counties of Scotland over the period 1975-1980. These data[3] have also been analyzed by Breslow and Clayton (1993) and Leyland (2001) among many others. The number of observed lip cancer cases, the expected number of cases and crude standardized mortality ratios (SMR) are presented in Table 11.7.

Table 11.7: Observed and expected numbers of lip cancer cases and various SMR estimates (in percent) for Scottish counties

County	#	Obs o_j	Exp e_j	Crude SMR	Norm.	NPML	Spatial IGAR Est	95% CI
Skye.Lochalsh	1	9	1.4	652.2	470.8	342.6	412.3	305.5, 492.2
Banf.Buchan	2	39	8.7	450.3	421.8	362.4	430.4	408.4, 444.3
Caithness	3	11	3.0	61.8	309.4	327.1	351.4	306.7, 394.9
Berwickshire	4	9	2.5	355.7	295.2	321.6	230.3	162.9, 281.8
Ross.Cromarty	5	15	4.3	352.1	308.5	327.6	321.8	277.2, 357.3
Orkney	6	8	2.4	333.3	272.1	311.1	332.4	283.4, 381.9
Moray	7	26	8.1	320.6	299.9	322.2	303.2	275.9, 324.4
Shetland	8	7	2.3	304.3	247.8	292.5	311.1	274.5, 353.5
Lochaber	9	6	2.0	303.0	238.9	280.1	231.2	190.1, 271.0
Gordon	10	20	6.6	301.7	279.1	319.9	285.5	261.3, 304.4
W.Isles	11	13	4.4	295.5	262.5	315.5	299.1	264.7, 335.6
Sutherland	12	5	1.8	279.3	219.2	254.3	304.1	261.7, 354.9
Nairn	13	3	1.1	277.8	198.4	222.7	266.2	215.0, 317.0
Wigtown	14	8	3.3	241.7	210.9	249.6	159.7	115.0, 194.6
NE.Fife	15	17	7.8	216.8	204.6	245.3	173.2	140.5, 194.0
Kincardine	16	9	4.6	197.8	178.9	171.4	190.9	167.9, 214.5

[3] The data can be downloaded from gllamm.org/books

Table 11.7: – continued

County	#	Obs o_j	Exp e_j	Crude SMR	Predicted SMRs			
					Norm.	NPML	Spatial IGAR Est	95% CI
Badenoch	17	2	1.1	186.9	151.9	163.2	207.7	169.7, 247.8
Ettrick	18	7	4.2	167.5	154.7	136.7	126.9	102.9, 153.5
Inverness	19	9	5.5	162.7	154.2	128.4	210.4	179.7, 252.3
Roxburgh	20	7	4.4	157.7	149.0	130.3	145.0	120.2, 172.6
Angus	21	16	10.5	153.0	147.8	117.1	138.8	125.2, 150.1
Aberdeen	22	31	22.7	136.7	135.0	116.4	145.0	139.0, 156.8
Argyll.Bute	23	11	8.8	125.4	123.3	116.4	111.5	96.5, 126.1
Clydesdale	24	7	5.6	124.6	122.9	116.7	79.7	63.7, 95.0
Kirkcaldy	25	19	15.5	122.8	121.6	116.4	123.9	115.1, 132.1
Dunfermline	26	15	12.5	120.1	119.1	116.3	108.4	94.6, 118.8
Nithsdale	27	7	6.0	115.9	116.3	115.5	99.8	79.9, 115.2
E.Lothian	28	10	9.0	111.6	111.4	115.9	108.7	94.7, 123.6
Perth.Kinross	29	16	14.4	111.3	111.1	116.3	116.9	105.4, 129.5
W.Lothian	30	11	10.2	107.8	108.5	115.9	78.9	64.9, 91.1
Cumnock-Doon	31	5	4.8	105.3	107.2	111.5	91.2	74.0, 108.4
Stewartry	32	3	2.9	104.2	109.1	109.4	107.4	82.6, 131.7
Midlothian	33	7	7.0	99.6	102.7	113.1	90.4	75.1, 104.3
Stirling	34	8	8.5	93.8	97.3	112.9	74.2	60.8, 85.7
Kyle.Carrick	35	11	12.3	89.3	92.2	114.1	91.1	79.1, 104.5
Inverclyde	36	9	10.1	89.1	92.5	112.4	84.4	73.7, 95.7
Cunninghame	37	11	12.7	86.8	89.7	113.3	81.9	71.4, 90.9
Monklands	38	8	9.4	85.6	89.5	109.4	63.2	50.1, 74.1
Dumbarton	39	6	7.2	83.3	89.4	104.9	78.1	66.0, 90.8
Clydebank	40	4	5.3	75.9	85.6	94.5	61.5	50.4, 73.0
Renfrew	41	10	18.8	53.3	59.1	40.6	56.6	49.6, 64.8
Falkirk	42	8	15.8	50.7	57.9	40.9	61.8	53.8, 70.7
Clackmannan	43	2	4.3	46.3	68.8	65.5	75.1	63.0, 89.8
Motherwell	44	6	14.6	41.0	50.6	37.5	50.1	43.5, 57.6
Edinburgh	45	19	50.7	37.5	40.8	36.2	46.6	41.8, 55.9
Kilmarnock	46	3	8.2	36.6	53.2	42.2	56.7	47.5, 67.6
E.Kilbride	47	2	5.6	35.8	57.9	49.7	51.2	40.6, 61.5
Hamilton	48	3	9.3	32.1	48.5	38.8	45.4	38.3, 54.8
Glasgow	49	28	88.7	31.6	33.8	36.2	37.0	33.8, 41.2
Dundee	50	6	19.6	30.6	39.8	36.2	57.4	46.3, 77.9
Cumbernauld	51	1	3.4	29.1	63.1	57.8	54.0	43.4, 66.8
Bearsden	52	1	3.6	27.6	61.1	55.4	48.3	37.9, 60.3
Eastwood	53	1	5.7	17.4	46.4	40.6	42.2	34.5, 52.7
Strathkelvin	54	1	7.0	14.2	40.8	37.8	43.2	33.9, 53.6
Tweeddale	55	0	4.2	0.0	43.2	40.8	73.0	55.1, 97.0
Annandale	56	0	1.8	0.0	64.9	60.6	71.3	56.0, 87.1

Source: Clayton and Kaldor (1987)

The expected number of lip cancer cases is based on the age-specific lip cancer rates for Scotland and the age distribution of the counties. The SMR for a county is defined as the ratio of the mortality rate to that expected if the age-specific mortality rates were equal to those of a reference population (e.g. Breslow and Day, 1987). The crude estimate of the SMR for county j is obtained using

$$\widehat{\text{SMR}}_j = \frac{o_j}{e_j},$$

where o_j is the observed number of cases and e_j the expected number. These estimates are shown in Table 11.7 under 'Crude SMR' and a map of the crude SMRs is shown in Figure 11.3.

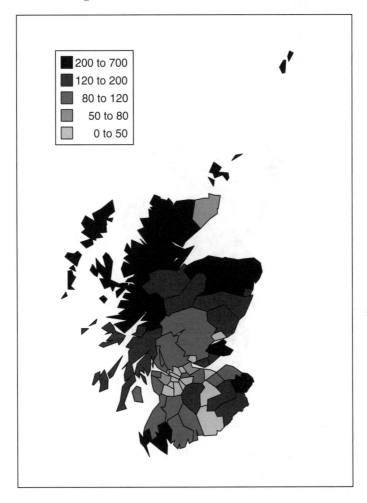

Figure 11.3 *Crude SMR in percent*

There are two important limitations of crude SMR's. First, crude SMR estimates for counties with small populations are very imprecise. Second, crude SMR's do not take into account that geographically close areas tend to have similar disease rates. The first problem can be addressed using random intercept Poisson models, yielding Bayesian or empirical Bayesian predictions of SMR's that are shrunken towards the overall SMR, thereby borrowing strength from other counties. The second problem can be addressed by allowing the random intercepts to be spatially correlated.

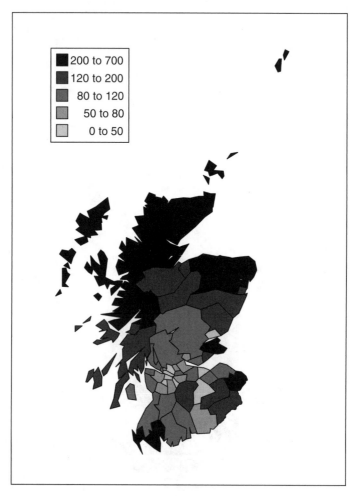

Figure 11.4 *SMR assuming normally distributed random effects (no covariate)*

11.4.2 Random intercept models

We first consider a simple random intercept model. The observed number of lip cancer cases in county j is be assumed to have a Poisson distribution with mean μ_j,

$$\ln(\mu_j) \;=\; \ln(e_j) + \beta_0 + \zeta_j,$$

where $\ln(e_j)$ is an offset (a covariate with regression coefficient set to 1), and ζ_j is a random intercept representing unobserved heterogeneity between counties. The empirical Bayes predictions of the SMRs,

$$\widetilde{\mathrm{SMR}}_j \;=\; \mathrm{E}_\zeta[\exp(\widehat{\beta}_0 + \zeta_j)|o_j, e_j],$$

will then be shrunken towards the average SMR, providing more stable values for counties with smaller populations.

The parameter estimates assuming $\zeta_j \sim \mathrm{N}(0, \psi)$ are given in Table 11.8 under 'Indep. Normal'. Empirical Bayes predictions of the SMRs assuming $\zeta_j \sim \mathrm{N}(0, \psi)$ are given under 'Norm.' in Table 11.7 and displayed as a map in Figure 11.4.

Table 11.8 *Estimates for different random intercept models for Scottish lip cancer data*

	Independent Normal		Independent NPML		Spatial IGAR	
	Est	(SE)	Est	(SE)	Est	(SE)
β_0 [Cons]	0.08	(0.12)	0.08	(0.12)	0.09	(0.07)
$\sqrt{\psi}$	0.76	(0.10)	0.80[†]	(–)	0.74	(0.14)

† derived from discrete distribution

Instead of assuming that the random intercept is normally distributed, we also used nonparametric maximum likelihood estimation (NPMLE) where ζ_j is discrete with locations $\zeta_j = e_c$, $c = 1, \ldots, C$, and probabilities π_c, and C is determined to maximize the likelihood (see Section 6.5). The NPML solution had $C = 4$ masses with locations $\widehat{\beta}_0 + \widehat{e}_c$ equal to -1.02, 0.15, 1.13 and 1.35 (corresponding to SMRs of 36.2%, 116.4%, 308.8% and 384.8%) and estimated probabilities $\widehat{\pi}_c$ equal to 27.5%, 47.9%, 18.5% and 6.2%. The parameter estimates are shown in Table 11.8 under 'Indep. NPML'. Predicted SMRs based on these NPML estimates are under given 'NPML' in Table 11.7.

11.4.3 Spatially correlated random intercepts

It is likely that unobserved risk factors, many of which may be environmental, are spatially correlated. We will therefore now consider spatial models where the random effects ζ_j are allowed to be correlated across neighboring counties. Clayton and Kaldor (1987) considered a conditional Gaussian autoregressive model (Besag, 1974), but this specification is problematic for the case of so-called irregular maps where the number of neighbors varies. For this common situation the intrinsic autoregressive Gaussian (IGAR) model (Besag et al., 1991; Bernardinelli and Montomoli, 1992) is deemed more appropriate. In this model the conditional distribution of a random intercept for a county, given the the random intercepts of its contiguous neighbors, does not depend on the random intercepts of non-neighboring counties. Such models are known as Markov random fields.

The model can be specified as

$$h(\zeta_j) \;\propto\; \exp\sum_{i\sim j}(\zeta_j - \zeta_i)^2/(2\psi),$$

where $i \sim j$ indexes counties contiguous to county j. Here the conditional expectation of ζ_j is the mean of ζ_i in the neighboring counties,

$$\mathrm{E}(\zeta_j|\zeta_i, i \neq j) \;=\; \frac{1}{m_j}\sum_{i\sim j}\zeta_i.$$

Note that the unconditional expectation of ζ_j is not specified. The conditional variance is inversely proportional to their number m_j,

$$\mathrm{Var}(\zeta_j|\zeta_i, i \neq j) \;=\; \frac{\psi}{m_j}.$$

The intrinsic autoregressive Gaussian model has been estimated for the lip cancer data using PQL-1 (Breslow and Clayton, 1993) and MCMC (Spiegelhalter *et al.*, 1996a). Since the random effects for all counties are correlated with each other (except for islands), methods based on numerical integration are not feasible due to the excessive dimensionality of the integrals. We therefore propose the following iterative algorithm which strongly resembles the AIP algorithm described in Section 6.11.5:

- Set initial offsets to zero,

$$a_j^0 = 0.$$

- In iteration k:

  1. Estimate a Poisson random intercept model with linear predictor

  $$\nu_j \;=\; \ln(e_j) + \beta_0 + a_j^k + u_j\frac{1}{\sqrt{m_j}}, \quad u_j \sim \mathrm{N}(0,\psi)$$

  giving parameter estimates $\widehat{\boldsymbol{\theta}}^k$ with estimated covariance matrix $\mathrm{Cov}(\widehat{\boldsymbol{\theta}})^k$.

  2. Sample parameters $\boldsymbol{\theta}^k$ from their approximate sampling distribution,

  $$\boldsymbol{\theta}^k \sim \mathrm{N}(\widehat{\boldsymbol{\theta}}^k, \mathrm{Cov}(\widehat{\boldsymbol{\theta}})^k).$$

  3. Obtain the posterior means and standard deviations of the random effects given the parameters $\boldsymbol{\theta}^k$,

  $$\widetilde{u}_j^k \;=\; \mathrm{E}(u_j|\mathbf{y}_j, \boldsymbol{\theta}^k)$$
  $$\tau_j^k \;=\; \mathrm{Var}(u_j|\mathbf{y}_j, \boldsymbol{\theta}^k).$$

  4. Sample the random effects from their approximate posterior distribution

  $$u_j^k \sim \mathrm{N}(\widetilde{u}_j^k, \tau_j^k).$$

5. Obtain ζ_j^k as

$$\zeta_j^k = a_j^k + u_j^k \frac{1}{\sqrt{m_j}}.$$

6. Calculate the mean of the neighbors' values of ζ_j^k

$$b_j^k = \frac{1}{m_j} \sum_{i \sim j} \zeta_i^k.$$

7. Update the offsets to

$$a_j^{k+1} = b_j^k - \frac{1}{J} \sum_{j=1}^{J} b_j^k.$$

Here the offset a_j^k represents the mean of the random effect ζ_j^k of cluster j for the current iteration, computed from the means of the neighbors' random effects from the previous iterations. (The mean-centering in step 7 is required to identify the intercept parameter β_0 since the intrinsic autoregressive Gaussian model does not specify a mean for the random effects.) In step 1, $\zeta_j = a_j^k + u_j \frac{1}{\sqrt{m_j}}$ has mean a_j^k and variance ψ/m_j as required. The random effect u_j is independently distributed, and the model can therefore be estimated using any software for random intercept models, here using adaptive quadrature in gllamm. In step 2, we sample from the approximate sampling distribution of the parameters to reflect the uncertainty of estimation. Hence, our approach is more elaborate than conventional empirical Bayes where parameter estimates are plugged in and this uncertainty is ignored. In steps 3 and 4, we sample the random effects from the normal approximation to the posterior distribution. In step 5 we form the corresponding ζ_j^k and combine them in step 6 to form the required offsets for the next iteration. In step 7 the offsets are mean-centered to allow estimation of the intercept β_0.

The algorithm was repeated until the estimates $\widehat{\theta}^k$ appear to come from a stationary distribution (the 'burn-in' period). This took only about 10 iterations. The algorithm was then repeated 500 times to obtain the mean and variance of the estimates as described in Section 6.11.5. The resulting estimates are shown in Table 11.8.

In step 3, the posterior mean SMRs were also obtained. Their means and 95% confidence intervals based on the 2.5 and 97.5 percentiles of the 500 replications are given in the last three columns of Table 11.7. The point estimates are shown in the map in Figure 11.5 and the confidence intervals in Figure 11.6. Here the hollow circles are the crude SMR estimates and the vertical line is the estimate of the mean SMR, $100 \exp(\widehat{\beta}_0)$, using the intrinsic Gaussian autoregressive model.

11.4.4 Introducing a county-level covariate

Breslow and Clayton (1993) suggest investigating the effect of the county-level covariate [Agric], the percent of the work force in each county employed in

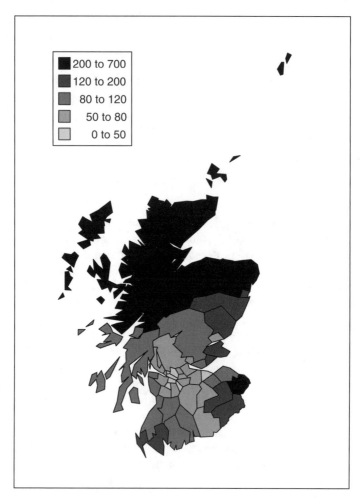

Figure 11.5 *SMRs and 95% CI based on intrinsic autoregressive Gaussian model without covariate*

agriculture, fishing or forestry divided by ten. [Agric] is believed to have an effect on lip cancer incidence since all three occupations involve outdoor work and exposure to sunlight is known to be the main risk factor for lip cancer (Kemp *et al.*, 1985). Following Breslow and Clayton (1993) we consider the model

$$\ln(\mu_j) \;=\; \ln(e_j) + \beta_0 + \beta_1 x_j + \zeta_j,$$

where [Agric] is represented by x_j.

Estimates for an intrinsic Gaussian autoregressive model including the effect of [Agric] using the data augmentation algorithm are given in Table 11.9. The estimates are very close to those using PQL-1 (Breslow and Clayton, 1993) and MCMC (from the BUGS Examples Manual (Spiegelhalter *et al.*, 1996c,

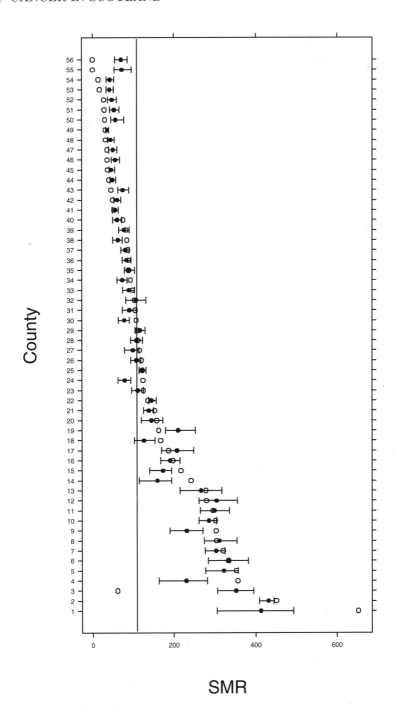

Figure 11.6 *Approximate 95% confidence intervals for SMRs based on intrinsic Gaussian autoregressive model without covariate*

Volume 2, Chapter 11) which are also given in the table. The priors for the
Bayesian model were $\beta_1 \sim N(0, 10^5)$ and $\zeta_j \sim N(0, \psi)$ and the hyperprior was
$\psi \sim IG(0.001, 0.001)$; see page 210 for the inverse gamma (IG). No constant
was included for the MCMC estimates because Spiegelhalter *et al.* (1996c)
did not set the mean of the random effects to zero. Although there appears
to be an effect of [Agric], conclusions regarding etiology should be made with
extreme caution when based on ecological or aggregated data such as these.

In Figure 11.7 we display a map of SMRs based on the intrinsic Gaussian autoregressive model with [Agric], estimated by the data augmentation
algorithm.

Table 11.9 *Estimates for intrinsic Gaussian autoregressive model with covariate for Scottish lip cancer data*

	PQL-1		MCMC		Data augmentation	
	Est	(SE)	Est	(SE)	Est	(SE)
β_0 [Cons]	−0.18	(0.12)	−		−0.21	(0.11)
β_1 [Agric]	0.35	(0.12)	0.37	(0.11)	0.36	(0.11)
$\sqrt{\psi}$	0.73	(0.13)	0.69	(0.12)	0.65	(0.14)

11.5 Summary and further reading

We have addressed the problem of overdispersion, using zero-inflated models as
well as random intercept models with both normal and nonparametric random
effects distributions. Notwithstanding their genesis in the statistical literature
(Lambert, 1992), recent research and applications of models for zero-inflated
count data have taken place mostly in econometrics (see Cameron and Trivedi
(1998) for references). Hall (1997) and Dobbie and Welsh (2001) discuss zero-
inflated Poisson models for clustered data.

We then used random intercept and random coefficient models for longitudinal epileptic seizure data and used diagnostics to check model assumptions. Comprehensive treatments of various types of count data modeling,
also including random effects, are provided in the econometric monographs of
Winkelmann (2003) and Cameron and Trivedi (1998).

Various models were used for mapping of lip cancer rates, including one with
spatially correlated random effects. An alternative model for spatial dependence is a conditional autoregressive model which conditions on the observed
counts in neighboring areas. Biggeri *et al.* (2000) include random effects in
such a model and use nonparametric maximum likelihood estimation. Knorr-
Held and Best (2001) propose a Bayesian model for two diseases where the
response model is essentially a common factor model. Application of small

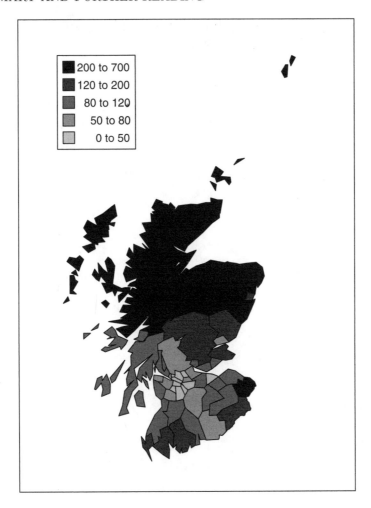

Figure 11.7 *SMRs using intrinsic autoregressive Gaussian model with* [Agric] *as covariate*

area estimation and mapping seems to be most common in medical research, with mapping of disease rates as the archetype. However, other applications include mapping of unemployment, crime rates and party preference. A nice review is given by Pfeffermann (2002). Lawson *et al.* (2003) discuss disease mapping using the MLwiN and WinBUGS softwares.

We have not used factor models in this chapter. For a discussion and application of factor models for counts we refer to Wedel *et al.* (2003). Arminger and Küsters (1988) consider a factor model for mixed responses including counts.

We return to an application involving counts in Section 14.5 of our chapter on multiple processes and mixed responses. Specifically, we model the effect of physician advice on the number of alcoholic beverages consumed, taking unobserved heterogeneity and endogeneity into account.

CHAPTER 12

Durations and survival

12.1 Introduction

In Section 2.5 we considered models for durations, distinguishing between continuous and discrete-time durations. In this chapter we discuss clustered survival or duration data often referred to as 'multivariate' duration data. It is useful to distinguish between 'single events' and 'multiple events' clustered duration data.

Single events clustered duration data comprise durations for one event (e.g. death from lung cancer) for each unit (e.g. subject) in different clusters (e.g. families). It is typically assumed that the event is 'absorbing' (can only occur once) and that the hazard for a unit is independent of the timing of events for other units in the cluster, although the hazards are dependent among units in a cluster.

In contrast, *multiple events clustered duration data* comprise durations for several events per cluster (e.g. subject). In this case it may be reasonable to expect that the hazard for one of the events for a particular subject could depend on the timing of other events for the subject. If the events are different and absorbing (e.g. death from different kinds of cancer) the events are of *multiple types*. The multiple events are called *recurrent* if the same event (e.g. occurrence and subsequent recurrence of colon cancer) is observed repeatedly. Complex duration data may of course include recurrent events of multiple types.

Various aspects of model specification for multiple events clustered duration data are discussed in Section 12.2. In Section 12.3 we apply models for discrete time single events clustered duration data. The data are on children clustered in schools and the event of interest is first experimentation with cigarettes. In Section 12.4 we apply models for multiple events clustered survival data in continuous time. A randomized clinical trial is considered where the responses are times to onset of angina in repeated exercise tests. Continuous random effects and factors are considered as well as discrete latent variables.

12.2 Modeling multiple events clustered duration data

The modeling of single events clustered duration data involves fairly straightforward inclusion of latent variables such as random effects and factors (possibly varying at several levels) in the duration models discussed in Section 2.5.

For multiple events clustered survival data, however, a number of additional considerations are important in choosing an appropriate model:

1. Are the events of different types?

2. Are the events ordered so that a unit becomes at risk for the next event only if the previous event has occurred? An example is HIV infection and AIDS. A less clear-cut example is malaria where the biological processes leading to the first infection are very different from subsequent recurrences.

3. Does the risk of an event depend on whether another event has already occurred for the same unit ('state dependence')?

4. Do the risks of all events evolve in parallel from a common time origin or do they evolve sequentially, starting from the previous event-time? For instance, risks of different types of infection may evolve from a common origin such as time of surgery. In contrast, for durations from repeated experiments such as time from starting exercise (after resting) to developing angina the risk would be expected to evolve from the beginning of each experiment (see Section 12.4 for an application).

Recall from Section 2.5 that both the proportional hazards model for continuous time durations and the discrete-time hazards models can be formulated in terms of risk sets. A difficulty with recurrent event data is that there are many different ways of constructing these risk sets; which are appropriate is determined by the considerations listed above.

Kelly and Lim (2000) suggest that the following aspects are useful to consider when constructing risk sets:

- Risk interval formulations

  - Total time: the clock continues to run from start of observation, undisturbed by event occurrences.
  - Counting process: the clock continues to run but a unit becomes at risk for its kth event only after having experienced the $k-1$th event.
  - Gap time: the clock is reset to zero at each event-time for a unit.

The difference between risk intervals in terms of total time, counting process and gap time is best illustrated using a simple example. Consider three subjects $j = A, B, C$ with possibly recurrent durations t_{ij}, from start of observation to the ith event ($\delta_{ij} = 1$) or censoring ($\delta_{ij} = 0$). The dataset is presented in Table 12.1, where we see for instance that A experiences the first event at time 2, the second at time 5 and is eventually censored at time 14. Risk intervals for the data are presented in Figure 12.1 for each of the three formulations.

The choice of risk interval formulation is guided by consideration 4 in the above list. Total time or the counting process formulation will typically be used if the risks evolve from a common origin and gap time used if the risks evolve sequentially.

Table 12.1 *Illustrative recurrent duration data*

j	i	t_{ij}	δ_{ij}
A	1	2	1
A	2	5	1
A	3	14	0
B	1	7	1
B	2	12	1
B	3	17	0
C	1	14	1

- Baseline hazard (within-unit stratification)

  - Common baseline hazard: the same baseline hazard $h^0(t_{ij})$ is assumed for all recurrent events
  - Event-specific baseline hazard: a different baseline hazard $h_i^0(t_{ij})$ is specified for each event i

  The choice of baseline hazard is guided by consideration 1 in the list above. A common baseline hazard is typically specified if the events are of the same type whereas event-specific baseline hazards are specified if the events are of different types.

- Risk set for the kth event (between-unit stratification)

  - Restricted: contributions to the kth risk-set are restricted to units having experienced $k-1$ events
  - Unrestricted: risk-sets include all units still at risk regardless of how many events they have previously experienced
  - Semi-restricted: contributions to the kth risk-set are restricted to units having experienced fewer than k events

  The choice is guided by consideration 2 in the above list (whether a unit is at risk for the next even only if the previous event has occurred).

Table 12.2 gives combinations of risk intervals, risk sets and baseline hazards and key references for continuous time proportional hazards modeling.

12.3 Onset of smoking: Discrete time frailty models

12.3.1 Introduction

We will analyze data on the smoking behavior of school children (Lader and Matheson, 1991) previously analyzed by Pickles *et al.* (2001) and Rabe-Hesketh *et al.* (2001d). Two cross-sectional surveys of school children aged 11 to 15 years were carried out in 1990 and 1993, giving a time-sequential design

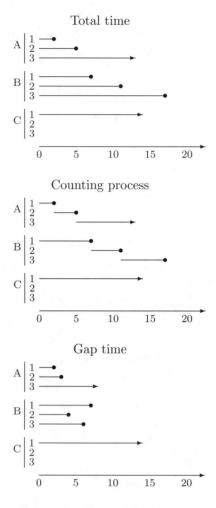

Figure 12.1 *Illustration of risk intervals for total time, counting process and gap time (Adapted from Kelly and Lim, 2000)*

(e.g. Appelbaum and McCall, 1983). Both studies followed similar two-stage sampling designs with schools as primary sampling units. The 1990 sample includes 3124 pupils from 125 schools and the 1993 sample includes 3140 different children from 110 schools. We assume that the sampling fraction for schools is sufficiently low so that the possibility of the same schools appearing in both samples can be ignored. The 5 years of classes sampled within each survey and the 3-year interval between surveys resulted in some age cohorts being sampled twice (e.g. 11-year-olds in 1990 are the same cohort as 14-year-olds in 1993).

The children were asked whether they had ever smoked a cigarette, and if so, how old they were the first time they smoked. In 2% of observations the

Table 12.2 *Combinations of types of risk interval, risk set and baseline hazards with key references*

	Risk set/ baseline hazard		
Risk interval	Unrestricted/ common	Semi-restricted/ event-specific	Restricted/ event-specific
Total time	Lee, Wei & Amato (1992)	Wei, Lin & Weissfeld (1989)	
Counting process	Andersen & Gill (1982)		Prentice, Williams & Peterson (1981)
Gap time		Not possible	Prentice, Williams & Peterson (1981)

Source: Kelly and Lim (2000)

child did not remember the age of onset (left censoring) and these observations were discarded. The children were surveyed after different periods of time since their last birthdays, giving those children interviewed immediately after their birthday little opportunity to have had a first cigarette at their current age. We will therefore analyze the data as if we had interviewed the children at the last birthday, treating the age of first experimentation as right-censored if it equals the current age.

There are six possible discrete-time durations $y = s$ corresponding to the possible age-ranges of onset, $\tau_{s-1} \leq T < \tau_s$, $s = 1, \ldots, 6$:

- $s = 1$: Before age 11, $\tau_0 = 0$, $\tau_1 = 11$
- $s = 2$: Age 11, $\tau_1 = 11$, $\tau_2 = 12$
- $s = 3$: Age 12, $\tau_2 = 12$, $\tau_3 = 13$
- $s = 4$: Age 13, $\tau_3 = 13$, $\tau_4 = 14$
- $s = 5$: Age 14, $\tau_4 = 14$, $\tau_5 = 15$
- $s = 6$: Age 15 and above, $\tau_5 = 15$, $\tau_6 = \infty$ (always censored)

Note that $s = 6$ is always censored because 15 was the oldest age and age of onset at the current age is treated as right-censored. The setup is presented in Table 12.3.

The explanatory variables that we consider as possible influences on age of first experimentation with smoking are:

- [Girl] a dummy for pupil a girl versus a boy
- [Boy] a dummy for pupil a boy versus a girl
- [Cohort2] a dummy for second versus first cohort
- [Parsmoke] a dummy for presence of a smoking parent at home.

Table 12.3 *Ages, possible times of onset and associated probabilities*

(a) Proportional Odds				
Age	11	12	13	14	15

	11	12	13	14	15
	$T < 11$	$T < 11$	$T < 11$	$T < 11$	$T < 11$
	(P_1)	(P_1)	(P_1)	(P_1)	(P_1)
Possible					
times	$11 \leq T$	$11 \leq T < 12$	$11 \leq T < 12$	$11 \leq T < 12$	$11 \leq T < 12$
of	$(1 - P_1)$	$(P_2 - P_1)$	$(P_2 - P_1)$	$(P_2 - P_1)$	$(P_2 - P_1)$
onset					
	$-$	$12 \leq T$	$12 \leq T < 13$	$12 \leq T < 13$	$12 \leq T < 13$
	$-$	$(1 - P_2)$	$(P_3 - P_2)$	$(P_3 - P_2)$	$(P_3 - P_2)$
	$-$	$-$	$13 \leq T$	$13 \leq T < 14$	$13 \leq T < 14$
	$-$	$-$	$(1 - P_3)$	$(P_4 - P_3)$	$(P_4 - P_3)$
	$-$	$-$	$-$	$14 \leq T$	$14 \leq T < 15$
	$-$	$-$	$-$	$(1 - P_4)$	$(P_5 - P_4)$
	$-$	$-$	$-$	$-$	$15 \leq T$
	$-$	$-$	$-$	$-$	$(1 - P_5)$

(b) Current Status				
Age	11	12	13	14	15

11	12	13	14	15
$T < 11$	$T < 12$	$T < 13$	$T < 14$	$T < 15$
(P_1)	(P_2)	(P_3)	(P_4)	(P_5)
$11 \leq T$	$12 \leq T$	$13 \leq T$	$14 \leq T$	$15 \leq T$
$(1 - P_1)$	$(1 - P_2)$	$(1 - P_3)$	$(1 - P_4)$	$(1 - P_5)$

Source: Rabe-Hesketh *et al.* (2001d)

12.3.2 Random intercept models

We consider two approaches to modeling discrete survival times, discrete time hazard models and the proportional odds model for ordinal data (with censoring).

Discrete time hazards models can be estimated by expanding the data as outlined in Display 2.2 on page 46. In the expanded dataset, a child with age of first experimentation in the sth interval has responses y_{ir}, $r = 1, \ldots, s$; a child who was interviewed during the sth interval but whose age of onset was not before his last birthday has responses y_{ir}, $r = 1, \ldots, s - 1$. The responses are equal to 1 if experimentation occurred in the corresponding interval and zero otherwise.

Logistic regression for the expanded data then corresponds to a continuation ratio model whereas complementary log-log regression corresponds to a proportional hazards model (see Section 2.5.2). In addition to the usual covariates, the models include a constant κ_s, $s = 1, \ldots, 5$ for each possible interval of first experimentation.

To model unobserved heterogeneity between schools, we can include a random intercept in the linear predictor for either of these models. The linear predictor for the rth response for the ith pupil in the jth school is therefore

$$\nu_{ijr} = \mathbf{x}'_{ij}\boldsymbol{\beta} + \kappa_r + \zeta_j, \tag{12.1}$$

where $\zeta_j \sim \mathrm{N}(0, \psi)$.

We also consider a censored proportional odds model with a random intercept. Here the response probabilities are as shown in Table 12.3 with cumulative probabilities

$$P_s \equiv \Pr(T_{ij} < \tau_s) = \frac{\exp(\mathbf{x}'_{ij}\boldsymbol{\beta} + \zeta_j + \kappa_s)}{1 + \exp(\mathbf{x}'_{ij}\boldsymbol{\beta} + \zeta_j + \kappa_s)} = g^{-1}(\mathbf{x}'_{ij}\boldsymbol{\beta} + \zeta_j + \kappa_s),$$

where g is the logit link. Note that the sign of the regression coefficients $\boldsymbol{\beta}$ is reversed compared with the usual proportional odds model so that positive coefficients imply an increased odds of *low* responses versus high responses.

For noncensored durations, the response probabilities have the form

$$\Pr(\tau_{s-1} \leq T < \tau_s) = g^{-1}(\mathbf{x}'_{ij}\boldsymbol{\beta} + \zeta_j + \kappa_s) - g^{-1}(\mathbf{x}'_{ij}\boldsymbol{\beta} + \zeta_j + \kappa_{s-1}).$$

The model can therefore be estimated using the composite link function discussed in Section 2.3.5. Alternatively, the model can be estimated by treating the responses of children of different current ages as distinct ordinal responses with different numbers of categories (see Table 12.3(a)). For example, those who were aged 11 when surveyed have an ordinal response with two possible categories and those who were aged 12 have a response with 3 categories, etc. The thresholds κ_s are then constrained equal across responses.

Parameter estimates are given in the first three columns of Table 12.4. All three models lead to essentially the same conclusions. Females are less at risk than males if no parent is smoking at home. A parent smoking at home increases the risk of smoking for boys and this effect is even greater for girls. There is an effect of cohort for girls with the risk of having a first cigarette earlier increasing over time. There does not appear to be a cohort effect for boys. Although the estimated ψ are quite small, there is significant heterogeneity between schools in the ages of onset. (For both continuation ratio and proportional hazards models the likelihood ratio tests give $p < 0.001$ using half the p-value based on the $\chi^2(1)$ distribution; see Section 8.3.4.)

Table 12.4 *Estimates and standard errors for random intercept discrete time duration models*

Parameters	Continuation ratio		Proportional hazards		Proportional odds		Current status		Telescoping	
	Est	(SE)	Est	(SE)	Est	(SE)	Est	(SE)	Est	(SE)
Fixed Effects:										
[Girl]	−0.34	(0.12)	−0.30	(0.11)	−0.45	(0.14)	−0.32	(0.08)	−0.31	(0.15)
[Parsmoke]	0.31	(0.07)	0.28	(0.06)	0.37	(0.08)	0.35	(0.08)	0.38	(0.08)
[Girl] × [Parsmoke]	0.29	(0.09)	0.26	(0.09)	0.33	(0.11)	0.38	(0.12)	0.36	(0.11)
[Boy] × [Cohort2]	0.02	(0.02)	0.02	(0.02)	0.02	(0.02)	0.04	(0.03)	0.03	(0.03)
[Girl] × [Cohort2]	0.08	(0.02)	0.07	(0.02)	0.08	(0.02)	0.09	(0.03)	0.08	(0.03)
Telescoping:										
[Boy]									−0.10	(0.04)
[Girl]									0.04	(0.04)
Thresholds or constants:										
κ_1	−2.17	(0.10)	−2.22	(0.09)	−2.18	(0.11)	−2.43	(0.20)	−2.37	(0.18)
κ_2	−2.41	(0.10)	−2.45	(0.09)	−1.52	(0.11)	−1.72	(0.17)	−1.68	(0.16)
κ_3	−1.91	(0.10)	−1.98	(0.09)	−0.89	(0.11)	−0.91	(0.15)	−1.01	(0.14)
κ_4	−1.38	(0.10)	−1.51	(0.09)	−0.22	(0.10)	−0.30	(0.13)	−0.31	(0.12)
κ_5	−1.23	(0.12)	−1.38	(0.10)	0.32	(0.11)	0.19	(0.12)	0.28	(0.12)
Random Effect:										
ψ	0.07	(0.02)	0.06	(0.02)	0.09	(0.03)	0.10	(0.03)	0.09	(0.03)
Log-likelihood	−6225.5		−6225.8		−6223.4		−3487.7		−5948.8	

Source: Rabe-Hesketh *et al.* (2001d)

12.3.3 Modeling 'telescoping' effects

One problem with the three models considered above is that they assume that the recalled ages of onset are reliable. Accounting for measurement error in age-of-onset data has received rather little attention in the literature. We consider two alternative approaches.

The first approach is to discard the timing element of the children's responses and simply model their current status (ever experimented) as a function of their current age, using a simple logistic regression model with the current smoking status indicator as the response variable, as indicated in Table 12.3(b). This gives the results in column 4 of Table 12.4 which are not very different from those for the proportional odds model.

Another approach is to model recall bias directly. It has been suggested that recall errors are characterized by an apparent shifting of events from the more distant past towards the time at which data collection is made (Sudman and Bradburn, 1973; Hobcraft and Murphy, 1987). This 'telescoping' could arise from an internal compression of the time scale so that an event that occurred a time t ago is reported as occurring a time γt ago with $0 < \gamma < 1$.

Telescoping could also result from heteroscedastic measurement error where the error variance increases with the lag between the event and the time of recollection even when the errors are symmetrically distributed. This is because more events from the distant past, that are typically recalled with larger errors, are shifted into the recent past than events in the recent past, that are typically recalled with smaller error, are shifted back into the distant past (Pickles et al., 1996). While Pickles et al. develop models to distinguish between these processes, here we will only consider systematic telescoping.

In the proportional-odds model, we assume that the log odds that the recalled age of onset is before a given age τ_s decreases linearly with the time that has passed since that age, $a_{ij} - \tau_s$, where a_{ij} is the child's current age:

$$\ln \frac{\Pr(T_{ij} < \tau_s)}{1 - \Pr(T_{ij} < \tau_s)} = \mathbf{x}'_{ij}\boldsymbol{\beta} + \zeta_j - (a_{ij} - \tau_s)\mathbf{w}'_{ij}\boldsymbol{\alpha} + \kappa_s.$$

Here, $\boldsymbol{\alpha}$ is a vector of coefficients and $\mathbf{w}_{ij}$ is a vector of explanatory variables that may predict the degree of telescoping (positive coefficients imply a compression of the time scale). If the proportional odds model is interpreted as a latent response model, telescoping corresponds to allowing the thresholds to depend on the time-lag, i.e., the thresholds are

$$-(a_{ij} - \tau_s)\mathbf{w}'_{ij}\boldsymbol{\alpha} + \kappa_s.$$

We assume that the degree of telescoping depends on sex only, giving the parameter estimates in the last column of Table 12.4. While there is little evidence of telescoping for girls, the boys tend to stretch the time scale (rather than compress it), perhaps 'showing off' with having experimented earlier than they actually did.

Note that separate identification of telescoping and cohort effects is possible here because some of the cohorts of children are represented in both surveys at different 'current' ages and therefore different time lags.

12.4 Exercise and angina: Proportional hazards random effects and factor models

12.4.1 Introduction

We analyze the dataset[1] published in Danahy *et al.* (1976) and previously considered by Pickles and Crouchley (1994, 1995). 21 subjects with coronary heart disease participated in a randomized crossover trial comparing the effect of the drug Isorbide dinitrate (ISDN) with placebo.

Before receiving the drug (or placebo), subjects were asked to exercise on exercise bikes to the onset of angina pectoris or, if angina did not occur, to exhaustion. The exercise time and outcome (angina or exhaustion) were recorded. The drug (or placebo) was then taken orally and the exercise test repeated one hour, three and five hours after drug (or placebo) administration. We therefore have repeated 'survival' times per subject pre and post administration of both an active drug and a placebo.

Each subject repeated the exercise test 4 times in the placebo condition and 4 times in the drug condition. The responses of interest are the durations to angina or exhaustion in the drug condition y_{ij} for exercise test i and subject j. d_{ij} is a censoring indicator taking the value 1 if angina occurred and 0 otherwise. The duration to angina t_{ij} in the corresponding placebo condition will be treated as a time-varying covariate (there were no censorings under the placebo). Unfortunately, the order in which placebo and drug were given was not reported in Danahy *et al.* (1976).

Since the subjects started each of the eight exercise tests at rest, so that the same processes leading to angina or exhaustion can be assumed to begin at the start of each exercise test, we will assume that the hazard functions for the tests are proportional. According to the classification of aspects involved in constructing risk sets on page 374 we specify:

Risk interval formulation: gap time with the time scale as starting at 0 at the beginning of each exercise test

Baseline hazard: common baseline hazard for each exercise test

Risk set for kth event: unrestricted because each event occurs in a separate exercise test.

This specification corresponds to assuming proportional hazards between tests and hence between pre and post drug administration. This is required to specify the treatment effect as a *hazard ratio*.

The following covariates were considered:

---

[1] The data can be downloaded from gllamm.org/books

- [Bypass] a dummy variable for previous coronary artery bypass graft surgery

$$x_{Bj} = \begin{cases} 1 \text{ if previous bypass surgery} \\ 0 \text{ otherwise} \end{cases}$$

- [TimeP] standardized time to angina in the placebo condition; t_{ij}
- [After] a dummy variable for drug administered, i.e.

$$x_{Tij} = \begin{cases} 0 \text{ if } i=1 \\ 1 \text{ if } i=2,3,4 \end{cases}$$

- [Lin] linear trend of drug effect

$$x_{Dij} = \begin{cases} 0 & \text{if } i=1 \\ i-3 & \text{if } i=2,3,4 \end{cases}$$

The definitions of the covariates that are varying over exercise tests [TimeP], [After], and [Lin] are summarized in Table 12.5.

Table 12.5 *Definitions of covariates varying over exercise tests i*

Test i	[TimeP] Placebo t_{ij}	[After] x_{Tij}	[Lin] x_{Dij}
1	t_{1j}	0	0
2	t_{2j}	1	-1
3	t_{3j}	1	0
4	t_{4j}	1	1

12.4.2 Cox regression

The Cox proportional hazards model can be written as (see also equation (2.22))

$$h_{ij}(t) = h^0(t) \exp(\nu_{ij}),$$

where $h^0(t)$ is the baseline hazard, the hazard when $\nu_{ij} = 0$. The linear predictor is

$$\nu_{ij} = \beta_1 x_{Bj} + \beta_2 t_{ij} + \beta_3 x_{Tij} + \beta_4 x_{Dij}, \tag{12.2}$$

where β_1 represents the effect of [Bypass], β_2 the effect of time to angina under placebo [TimeP], β_3 the difference in treatment effect between the mid post-treatment test $i=3$ and the baseline $i=1$ [After] and β_4 the linear change in treatment effect over the three post-treatment tests [Lin].

It was pointed out in Section 2.5.1 that including a dummy variable for each risk set in Poisson regression and using the log interval length between failure times as an offset leads to a likelihood proportional to the partial likelihood. A problem with this approach is that there will often be a very large number of risk sets and therefore an excessive number of parameters to be estimated. For instance, as many as 64 dummies are required for the present small dataset. This suggests approximating the baseline hazard with a parsimonious smooth function of time. Specifically, we consider an orthogonal polynomial of degree 6,

$$\ln h^0(t) = \alpha_0 + \alpha_1 f_1(t) + \alpha_2 f_2(t) + \alpha_3 f_3(t) + \alpha_4 f_4(t) + \alpha_5 f_5(t) + \alpha_6 f_6(t), \quad (12.3)$$

where $f_k(t)$ is the kth order term. Note that although the baseline hazard is modeled as a smooth function of time, the piecewise exponential formulation implies that the hazard is constant between successive events giving steps whose heights are determined by the smooth function.

Estimates and standard errors for the effects of [Bypass], [TimeP], [After] and [Lin] are reported in Table 12.6 for partial likelihood (Cox regression), maximum likelihood Poisson regression with dummies for the risk-sets and maximum likelihood Poisson regression with a 6th order orthogonal polynomial of time (orthpol-6). For each implementation, we report the estimated β's, standard errors (SE) and hazard ratios $\exp(\beta)$ (denoted HR). Note that

Table 12.6 *Estimates for different implementations of Cox regression*

Parameter	Partial likelihood			ML Poisson dummies			ML Poisson orthpol-6		
	Est	(SE)	HR	Est	(SE)	HR	Est	(SE)	HR
β_1 [Bypass]	0.88	(0.29)	2.42	0.88	(0.29)	2.42	0.90	(0.30)	2.45
β_2 [TimeP]	−1.17	(0.20)	0.31	−1.17	(0.20)	0.31	−1.12	(0.20)	0.33
β_3 [After]	−1.02	(0.29)	0.36	−1.02	(0.29)	0.36	−1.00	(0.29)	0.37
β_4 [Lin]	0.62	(0.19)	1.85	0.62	(0.19)	1.85	0.59	(0.19)	1.81
Log-likelihood	−224.76			−284.14			−320.22		

the estimates and standard errors are identical under partial likelihood and maximum likelihood Poisson regression with dummies as prescribed by theory. Also note that the estimates from Cox regression are well recovered by modeling the log baseline hazard as a 6th order orthogonal polynomial of time. This approach is much more parsimonious than Poisson regression with a dummy variable for each risk set, which requires 57 additional parameters in the present example. For larger datasets, the number of additional parameters will simply be unmanageable. We will therefore rely on the orthogonal polynomial approximation to the log baseline hazard $\ln[h^0(t)]$ for the duration models with latent variables considered subsequently.

Figure 12.2 shows the estimated constants and their 95% confidence intervals for the Poisson model with dummies for the risk sets. The smooth curve is the estimated log baseline hazard for the Poisson model with a sixth order polynomial of time. The polynomial does not seem to oversmooth the log baseline hazard since the curve falls well within the confidence intervals for the constants.

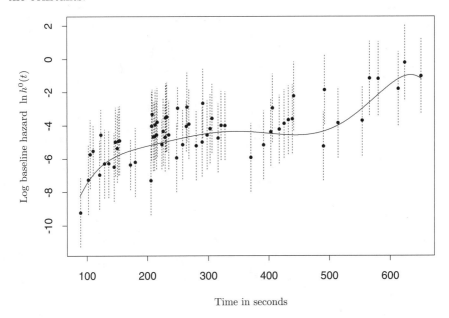

Figure 12.2 *Log baseline hazard for angina data. Points represent constants estimated for each risk set with 95% confidence intervals shown as dotted lines. Curve represents the corresponding sixth order polynomial*

12.4.3 Proportional hazards regression with multinormal latent variables

We first consider the conventional *frailty model*, a simple random intercept model with linear predictor

$$\nu_{ij} = \beta_1 x_{Bj} + \beta_2 t_{ij} + \beta_3 x_{Tij} + \beta_4 x_{Dij} + \zeta_{1j},$$

where $\zeta_{1j} \sim N(0, \psi_{11})$. The exponential of the random intercept, $\exp(\zeta_{1j})$, is often called the *frailty*, and we are hence specifying a log-normal frailty (e.g. McGilchrist and Aisbett, 1991). Alternative distributions for the frailty include the one-parameter gamma (e.g. Clayton, 1978), the positive stable (e.g. Hougaard, 1986b) and the inverse Gaussian (e.g. Hougaard, 1986a).

Estimates and standard errors for the random intercept model are given in the second column of Table 12.7. The estimates for the fixed part are also given in exponentiated form, i.e. as hazard ratios HR.

To allow for a less restrictive dependence structure, we depart from the conventional frailty model and consider a *factor model*

$$\nu_{ij} = \beta_1 x_{Bj} + \beta_2 t_{ij} + \beta_3 x_{Tij} + \beta_4 x_{Dij} + \lambda_i \zeta_{1j}. \tag{12.4}$$

Table 12.7 *Estimates for random intercept (frailty) and factor structured proportional hazards models*

| | Frailty | | | Factor | | |
Parameter	Est	(SE)	HR	Est	(SE)	HR
Fixed part:						
β_1 [Bypass]	1.09	(0.66)	2.96	-0.03	(0.60)	0.97
β_2 [TimeP]	-1.44	(0.30)	0.24	-1.86	(0.34)	0.16
β_3 [After]	-0.92	(0.31)	0.40	-1.41	(0.51)	0.24
β_4 [Lin]	0.81	(0.21)	2.24	1.14	(0.33)	3.11
Random part:						
ψ_{11}	1.32	(0.69)		8.86	(4.97)	
λ_1				0.19	(0.10)	
λ_2				1		
λ_3				0.71	(0.14)	
λ_4				0.50	(0.14)	
Log-likelihood		-311.19			-301.41	

Here, $\zeta_{1j} \sim \mathrm{N}(0, \psi_{11})$ and we have identified the model by restricting one of the factor loadings, $\lambda_2 = 1$. Note that the frailty model is obtained if the restrictions $\lambda_1 = \lambda_2 = \lambda_3 = \lambda_4 = 1$ are imposed in the factor model. Estimates and standard errors for the factor model are given in the third column of Table 12.7. We prefer this model to the random intercept model since the log-likelihood increases quite considerably.

Another candidate is a *random treatment model* where [After] has an associated random coefficient ζ_{2j} but there is no random intercept,

$$\nu_{ij} = \beta_1 x_{Bj} + \beta_2 t_{ij} + (\beta_3 + \zeta_{2j}) x_{Tij} + \beta_3 x_{Dij}, \tag{12.5}$$

where $\zeta_{2j} \sim \mathrm{N}(0, \psi_{22})$. Estimates and confidence intervals for the random treatment model are given in the second column of Table 12.8. Note that this model is a special case of the factor model with $\lambda_1 = 0$ and $\lambda_2 = \lambda_3 = \lambda_4 = 1$. The decrease in the log-likelihood from imposing these restrictions is moderate, which could have been surmised from inspecting the estimated factor loadings. Figure 12.3 shows the log-hazards for the third exercise test for hypothetical individuals with [TimeP] equal to the mean (240.5) and log frailties of -2, -1, 0, 1 and 2 standard deviations.

Finally, we consider a *random intercept and treatment model*. This model has two random effects, a random intercept ζ_{1j} correlated with a random treatment effect ζ_{2j},

$$\nu_{ij} = \zeta_{1j} + \beta_1 x_{Bj} + \beta_2 t_{ij} + (\beta_3 + \zeta_{2i}) x_{Tij} + \beta_4 x_{Dij},$$

where (ζ_{1j}, ζ_{2j}) are assumed to have a bivariate normal distribution, with vari-

Table 12.8 *Estimates for random coefficient proportional hazards models*

	Random treatment			Random intercept & treatment		
Parameter	Est	(SE)	HR	Est	(SE)	HR
Fixed part:						
β_1 [Bypass]	0.58	(0.44)	1.78	0.47	(0.54)	1.61
β_2 [TimeP]	−1.85	(0.32)	0.15	−1.75	(0.33)	0.17
β_3 [After]	−1.09	(0.46)	0.34	−1.16	(0.42)	0.32
β_4 [Lin]	0.78	(0.21)	2.18	0.83	(0.22)	2.30
Random part:						
ψ_{11}				0.20	(0.30)	
ψ_{22}	2.62	(1.24)		1.74	(1.09)	
ψ_{21}				†0.59	(0.42)	
Log-likelihood		−305.48			−304.53	

†boundary solution

ances ψ_{11} and ψ_{22} and covariance ψ_{21}. The multivariate normal distribution is a convenient choice for models with several latent variables since it is not obvious how to generalize the other (univariate) distributions often used for frailty models. Estimates and standard errors for this model are given in the second column of Table 12.8. The estimated correlation between the random effects is virtually 1, and this boundary solution might indicate that one random effect would suffice for this application. This is also reflected in the very small increase in log-likelihood compared to the random treatment model. We therefore abandon the random intercept and treatment model.

The estimated hazard ratios HR, reported in the fixed part in Tables 12.7 and 12.8, are conditional on the latent variables. We observe that the estimates of the hazard ratio for [Bypass] are larger than 1 for all models apart from the factor model. However, the estimates are quite imprecise and vary considerably across models, from 2.96 to 0.96. As would be expected, longer durations under placebo seem to reduce the hazard under the drug condition. The estimates of the hazard ratio for [After] suggest that the drug reduces the hazard at the mid post-treatment exercise test compared with baseline. We also note that the log hazard increases over the post-treatment tests, since β_4 is estimated as positive.

Regarding the random part, we note that a likelihood ratio test for the random intercept variance in the random intercept model strongly indicates that it is needed, which contradicts the corresponding Wald test. We recommend using the likelihood-ratio test with halved p-value for testing the null hypothesis that the variance of a latent variable is zero; see Section 8.3.4.

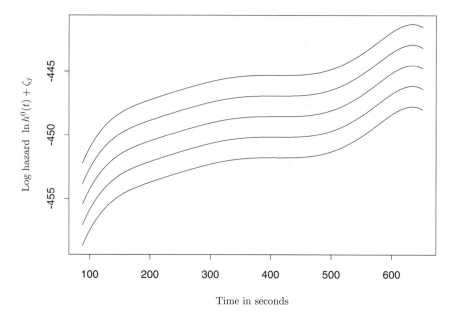

Figure 12.3 *Log hazards for exercise test 3 for hypothetical individuals with log frail-ties ζ_j equal to $-2, -1,\ 0,\ 1$ and 2 estimated standard deviations $\widehat{\psi}^{1/2}$*

12.4.4 Latent class proportional hazards models

Sometimes random effects are assumed to be truly discrete leading to latent class models. Based on the results reported for continuous random effects, we will concentrate on the random treatment and factor models in this section. The models take the same form as in the continuous case, with the vital difference that the random effects ζ_j are now discrete with locations

$$\zeta_j = e_c, \quad c = 1, \ldots, C,$$

and probabilities π_c instead of normally distributed. We also confine the discussion to two-class models $C = 2$.

Estimates for the two-class random treatment model in (12.5) are reported in the second column of Table 12.9. The location -2.59 of the first class for the random treatment model suggests that there is a class of patients who responds remarkably well to the drug. Moreover, this class comprises 21% of the population.

Regarding the fixed part estimates, the results for both latent class models are quite similar to the estimates for the normal random treatment and factor models reported in Table 12.7. The loadings in the factor model have the same pattern as for the model assuming normality.

Table 12.9 *Estimates for 2-class random treatment and factor models*

Parameter	Random treatment			Factor		
	Est	(SE)	HR	Est	(SE)	HR
Fixed part:						
β_1 [Bypass]	0.28	(0.29)	1.32	0.28	(0.30)	1.33
β_2 [TimeP]	−1.95	(0.26)	0.14	−1.97	(0.26)	0.14
β_3 [After]	−1.22	(0.41)	0.30	−1.39	(0.47)	0.25
β_4 [Lin]	0.85	(0.20)	2.33	1.13	(0.34)	3.08
Random part:						
$\mathrm{Var}(\eta_j)$	1.75			5.52		
λ_1				0.07	(0.10)	
λ_2				1		
λ_3				0.68	(0.29)	
λ_4				0.40	(0.20)	
Locations:						
e_1	−2.59			−4.63		
e_2	0.68			1.19		
Class probabilities:						
$\mathrm{Pr}(e_1)$	0.21			0.21		
$\mathrm{Pr}(e_2)$	0.79			0.79		
Log-likelihood	−303.82			−301.42		

12.4.5 Nonparametric maximum likelihood estimation

Assuming parametric continuous distributions for the random effects, such as the multivariate normal used above, has caused consternation among some researchers. Admittedly, it is problematic if results are highly sensitive to the choice of parametric distributions whose main motivation is convenience. An alternative is to use models with latent classes as described above, but it often seems unreasonable to assume that the population consists of truly discrete classes. This has led to the development of duration models where the latent variable distribution may be continuous and/or discrete, not requiring any prior specification. Both locations and mass-points for the latent variables are then estimated, and nonparametric maximum likelihood estimates (NPMLE) are obtained when no further improvement in the likelihood can be achieved by introducing additional mass-points (e.g. Heckman and Singer, 1982, 1984); see Section 6.5.

NPML estimates for the random treatment model are reported in the second column of Table 12.10. The location and mass estimates of the obtained

Table 12.10 *NPMLE for random treatment and factor models*

Parameter	Random treatment 3 masses			Factor 5 masses		
	Est	(SE)	HR	Est	(SE)	HR
Fixed part:						
β_1 [Bypass]	0.33	(0.30)	1.39	−0.22	(0.36)	0.81
β_2 [TimeP]	−2.18	(0.29)	0.11	−2.33	(0.34)	0.10
β_3 [After]	−1.07	(0.48)	0.34	−2.18	(1.12)	0.11
β_4 [Lin]	0.84	(0.21)	2.34	1.94	(0.94)	6.93
Random part:						
$\mathrm{Var}(\zeta_j)$	2.92	(–)		47.24	(–)	
λ_1				0.08	(0.05)	
λ_2				1		
λ_3				0.54	(0.11)	
λ_4				0.23	(0.13)	
Log-likelihood	−301.34			−295.59		

3 mass solution are $(-2.91, 0.32, 2.84)$ and $(0.21, 0.66, 0.14)$, respectively. NPML estimates for the factor model in (12.4) are given in the third column of Table 12.10. A five mass solution was obtained with locations estimated as $(-18.10, -5.51, 0.74, 3.88, 6.91)$ and corresponding masses as $(0.10, 0.10, 0.28, 0.44, 0.08)$. Note that the first class has a frailty approaching zero, indicating a boundary solution. This is reflected in the large estimated $\mathrm{Var}(\zeta_j)$. The estimates are close to those assuming a normally distributed factor, notwithstanding that the nonparametric distribution of the factor appears to be quite skewed. Thus, the factor model appears to be robust against misspecification of the factor distribution in this application.

12.5 Summary and further reading

We initially discussed special issues arising in constructing risk sets for multiple events clustered duration data, pointing out that appropriate modeling of this case is more complex than modeling of single events clustered duration data.

The first application used discrete time frailty models for the onset of smoking among adolescents. Approaches to handling recall bias for age of onset data were discussed and a model with 'systematic telescoping' was estimated.

We then considered proportional hazards models for onset of angina pectoris in a randomized clinical trial. Different kinds of latent variables were introduced to model the dependence among repeated durations within each patient. Models with continuous latent variables such as factor models and various random effects models as well as discrete latent variables were estimated. We also left the latent variable distribution unspecified by using nonparametric maximum likelihood.

Hougaard (2000) provides an extensive treatment of multivariate frailty models, concentrating on models with positive stable frailty distributions. Vermunt (1997) gives an easily accessible discussion of survival models with discrete latent variables. Useful papers on survival or duration modeling with frailty and latent variables include Clayton (1978), Vaupel *et al.* (1979), Clayton and Cuzick (1985), Aalen (1988), Pickles and Crouchley (1994, 1995) and Vaida and Xu (2000).

There has recently been considerable interest in joint modeling of duration and other responses, see Hogan and Laird (1997ab) for an overview. We consider a joint model for survival of liver cirrhosis and a marker process in Section 14.6, estimating both direct and indirect effects of treatment.

CHAPTER 13

Comparative responses: Polytomous responses and rankings

13.1 Introduction

Polytomous responses (also known as 'first choices' or 'discrete choices'), pairwise comparisons and rankings were introduced in Section 2.4.3. A distinguishing feature of the latent response formulation for these processes is that the response is a *vector* of utilities.

It is unrealistic to assume that the utilities for different alternatives are uncorrelated. For the multinomial logit model this would imply the unrealistic property known as 'independence from irrelevant alternatives' to be discussed in Section 13.2. To model the dependence, it is tempting to simply introduce factors and/or random coefficients in the same way as in multivariate regression models. However, in contrast to the case of continuous responses, we do not observe the multivariate vector of utilities directly because the response processes are comparative. Identification thus becomes more difficult as discussed in Bunch (1991), Keane (1992) and Skrondal and Rabe-Hesketh (2003b). In Section 13.3 we describe how the fixed and random parts of the model are usually structured.

We consider multinomial logit instead of probit models with latent variables in this chapter. The reason is pragmatic; the models are very similar but the logit versions are more convenient computationally. Specifically, the conditional probability of a response given the latent variables can be expressed in closed form for multinomial logit models in contrast to probit models. As was pointed out in Section 2.4.3, obtaining response probabilities in multinomial probit models requires integration even when the model is void of latent variables.

The applications considered in this chapter are from political science and marketing. We discuss models for rankings as well as discrete choices, both with continuous and discrete latent variables.

13.2 Heterogeneity and 'Independence from Irrelevant Alternatives'

It follows from the logistic discrete choice model (2.18) presented on page 37 that the odds for two alternatives a and b for unit i are

$$\frac{\Pr(a)}{\Pr(b)} = \exp(\nu_i^a - \nu_i^b), \tag{13.1}$$

which only depend on the linear predictors for the two alternatives. Hence, the odds do not depend on which other alternatives there are in the alternative set. In the ranking case it furthermore follows that the odds do not depend on which other alternatives have already been chosen, the number of alternatives already chosen or the order in which those alternatives were chosen. Luce (1959) denoted this property 'Independence from Irrelevant Alternatives' (IIA).

The problem associated with the IIA property can be illustrated by adapting the red bus - blue bus example of McFadden (1973). Let there be three political parties, Labour I, Labour II and Conservatives. The first two parties are indistinguishable and have the same linear predictor ν_i^{lab} whereas the Conservative party has linear predictor ν_i^{con}. The probability of voting for either Labour I or Labour II is

$$\frac{2\exp(\nu_i^{\text{lab}})}{2\exp(\nu_i^{\text{lab}}) + \exp(\nu_i^{\text{con}})}.$$

Considering the scenario that the two Labour parties merge to form a single Labour party, the probability of voting Labour decreases to

$$\frac{\exp(\nu_i^{\text{lab}})}{\exp(\nu_i^{\text{lab}}) + \exp(\nu_i^{\text{con}})},$$

which is clearly counterintuitive. The model is thus often deemed unduly restrictive (e.g. Takane, 1987 and the references therein).

The consequences of IIA are most pronounced for an indifferent voter with $\nu_i^{\text{lab}} - \nu_i^{\text{con}} = 0$. The merger would in this case reduce the probability of voting Labour from 0.67 to 0.50, consistent with an equiprobable choice among initially three and then two available parties. However, in reality almost all voters will have a party preference and there will be heterogeneity in party preference among voters. This heterogeneity, observed and unobserved, ensures that the model does not imply a substantial reduction in the share of the Labour vote due to the merger.

To illustrate the effect of observed heterogeneity, consider a fixed effect of gender giving $\nu_i^{\text{lab}} - \nu_i^{\text{con}} = -1.2$ for men and $\nu_i^{\text{lab}} - \nu_i^{\text{con}} = 2.8$ for women. For a population with 50% men, the merger reduces the marginal probability for Labour from 0.67 to 0.59 and, marginally to the observed covariates, IIA is therefore violated. In practice, observed covariates cannot explain all variability in individual party preferences. The remaining unobserved heterogeneity can be modeled by including a shared random effect for the two Labour parties in the linear predictor. For example, if $\nu_i^{\text{lab}} - \nu_i^{\text{con}} = -0.8 + \zeta_i$ for men and $\nu_i^{\text{lab}} - \nu_i^{\text{con}} = 3.2 + \zeta_i$ for women, where ζ_i is normally distributed random intercept with variance 16, the marginal (with respect to ζ_i) probabilities for Labour become 0.67 before the merger and 0.63 after. Increasing the magnitude of the fixed effects or the random effects variances decreases the difference even further.

A consequence of failing to include random effects would be that the unrealistic IIA property still holds conditional on the observed covariates, e.g. among men and among women if gender is the only observed covariate. Even with voter-specific random effects IIA still holds for a given voter. This could be relaxed by including voting-occasion-specific random effects if the resulting model is identified.

13.3 Model structure

We give a brief description of the model structure suggested by Skrondal and Rabe-Hesketh (2003b) and refer to this work for details. For simplicity, we consider two-level models with units i (level 1) nested in clusters j (level 2).

It is useful to formulate multilevel models for nominal responses in terms of latent utilities u_{ij}^s with

$$u_{ij}^s = \nu_{ij}^s + \epsilon_{ij}^s.$$

Recall from Section 2.4.3 that multinomial logit models result if the ϵ_{ij}^s are specified as independently Gumbel distributed. The linear predictor ν_{ij}^s represents the mean utility for category or alternative s and contains a fixed part f_{ij}^s and random parts $\delta_{ij}^{s(1)}$ and $\delta_{ij}^{s(2)}$ at levels 1 and 2, respectively,

$$u_{ij}^s = f_{ij}^s + \delta_{ij}^{s(1)} + \delta_{ij}^{s(2)} + \epsilon_{ij}^s.$$

For polytomous responses, it is assumed that the alternative with the greatest utility is chosen and for rankings the alternatives are ranked according to the ordering of the utilities.

Let us now describe the structure of the fixed and random parts of the model.

13.3.1 Fixed part

The fixed part of the model is structured as

$$f_{ij}^s = m^s + \mathbf{x}_{ij}^{s\prime}\mathbf{b} + \mathbf{x}_{ij}'\mathbf{g}^s, \tag{13.2}$$

where m^s is a constant, $\mathbf{x}_{ij}^s$ is a vector which varies over alternatives and may also vary over units and/or clusters, whereas the vector $\mathbf{x}_{ij}$ varies over units and/or clusters but not alternatives. The corresponding fixed coefficient vectors are $\mathbf{b}$ and $\mathbf{g}^s$, respectively. Note that the effect $\mathbf{b}$ is assumed to be the same on all utilities, so that this part of the model simply represents a linear relationship between the utilities and alternative (and possibly unit and/or cluster)-specific covariates. For instance, in the election example considered in Section 13.4, we will assume a linear relationship between a measure of distance on the left-right political dimension between the sth party and the ith voter and the voter's utilities for the party. For some alternative and unit and/or cluster-specific variables the effect may differ between alternatives. Such effects can be accommodated by including interactions between these variables and dummy variables for the alternatives in $\mathbf{x}_{ij}^s$.

Discrete choice models only including fixed effects have been considered by

Train (1986) and ranking versions by Chapman and Staelin (1982) and Allison and Christakis (1994). The *conditional logit model*, standard in econometrics, arises as the special case where $\mathbf{x}'_i\mathbf{g}^s$ and m^s are omitted in (13.2). The model is used for discrete choices (e.g. McFadden, 1973), rankings (e.g. Hausman and Ruud, 1987) and paired comparisons (e.g. Bradley and Terry, 1952). The *polytomous logistic regression model*, a standard model for discrete choices in for instance biostatistics (e.g. Hosmer and Lemeshow, 2000), results as the special case where $\mathbf{x}^{s\prime}_{ij}\mathbf{b}$ is omitted in (13.2).

13.3.2 Level-1 random part

The component $\delta^{s(1)}_{ij}$, inducing dependence between alternatives within units, is structured as

$$\delta^{s(1)}_{ij} = \mathbf{z}^{s(1)\prime}_{ij}\boldsymbol{\xi}^{(1)}_{ij} + \boldsymbol{\lambda}^{s(1)\prime}\boldsymbol{\eta}^{(1)}_{ij}, \tag{13.3}$$

where $\boldsymbol{\xi}^{(1)}_{ij}$ are random coefficients allowing the effects of alternative-specific covariates $\mathbf{z}^{s(1)}_{ij}$ to vary between units i and $\boldsymbol{\eta}^{(1)}_{ij}$ are factors representing unobserved variables having effects $\boldsymbol{\lambda}^{s(1)}$ on u^s_{ij}. Alternatively, $\boldsymbol{\lambda}^{s(1)}$ can be interpreted as unobserved attributes of alternative s and $\boldsymbol{\eta}^{(1)}_{ij}$ as random effects on the utilities. In the election example, letting $z^{s(1)}_{ij} = x^{s(1)}_{ij}$ represent the distance on the left-right political dimension, the random slope $\xi^{(1)}_{ij}$ allows the effect of political distance to vary between voters.

A model of the above type was discussed by McFadden and Train (2000) for the case of a multinomial logit. A special case including only random coefficients was considered by Hausman and Wise (1978). Models only including common factors were suggested for paired comparisons by Bloxom (1972) and Arbuckle and Nugent (1973) and for rankings by Brady (1989), Böckenholt (1993) and Chan and Bentler (1998). It should be noted that unit-level factor models are not identified for discrete choices unless unit and alternative-specific covariates are included in the fixed part (e.g. Keane, 1992).

13.3.3 Level-2 random part

The component $\delta^{s(2)}_{ij}$, inducing dependence among utilities within clusters, is structured as

$$\delta^{s(2)}_{ij} = \mathbf{z}^{s(2)\prime}_{ij}\boldsymbol{\xi}^{(2)}_j + \mathbf{z}^{(2)\prime}_{ij}\boldsymbol{\gamma}^{s(2)}_j + \boldsymbol{\lambda}^{s(2)\prime}\boldsymbol{\eta}^{(2)}_j, \tag{13.4}$$

where $\boldsymbol{\xi}^{(2)}_j$ and $\boldsymbol{\gamma}^{s(2)}_j$ are vectors of random coefficients with corresponding variable vectors $\mathbf{z}^{s(2)}_{ij}$ and $\mathbf{z}^{(2)}_{ij}$. $\boldsymbol{\xi}^{(2)}_j$ allows the effects of alternative-specific covariates to vary between clusters j, whereas $\boldsymbol{\gamma}^{s(2)}_j$ represents the variability in the effect of unit-specific covariates $\mathbf{z}^{(2)}_{ij}$ between clusters. Note that there is in this case a random coefficient for each alternative. In the context of the election example $\xi^{(2)}_j$ could represent heterogeneity in the distance effect between constituencies. $\boldsymbol{\gamma}_j^{s(2)}$ are random intercepts when $\mathbf{z}_{ij}^{(2)} = 1$ and could more generally represent the random coefficients (varying over constituencies) of a covariate that varies within constituencies (e.g. age). $\boldsymbol{\eta}_j^{(2)}$ are factors representing unobserved variables at the cluster level having effects $\boldsymbol{\lambda}^{s(2)}$ on u_{ij}^s. The random terms at levels 1 and 2 are analogous except that there is no term corresponding to $\gamma_j^{s(2)}$ at level 1, since identification would then be fragile.

The special case including only random coefficients for unit-specific covariates has been considered for discrete choices by Hedeker (2003) and Daniels and Gatsonis (1997) and for paired-comparisons by Böckenholt (2001a). Revelt and Train (1998) specified the special case where there are only random coefficients for alternative-specific or alternative and unit-specific covariates. Random coefficients for alternative-specific covariates were used in a conjoint choice experiment by Haaijer *et al.* (1998). The special case including only common factors was considered by Elrod and Keane (1995). McFadden and Train (2000), among others, specify 'mixed logit models' containing both random coefficient and factor structures. Böckenholt (2001b) used this model for rankings, also including discrete random intercepts $\gamma_j^{s(2)} = e_{jc}^s$, $c = 1, \cdots, C$.

Latent class models, containing only a discrete random intercept, have been considered for discrete choice data by Kamakura *et al.* (1996), for rankings by Croon (1989) and for paired comparisons by Dillon *et al.* (1993), among others. We will give examples of latent class models for rankings and discrete choices in Sections 13.5 and 13.6, respectively.

Allenby and Lenk (1994) and Skrondal and Rabe-Hesketh (2003b) are the only contributions we are aware of modeling dependence at both unit and cluster levels although the former do not include explicit terms representing unit-level heterogeneity.

13.4 British general elections: Multilevel models for discrete choice and rankings

Skrondal and Rabe-Hesketh (2003ab) modeled data from the 1987-1992 panel of the British Election Study (Heath *et al.*, 1991, 1993, 1994).[1] 1608 respondents participated in the panel. Voting occasions with missing covariates and where the voters did not vote for candidates from the major parties were excluded. The resulting data comprised 2458 voting occasions, 1344 voters and 249 constituencies.

The alternatives Conservative, Labour, and Liberal (Alliance) are sometimes labeled as con, lab, lib, corresponding to $s = 1, 2, 3$. The voters were not explicitly asked to rank order the alternatives, but the first or discrete choice clearly corresponds to rank 1. The voters also rated the parties on a five point scale from 'strongly against' to 'strongly in favour'. We used these

---

[1] The data were made available by the UK Data Archive and the subset analyzed here can be downloaded from gllamm.org/books, courtesy of Anthony Heath.

ratings to assign ranks to the remaining alternatives, ordering the parties in terms of their rating. Tied second and third choices were observed for 394 voting occasions yielding top-rankings (first choices only).

The covariates considered are

- [LRdist] the distance between a voter's position on the left-right political dimension and the mean position of the party voted for. The mean positions of the parties over voters were used to avoid rationalization problems (e.g. Brody and Page, 1972). The placements were constructed from four scales where respondents located themselves and each of the parties on a 11 point scale anchored by two contrasting statements (priority should be unemployment versus inflation, increase government services versus cut taxation, nationalization versus privatization, more effort to redistribute wealth versus less effort).

- [1987] a dummy variable for the 1987 national elections
- [1992] a dummy variable for the 1992 national elections
- [Male] a dummy for the voter being male
- [Age] age of the voter in 10 year units
- [Manual] a dummy for father of voter a manual worker
- [Inflation] rating of perceived inflation since the last election on a five point scale

The data have a hierarchical structure with elections i (level 1) nested within voters j (level 2) and voters nested within constituencies k (level 3). The variable [LRdist] is alternative and election-specific and will be denoted x^s_{ijk}. The variables [1987], [1992] and [Inflation] are election-specific, denoted $\mathbf{x}_{ijk}$, whereas [Male], [Age] and [Manual] are voter-specific, denoted $\mathbf{x}_{jk}$.

The fixed part of the model includes all covariates and is of the same form as (13.2) with a constant effect b for [LRdist] and party-specific effects $\mathbf{g}^s$ for all other covariates. The constant m^s is not needed since the coefficients of [1987] and [1992] represent election-specific constants.

We first estimated the conventional logistic model $\mathcal{M}0$ without latent variables for both discrete choices and rankings and the estimates are reported in Table 13.1 We see that the estimated effects of the election and/or voter-specific covariates are in accordance with previous research on British elections. Being male and older increases the probability of voting Conservative, whereas a perceived high inflation since the last election harms the incumbent party (the Conservatives). The impact of social class is indicated by the higher probability of voting Labour among voters with a father who is/was a manual worker. Regarding our election, voter and *alternative*-specific covariate [RLdist], the estimate also makes sense: the larger the political distance between voter and party, the less likely it is that the voter will vote for the party.

We consider three types of models for the random part:

(a) a random coefficient model for political distance [LRdist], inducing dependence and allowing the effect of x^s_{ijk} to vary over elections: $x^s_{ijk}\xi^{(1)}_{ijk}$,

Table 13.1 *Estimates for the conventional logistic model* $\mathcal{M}0$

	Ranking		Dicrete Choice	
	Lab vs. Con Est (SE)	Lib vs. Con Est (SE)	Lab vs. Con Est (SE)	Lib vs. Con Est (SE)
g_1^s [1987]	0.38 (0.20)	0.12 (0.17)	0.51 (0.23)	0.13 (0.22)
g_2^s [1992]	0.51 (0.20)	0.13 (0.18)	0.63 (0.24)	−0.13 (0.23)
g_3^s [Male]	−0.79 (0.11)	−0.53 (0.09)	−0.77 (0.13)	−0.67 (0.12)
g_4^s [Age]	−0.37 (0.04)	−0.18 (0.03)	−0.34 (0.04)	−0.20 (0.04)
g_5^s [Manual]	0.65 (0.11)	−0.05 (0.10)	0.69 (0.13)	−0.10 (0.12)
g_6^s [Inflation]	0.87 (0.09)	0.18 (0.03)	0.76 (0.10)	0.57 (0.09)
b [LRdist]	−0.62 (0.02)		−0.54 (0.02)	
Log-likelihood	−2963.68		−1957.91	

Source: Skrondal and Rabe-Hesketh (2003b)

over voters: $x_{ijk}^s\xi_{jk}^{(2)}$ and over constituencies: $x_{ijk}^s\xi_k^{(3)}$. Note that $z_{ijk}^s=x_{ijk}^s$ in this application.

(b) a one-factor model, inducing dependence within elections: $\lambda^{s(1)}\eta_{ijk}^{(1)}$, within voters: $\lambda^{s(2)}\eta_{jk}^{(2)}$ and within constituencies: $\lambda^{s(3)}\eta_k^{(3)}$. At the election level, this is a common factor model because the ϵ_{ijk}^s can be interpreted as unique factors at that level. However, at the higher levels we have factor models without unique factors.

(c) a correlated alternative-specific random intercept model, inducing dependence within voters: $\gamma_{jk}^{s(2)}$ and within constituencies: $\gamma_k^{s(3)}$.
We do not consider correlated alternative-specific random intercept models at the election level since they would be extremely fragile in terms of identification.

At a given level, e.g. the voter level, the model in (b) with random terms $\lambda^{s(2)}\eta_{jk}^{(2)}$, $s = 2,3$ is nested in the random coefficient model (c) with random terms $\gamma_{jk}^{s(2)}$, $s = 2,3$ since the variances of the random terms are unconstrained in both cases whereas the covariance is fixed at one in the factor model and unconstrained in the random coefficient model.

When dependence is modeled at several levels, we use the same kind of model (e.g. (a), (b) or (c)) at the different levels in order to limit the set of models considered. Note that this is a practical consideration; combinations of models can be used at the same level and different models specified at different levels. Furthermore, no parameter restrictions are imposed across levels. The multilevel models are referred to by numbers indicating the levels followed by letters in parentheses for the model type, for instance $\mathcal{M}12$(b) for a one-factor model specified at the election and voter levels.

The sequence of fitted models, their number of parameters (# Par) and log-likelihoods are reported in Table 13.2 for rankings (10-point adaptive quadrature was used).

Table 13.2 *Estimated models for rankings (the fixed part includes all covariates)*

	Random Part				
	Election	Voter	Constit.	# Par	Log-likelihood
$\mathcal{M}0$				13	-2963.68
$\mathcal{M}1$(a)	$z^s_{ijk}\beta^{(1)}_{ijk}$			14	-2945.83
$\mathcal{M}1$(b)	$\lambda^{s(1)}\eta^{(1)}_{ijk}$			15	-2842.73
$\mathcal{M}2$(a)		$z^s_{ijk}\beta^{(2)}_{jk}$		14	-2893.19
$\mathcal{M}2$(b)		$\lambda^{s(2)}\eta^{(2)}_{jk}$		15	-2693.78
$\mathcal{M}2$(c)		$\gamma^{s(2)}_{jk}$		16	-2645.99
$\mathcal{M}3$(a)			$z^s_{ijk}\beta^{(3)}_{k}$	14	-2948.44
$\mathcal{M}3$(b)			$\lambda^{s(3)}\eta^{(3)}_{k}$	15	-2846.26
$\mathcal{M}3$(c)			$\gamma^{s(3)}_{k}$	16	-2844.41
$\mathcal{M}12$(a)	$z^s_{ijk}\beta^{(1)}_{ijk}$	$z^s_{ijk}\beta^{(2)}_{jk}$		15	-2893.19
$\mathcal{M}12$(b)	$\lambda^{s(1)}\eta^{(1)}_{ijk}$	$\lambda^{s(2)}\eta^{(2)}_{jk}$		17	-2691.97
$\mathcal{M}23$(a)		$z^s_{ijk}\beta^{(2)}_{jk}$	$z^s_{ijk}\beta^{(3)}_{k}$	15	-2892.87
$\mathcal{M}23$(b)		$\lambda^{s(2)}\eta^{(2)}_{jk}$	$\lambda^{s(3)}\eta^{(3)}_{k}$	17	-2630.12
$\mathcal{M}23$(c)		$\gamma^{s(2)}_{jk}$	$\gamma^{s(3)}_{k}$	19	-2601.33

Here $x^s_{ijk} = z^s_{ijk}$ is the distance on the left-right political dimension.
Source: Skrondal and Rabe-Hesketh (2003b)

We first introduce latent variables only at the election level in $\mathcal{M}1$(a) and $\mathcal{M}1$(b) and see from the table that the fit is considerably improved compared to the conventional model $\mathcal{M}0$, indicating that there is cross-sectional dependence among the utilities at a given election (given the covariates in the fixed part).

Latent variables are then introduced only at the voter level in models $\mathcal{M}2$(a), $\mathcal{M}2$(b) and $\mathcal{M}2$(c). The fit is much improved compared to the conventional model, indicating that there is unobserved heterogeneity at the voter level inducing longitudinal dependence within voters. We note in passing that models akin to $\mathcal{M}2$(a) have attracted considerable interest in political science (e.g. Rivers, 1988).

We next incorporate latent variables only at the constituency level in $\mathcal{M}3$(a), $\mathcal{M}3$(b) and $\mathcal{M}3$(c), and see that the fit is once more improved. Importantly, latent variables at a given level will not only induce dependence at that level but also induce dependence at all lower levels. Hence, it is possible that latent variables at the election level are superfluous once latent variables are included at the voter level.

To resolve this issue we include latent variables at both election and voter levels in $\mathcal{M}12(a)$ and $\mathcal{M}12(b)$. The improvement in fit achieved by including the additional election level latent variables is negligible suggesting that the cross-sectional dependence within elections is largely due to subject level heterogeneity. We therefore do not need to include latent variables at the election level in the subsequent analyses as long as latent variables at the voter level are included. It would be surprising if the latent variables at the voter level were not needed when latent variables are specified at the constituency level since this would imply conditional independence between a voter's utilities at the two elections given the constituency level effects. As expected, therefore, the $\mathcal{M}23$ models fit considerably better than the $\mathcal{M}3$ models confirming that latent variables are needed at both voter and constituency levels. Regarding the choice between the $\mathcal{M}23$ models, we observe that the fit of the random coefficient model $\mathcal{M}23(a)$ is inferior to the factor model $\mathcal{M}23(b)$ and the random intercepts models $\mathcal{M}23(c)$. The choice between the latter two models, which are nested, suggests that the correlated random intercepts model is the preferred model.

Estimates for our retained model $\mathcal{M}23(c)$ are reported in Table 13.3 from rankings and discrete choices in the left and right panels, respectively. The estimates of the fixed regression coefficients are greater than those shown in Table 13.1 for the conventional model as expected. The random effects variances at the voter level are larger than at the constituency level consistent with a greater residual variability between voters within constituencies than between constituencies as would be expected.

The variance of the random effect for Labour, representing residual variability in the utility differences between Labour and Conservatives, is particularly large reflecting the presence of a mixture of people with strong residual (unexplained) support for the Labour or Conservative parties. There is a positive correlation between the random effects for the Labour and Liberal parties suggesting that those who prefer Labour to the Conservatives, after conditioning on the covariates, also tend to prefer the Liberal party to the Conservatives. This is consistent with the Liberal party being placed between the Labour and Conservative parties and suggests that the [LRdist] covariate has not fully captured this ordering.

To further interpret the estimates for the random part of the model, we present the model-implied residual (conditional on the covariates) correlation matrices for the utility differences in Table 13.4. To derive the residual correlations between utility-differences for a voter at a given election, write the

Table 13.3 *Estimates for correlated alternative-specific random intercepts model at voter and constituency levels* $\mathcal{M}23(c)$

	Ranking		Discrete Choice	
	Lab vs. Con	Lib vs. Con	Lab vs. Con	Lib vs. Con
	Est (SE)	Est (SE)	Est (SE)	Est (SE)
Fixed part				
g_1^s [1987]	0.77 (0.56)	0.75 (0.37)	0.95 (0.52)	0.13 (0.52)
g_2^s [1992]	1.28 (0.59)	0.78 (0.39)	1.32 (0.54)	-0.30 (0.55)
g_3^s [Male]	-0.99 (0.31)	-0.71 (0.20)	-1.15 (0.28)	-0.96 (0.27)
g_4^s [Age]	-0.74 (0.11)	-0.37 (0.07)	-0.61 (0.10)	-0.36 (0.09)
g_5^s [Manual]	1.57 (0.34)	0.10 (0.22)	1.31 (0.31)	0.04 (0.29)
g_6^s [Inflation]	1.31 (0.18)	0.74 (0.13)	1.17 (0.19)	0.97 (0.18)
b [LRdist]	-0.79 (0.04)		-0.87 (0.05)	
Random part				
Voter-level				
$\psi_{\gamma^s}^{(2)}$	16.13 (2.05)	6.03 (0.90)	7.43 (1.62)	9.11 (1.61)
$\psi_{\gamma^2,\gamma^3}^{(2)}$	8.53 (1.15)		5.90 (1.30)	
Const.-level				
$\psi_{\gamma^s}^{(3)}$	4.91 (1.12)	0.60 (0.29)	3.12 (0.86)	1.74 (0.60)
$\psi_{\gamma^2,\gamma^3}^{(3)}$	1.21 (0.48)		1.11 (0.60)	
Log-likelihood	-2600.90		-1748.95	

Source: Skrondal and Rabe-Hesketh (2003b)

model for the three-dimensional vector of utility residuals as

$$
\underbrace{\begin{bmatrix} u_{ijk}^{con} \\ u_{ijk}^{lab} \\ u_{ijk}^{lib} \end{bmatrix}}_{\mathbf{u}_{ijk}} - \underbrace{\begin{bmatrix} f_{ijk}^{con} \\ f_{ijk}^{lab} \\ f_{ijk}^{lib} \end{bmatrix}}_{\mathbf{f}_{ijk}} = \underbrace{\begin{bmatrix} 0 & 0 \\ 1 & 0 \\ 0 & 1 \end{bmatrix}}_{\mathbf{Z}} \underbrace{\begin{bmatrix} \gamma_{jk}^{lab(2)} \\ \gamma_{jk}^{lib(2)} \end{bmatrix}}_{\boldsymbol{\gamma}_{jk}^{(2)}} + \underbrace{\begin{bmatrix} 0 & 0 \\ 1 & 0 \\ 0 & 1 \end{bmatrix}}_{\mathbf{Z}} \underbrace{\begin{bmatrix} \gamma_k^{lab(3)} \\ \gamma_k^{lib(3)} \end{bmatrix}}_{\boldsymbol{\gamma}_k^{(3)}}.
$$

The covariance matrix of this vector of utility residuals becomes

$$
\mathrm{Cov}(\mathbf{u}_{ijk} - \mathbf{f}_{ijk}) = \mathbf{Z}\boldsymbol{\Psi}^{(2)}\mathbf{Z}' + \mathbf{Z}\boldsymbol{\Psi}^{(3)}\mathbf{Z}' + \mathbf{I}_3\pi^2/3.
$$

Consider the vector of differences in utility residuals,

$$
\begin{bmatrix} (u_{ijk}^{lab} - f_{ijk}^{lab}) - (u_{ijk}^{con} - f_{ijk}^{con}) \\ (u_{ijk}^{lib} - f_{ijk}^{lib}) - (u_{ijk}^{con} - f_{ijk}^{con}) \\ (u_{ijk}^{lab} - f_{ijk}^{lab}) - (u_{ijk}^{lib} - f_{ijk}^{lib}) \end{bmatrix}.
$$

Table 13.4 *Residual within-constituency correlation matrix implied by $M23(c)$ for the utility differences (subscript k omitted)*

Overview

		voter j		voter j'	
		1987	1992	1987	1992
voter j	1987	A			
	1992	B	A		
voter j'	1987	C	C	A	
	1992	C	C	B	A

A: Within voter and election

	$u_{ij}^{lab} - u_{ij}^{con}$	$u_{ij}^{lib} - u_{ij}^{con}$	$u_{ij}^{lab} - u_{ij}^{lib}$
$u_{ij}^{lab} - u_{ij}^{con}$	1		
$u_{ij}^{lib} - u_{ij}^{con}$	0.73	1	
$u_{ij}^{lab} - u_{ij}^{lib}$	0.78	0.14	1

B: Within voter between elections

	$u_{i'j}^{lab} - u_{i'j}^{con}$	$u_{i'j}^{lib} - u_{i'j}^{con}$	$u_{i'j}^{lab} - u_{i'j}^{lib}$
$u_{ij}^{lab} - u_{ij}^{con}$	0.86		
$u_{ij}^{lib} - u_{ij}^{con}$	0.63	0.67	
$u_{ij}^{lab} - u_{ij}^{lib}$	0.68	0.29	0.71

C: Between voters (within or between elections)

	$u_{i'j'}^{lab} - u_{i'j'}^{con}$	$u_{i'j'}^{lib} - u_{i'j'}^{con}$	$u_{i'j'}^{lab} - u_{i'j'}^{lib}$
$u_{ij}^{lab} - u_{ij}^{con}$	0.21		
$u_{ij}^{lib} - u_{ij}^{con}$	0.09	0.08	
$u_{ij}^{lab} - u_{ij}^{lib}$	0.22	0.06	0.27

Source: Skrondal and Rabe-Hesketh (2003b)

Defining the comparison matrix

$$\mathbf{H} = \begin{bmatrix} -1 & 1 & 0 \\ -1 & 0 & 1 \\ 0 & 1 & -1 \end{bmatrix},$$

the covariance matrix of the differences in utility residuals becomes

$$\mathrm{Cov}(\mathbf{H}\,(\mathbf{u}_{ijk} - \mathbf{f}_{ijk})) = \mathbf{H}\,\mathrm{Cov}(\mathbf{u}_{ijk} - \mathbf{f}_{ijk})\,\mathbf{H}'.$$

Here the signs of the utility differences are such that we expect positive correlations if the Liberal party is positioned between the Conservative and Labour parties conditional on the covariates. For example, those who prefer Labour to Conservative (positive $u_{ijk}^{\mathrm{lab}} - u_{ijk}^{\mathrm{con}}$) are likely to also prefer Liberal to Conservative (positive $u_{ijk}^{\mathrm{lib}} - u_{ijk}^{\mathrm{con}}$).

As expected, the implied cross-sectional correlations at a given election (A) are larger than those implied from the fixed effects model (0.5 in column 1 and -0.5 in column 2). The longitudinal correlations across elections within voters (B) are all positive (the fixed effects model implies zero correlations), the difference in utilities between the Labour and Conservative parties being the most highly correlated across elections. As would be expected, the correlations between different voters in the same constituency (C) are much lower than between elections for the same voter. The correlation involving the Liberal-Conservative differences tend to be lower than the others, suggesting that these parties were less distinguishable from each other after adjusting for the covariates than the other pairs of parties.

13.5 Post-materialism: A latent class model for rankings

In 'The Silent Revolution' Inglehart (1977) contended that a transformation was taking place in the political cultures of advanced industrial societies. Materialist goals of economic and national security were fading from people's basic value priorities. In their wake was a growing wave of *post-materialist values* - values which emphasize such goals as protecting the freedom of speech and giving people more say in important political decisions. With this change in values, the theory continues, came changes in the salient political issues, changes in political cleavages and changes in political participation.

Inglehart argued that post-materialism is best measured by asking respondents to rank materialistic and post-materialistic values (instead of using for instance rating scales). In the eight nation survey conducted in 1974/1975 (Wieken-Mayser *et al.*, 1979; see also Barnes *et al.*, 1979), also analyzed in Section 10.3, respondents were therefore asked to rank the following four political values according to their desirability:

1. [Order] maintain order in the nation

2. [Say] give people more say in decisions of the government

3. [Prices] fight rising prices

4. [Freedom] protect freedom of speech

Materialists would be expected to give preference to [Order] and [Prices] whereas post-materialists would be expected to prefer [Say] and [Freedom]. Following Croon (1989), we consider data[2,3] on the 2262 German respondents given in Table 13.5. The alternatives are numbered as above so that the first ranking in the table, 1234, represents the rank order [Order], [Say], [Prices], [Freedom].

The heterogeneity in value orientations can be modeled by assuming that respondents' 'utilities' for the political values vary randomly from the overall mean. Croon (1989) describes an exploratory latent class analysis of these rankings. The linear predictor for value s and latent class c can be parameterized as

$$\nu_{jc}^s = e_c^s, \quad s = 1, 2, 3, 4, \quad c = 1, \ldots, C, \qquad (13.5)$$

where

$$e_c^4 = 0.$$

Here we estimate the locations for all latent classes instead of setting their mean to zero. The parameter estimates are shown in Table 13.6 for one, two and three latent classes.

From the one-class model it is clear that, on average, the materialistic values were preferred since the estimated log-odds $\hat{e}^1$ and $\hat{e}^3$ for [Order] and [Prices] are larger than $\hat{e}^2$ and $\hat{e}^4$ ($=0$) for [Say] and [Freedom]. The two-class solution suggests that about 21% of the population is post-materialistic (class 2) with negative log odds for values 1 and 3, whereas about 79% of the population is materialistic (class 1). In the three-class model, the materialists are split into 32% who value [Order] most (class 3) and 45% who value [Prices] most (class 1). The prevalence of post-materialism is now estimated as 23%.

Table 13.5 shows the posterior probabilities for the three-class solution with the highest probability in bold. Class 1 has high posterior probability when [Prices] (3) is ranked high, class 2 if [Say] (2) and [Freedom] (4) are ranked high and class 3 if [Order] (1) is ranked high. Croon's three-class estimates are given in his Table 3 where he uses a different parameterization, imposing the constraint $\sum_s e_c^s = 0$ instead of fixing $e_c^4 = 0$. For the largest class with probability 0.45, this gives the locations 0.60, -1.07, 1.71 and -1.24 which (almost) agrees with Croon's result of $0.59, -1.07, 1.73, -1.25$.

Croon (1989) assesses the adequacy of different numbers of latent classes using the deviance defined as twice the difference in log-likelihoods between a given model and the full or saturated model. The log-likelihood for the saturated model can be obtained from the original data as

$$\sum n_R \ln p_R = -6269.52,$$

[2] The data used in this section were compiled by S.H.Barnes, R.Inglehart, M.K.Jennings and B.Farah and made available by the Norwegian Social Science Data Services (NSD). Neither the original collectors nor NSD are responsible for the analysis reported here.
[3] The data in Table 13.5 are available at gllamm.org/books

Table 13.5 *Materialism data and posterior probabilities*

Data					Results for 3 class model				
							Posterior prob. (%)		
					Pred.	Class:	1	2	3
Ranking				Freq.	freq.	Prior:	0.45	0.23	0.32
1	2	3	4	137	126		10	10	**80**
1	2	4	3	29	46		1	31	**67**
1	3	2	4	309	315		36	3	**61**
1	3	4	2	255	257		37	2	**61**
1	4	2	3	52	40		2	26	**73**
1	4	3	2	93	93		11	6	**83**
2	1	3	4	48	50		20	**40**	40
2	1	4	3	23	29		2	**77**	22
2	3	1	4	61	57		**48**	46	6
2	3	4	1	55	61		7	**93**	0
2	4	1	3	33	32		1	**96**	3
2	4	3	1	59	61		2	**98**	0
3	1	2	4	330	339		**85**	3	12
3	1	4	2	294	281		**86**	2	12
3	2	1	4	117	109		**79**	18	4
3	2	4	1	69	56		25	**75**	0
3	4	1	2	70	81		**87**	9	4
3	4	2	1	34	41		32	**67**	0
4	1	2	3	21	18		3	**66**	32
4	1	3	2	30	30		27	21	**51**
4	2	1	3	29	25		2	**94**	4
4	2	3	1	52	47		3	**97**	0
4	3	1	2	35	33		**68**	23	9
4	3	2	1	27	33		13	**87**	0

Source: Croon (1989)

where n_R and p_R are the absolute and relative observed frequencies of ranking R and the sum is over all 24 observed rankings. The deviances are given in Table 13.6.

We will now extend Croon's analysis by allowing the class probabilities to depend on the following covariates:

- [Female] a dummy variable for females

- [Age] age categories: 15-30 (reference), 31-45, 46-60 and above 60

- [Education] education categories: compulsory school (reference), middle level or academic level

Table 13.6 *Parameter estimates for Croon's latent class model*

	One class	Two classes	Three classes
Class 1			
Probability π_1	1	0.79	0.45
Locations			
e_1^1 [Order]	1.16 (0.04)	1.94 (0.09)	1.84 (0.15)
e_1^2 [Say]	0.21 (0.04)	0.21 (0.05)	0.17 (0.09)
e_1^3 [Prices]	1.28 (0.04)	1.87 (0.09)	2.96 (0.31)
e_1^4 [Freedom]	0	0	0
Class 2			
Probability π_2		0.21	0.23
Locations			
e_2^1 [Order]		−0.87 (0.09)	−0.76 (0.26)
e_2^2 [Say]		0.44 (0.12)	0.56 (0.12)
e_2^3 [Prices]		−0.21 (0.16)	−0.09 (0.19)
e_2^4 [Freedom]		0	0
Class 3			
Probability π_3			0.32
Locations			
e_3^1 [Order]			3.14 (0.40)
e_3^2 [Say]			0.21 (0.10)
e_3^3 [Prices]			1.18 (0.16)
e_3^4 [Freedom]			0
Log-likelihood	−6427.05	−6311.69	−6281.36
Deviance	315.05	84.32	23.58
Degrees of freedom	20	16	12

We specify a structural model (see Section 4.3.2) as

$$\pi_{jc} = \frac{\exp(\mathbf{w}_j' \boldsymbol{\varrho}^c)}{1 + \exp(\mathbf{w}_j' \boldsymbol{\varrho}^c)}.$$

Full covariate information was available on 2246 of the 2262 subjects. Re-estimating the three-class model without covariates on this subsample gave a log-likelihood of −6239.58. The estimates are shown under $\mathcal{M}_0$ in Table 13.7. Allowing the class probabilities to depend on the covariates increased the log-likelihood to −6043.82 (for a loss of 12 degrees of freedom) with estimates shown under $\mathcal{M}_1$ in Table 13.7. Constraining the effect of age and education to be linear across categories (scored 1,2,...) decreased the log-likelihood by only 2.79 (6 degrees of freedom) so that this model $\mathcal{M}_2$, with estimates shown in Table 13.7, is preferred.

For models $\mathcal{M}_1$ and $\mathcal{M}_2$ with covariates, the interpretation of the latent classes remains as for model $\mathcal{M}_0$ without covariates with only small changes in the estimated parameters of the response model. Interestingly, being female reduces the probability of being post-materialistic (class 2) as does [Age], whereas [Education] increases the probability. [Education] and [Age] both increase the probability of valuing [Order] (class 3) over [Prices] whereas [Female] has little effect.

For models $\mathcal{M}_0$ and $\mathcal{M}_2$, the proportions of classification errors (or misclassification rate) $\bar{f}_j$ estimated as described on page 237 in Section 7.4 are given in Table 13.8. If the ranking response has been observed, respondents can be classified by assigning them to the class with the highest posterior probability $\omega(e_c | \mathbf{y}_j, \mathbf{X}_j; \widehat{\boldsymbol{\theta}})$. The corresponding estimated misclassification rates are given in rows 2 and 4 of the table for models $\mathcal{M}_0$ and $\mathcal{M}_2$, respectively. Model $\mathcal{M}_2$ has a slightly lower misclassification rate of 0.217 compared with 0.202 for model $\mathcal{M}_0$ because it uses covariate information.

If the response has not been observed, we must base classification on the prior probabilities π_{jc}. In model $\mathcal{M}_0$ this amounts to assigning everyone to class 1 (with modal probability $\pi_{j1} = 0.460$), giving a misclassification rate of $f_j = 0.540$. In model $\mathcal{M}_2$ the prior probability uses covariate information so that the proportion of classification errors is reduced to 0.466; see rows 1 and 3 of Table 13.8. Such classification based on covariates or 'concomitant' variables only is for instance sometimes required for targeted marketing where rankings, choices or ratings of products are available only for a small 'training set' whereas covariate information is also available for future potential customers (e.g. Wedel, 2002b).

Classification accuracy can also be expressed in terms of the proportional reduction in error (PRE), here relative to model $\mathcal{M}_0$ when the response has not been observed (row 1). These PREs are given in the last column of Table 13.8. The PRE for model $\mathcal{M}_2$ when the response is observed is 0.63.

13.6 Consumer preferences for coffee makers: A conjoint choice model

Conjoint analysis is a marketing research technique that can provide valuable information for market segmentation, new product development, forecasting and pricing decisions (e.g. Wedel and Kamakura, 2000).

In a real purchase situation, shoppers examine and evaluate a range of features or attributes in making their final purchase choice. Conjoint analysis examines these trade-offs to determine what features are most valued by purchasers. Once data are collected the researcher can conduct a number of 'choice simulations' to estimate market share for products with different attributes/features. This gives the researcher some idea which products or services are likely to be successful before they are introduced to the market.

Table 13.7 *Parameter estimates for three-class model with and without covariates*

	$\mathcal{M}_0$ Without covariates	$\mathcal{M}_1$ With covariates (Categorical)	$\mathcal{M}_2$ With covariates (Continuous)
Response model			
Class 1			
e_1^1 [Order]	1.83 (0.15)	1.70 (0.14)	1.71 (0.14)
e_1^2 [Say]	0.18 (0.08)	0.23 (0.08)	0.22 (0.09)
e_1^3 [Prices]	2.93 (0.30)	3.11 (0.25)	3.18 (0.26)
Class 2			
e_2^1 [Order]	−0.80 (0.25)	−0.65 (0.15)	−0.64 (0.15)
e_2^2 [Say]	0.54 (0.12)	0.36 (0.11)	0.37 (0.10)
e_2^3 [Prices]	−0.12 (0.19)	−0.17 (0.13)	−0.15 (0.13)
Class 3			
e_3^1 [Order]	3.14 (0.40)	3.18 (0.29)	3.17 (0.29)
e_3^2 [Say]	0.19 (0.10)	0.20 (0.09)	0.20 (0.09)
e_3^3 [Prices]	1.16 (0.17)	1.40 (0.13)	1.42 (0.13)
Structural model			
Class 2			
ϱ_0^2 [Cons]	−0.37 (0.34)	0.31 (0.27)	0.41 (0.24)
ϱ_1^2 [Female]		−0.90 (0.19)	−0.88 (0.19)
ϱ_2^2 [Age]†			−0.94 (0.11)
ϱ_3^2 15-30		0	
ϱ_4^2 31-45		−0.97 (0.23)	
ϱ_5^2 46-60		−1.57 (0.28)	
ϱ_6^2 > 60		−3.15 (0.51)	
ϱ_7^2 [Education]†			1.51 (0.19)
ϱ_8^2 Compulsory		0	
ϱ_9^2 Middle		1.65 (0.23)	
ϱ_{10}^2 Academic		2.64 (0.47)	
Class 3			
ϱ_0^3 [Cons]	−0.73 (0.23)	−0.89 (0.37)	−0.69 (0.29)
ϱ_1^3 [Female]		−0.07 (0.16)	−0.07 (0.16)
ϱ_2^3 [Age]†			0.26 (0.08)
ϱ_3^3 15-30		0	
ϱ_4^3 31-45		0.39 (0.31)	
ϱ_5^3 46-60		0.90 (0.31)	
ϱ_6^3 > 60		0.82 (0.31)	
ϱ_7^3 [Education]†			0.59 (0.19)
ϱ_8^3 Compulsory		0	
ϱ_9^3 Middle		0.64 (0.22)	
ϱ_{10}^3 Academic		0.84 (0.57)	
Log-likelihood	−6239.58	−6043.82	−6046.79

† linear effect across ordered categories

Table 13.8 *Misclassification rate and proportional reduction in error* (PRE)

Information	Model	$\bar{f}_j$	PRE
No information	$\mathcal{M}_0$	0.540	
Response only	$\mathcal{M}_0$	0.217	0.60
Covariates only	$\mathcal{M}_2$	0.466	0.14
Response and covariates	$\mathcal{M}_2$	0.202	0.63

As an example we will consider data[4,5] on a conjoint choice experiment for coffee makers. After in-depth discussions with experts and consumers, hypothetical coffee-makers were defined using the following five attributes:

- [Brand] brand-name: Philips, Braun, Moulinex
- [Capacity] number of cups: 6, 10, 15
- [Price] price in Dutch Guilders f: 39, 69, 99
- [Thermos] presence of a thermos flask: yes, no
- [Filter] presence of a special filter: yes, no

A total of sixteen profiles were constructed from combinations of the levels of these attributes using an incomplete design (excluding unrealistic combinations such as a coffee maker with all the features costing only 39f). In the choice experiment, respondents were then asked to make choices out of sets of three profiles, each set containing the same base alternative.

There are several advantages of conjoint choice experiments as compared to conventional conjoint analysis based on rating scales. One advantage is that choices may be more realistic since they resemble the purchasing situation, another that the problem of individual differences in interpreting rating scales is avoided.

To construct the choice sets, the profiles were divided in two different ways into eight sets of two alternatives. A base alternative was added to each set, resulting in two groups of eight choice sets with three alternatives shown in Table 13.9. 185 respondents were recruited at a large shopping mall in the Netherlands. These respondents were randomly divided into two groups of 94 and 91 subjects and each group was administered one of the groups of eight choice sets.

The multinomial logit model containing only fixed effects of the vector of attributes $\mathbf{x}^s$ for alternative (or profile) s has linear predictor

$$\nu_j^s = \mathbf{x}^{s\prime}\mathbf{b}$$

for subject j. Note that the intercepts m^s are omitted to identify the model

---

[4] We thank Michel Wedel for providing us with this dataset which accompanies the GLIMMIX program (Wedel, 2002b).

[5] The data can be downloaded from `gllamm.org/books`

Table 13.9 *Choice sets for conjoint choice experiment*

	Alternative 1					Alternative 2			
Set	Brand	Cap.	Pr.	Th.	Fi.	Brand	Cap.	Pr.	Th.	Fi.
			Choice sets for group 1							
1	Philips	10	69	–	Fi	Braun	15	69	Th	Fi
2	Braun	6	69	–	–	Moulinex	10	69	Th	Fi
3	Braun	10	39	–	–	Braun	10	99	Th	Fi
4	Philips	6	39	Th	Fi	Braun	10	39	Th	–
5	Philips	10	69	Th	–	Moulinex	15	39	–	Fi
6	Braun	6	69	–	Fi	Moulinex	10	69	–	–
7	Philips	15	99	–	–	Moulinex	6	99	Th	–
8	Braun	15	69	Th	–	Braun	10	99	–	Fi
			Choice sets for group 2							
1	Philips	10	69	Th	–	Moulinex	10	69	Th	Fi
2	Philips	15	99	–	–	Braun	15	69	Th	Fi
3	Braun	10	39	–	–	Moulinex	15	39	–	Fi
4	Braun	15	69	Th	–	Braun	10	99	–	Fi
5	Philips	10	69	–	Fi	Moulinex	6	99	Th	–
6	Braun	6	69	–	–	Braun	10	99	Th	Fi
7	Braun	6	69	–	–	Moulinex	10	69	–	–
8	Philips	6	39	Th	Fi	Braun	10	39	Th	–

	Alternative 3 (base)				
Set	Brand	Cap.	Pr.	Th.	Fi.
1	Philips	6	69	–	Th

because the covariates $\mathbf{x}^s$ do not vary between subjects. Excluding a constant also has the advantage that we can make predictions involving profiles not included in the data.

This model is unrealistic since it assumes that all subjects have the same mean utilities for the coffee makers. A more realistic model allows subjects to differ in their coefficients for the attributes, reflecting different preferences or 'tastes'. The linear predictor becomes

$$\nu_j^s = \mathbf{x}^{s\prime}\mathbf{b} + \mathbf{x}^{s\prime}\boldsymbol{\gamma}_j.$$

If the market is believed to consist of different 'segments' that are homogeneous in their preferences, a latent class model can be specified using a discrete random coefficient vector

$$\boldsymbol{\gamma}_j = \mathbf{e}_c, \quad c = 1, \cdots, C,$$

with probabilities π_c, where c labels the market segments. This model is often referred to as a mixture regression model. We will exclude the fixed part $\mathbf{x}^s$ from the model instead of imposing the constraint $\sum_c \pi_c \mathbf{e}_c = \mathbf{0}$, so that the linear predictor for class c becomes

$$\nu_{jc}^s = \mathbf{x}^{s\prime}\mathbf{e}_c.$$

Alternatively, the random tastes $\boldsymbol{\gamma}_j$ can be assumed to be continuous. Haaijer *et al.* (1998) develop a random coefficient multinomial probit model where $\boldsymbol{\gamma}_j$ is multivariate normal with covariance structure

$$\mathrm{Cov}(\boldsymbol{\gamma}_j) = \mathbf{Q}\mathbf{Q}'.$$

Here, $\mathbf{Q}$ is a $8 \times T$ matrix and some constraints on $\mathbf{Q}$ are necessary for identification. Even with these constraints, they found identification to be fragile for $T > 1$. With $T = 1$, the multinomial logit version of the model can be written in GRC notation as

$$\nu_j^s = \mathbf{x}^{s\prime}\mathbf{b} + \eta_j \mathbf{x}^{s\prime}\boldsymbol{\lambda}, \quad \mathrm{Var}(\eta_j) = 1.$$

Here we have set $\boldsymbol{\gamma}_j = \eta_j \boldsymbol{\lambda}$, thereby reducing the number of dimensions from eight to one and greatly simplifying estimation.

The estimates for the one-class, two-class and random coefficient models are given in Table 13.10. The coefficients in the one-class model suggest that Braun is the least popular brand, a capacity of 10 cups is most desirable, followed by 15 cups and then 6 cups, cheaper coffee makers (39 or 69f) are preferred to the most expensive ones (99f) and coffee makers with filters and thermos flasks are preferred to coffee makers without these features.

The two-class model fits considerably better than the one-class model (difference in log-likelihoods is 188). The size of the first market segment is estimated as 72% and that of the second segment as 28%. The first market segment cares little about brands whereas the second strongly prefers Philips and dislikes Braun. The first segment has a more marked preference for 10 cups and cares more about prices, as well as having a stronger preference for thermos flasks than the second segment.

The random coefficient model suggests that subjects vary mostly in the degree to which they prefer Philips over Braun, the extent of dislike of a 6-cup capacity and their price sensitivity. It is interesting to note that the log-likelihood for the independent multinomial logit model is very close to that found by Haaijer *et al.* (1998) for the independent multinomial probit model using simulated maximum likelihood (-1298.7 and -1299.9, respectively). The log-likelihoods for the logit and probit versions of the random coefficient model are also very similar (-1086.1 and -1086.6, respectively).

Haaijer *et al.* (1998) also consider choice simulations for different product

Table 13.10 *Estimates for conjoint choice analysis*

| | | Latent class models | | | Random coeff. model | | |
| | | $\mathcal{M}1$ One class | | $\mathcal{M}2$ Two classes | | $\mathcal{M}3$ | | |
Variable	Par.	Est (SE)		Est (SE)	Par.	Est (SE)	
[Brand]							
Philips	e_1^1	−0.25 (0.11)		−0.37 (0.17)	b_1	−0.31	(0.15)
Braun	e_2^1	−0.62 (0.11)		−0.40 (0.16)	b_2	−0.65	(0.15)
Moulinex	e_3^1	0		0	b_3	0	
[Capacity]							
6	e_4^1	−1.54 (0.12)		−2.48 (0.21)	b_4	−2.08	(0.20)
10	e_5^1	−0.03 (0.10)		0.06 (0.14)	b_5	0.13	(0.13)
15	e_6^1	0		0	b_6	0	
[Price]							
39	e_7^1	1.00 (0.15)		1.97 (0.34)	b_7	1.52	(0.27)
69	e_8^1	1.06 (0.11)		1.48 (0.17)	b_8	1.21	(0.16)
99	e_9^1	0		0	b_9	0	
[Thermos]							
yes	e_{10}^1	0.62 (0.09)		1.14 (0.18)	b_{10}	1.09	(0.15)
no	e_{11}^1	0		0	b_{11}	0	
[Filter]							
yes	e_{12}^1	0.68 (0.08)		0.92 (0.12)	b_{12}	0.97	(0.11)
no	e_{13}^1	0		0	b_{13}	0	
$\ln[\pi_1/(1-\pi_1)]$	ϱ_0^1	0		0.92 (0.21)			
[Brand]							
Philips	e_1^2			0.12 (0.21)	λ_1	0.51	(0.18)
Braun	e_2^2			−1.43 (0.31)	λ_2	−0.51	(0.20)
Moulinex	e_3^2			0	λ_3	0	
[Capacity]							
6	e_4^2			−0.25 (0.26)	λ_4	1.72	(0.22)
10	e_5^2			0.07 (0.25)	λ_5	0.19	(0.21)
15	e_6^2			0	λ_6	0	
[Price]							
39	e_7^2			−0.49 (0.32)	λ_7	−1.71	(0.30)
69	e_8^2			−0.04 (0.22)	λ_8	−1.11	(0.21)
99	e_9^2			0	λ_9	0	
[Thermos]							
yes	e_{10}^2			0.35 (0.20)	λ_{10}	−0.60	(0.19)
no	e_{11}^2			0	λ_{11}	0	
[Filter]							
yes	e_{12}^2			1.00 (0.20)	λ_{12}	0.08	(0.14)
no	e_{13}^2			0	λ_{13}	0	
Log-likelihood		−1298.71		−1110.63		−1086.07	

introductions. They use the four profiles listed in Table 13.11 to generate three 'managerially relevant' situations:

Table 13.11 *Profiles for market simulations*

Prof.	Brand	Cap.	Pr.	Th.	Fi.	Prof.	Brand	Cap.	Pr.	Th.	Fi.
P1	Philips	10	39	–	–	**B2**	Braun	15	69	Th	–
M3	Moulinex	15	69	–	–	**P4**	Philips	10	69	–	Fi

- *Product modification*: The current market consist of two products, Philips (P1) and Braun (B2). Philips modifies its existing product (from P1 to P4) by introducing a special filter and increasing the price.
- *Product line extension*: The current market consist of two products, Philips (P1) and Braun (B2). Philips introduces a new product (P4) in addition to its existing product.
- *Introduction of a 'me-too' brand*: The current market consist of two products, Philips (P1) and Braun (B2). A third brand, Moulinex, introduces a product (M3), close to the existing product of the current market-leader Braun, but without the thermos-flask.

The market-share predictions for these three scenarios are given in Table 13.12 for each of the models. Product modification leads to a greater increase in market share for the independent and two-class models than for the random coefficient model. Product-line extension by Philips leads to a greater decrease in Braun's market share for the independent and two-class models than for the random coefficient model. Similarly, the introduction of a new brand ('me-too') leads to a greater decrease in Braun's market share for the independent and two-class models than for the random coefficient model. These simulations illustrate how assuming different dependence structures can have important implications for predictions.

Table 13.12 *Predicted market shares in percent*

Product modification				Product-line extension				Intro. 'me-too' brand			
Pr.	$\mathcal{M}1$	$\mathcal{M}2$	$\mathcal{M}3$	Pr.	$\mathcal{M}1$	$\mathcal{M}2$	$\mathcal{M}3$	Pr.	$\mathcal{M}1$	$\mathcal{M}2$	$\mathcal{M}3$
Before:											
P1	41.5	45.8	43.9	P1	41.5	45.8	43.9	P1	41.5	45.8	43.9
B2	58.5	54.2	56.1	B2	58.5	54.2	56.1	B2	58.5	54.2	56.1
After:											
P4	59.8	59.4	55.2	P1	22.2	21.6	18.6	P1	26.2	30.3	30.4
B2	40.2	40.6	44.8	B2	31.3	31.4	37.1	B2	37.0	39.4	43.8
				P4	46.5	47.1	44.3	M3	36.8	30.3	25.7

13.7 Summary and further reading

We have considered models for comparative responses in this chapter, basing all applications on logistic regression models with continuous or discrete latent variables. The first application considered panel or longitudinal data on discrete choices and rankings from British general elections. The model included both alternative and unit-specific covariates and latent variables at different hierarchical levels. The second application concerned rankings on materialistic and post-materialistic values among Germans. Latent class models previously suggested for this application were extended by including covariates. The third application was based on data from a conjoint choice experiment where respondents were asked to choose among coffee makers. The profiles of the coffee makers making up the choice sets were constructed according to an experimental design, helping the market researcher to gauge which product will be successful before introducing it to the market. See Wedel and Kamakura (2000) for a modern treatment of market segmentation.

In all applications discussed in this chapter, the responses correspond to decisions, but this need not be the case. Other examples of polytomous responses are eye-color or diagnosis. Yang (2001) and Skrondal and Rabe-Hesketh (2003c) use a three-level multinomial logit model to analyze the quality of physicians' treatment decisions. Pairwise comparison data also arise in tournaments such as chess or football, whereas rankings could be the finishing order in a horse-race.

In addition to the numerous references given in the introduction of this chapter, see Hartzel *et al.* (2001) and Train (2003) for useful overviews. For a review of latent variable models for polytomous responses and rankings, see Skrondal and Rabe-Hesketh (2003b). A more introductory treatment is given in Skrondal and Rabe-Hesketh (2003a). Takane (1987) and Böckenholt (2001a) discuss latent variable models for pairwise comparisons.

CHAPTER 14

Multiple processes and mixed responses

14.1 Introduction

In the previous application chapters we considered responses of particular types. In this chapter we exploit the generality of the general model framework and discuss applications where the responses are from multiple processes and possibly of mixed type. Combinations treated here are continuous and dichotomous, dichotomous and counts, continuous and continuous time survival and dichotomous and continuous time survival. Importantly, we will see that it is often not permissible to simply decompose such problems, that is by separately modeling the different processes. As in other application chapters, we will use continuous latent variables with parametric distributions and discrete latent variables, interpretable as latent classes or as a 'nonparametric' estimator of an unspecified distribution. The usefulness of structural models, regressing latent variables on other variables, will be demonstrated.

14.2 Diet and heart disease: A covariate measurement error model

14.2.1 Introduction

We consider estimating the effect of dietary fiber intake on coronary heart disease (CHD), following the analysis in Rabe-Hesketh *et al.* (2003ab).

The dataset is on 337 middle aged men, recruited between 1956 and 1966 and followed until 1976 (Morris *et al.*, 1977)[1]. There were two occupational groups, bank staff and London Transport staff (drivers and conductors). At the time of recruitment, the men were asked to weigh their food over a seven-day period from which the total number of calories were derived as well as the amount of fat and fiber. Seventy-six bank staff had their diet measured in the same way again six months later. Coronary heart disease was determined from personnel records and, after retirement, by direct communication with the retired men and by 'tagging' them at the Registrar General's Office.

The explanatory variables used in the analysis are

- [Age] age in years

- [Transp] dummy variable for London Transport staff versus bank staff

---

[1] We thank David Clayton for providing us with these data.

14.2.2 Logistic regression with covariate measurement error

We will estimate the association between fiber intake (exposure) and heart disease using logistic regression, taking into account that exposure measurement is imperfect and that replicate measurements are available for a subsample. This may be accomplished by introducing a latent variable for unobserved true exposure and specifying three submodels (following Clayton's (1992) terminology): an exposure model, a measurement model and a disease model.

Exposure model

True fiber intake η_j for subject j is modeled using the structural model

$$\eta_j \;=\; \mathbf{x}'_j \boldsymbol{\gamma} + \zeta_j, \tag{14.1}$$

where the covariates $\mathbf{x}_j$ are [Age], [Transp] and their interaction. Traditionally, a normal exposure distribution is assumed with $\zeta_j \sim N(0, \psi)$, but we will also consider a nonparametric exposure distribution of the kind introduced in Section 4.4.2.

Measurement model

The classical measurement model assumes that the ith fiber measurement for subject j, y_{ij}, differs from true fiber intake η_j by a normally distributed measurement error ϵ_{ij},

$$y_{ij} \;=\; \eta_j + \epsilon_{ij}, \qquad \epsilon_{ij} \sim N(0, \theta),$$

where ϵ_{ij} is independent from η_j (and ζ_j in the exposure model). Two measurements were available for a subsample, but information from subjects only providing one measurement is also used. We will also allow for a 'drift' in the fiber measurements by including a dummy variable [Drift] for the second measurement.

Disease model

The disease model specifies a logistic regression of heart disease D_j on true fiber intake

$$\mathrm{logit}[\Pr(D_j{=}1|\eta_j)] \;=\; \mathbf{x}'_j \boldsymbol{\beta} + \lambda \eta_j.$$

Here, the factor loading λ represents the effect of true exposure on the log-odds of disease. The same covariates $\mathbf{x}_j$ are used in both the disease and exposure models to allow for both direct and indirect effects of these covariates on disease. We also consider a model with $\boldsymbol{\beta} = \mathbf{0}$, thus specifying only indirect effects of the covariates on heart disease via true fiber intake. Both kinds of models are shown in Figure 14.1.

Joint model

Let $i = 1, 2$ index the fiber measurements and $i = 3$ the disease response, and define the corresponding dummy variables d_{1i}, d_{2i} and d_{3i}. The joint response

Direct and indirect effects

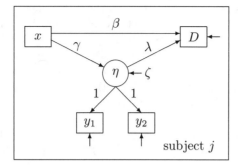

Indirect effects only

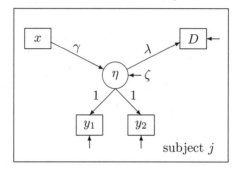

Figure 14.1 *Path diagram for covariate measurement error models*

model for measurement and disease can then be written in GRC formulation as

$$\begin{aligned}
\nu_{ij} &= (d_{1i} + d_{2i})\eta_j + d_{3i}[\mathbf{x}_j'\boldsymbol{\beta} + \lambda\eta_j] \\
&= d_{3i}\mathbf{x}_j'\boldsymbol{\beta} + \eta_j[(d_{1i} + d_{2i}) + \lambda d_{3i}],
\end{aligned}$$

and the structural model for exposure remains as in (14.1).

Note that measurement error is assumed to be *nondifferential* since it is conditionally independent of disease status D_j given true exposure η_j (no direct path between y_{ij} and D_j in the path diagrams). The joint model is a generalization of the MIMIC model discussed in Section 3.5, allowing responses of mixed type (diet measurements are continuous and heart disease is dichotomous) and direct effects of covariates on the responses.

The parameter estimates, based on a normal and nonparametric exposure distribution and including and excluding direct effects of the covariates on heart disease, are given in Table 14.1. Similar estimates of the effect of true fiber on heart disease are obtained for all four cases considered, with an odds ratio of about 0.87 per gram of fiber per day. Fiber intake therefore appears to have a remarkable protective effect on heart disease. However, we have not adjusted for exercise, an important confounder which both increases total food intake (including fiber) and reduces the risk of heart disease (Morris *et al.*, 1977). London Transport staff eat less fiber than bank staff, fiber intake decreases with age and there is no substantial interaction between occupation and age. Excluding the direct effect of these variables on heart disease leads to a very small decrease in the log-likelihood, both for the normal and non-parametric exposure distributions. Thus the covariates appear to affect the risk of heart disease only indirectly via fiber intake as in the lower panel of Figure 14.1.

For the model with indirect effects only, the reliability of the fiber measurements given the covariates is estimated as 0.77 when exposure is assumed to have a normal distribution and as 0.80 using NPMLE. The somewhat higher estimate for NPMLE is consistent with simulations reported in Rabe-Hesketh *et al.* (2003a) and Hu *et al.* (1998).

Table 14.1 *Parameter estimates for heart disease data*

| | Indirect effects only | | Direct and indirect effects | |
| | Quadrature | NPMLE | Quadrature | NPMLE |
	Est (SE)	Est (SE)	Est (SE)	Est (SE)
Exposure model				
γ_1 [Transp]	-1.66 (0.64)	-1.12 (0.44)	-1.68 (0.64)	-1.14 (0.44)
γ_2 [Age]	-0.21 (0.10)	-0.29 (0.06)	-0.21 (0.10)	-0.28 (0.06)
γ_3 [Age] $\times$[Transp]	0.17 (0.11)	0.22 (0.07)	0.17 (0.11)	0.22 (0.07)
Var(ζ_i)	23.66 (2.53)	24.94 (–)	23.64 (2.52)	24.98 (–)
Measurement model				
α_0 [Const]	17.93 (0.49)	17.58 (0.40)	17.95 (0.49)	17.60 (0.40)
α_1 [Drift]	0.24 (0.42)	0.16 (0.38)	0.23 (0.42)	0.15 (0.38)
θ	6.95 (1.14)	6.13 (0.85)	6.95 (1.14)	6.13 (0.85)
Disease model				
λ	-0.13 (0.05)	-0.15 (0.06)	-0.13 (0.05)	-0.15 (0.06)
β_0 [Const]	-2.08 (0.21)	-2.07 (0.21)	-1.92 (0.28)	-1.90 (0.28)
β_1 [Transp]			-0.26 (0.34)	-0.27 (0.34)
β_2 [Age]			0.04 (0.06)	0.04 (0.06)
β_3 [Age] $\times$[Transp]			-0.03 (0.06)	-0.03 (0.07)
Log likelihood	-1373.33	-1320.25	-1372.35	-1319.79

The NPMLE solutions required six masses which are displayed for the case of indirect effects only in the upper panel of Figure 14.2 (the distribution for the model also including direct effects was very similar). As might be expected,

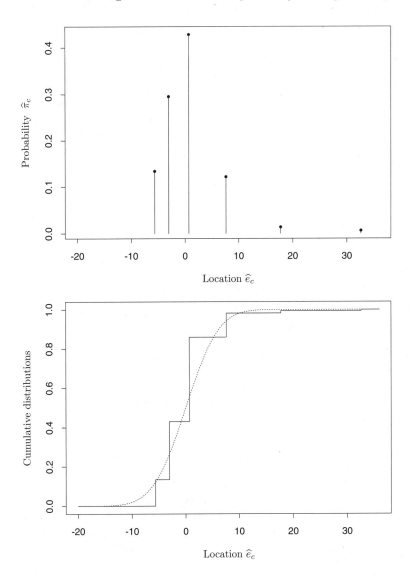

Figure 14.2 *NPMLE probability masses and cumulative distribution for NPMLE and normal distributions*

the distribution is positively skewed. Another indication of nonnormality of the true fiber intake distribution is the considerable increase in log-likelihood when relaxing the normality assumption. (The NPMLE solutions have nine extra parameters for a change in log-likelihood of about 53.) The lower panel of the figure shows the estimated cumulative distribution both for normal and nonparametric true fiber intake distributions.

The empirical Bayes predictions of the disturbances ζ_j in the exposure model are shown in Figure 14.3 for both NPMLE and normal exposure. The discrepancy appears to be greater for larger predictions where NPMLE

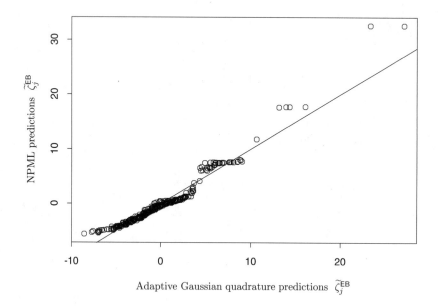

Figure 14.3 *Empirical Bayes predictions of true residual fiber for normal and non-parametric exposure distributions.*

produces larger values. Simulations reported in Rabe-Hesketh *et al.* (2003a) showed that NPMLE predictions are superior to those assuming normality if the true distribution is highly skewed. In particular, the parametric empirical Bayes predictions are too severely shrunken when the true values are large.

14.3 Herpes and cervical cancer: A latent class covariate measurement error model for a case-control study

14.3.1 Introduction

Sampling is sometimes conducted stratified on a dichotomous response. Such retrospective designs are called case-control designs in epidemiology (e.g. Breslow and Day, 1980; Schlesselman, 1982) and choice-based sampling in economics (e.g. Manski, 1981). Sampling typically proceeds by including all 'cases' (or ill persons in epidemiology) and a random sample of roughly as many 'controls' (healthy persons) as there are cases. Importantly, when the case prevalence is low, the loss of efficiency compared to a prospective cohort design is small. An important merit is that more accurate covariate information may be potentially obtained than in a prospective cohort design since measurements are only required for a sample of controls.

It has been shown that logistic regression for this retrospective design can be performed as if the design were prospective (e.g. Farewell, 1979), giving appropriate estimates except for the intercept. Importantly, Carroll et al. (1995b) point out that prospective analysis of case-control studies with covariate measurement error also typically produces consistent estimators and asymptotically correct standard errors. Specifically, for the case of a categorical exposure and nondifferential measurement error, Satten and Kupper (1993) and Carroll et al. (1995b) demonstrate that maximizing the prospective likelihood produces correct inferences. We adopt this approach in this section, effectively ignoring the retrospective design in specifying the likelihood.

Table 14.2 *Data for case control study of cervical cancer*

[Case]	True exposure [TrueE]	Measured exposure [MeasE]	Count
Validation sample			
1	0	0	13
1	0	1	3
1	1	0	5
1	1	1	18
0	0	0	33
0	0	1	11
0	1	0	16
0	1	1	16
Incomplete sample			
1		0	318
1		1	375
0		0	701
0		1	535

Source: Carroll et al. (1993)

Hildesheim et al. (1991) conducted a case-control study to examine a potential association between exposure to herpes simplex virus type 2 and invasive cervical cancer. The exposure was measured for cases and controls using an inaccurate western blot procedure. To investigate misclassification, a gold standard measurement using a refined western blot procedure was obtained for a random sub-sample of women from both groups, the 'validation sample'.

We will treat this measurement as 'true exposure' [TrueE] and refer to the inaccurate western blot as 'measured exposure' [MeasE]. The data[2] have been analyzed by Carroll *et al.* (1993) and others and are given in Table 14.2.

We can of course validly estimate the odds ratio for true exposure based on just the validation sample. However, this approach would be inefficient since the information from the large incomplete sample is discarded.

14.3.2 Latent class logistic regression

We can estimate the odds ratio for true exposure based on all information by specifying three component models as in the previous section: an exposure model, a measurement model and a disease model.

Exposure model

Let X_{1j} denote true exposure in the validation sample (1 for exposed and 0 for unexposed). We can treat the missing exposure in the incomplete data sample as a dichotomous latent variable η_j taking the values $\eta_j = e_c$, $c = 1, 2$, $e_1 = 1$, $e_2 = 0$. The exposure models for the two samples are simply

$$\text{logit}[\Pr(X_{1j} = 1)] = \varrho_0,$$

and

$$\text{logit}[\pi_1] = \varrho_0,$$

where π_1 is the probability that a subject in the incomplete data sample is in latent class 1.

Measurement model

An assumption often made in covariate measurement error problems, for instance in Section 14.2.2, is that there is nondifferential measurement error, i.e. that measured exposure is conditionally independent of disease status given true exposure. Let W_{1j} and W_{0j} denote measured exposure in the validation and incomplete data samples, respectively. Also let D_{1j} and D_{0j} denote disease status (1 for cases and 0 for controls) in the validation sample and incomplete data samples, respectively. A *nondifferential* measurement error model can be specified as

$$\text{logit}[\Pr(W_{1j}|X_{1j}, D_{1j})] = \alpha_0 + \alpha_1 X_{1j},$$

for the validation sample and

$$\text{logit}[\Pr(W_{0j}|\eta_j, D_{0j})] = \alpha_0 + \alpha_1 \eta_j,$$

for the incomplete data sample.

Carroll *et al.* (1993) point out that there is evidence for differential measurement error in the validation sample with cases having a higher estimated

---

[2] The data can be downloaded from `gllamm.org/books`

sensitivity ($18/23 = 0.78$) than controls ($16/32 = 0.50$). The saturated *differential* measurement error model includes the effects of true exposure, disease status and their interaction,

$$\text{logit}[\Pr(W_{1j}|X_{1j}, D_{1j})] = \alpha_0 + \alpha_1 X_{1j} + \alpha_2 D_{1j} + \alpha_3 D_{1j} X_{1j},$$

and

$$\text{logit}[\Pr(W_{0j}|\eta_j, D_{0j})] = \alpha_0 + \alpha_1 \eta_j + \alpha_2 D_{0j} + \alpha_3 D_{0j} \eta_j.$$

Disease model

For the validation sample, we can specify a logistic regression model for disease given true exposure

$$\text{logit}[\Pr(D_{1j} = 1|X_{1j})] = \beta_0 + \beta_1 X_{1j}.$$

For the remainder of the sample, true exposure status is missing and represented by latent classes. The model for disease status D_{0j} in the incomplete data sample then is

$$\text{logit}[\Pr(D_{0j} = 1|\eta_j)] = \beta_0 + \beta_1 \eta_j.$$

Joint model

Let y_{ij} denote the three responses, exposure ($i = 1$), measurement ($i = 2$) and disease ($i = 3$) with dummy variables $d_{ri} = 1$ if $r = i$ and 0 otherwise. Let $v_j = 1$ if subject j is in the validation sample and 0 otherwise. The joint model (allowing for differential measurement error) can be written in the GRC formulation as

$$
\begin{aligned}
\nu_{ij} &= d_{1i}[\varrho_0] + d_{2i}[\alpha_0 + \alpha_1 X_{1j} v_j + \alpha_1 \eta_j (1 - v_j) \\
&\quad + \alpha_2 D_{1j} v_j + \alpha_2 D_{0j}(1 - v_j) + \alpha_3 X_{1j} D_{1j} v_j + \alpha_3 \eta_j D_{0j}(1 - v_j)] \\
&\quad + d_{3i}[\beta_0 + \beta_1 X_{1j} v_j + \beta_1 \eta_j (1 - v_j)] \\
&= \varrho_0 d_{1i} + \alpha_0 d_{2i} + \alpha_1 X_{1j} v_j d_{2i} + \alpha_2 D_{1j} v_j d_{2i} + \alpha_2 D_{0j}(1 - v_j) d_{2i} \\
&\quad + \alpha_3 X_{1j} D_{1j} v_j d_{2i} + \beta_0 d_{3i} + \beta_1 X_{1j} v_j d_{3i} \\
&\quad + \eta_j[\alpha_1(1 - v_j) d_{2i} + \alpha_3 D_{0j}(1 - v_j) d_{2i} + \beta_1(1 - v_j) d_{3i}],
\end{aligned}
$$

and

$$\text{logit}[\pi_1] = \varrho_0.$$

Here α_1, α_3 and β_1 are factor loadings constrained equal to the corresponding regression coefficients for the validation sample.

The parameter estimates for the models assuming differential and nondifferential measurement error are given in Table 14.3. We can use the estimates for the measurement model to estimate the conditional probabilities $\Pr(W_j = 1|X_j, D_j)$, providing information on the sensitivity and specificity of the measurements, and these are given in Table 14.4.

The likelihood ratio test statistic for comparing the two models is 5.14 for 2

Table 14.3 *Estimates for cervical cancer data*

Parameter	Differential measurement error		Nondifferential measurement error	
	Est	(SE)	Est	(SE)
Exposure model				
ϱ_0 [Cons]	−0.023	(0.171)	0.006	(0.165)
Measurement model				
α_0 [Cons]	−0.791	(0.259)	−1.063	(0.224)
α_1 [TrueE]	1.099	(0.496)	1.813	(0.365)
α_2 [Case]	−0.665	(0.616)		
α_3 [TrueE]×[Case]	1.649	(0.955)		
Disease model				
β_0 [Cons]	−0.897	(0.192)	−1.097	(0.149)
β_1 [TrueE]	0.608	(0.350)	0.958	(0.237)
Log likelihood	−2802.63		−2805.19	

Table 14.4 *Conditional probabilities* $\Pr(W_j = 1 | X_j, D_j)$ *of measured exposure*

X_j [TrueE]	D_j [Case]	Differential meas. error	Nondifferential meas. error
		1–Specificity	
0	0	0.31	0.26
0	1	0.19	0.26
		Sensitivity	
1	0	0.58	0.68
1	1	0.79	0.68

degrees of freedom, so that the simpler model assuming nondifferential measurement error appears to be adequate. The log odds ratios for true exposure are estimated as 0.608 (0.350) assuming differential measurement error and 0.958 (0.237) assuming nondifferential measurement error. Using just the validation sample, the odds ratio is estimated as 0.681 (0.400). Note that we have gained very little in terms of precision by analyzing the full sample, probably because of the low sensitivity and specificity of measured exposure. Using the inaccurate measurement of exposure instead of the gold standard for the full sample gives an estimate of 0.453 (0.093). As expected, this estimate is attenuated compared with the other estimates due to regression dilution. Our estimates based on a prospective likelihood agree very closely with the estimates obtained by Carroll *et al.* (1993) using a retrospective likelihood.

For the incomplete data, we can predict true exposure to the herpes virus using the posterior probabilities of true exposure given measured exposure and disease status. These posterior probabilities, providing information on positive predictive values (PPV) and negative predictive values (NPV), are given in Table 14.5.

Table 14.5 *Posterior probabilities* $\Pr(X_j = 1 | W_j, D_j)$ *of true exposure*

W_j [MeasE]	D_j [Case]	Differential meas. error	Nondifferential meas. error
		1–NPV	
0	0	0.33	0.24
0	1	0.28	0.45
		PPV	
1	0	0.59	0.65
1	1	0.86	0.83

14.4 Job training and depression: A complier average causal effect model

14.4.1 Introduction

Little and Yau (1998) analyzed data from the JOBS II intervention trial described in Vinokur *et al.* (1995)[3]. Unemployed individuals who had lost their jobs within the last 13 weeks and were looking for a job where randomized to receive either five half-day sessions of job training plus a booklet briefly describing search methods and tips (treatment group) or just the booklet (control group). The aim of the intervention was to prevent poor mental health and promote high-quality re-employment. Noncompliance was a problem in the intervention group with 46% never attending any of the seminars.

14.4.2 The compliance problem

Consider a randomized study with two treatment arms. For simplicity, we will refer to these as the treatment (or active treatment) group and the control group, although the control treatment need not be a placebo.

Imbens and Rubin (1997ab) classified study participants into four types

---

[3] These data can be downloaded from `gllamm.org/books` or the from ICPSR at the Institute for Social Research of the University of Michigan at
`http://www.icpsr.umich.edu/access/index.html` (under JOBS #2739).

of compliance status: *compliers* (adhere to their assigned treatment), *always-takers* (take active treatment regardless of assignment), *never-takers* (never take active treatment regardless of assignment) or *defiers* (take opposite treatment to assigned).

Importantly, compliance status cannot be observed in a parallel trial since whether or not a subject takes the treatment can only be observed in the assigned treatment group (and not the other group). Consider first a subject assigned to the active treatment. If the subject takes the active treatment he may either be a complier or an always-taker. If he fails to take it he may either be a never-taker or a defier. Consider then a subject assigned to the control treatment. If the subject takes the control treatment he may either be a complier or a never-taker. If he fails to take it he may either be an always-taker or a defier. It might be helpful to have a look at the graphical summary of this setup provided in Figure 14.4.

It often seems reasonable to assume that there are no defiers, sometimes called the 'monotonicity' assumption. This is because participation in a trial is generally voluntary and nonadherence to the assigned treatment is usually due to prior preference for one of the treatments. Thus, we will henceforth assume that there are no defiers. As has been demonstrated by Imbens and Rubin (1997b), this assumption is useful for identification. Specifically, if the probabilities of actually taking the assigned treatments in both groups can be estimated, the probabilities of being an always-taker, never-taker and complier become identified.

This can be seen as follows (consulting Figure 14.4 may once more be helpful): In the treatment group, the probability of being a never-taker is equal to the probability of not taking the treatment since there are no defiers. Due to randomization, the probability of being a never-taker is the same in the control group. In the control group, the probability of being a complier can then be obtained by subtracting the probability of being a never-taker from the probability of taking the control treatment. Due to randomization, the probability of being a complier is the same in the treatment group. In the treatment group, the probability of being an always-taker can now be obtained by subtracting the probability of being a complier from the probability of taking the treatment. Finally, randomization once more ensures that the probability of being an always-taker is the same in the control group.

Conventionally, three different kinds of analysis have been conducted when there is noncompliance:

1. An 'as-randomized analysis' or intention to treat (ITT) analysis compares the outcomes of participants by assigned group as shown in Figure 14.4. The ITT effect is the effect of treatment *assignment* rather than the effect of treatment taken (often called 'effectiveness' as opposed to 'efficacy'). Importantly, the standard ITT estimator is protected from selection bias by randomized treatment assignment.

2. An 'as-treated analysis' compares the outcomes of participants by treat-

ment actually taken. Here, randomization is violated and causal interpretations of treatment effects may be dubious.

3. A 'per-protocol analysis' compares subjects who adhered to the assigned treatments, excluding those not adhering to assigned treatment, as shown in Figure 14.4. This does not correspond to any useful summary of individual causal effects.

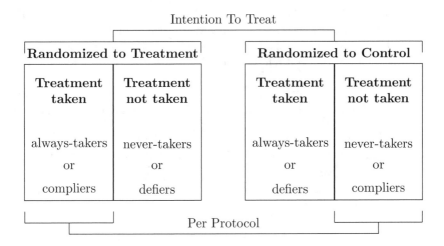

Figure 14.4 *Compliance status and different ways of handling noncompliance*

An alternative treatment effect was introduced by Imbens and Rubin (1997b). Their 'complier average causal effect' (CACE) is the treatment effect among true compliers; the mean difference in outcome between compliers in the treatment group and those controls who would have complied with treatment had they been randomized to the treatment group. Thus, the CACE may be viewed as a measure of 'efficacy' as opposed to 'effectiveness'. The crux of the CACE formulation is the distinction between a 'true complier' (complier under *both* treatments) and an 'observed complier' (taking the assigned treatment). Hence, the challenge in modeling CACE is that the true compliance status of subjects is generally unknown. Formally, the complier average causal effect (denoted δ_c) is defined as

$$\delta_c \equiv \mu_{1c} - \mu_{0c},$$

where μ_{1c} is the mean outcome of compliers in the treatment group and μ_{0c} the mean outcome of compliers in the control group.

In the JOBS II trial, the job training treatment was not available to those assigned to the control group. It is nevertheless plausible that there are always-takers who would take part in training if allowed regardless of assigned treatment. However, we cannot estimate the probability of being an always-taker since always-takers in the control group cannot act as always-takers and always-takers in the treatment group are indistinguishable from compliers. One approach is to combine always-takers and compliers into a single group (e.g. Little and Yau, 1998) and another is to assume that there are no always-takers (e.g. Jo, 2002). We prefer the former approach and will loosely refer to the combined group of compliers and always-takers as 'compliers'. The resulting 'complier' average causal effect is still meaningful since it represents the effect of training on those who will participate given the opportunity. The design for the JOBS II trial and the meaning of the CACE in this setting is shown in Figure 14.5, where the broken line indicates that we do not know who would take the treatment if offered and who wouldn't in the control group.

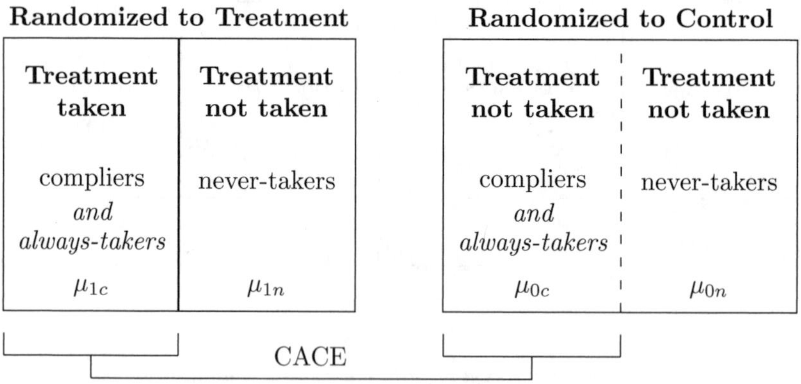

Randomized to Treatment   **Randomized to Control**

Treatment taken	Treatment not taken		Treatment not taken	Treatment not taken
compliers *and* *always-takers* μ_{1c}	never-takers μ_{1n}		compliers *and* *always-takers* μ_{0c}	never-takers μ_{0n}

CACE

Figure 14.5 *Compliance Causal Average Effect (CACE) in job training trial*

We also assume that never-takers have the same mean in the treatment and control groups,

$$\mu_{0n} = \mu_{1n}$$

(often called an 'exclusion restriction').

14.4.3 CACE modeling

Models for CACE are discussed in Angrist *et al.* (1996), Imbens and Rubin (1997ab), Little and Yau (1998) and Jo (2002). Formulation of this model as a latent class model with training data is described in Muthén (2002) and we use this approach here.

Compliance model

Let c_j be a dummy variable for compliers (or always-takers) versus never-takers and r_j a dummy for being randomized to the treatment versus control group.

Compliance status is not observable in the control group and is thus represented by η_j, a discrete latent variable with two values, $e_1 = 1$, $e_2 = 0$. Because of the randomization, we can set the parameters in the model for latent compliance in the control group equal to those for observed compliance in the treatment group,

$$\text{logit}[\pi_{1j}] \equiv \text{logit}[\Pr(\eta_j = 1 | r_j = 0)] = \mathbf{w}_j' \boldsymbol{\varrho} = \text{logit}[\Pr(c_j = 1 | r_j = 1)].$$

The covariates for the compliance model are

- [Age] age in years
- [Motivate] motivation to attend
- [Educ] school grade completed
- [Assert] assertiveness
- [Single] dummy for being single
- [Econ] economic hardship
- [Nonwhite] dummy variable for not being white versus white

Depression model

The outcome considered is the change in depression score (11-item subscale of Hopkins Symptom Checklist) from baseline to six months after the training seminars.

If compliance were known in the control group, we could model depression as

$$y_j = \beta_0 + \beta_1 c_j (1 - r_j) + \beta_2 c_j r_j + \epsilon_j,$$

where $\epsilon_j \sim \text{N}(0, \theta)$, so that $\mu_{0n} = \mu_{1n} = \beta_0$, $\mu_{0c} = \beta_0 + \beta_1$, $\mu_{1c} = \beta_0 + \beta_2$ and $\delta_c = \beta_2 - \beta_1$. However, c_j in the second term (control group) is never observed. We therefore write the model in terms of latent compliance η_j as

$$y_j = \beta_0 + \beta_1 \eta_j (1 - r_j) + \beta_2 c_j r_j + \epsilon_j.$$

We can add covariates $\mathbf{x}_j$ with constant effects $\boldsymbol{\alpha}$ across treatment groups by specifying

$$y_j = \beta_0 + \mathbf{x}_j' \boldsymbol{\alpha} + \beta_1 \eta_j (1 - r_j) + \beta_2 c_j r_j + \epsilon_j.$$

The covariates considered for the depression model are

- [Basedep] baseline depression score
- [Risk] baseline risk score; an index based on depression, financial strain and assertiveness

Joint model

In CACE modeling we have two responses, compliance c_j for the treatment group and depression y_j, for the entire sample. We denote these responses y_{1j} and y_{2j}, respectively, and define corresponding dummy variables $d_{ri} = 1$ if $r = i$ and zero otherwise. The response model can then be written in the GRC formulation as

$$\nu_{ij} = d_{i1}\mathbf{w}_j'\boldsymbol{\varrho} + \beta_0 d_{2i} + d_{2i}\mathbf{x}_j'\boldsymbol{\alpha} + \eta_j\beta_1(1 - r_j)d_{2i} + \beta_2 c_j r_j d_{2i}$$

where a logit link and Bernouilli distribution are specified when $i = 1$ whereas an identity link and normal density are used when $i = 2$. The structural model is

$$\text{logit}[\pi_{1j}] = \mathbf{w}_j'\boldsymbol{\varrho}.$$

Here we will replicate the analysis of the 'high risk' group presented in Little and Yau (1998). This group consisted of 335 subjects randomized to job training and 167 subjects randomized to the control group. Only 183 (55%) out of the 335 subjects randomized to job training actually participated. Parameter estimates and standard errors for the CACE model with and without covariates in the compliance model are given in Table 14.6.

Increasing age, motivation and education are associated with higher probability of compliance while assertiveness reduces the probability. There seems to be a greater reduction in depression amongst those attending job training. The complier average causal effect is estimated as -0.14 (0.14) when there are no covariates for compliance. A more pronounced complier average causal effect of -0.31 (0.12) is obtained for the model with covariates. For comparison, the conventional estimates of treatment effect, using baseline depression and risk as covariates for depression as in the CACE model, are

1. Intention to treat: -0.15 (0.07)
2. As treated (combining nonparticipants with control group): -0.18 (0.07)
3. Per protocol (excluding nonparticipants): -0.21 (0.08)

As expected, the intention to treat effect is the smallest since inclusion of nonparticipants in the treatment group dilutes the treatment effect. It is interesting to compare the CACE and per protocol effects since both exclude never-takers from the treatment group, whereas only CACE also excludes them from the control group. If never-takers had a worse outcome, the per protocol effect would be higher, but somewhat surprisingly this is not the case ($\widehat{\beta}_1$ is positive).

14.5 Physician advice and drinking: An endogenous treatment model

14.5.1 Introduction

Alcohol abuse is a significant public health concern, not only leading to health problems but also to for instance alcohol related traffic accidents. One approach to reducing alcohol related problems is through physicians advising

Table 14.6 *Parameter estimates for CACE model with and without covariates for compliance model*

Parameter	No covariates Est	No covariates (SE)	Covariates Est	Covariates (SE)
Compliance model				
ϱ_0 [Cons]	−0.19	(0.11)	−8.74	(1.58)
ϱ_1 [Age]			0.08	(0.01)
ϱ_2 [Motivate]			0.67	(0.16)
ϱ_3 [Educ]			0.30	(0.07)
ϱ_4 [Assert]			−0.38	(0.15)
ϱ_5 [Single]			0.54	(0.28)
ϱ_6 [Econ]			−0.16	(0.16)
ϱ_6 [Nonwhite]			−0.50	(0.31)
Depression model				
β_0 [Cons]	−0.39	(0.07)	1.63	(0.28)
β_1	0.02	(0.17)	0.18	(0.13)
β_2	−0.12	(0.09)	−0.13	(0.08)
$\delta_c = \beta_2 - \beta_1$	−0.14	(0.14)	−0.31	(0.12)
α_1 [Basedep]			−1.46	(0.18)
α_1 [Risk]			0.91	(0.26)
θ	0.60	(0.04)	0.51	(0.03)
Log-likelihood	−815.15		−729.41	

problem drinkers to reduce their consumption. The *efficacy* of physician advice in reducing drinking has been demonstrated in controlled clinical trials. However, studies of the effect of physician advice based on observational data are required, since efficacy does not necessarily translate into *effectiveness* in everyday practice.

Kenkel and Terza (2001) analyzed data from the 1990 National Health Interview Survey core questionnaire and special supplements. The data comprise a sub-sample of 2467 males who are current drinkers and have been told that they have hypertension[4]. The response variable is the number of alcoholic beverages consumed in the last two weeks, i.e. a count variable. 28 percent of the drinkers report having been advised to reduce drinking. The objective is to estimate the 'treatment effect' of physician advice.

In contrast to randomized studies, a major problem in estimating treatment effects from observational studies is that the treatment is often 'endogenous', in the sense that the treatment is correlated with unobserved heterogeneity.

---

[4] These data can be downloaded from the data archive of *Journal of Applied Econometrics* at http://qed.econ.queensu.ca/jae/2001-v16.2/kenkel-terza/.

For instance, subjects with a poor prognosis (unobserved heterogeneity) may self-select into the treatment perceived to be the best. It is hardly surprising that conventional modeling in this case can produce severely biased estimates of the treatment effect. Hence, models attempting to correct for selection bias, often called endogenous treatment models, have been suggested in econometrics (e.g. Heckman, 1978). In this section we apply a class of endogenous treatment models for counts as outcome discussed by Terza and colleagues (e.g. Terza, 1998; Kenkel and Terza, 2001).

14.5.2 Drinking model

The main objective is to study the effect of

- [Advice] answer to the question 'Have you ever been told by a physician to drink less?', treated as a dummy variable T_j for patient j.

In addition, covariates included in $\mathbf{x}_j$ for the drinking process are:

- [HiEduc] dummy for high education (> 12 years vs. ≤ 12 years)
- [Black] dummy for race (black vs. nonblack).

For simplicity, this is a subset of the covariates used in the analyses reported by Kenkel and Terza (2001).

Since the number of beverages consumed y_j is a count, a natural approach is to consider a Poisson regression model

$$\Pr(y_j; \mu_j) = \frac{\mu_j^{y_j} \exp(-\mu_j)}{y_j!},$$

where the expectation μ_j is structured as a log-linear model

$$\log(\mu_j) = \alpha T_j + \mathbf{x}_j'\boldsymbol{\beta}.$$

Note that the treatment effect of [Advice] is represented by α in this model. Estimates for the model are given in the second column of Table 14.7.

Since a zero response occurred for as many as 21 percent of the patients, we might consider models inducing overdispersion, inflating the number of zeros as compared to Poisson regression. A zero-inflated Poisson (ZIP) model of the kind discussed in Section 11.2 does not seem appropriate for the present application since abstainers are excluded from the sample. Thus, we consider a random intercept Poisson regression to model overdispersion,

$$\log(\mu_j) = \alpha T_j + \mathbf{x}_j'\boldsymbol{\beta} + \zeta_j,$$

where $\zeta_j \sim \mathrm{N}(0, \psi)$. Estimates for this model are presented in the third column of Table 14.7.

For both Poisson models [HiEduc] and [Black] have negative effects on drinking, which does not seem unreasonable. The important point to note, however, is that [Advice] has a rather perverse *positive* effect on drinking! Moreover, the sign does not change when numerous other potential confounders are included (see Kenkel and Terza, 2001). Physician advice regarding problem drinking

Table 14.7 *Estimates and standard errors for drinking and advice models*

Parameter	Poisson Est	(SE)	Overdisp. Poisson Est	(SE)	Probit Est	(SE)	Endog. Treatment Est	(SE)
Fixed part								
Drinking model								
α [Advice]	**0.47**	(0.01)	**0.59**	(0.08)			−2.42	(0.23)
β_0 [Cons]	2.65	(0.01)	1.43	(0.06)			2.32	(0.09)
β_1 [HiEduc]	−0.18	(0.01)	0.02	(0.07)			−0.29	(0.10)
β_2 [Black]	−0.31	(0.02)	−0.29	(0.11)			0.20	(0.11)
Advice model								
γ_0 [Cons]					−0.48	(0.08)	−1.13	(0.16)
γ_1 [HiEduc]					−0.25	(0.06)	−0.40	(0.10)
γ_2 [Black]					0.30	(0.08)	0.60	(0.15)
γ_3 [HlthIns]					−0.27	(0.07)	−0.33	(0.10)
γ_4 [RegMed]					0.18	(0.07)	0.39	(0.10)
γ_5 [Heart]					0.17	(0.08)	0.51	(0.11)
Random part								
Variance								
ψ			2.90	(0.11)			2.50	(0.69)
Loading								
λ							1.43	(0.15)
Log-likelihood	−32939.15		−8857.85		−1419.90		−10254.02	

should thus be abandoned if we were to take these results seriously. Note that there is considerable overdispersion with an estimated random intercept variance of 2.90.

14.5.3 Advice model

In order to gain some insight into the advice process, we specify and estimate a probit model for the treatment [Advice]. It is useful to formulate the model as a latent response model

$$T_j^* = \mathbf{w}_j'\boldsymbol{\gamma} + \epsilon_j,$$

where $\epsilon_j \sim N(0,1)$. The treatment T_j is generated from the threshold model

$$T_j = \begin{cases} 1 & \text{if } T_j^* > 0 \\ 0 & \text{otherwise.} \end{cases}$$

The covariate vector $\mathbf{w}_j$ contains three dichotomous health service utilization variables in addition to the covariates already introduced:

- [HlthIns] dummy for health insurance

- [RegMed] dummy for registered source of medical care
- [Heart] dummy for heart condition.

Once more, we have for simplicity included a subset of the covariates used in Kenkel and Terza (2001).

Estimates for this model are presented in the third column of Table 14.7, where we see that [Black], [RegMed] and [Heart] have positive effects on [Advice], whereas [HiEduc] and [HlthIns] have negative effects.

14.5.4 Joint model for drinking and advice: advice as endogenous treatment

It is likely that there could be shared unobserved heterogeneity for the drinking and advice processes, since physicians may be more likely to give advice to patients considered at risk. Hence, we consider a simultaneous model for drinking and advice, a so-called endogenous treatment model

$$\log(\mu_j) = \alpha T_j + \mathbf{x}'_j\boldsymbol{\beta} + \lambda\zeta_j,$$
$$T^*_j = \mathbf{w}'_j\boldsymbol{\gamma} + \zeta_j + \epsilon_j,$$

where $\zeta_j \sim N(0, \psi)$ is a factor representing shared unobserved heterogeneity, λ is a factor loading and $\epsilon_j \sim N(0,1)$. Note that this factor model is simply a reparametrization of the model considered by Terza (1998), reducing the dimension of integration from 2 to 1.

It follows from the endogenous treatment model that

$$\text{Var}[\log(\mu_j)|T_j, \mathbf{x}_j] = \lambda^2\psi \geq 0,$$

thus allowing for overdispersion,

$$\text{Var}[T^*_j|\mathbf{w}_j] = \psi + 1,$$

and

$$\text{Cov}[\log(\mu_j), T^*_j|\mathbf{w}_j] = \lambda\psi.$$

Note that due to the increased residual variance in the probit model, the coefficients $\boldsymbol{\gamma}$ are expected to increase by a factor of about $\sqrt{\psi+1}$.

Viewing the responses $i = 1$ for drinking and $i = 2$ for advice as clustered within patients j, we define the dummy variables $d_{1i} = 1$ for drinking and $d_{2i} = 1$ for advice. Using the GRC notation, the endogenous treatment model can be written as

$$\begin{aligned}
\nu_{ij} &= d_{1i}[\alpha T_j + \mathbf{x}_j'\boldsymbol{\beta} + \lambda\zeta_j] + d_{2i}[\mathbf{w}'_j\boldsymbol{\gamma} + \zeta_j] \\
&= \alpha d_{1i}T_j + d_{1i}\mathbf{x}_j'\boldsymbol{\beta} + d_{2i}\mathbf{w}'_j\boldsymbol{\gamma} + \zeta_j[d_{1i}\lambda + d_{2i}],
\end{aligned}$$

with log link and Poisson distribution for $i = 1$ and probit link for $i = 2$. A path diagram of this model is shown in Figure 14.6, where $\mathbf{w}'_j = (\mathbf{x}'_j, \mathbf{z}'_j)$.

An important feature of the advice model is that T^*_j and ζ_j are dependent, which implies that T_j and ζ_j become dependent in the drinking model. [Advice] is thus 'endogenous' in the drinking model and valid inference regarding the treatment effect α must in general be based on the simultaneous model

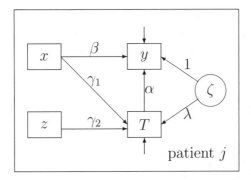

Figure 14.6 *Path diagram for endogenous treatment model*

including both drinking and advice as response or 'endogenous' variables. In contrast, the restriction $\lambda = 0$ decomposes the likelihood into a component for the conventional Poisson model for drinking and a probit component for advice (equal to that considered above if $\psi = 0$).

Importantly, randomization of patients to [Advice] would render the treatment exogenous, since T_j would become independent of the unobserved heterogeneity ζ_j. In this case [Advice] becomes 'exogenous' and valid inference regarding α can be obtained from the conventional Poisson regression model. Observe that λ governs the sign of the covariance and that a shared random intercept model (for drinking and advice) is obtained if $\lambda = 1$.

The *switching regime model* is a generalization of the endogenous treatment model. Here, a probit model governs whether a patient is allocated to treatment or nontreatment,

$$y_j^* = \mathbf{w}_j'\boldsymbol{\gamma} + \zeta_{1j} + \zeta_{2j} + \epsilon_j,$$

where $\zeta_{1j} \sim N(0, \psi_{11})$ and $\zeta_{2j} \sim N(0, \psi_{22})$, independently distributed, and independent from ϵ_j. In the treatment regime drinking is modeled as

$$\log(\mu_{2j}) = \beta_{20} + \mathbf{x}_j'\boldsymbol{\beta}_2 + \lambda_2\zeta_{2j},$$

and in the nontreatment regime as

$$\log(\mu_{1j}) = \beta_{10} + \mathbf{x}_j'\boldsymbol{\beta}_1 + \lambda_1\zeta_{1j},$$

where $\mathbf{x}_j$ no longer includes a constant. The treatment effect now becomes $\alpha = \beta_{20} - \beta_{10}$. Note that a patient is only allocated to one of the regimes and never both, the responses in the regimes thus represent *potential outcomes*. Unfortunately, attempts to obtain reliable estimates for the switching regimes model failed for this application.

Estimates for the endogenous treatment model are presented in the fourth column of Table 14.7. The main point to note is that the effect of physician advice is now reversed. Receiving advice leads to reduced drinking, consistent with the results from controlled clinical trials. Introducing further covariates (see Kenkel and Terza, 2001) does not alter this conclusion. Since $\widehat{\lambda}$ is positive, there appears to be a positive correlation between the unobserved heterogeneity for the drinking and advice processes. Thus, conditional on the covariates (the observed heterogeneity) the patients most prone to be heavy drinkers are also most likely to receive drinking advice. As expected, including unobserved heterogeneity furthermore increases the magnitude of the estimates in the advice model.

Note that the endogenous treatment model imposes so-called exclusion restrictions (different from the exclusion restriction in CACE) since the effects of the health service utilization variables are implicitly set to zero in the drinking part of the model. According to Kenkel and Terza (2001), the restrictions can be motivated from standard models for demand for alcohol as a consumer good. Although beneficial for identification, the restrictions are not necessary for identification. Relaxing the restrictions did not alter the main results appreciably. Results should be interpreted with extreme caution when this is not the case.

It is useful to note that a *sample selection model* for counts is obtained as the special case of the endogenous treatment model where $\alpha = 0$ and T_j plays the role of sample selection indicator, taking the value 1 if subject j is included in the sample and 0 otherwise. This model is a generalization of the sample selection model originally suggested by Heckman (e.g. Heckman, 1979) for continuous response.

14.6 Treatment of liver cirrhosis: A joint survival and marker model

14.6.1 Introduction

Andersen *et al.* (1993) analyzed a randomized controlled trial with Danish patients suffering from histologically verified liver cirrhosis (severely damaged liver, often caused by alcohol abuse). The patients were randomized to either treatment with the hormone prednisone or to placebo. 488 patients were considered for whom the initial biopsy could be reevaluated using more restrictive criteria, of whom 251 received prednisone and 237 placebo[5,6]. Treatment with prednisone is denoted [Treat] and the corresponding dummy variable as T in the sequel.

The main purpose of the trial was to ascertain whether prednisone reduces the death hazard for cirrhosis patients. Patients were considered censored if lost to follow-up or alive at the end of the observation period. Repeated measurements of prothrobin, a biochemical marker of liver functioning, were also obtained. A prothrobin index was based on a blood test of coagulation factors II, VII and X produced by the liver (we divide the original index by 10 to avoid very small regression coefficients). We also consider a dichotomized

---

[5] We thank Per Kragh Andersen for providing us with these data.
[6] The data can be downloaded from `gllamm.org/books`

version of the index, defining a value as normal if the original index is higher than 70 and otherwise as abnormal.

The measurements of prothrobin were scheduled to take place 3, 6 and 12 months after randomization and thereafter once a year but the actual follow-up times were irregular. Thus, one problem is a highly unbalanced design and missing measurements for the marker.

Another problem, specific for the survival setting, is that values of the time-varying covariate are required at each failure time for each subject surviving beyond this time (see Display 2.1 on page 43 for risk-set expansion of data). However, measurements of the marker are not available at each failure time so some kind of interpolation method must be used. Christensen *et al.* (1986) made the somewhat unrealistic assumption that time-varying covariates were constant between follow-up occasions. Instead, we propose using a growth curve model for the marker process.

It also seems plausible that liver functioning could be an intervening variable, implying that prednisone could have an *indirect* effect on the death hazard via liver functioning in addition to a *direct* treatment effect. Disentangling these effects may first of all improve our understanding of how the treatment works. Furthermore, it may shed light on whether the marker can be regarded as a 'surrogate' for survival. If there is no direct effect, survival is conditionally independent of treatment given the marker. Thus, survival information conveys no additional information on the treatment effect once the marker is known and the marker is called a 'perfect surrogate' (e.g. Prentice, 1989). The marker may then be used as response variable instead of survival, which can be beneficial for various practical reasons, for instance by reducing the required follow-up time.

To address these questions, we specify a structural equation survival model, composed of a latent marker model and a hazard model with a latent covariate.

14.6.2 Latent marker model

Let t_{ij} be the time of the ith measurement occasion for patient j. The observed marker is denoted y_{ij} at t_{ij} and is related to the latent or 'true' marker η_{ij} via the measurement model

$$y_{ij} = \beta_0 + \eta_{ij}^{(2)} + \epsilon_{ij},$$

where $\epsilon_{ij} \sim N(0, \theta)$.

A structural model for the latent marker is specified as

$$\eta_{ij}^{(2)} = \gamma_1 t_{ij} + \gamma_2 T_j + \eta_j^{(3)}, \tag{14.2}$$

where $\eta_j^{(3)} \sim N(0, \psi)$ is independently distributed from ϵ_{ij}. Note that there is no disturbance $\zeta_{ij}^{(2)}$ in the structural model or equivalently $\text{Var}(\zeta_{ij}^{(2)}) = 0$. We also considered a more general structural model, including a random slope for time in addition to the random intercept (bivariate normal random intercept and slope), but the log-likelihood did not increase appreciably.

Substituting for the latent marker $\eta_{ij}^{(2)}$ in the measurement model, the reduced form marker model becomes

$$y_{ij} = \beta_0 + \gamma_1 t_{ij} + \gamma_2 T_j + \eta_j^{(3)} + \epsilon_{ij},$$

a random intercept linear growth model. Note that ϵ_{ij} is often interpreted as measurement error (e.g. Faucett and Thomas, 1996; Wulfsohn and Tsiatis, 1997; Xu and Zeger, 2001), although this is problematic since the term also incorporates the deviation of the true marker from the linear time trend. The two error components cannot be separated since there is only a single measurement per patient at each time point.

We will also consider a probit model for the dichotomous marker variable 'normal versus abnormal prothrobin', using the 70% cut off for the prothrobin index. The marker model is similar to that outlined above, with the latent response y_{ij}^* taking the place of y_{ij}. The dichotomous measurements are generated from the threshold model

$$y_{ij} = \begin{cases} 1 & \text{if } y_{ij}^* > 0 \\ 0 & \text{otherwise.} \end{cases}$$

14.6.3 Hazard model with latent covariate

Let t_{rj} be the rth death time survived by patient j and let the hazard at t_{rj} be denoted h_{rj}. A proportional hazards model with the latent marker $\eta_{rj}^{(2)}$ (at time t_{rj}) as covariate is specified as:

$$\ln h_{rj} = \ln h_{rj}^0 + \lambda \eta_{rj}^{(2)} + \alpha_4 T_j$$

with baseline hazard modeled as

$$\ln h_{rj}^0 = \alpha_0 + \alpha_1 t_{rj} + \alpha_2 t_{rj}^2 + \alpha_3 t_{rj}^3.$$

This third degree polynomial was the chosen model from a systematic search among third degree fractional polynomials (e.g. Royston and Altman, 1994).

We then interpolate the latent marker to the required death times t_{rj} by substituting for $\eta_{rj}^{(2)}$ from the structural marker model (14.2), obtaining the reduced form hazard model

$$\begin{aligned} \ln h_{rj} &= \ln h_{rj}^0 + \lambda(\gamma_1 t_{rj} + \gamma_2 T_j + \eta_j^{(3)}) + \alpha_4 T_j \\ &= \ln \bar{h}_{rj}^0 + [\lambda\gamma_2 + \alpha_4]T_j + \lambda\eta_j^{(3)}, \end{aligned}$$

where

$$\ln \bar{h}_{rj}^0 = \alpha_0 + [\lambda\gamma_1 + \alpha_1]t_{rj} + \alpha_2 t_{rj}^2 + \alpha_3 t_{rj}^3$$

is the reduced form baseline hazard.

We see that the total treatment effect $\lambda\gamma_2 + \alpha_4$ on the log hazard is decomposed into the indirect effect $\lambda\gamma_2$ mediated through the latent marker and a direct effect α_4. If there is no direct effect ($\alpha_4 = 0$), the log hazard is conditionally independent of treatment given the marker. In this case, survival information conveys no additional information on the treatment effect if the marker is known, and the marker $\eta_{tj}^{(2)}$ is a 'perfect surrogate'.

A path diagram of the model is shown in Figure 14.7, where the letters i, r and j represent indices over which the variables within each frame vary and D_{rj} is a dummy variable taking the value 1 if patient j dies in risk set r and 0 otherwise (the response for the survival model).

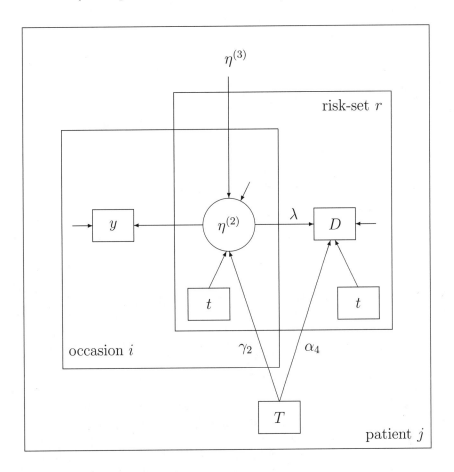

Figure 14.7 *Path diagram for structural equation hazards model*

Estimates and standard errors for the models with continuous and dichotomous markers are shown in Table 14.8. Note that we have fixed θ, which is not identified in the probit marker model, to the estimate for the continuous marker model to ease comparison of the estimates and their standard errors. For the marker model, we note that the estimates for the continuous and dichotomous case are quite similar. The exception is [Cons] which takes on different roles in the models, being closely related to a threshold in the dichoto-

Table 14.8 *Estimates for structural equation hazard models, using both continuous and dichotomous versions of the marker*

	Continuous		Dichotomous	
	Est	(SE)	Est	(SE)
Marker model				
β_0 [Cons]	7.88	(0.12)	0.84	(0.19)
γ_1 [Time]	0.15	(0.02)	0.11	(0.03)
γ_2 [Treat]	−0.64	(0.19)	−0.57	(0.27)
ψ	3.71	(0.26)	4.66	(0.50)
θ	3.43	(0.97)	3.43	
Hazard model				
α_0 [Cons]	−7.96	(0.12)	−7.97	(0.12)
α_1 [Time]	−0.15	(0.05)	−0.16	(0.05)
α_2 [Time]^2	0.10	(0.02)	0.10	(0.02)
α_3 [Time]^3	−0.01	(0.00)	−0.01	(0.00)
α_4 [Treat]	−0.18	(0.12)	−0.10	(0.13)
λ	−0.38	(0.04)	−0.34	(0.04)
Log-likelihood	−8422.87		−3527.28	

mous case. As expected, the standard errors are higher for the probit, due to the loss of information incurred from dichotomization. The estimates suggest that the marker increases over time but is reduced by [Treat]. The residual intraclass correlations for the continuous observed marker and the underlying variable for the dichotomous marker are 0.52 and 0.58, respectively.

Regarding the hazard model, both estimates and standard errors are similar, whether the observed marker was dichotomized or not. The exception concerns the negative estimate for [Treat], which is lower in absolute value in the dichotomous case. The treatment effect on liver functioning (the latent marker) is estimated as $\hat{\gamma}_2 = -0.64$, with 95% confidence interval $(-1.00, -0.28)$. Thus, the hormone prednisone has a negative 'side-effect' on liver functioning. The estimated effect of liver functioning on death hazard is $\hat{\lambda} = -0.38$ with 95% confidence interval $(-0.46, -0.31)$. The corresponding hazard ratio is $\exp(\hat{\lambda}) = 0.68$. As expected, good liver functioning reduces the death hazard.

The direct treatment effect on the death hazard was estimated as $\hat{\alpha}_4 = -0.18$ (hazard ratio 0.84) with 95% confidence interval $(-0.42, 0.06)$. Thus, the direct effect of treatment appears to reduce the hazard a bit, but the estimate is imprecise. The indirect treatment effect is estimated as $\hat{\lambda}\hat{\gamma}_2 = 0.25$ (hazard ratio 1.28) with 95% confidence interval $(0.08, 0.41)$. This increased death hazard is due to the negative side-effect of treatment on liver functioning, which in turn transmits to an increased hazard. The total treatment effect, the sum of the direct and indirect effects, is estimated as $\widehat{\lambda}\widehat{\gamma}_2 + \widehat{\alpha}_4 = 0.07$ (hazard ratio 1.07) with 95% confidence interval $(-0.22, 0.35)$. Hence, there is no evidence for a beneficial treatment effect of prednisone on the death hazard for cirrhosis patients. Interestingly, prednisone is no longer used as a treatment of liver cirrhosis.

Regarding surrogacy, the direct effect α_4 does not differ significantly from 0 at the 5% level, so we might conclude that the marker is a perfect surrogate. However, this does not appear to make sense since the indirect effect of the marker is detrimental. Furthermore, even a perfect surrogate is not as informative as survival times, if our aim is to estimate the size of the treatment effect on survival, rather than just testing for a treatment effect.

14.7 Summary and further reading

We discussed covariate measurement error models for both continuous and discrete true covariates in both prospective and retrospective studies, for the case of repeated measurements and validation samples. Covariate measurement error models are discussed in Carroll *et al.* (1995a) and Gustafson (2004). Roeder *et al.* (1996), Schafer (2001), Aitkin and Rocci (2002) and Rabe-Hesketh *et al.* (2003a) also consider nonparametric maximum likelihood estimation in this context.

This chapter illustrates how mathematically very similar models can be used for very different problems. For example, the model structure used in the joint modeling of the marker and survival process is similar to that discussed for the diet and heart disease application in Section 14.2. If the random intercept is removed in the marker model, the models in both applications have the same structure apart from the different response types of the outcome (durations in marker and survival example and dichotomous responses in the diet and heart disease example). However, the interpretation of the models is quite different.

The models for the job training and cervical cancer applications also have very similar structures. In both examples, a latent class model is used for a dichotomous covariate (compliance and exposure to herpes virus) that is perfectly observed in a subsample. In both models, the prevalence of the true covariate categories and relationship between true covariate and outcome are assumed to be the same in both subsamples. In the job training example, further information on latent class membership comes from covariates for compliance, whereas in the cervical cancer application, further information comes from imperfect measurements of exposure to herpes virus available in both subsamples.

A very interesting application of multiple process modeling is given in Lillard (1993), who considered a simultaneous equations survival model for marriage duration and fertility. We are currently only at the beginning of comprehending the scope of latent variable modeling of multiple processes and mixed responses. However, we expect that there will be exciting future developments in this area.

References

Aalen, O. O. 1988. Heterogeneity in survival analysis. *Statistics in Medicine*, **7**, 1121–1137.

Abramson, P. R. 1983. *Political Attitudes in America*. San Francisco, CA: Freeman.

Adams, R. C. 1975. On the use of a marker variable in lieu of its factor-analytic construct. *Perceptual and Motor Skills*, **41**, 665–666.

Agresti, A. 1994. Simple capture-recapture models permitting unequal catchability and variable sampling effort. *Biometrics*, **50**, 494–500.

Agresti, A. 1996. *Introduction to Categorical Data Analysis*. New York: Wiley.

Agresti, A. 2002. *Categorical Data Analysis (Second Edition)*. Hoboken, NJ: Wiley.

Agresti, A., & Hartzel, J. 2000. Tutorial in biostatistics. Strategies for comparing treatments on a binary response with multi-centre data. *Statistics in Medicine*, **19**, 1115–1139.

Agresti, A., & Natarajan, R. 2001. Modeling clustered ordered categorical data: A survey. *International Statistical Review*, **69**, 345–371.

Aish, A.-M., & Jöreskog, K. G. 1989. *Political Efficacy: Measurement and Dimensionality*. Tech. rept. 89-4. Department of Statistics, Uppsala University.

Aitchison, J., & Silvey, S. 1957. The generalization of probit analysis to the case of multiple responses. *Biometrika*, **44**, 131–140.

Aitkin, M. 1996. A general maximum likelihood analysis of overdispersion in generalised linear models. *Statistics and Computing*, **6**, 251–262.

Aitkin, M. 1999a. A general maximum likelihood analysis of variance components in generalized linear models. *Biometrics*, **55**, 117–128.

Aitkin, M. 1999b. Meta-analysis by random effect modelling in generalised linear models. *Statistics in Medicine*, **18**, 2343–2351.

Aitkin, M., & Longford, N. T. 1986. Statistical modeling issues in school effectiveness studies. *Journal of the Royal Statistical Society, Series A*, **149**, 1–43.

Aitkin, M., & Rocci, R. 2002. A general maximum likelihood analysis of measurement error in generalised linear models. *Statistics and Computing*, **12**, 163–174.

Aitkin, M., & Rubin, D. B. 1985. Estimation and hypothesis testing in finite mixture distributions. *Journal of the Royal Statistical Society, Series B*, **47**, 67–75.

Aitkin, M., Anderson, D., & Hinde, J. 1981. Statistical modeling of data on teaching styles. *Journal of the Royal Statistical Society, Series A*, **144**, 419–461.

Aitkin, M., Anderson, D., Francis, B. J., & Hinde, J. 1989. *Statistical Modelling in GLIM*. Oxford: Oxford University Press.

Aitkin, M., Anderson, D., Francis, B. J., & Hinde, J. 2004. *Statistical Modelling in GLIM (Second Edition)*. Oxford: Oxford University Press.

Akaike, H. 1973. Information theory and an extension of the maximum likelihood principle. *Pages 267–281 of:* Petrov, B. N., & Csaki, F. (eds), *Second International Symposium on Information Theory*. Budapest: Academiai Kiado.

Akaike, H. 1987. Factor analysis and AIC. *Psychometrika*, **52**, 317–332.

Albert, J. H. 1992. Bayesian estimation of normal ogive item response curves using Gibbs sampling. *Journal of Educational Statistics*, **17**, 251–269.

Albert, J. H., & Chib, S. 1993. Bayesian analysis of binary and polychotomous response data. *Journal of the American Statistical Association*, **88**, 669–679.

Albert, J. H., & Chib, S. 1995. Bayesian residual analysis for binary response regression models. *Biometrika*, **82**, 747–759.

Albert, J. H., & Chib, S. 1997. Bayesian tests and model diagnostics in conditionally independent hierarchical models. *Journal of the American Statistical Association*, **92**, 916–925.

Albert, P. S. 1999. Tutorial in biostatistics. Longitudinal data analysis (repeated measures) in clinical trials. *Statistics in Medicine*, **18**, 1707–1732.

Albert, P. S., & Follmann, D. A. 2000. Modeling repeated count data subject to informative dropout. *Biometrics*, **56**, 667–677.

Allenby, G. M., & Lenk, P. J. 1994. Modeling household purchase behavior with logistic normal regression. *Journal of the American Statistical Association*, **89**, 1218–1231.

Allenby, G. M., Arora, N., & Ginter, J .L. 1998. On the heterogeneity of demand. *Journal of Marketing Research*, **35**, 384–389.

Allison, P. D. 1984. *Event History Analysis. Regression for Longitudinal Event Data.* Sage University Paper Series on Quantitative Applications in the Social Sciences. Newbury Park, CA: Sage.

Allison, P. D., & Christakis, N. A. 1994. Logit models for sets of ranked items. *Pages 199–228 of:* Marsden, P. V. (ed), *Sociological Methodology 1994.* Oxford: Blackwell.

Alwin, D. F. 1989. Problems in the estimation and interpretation of the reliability of survey data. *Quality and Quantity*, **23**, 277–331.

Alwin, D. F., & Krosnick, J. A. 1985. The measurement of values in surveys: A comparison of ratings and rankings. *Public Opinion Quarterly*, **49**, 535–552.

American Psychological Association (APA), American Educational Research Association (AERA), & National Council on Measurement in Education (NCME). 1974. *Standards for Educational and Psychological Tests.* Washington, DC: American Psychological Association.

Andersen, E. B. 1970. Asymptotic properties of conditional maximum likelihood estimators. *Journal of the Royal Statistical Society, Series B*, **32**, 283–301.

Andersen, E. B. 1973. *Conditional Inference and Models for Measuring.* Copenhagen: Mentalhygiejnisk Forsknings Institut.

Andersen, E. B. 1980. *Discrete Statistical Models with Social Science Applications.* Amsterdam: North-Holland.

Andersen, P. K., & Gill, R. D. 1982. Cox's regression model for counting processes: A large sample study. *The Annals of Statistics*, **10**, 1100–1120.

Andersen, P. K., Borgan, Ø., Gill, R. D., & Keiding, N. 1993. *Statistical Models Based on Counting Processes.* New York: Springer.

Anderson, J. A. 1984. Regression and ordered categorical variables. *Journal of the Royal Statistical Society, Series B*, **46**, 1–30.

Anderson, T. W. 1963. The use of factor analysis in the statistical analysis of multiple time series. *Psychometrika*, **28**, 1–25.

Anderson, T. W. 2003. *Introduction to Multivariate Statistical Analysis (Third Edition).* New York: Wiley.

Anderson, T. W., & Rubin, H. 1956. Statistical inference in factor analysis. *Pages 111–150 of:* Neyman, J. (ed), *Proceedings of the Third Berkeley Symposium On Mathematical Statistics and Probability.* Berkeley: University of California Press.

Andrews, F. M., & Withey, S. B. 1976. *Social Indicators of Well-Being: Americans' Perceptions of Life Quality.* New York: Plenum.

Andrich, D. 1978. A rating formulation for ordered response categories. *Psychometrika*, **43**, 561–573.

Angrist, J. D., Imbens, G. W., & Rubin, D. B. 1996. Identification of causal effects using instrumental variables. *Journal of the American Statistical Association*, **91**, 444–455.

Ansari, A., & Jedidi, K. 2000. Bayesian factor analysis for multilevel binary observations. *Psychometrika*, **65**, 475–496.

Appelbaum, M. I., & McCall, R. B. 1983. Design and analysis in developmental psychology. *Pages 415–476 of:* Mussen, P. H. (ed), *Handbook of Child Psychology*, vol. I. New York: Wiley.

Arbuckle, J. L., & Nugent, J. H. 1973. A general procedure for parameter estimation for the law of comparative judgment. *British Journal of Mathematical and Statistical Psychology*, **26**, 240–260.

Arbuckle, J. L., & Wothke, W. 1999. *Amos 4.0 Users' Guide*. Chicago, IL: SmallWaters Corporation.

Arbuckle, J. L., & Wothke, W. 2003. *Amos 5.0 Update to the Amos User's Guide*. Chicago, IL: SmallWaters Corporation.

Arminger, G., & Küsters, U. 1988. Latent trait models with indicators of mixed measurement levels. *Pages 51–73 of:* Langeheine, R., & Rost, J. (eds), *Latent Trait and Latent Class Models*. New York: Plenum.

Arminger, G., & Küsters, U. 1989. Construction principles for latent trait models. *Pages 369–393 of:* Clogg, C. C. (ed), *Sociological Methodology 1989*. Oxford: Blackwell.

Arminger, G., & Muthén, B. O. 1998. A Bayesian approach to nonlinear latent variable models using the Gibbs sampler and the Metropolis-Hastings algorithm. *Psychometrika*, **63**, 271–300.

Arminger, G., Wittenberg, J., & Schepers, A. 1996. *MECOSA 3 User Guide*. Friedrichsdorf: ADDITIVE.

Arminger, G., Stein, P., & Wittenberg, J. 1999. Mixtures of conditional mean- and covariance-structure models. *Psychometrika*, **64**, 475–494.

Armstrong, J. S. 1967. Derivation of theory by means of factor analysis or Tom Swift and his electric factor analysis machine. *The American Statistician*, **21**, 17–21.

Bagozzi, R. P. 1980. *Causal Models in Marketing*. New York: Wiley.

Bagozzi, R. P. 1981. An examination of the validity of two models of attitude. *Multivariate Behavioral Research*, **16**, 323–359.

Bahadur, R. R. 1961. A representation of the joint distribution of responses to n dichotomous items. *Pages 158–168 of:* Solomon, H. (ed), *Studies in Item Analysis and Prediction*. Stanford Mathematical Studies in the Social Sciences, vol. VI. Stanford, CA: Stanford University Press.

Bailey, K. R. 1987. Inter-study differences: How should they influence the interpretation and analysis of results? *Statistics in Medicine*, **6**, 351–358.

Baltagi, B. 2001. *The Econometrics of Panel Data (Second Edition)*. London: Wiley.

Banfield, J. D., & Raftery, A. E. 1993. Model-based Gaussian and non-Gaussian clustering. *Biometrics*, **49**, 803–821.

Barndorff-Nielsen, O. E. 1978. *Information and Exponential Families in Statistical Theory*. New York: Wiley.

Barnes, S. H., Kaase, M., Allerbeck, K. R., Farah, B. J., Heunks, F., Inglehart, R., Jennings, M. K., Klingemann, H. D., Marsh, H., & Rosenmayr, L. 1979. *Political Action. Mass Participation in Five Western Democracies*. Beverly Hills, CA: Sage.

Barnett, V., & Lewis, T. 1984. *Outliers in Statistical Data*. London: Wiley.

Baron, R. M., & Kenny, D. A. 1986. The moderator-mediator variable distinction in social psychological research: Conceptual, strategic and statistical considerations. *Journal of Personality and Social Psychology*, **51**, 1173–1182.

Bartholomew, D. J. 1981. Posterior analysis of the factor model. *British Journal of Mathematical and Statistical Psychology*, **34**, 93–99.

Bartholomew, D. J. 1987. *Latent Variable Models and Factor Analysis*. Oxford: Oxford University Press.

Bartholomew, D. J. 1988. The sensitivity of latent trait analysis to the choice of prior distribution. *British Journal of Mathematical and Statistical Psychology*, **41**, 101–107.

Bartholomew, D. J. 1991. Latent variable methods. *In:* Lovie, P., & Lovie, A. D. (eds), *New developments in statistics for psychology and the social sciences*, vol. 2. London: BPS Books and Routledge.

Bartholomew, D. J. 1993. Estimating relationships between latent variables. *Sankhya*, **55**, 409–419.

Bartholomew, D. J. 1994. Bayes' theorem in latent variable modelling. *Pages 41–50 of:* Freeman, P. R., & Smith, A. F. M. (eds), *Aspects of uncertainty: A tribute to D. V. Lindley.* London: Wiley.

Bartholomew, D. J., & Knott, M. 1999. *Latent Variable Models and Factor Analysis.* London: Arnold.

Bartholomew, D. J., & Leung, S. O. 2002. A goodness of fit test for sparse 2^p contingency tables. *British Journal of Mathematical and Statistical Psychology,* **55**, 1–15.

Bartlett, M. S. 1937. The statistical conception of mental factors. *British Journal of Psychology,* **28**, 97–104.

Bartlett, M. S. 1938. Methods of estimating mental factors. *Nature,* **141**, 609–610.

Bayarri, M. J., & Berger, J. O. 2000. P-values for composite null models. *Journal of the American Statistical Association,* **95**, 1127–1142.

Bechger, T. M., Verhelst, N. D., & Verstralen, H. H. F. M. 2001. Identifiability of nonlinear logistic test models. *Psychometrika,* **66**, 357–372.

Bechger, T. M., Verstralen, H. H. F. M., & Verhelst, N. D. 2002. Equivalent linear logistic test models. *Psychometrika,* **67**, 123–136.

Becker, M. P., Yang, I., & Lange, K. 1997. EM algorithms without missing data. *Statistical Methods in Medical Research,* **6**, 37–53.

Beggs, S., Cardell, S., & Hausman, J. A. 1981. Assessing the potential demand for electric cars. *Journal of Econometrics,* **16**, 1–19.

Bekker, P. A., Merckens, A., & Wansbeek, T. J. 1994. *Identification, Equivalent Models and Computer Algebra.* Boston: Academic Press.

Belsley, D. A., Kuh, E., & Welsch, R. E. 1980. *Regression Diagnostics.* London: Wiley.

Bennett, S. 1983. Analysis of survival data by the proportional odds model. *Statistics in Medicine,* **2**, 273–277.

Bensmail, H., Celeux, G., Raftery, A. E., & Robert, C. P. 1997. Inference in model-based cluster analysis. *Statistics and Computing,* **7**, 1–10.

Bentler, P. M. 1978. The interdependence of theory, methodology and empirical data. *In:* Kandel, D. (ed), *Longitudinal Research On Drug Use. Empirical Findings And Methodological Issues.* Washington: Hemisphere.

Bentler, P. M. 1980. Multivariate analysis with latent variables: Causal Modelling. *Annual Review of Psychology,* **31**, 419–456.

Bentler, P. M. 1982. Linear systems with multiple levels and types of latent variables. *Pages 101–130 of:* Jöreskog, K. G., & Wold, H. (eds), *Systems Under Indirect Observation - Causality ⋆ Structure ⋆ Prediction - Part I.* Amsterdam: North-Holland.

Bentler, P. M. 1995. *EQS Structural Equation Program Manual.* Encino, CA: Multivariate Software.

Bentler, P. M., & Bonett, D. G. 1980. Significance tests and goodness of fit in the analysis of covariance structures. *Psychological Bulletin,* **88**, 588–606.

Bergan, J. R. 1988. Latent variable techniques for measuring development. *Pages 233–261 of:* Langeheine, R., & Rost, J. (eds), *Latent Class and Latent Trait Models.* New York: Plenum Press.

Berger, J. O., & Delampady, M. 1987. Testing precise hypotheses (with discussion). *Statistical Science,* **2**, 317–352.

Berger, J. O., & Sellke, T. 1987. Testing a point null hypothesis: The irreconcilability of p values and evidence (with discussion). *Journal of the American Statistical Association,* **82**, 112–139.

Berkey, C. S., Hoaglin, D. C., Mosteller, F., & Colditz, G. A. 1995. Random effects regression model for meta-analysis. *Statistics in Medicine,* **14**, 396–411.

Berkhof, J., & Snijders, T. A. B. 2001. Variance component testing in multilevel models. *Journal of Educational and Behavioral Statistics,* **26**, 132–152.

Berkson, J. 1950. Are there two regressions? *Journal of the American Statistical Association,* **45**, 164–80.

Bernardinelli, L., & Montomoli, C. 1992. Empirical Bayes versus fully Bayesian analysis of geographical variation in disease risk. *Statistics in Medicine,* **11**, 983–1007.

Berndt, E. R., Hall, B. H., Hall, R. E., & Hausman, J. A. 1974. Estimation and inference in nonlinear structural models. *Annals of Economic and Social Measurement*, **3**, 653–666.

Besag, J. 1974. Spatial interaction and the statistical analysis of lattice systems (with discussion). *Journal of the Royal Statistical Society, Series B*, **36**, 192–236.

Besag, J. 1986. On the statistical analysis of dirty pictures (with discussion). *Journal of the Royal Statistical Society, Series B*, **48**, 259–302.

Besag, J., York, J., & Mollie, A. 1991. Bayesian image restoration, with two applications in spatial statistics. *Annals of the Institute of Statistical Mathematics*, **43**, 1–21.

Biggeri, A., Marchi, M., Lagazio, C., Martuzzi, M., & Böhning, D. 2000. Non-parametric maximum likelihood estimators for disease mapping. *Statistics in Medicine*, **19**, 2539–2554.

Binder, D. A. 1983. On the variances of asymptotically normal estimators from complex surveys. *International Statistical Review*, **51**, 279–292.

Birnbaum, A. 1968. Test scores, sufficient statistics, and the information structures of tests. *Pages 425–435 of:* Lord, F. M., & Novick, M. R. (eds), *Statistical Theories of Mental Test Scores*. Reading: Addison-Wesley.

Bishop, Y. M. M., Fienberg, S. E., & Holland, P. W. 1975. *Discrete Multivariate Analysis. Theory and Practice*. Cambridge: MIT Press.

Blåfield, E. 1980. *Clustering of Observation from Finite Mixtures with Structural Information*. Ph.D. thesis, Jyväskyla studies in computer science, economics, and statistics, Jyväskyla.

Blalock, H. M. 1960. *Social Statistics*. New York: McGraw-Hill.

Bliss, C. I. 1934. The method of probits. *Science*, **79**, 38–39.

Bloxom, B. 1972. The simplex in pair comparisons. *Psychometrika*, **37**, 119–136.

Bock, R. D. 1972. Estimating item parameters and latent ability when responses are scored in two or more nominal categories. *Psychometrika*, **37**, 29–51.

Bock, R. D. 1983. The discrete Bayesian. *Pages 103–115 of:* Wainer, H., & Messick, S. (eds), *Principals of Modern Psychological Measurement*. Hillsdale, NJ: Erlbaum.

Bock, R. D. 1985. Contributions of empirical Bayes and marginal maximum likelihood methods to the measurement of individual differences. *Pages 75–99 of:* Roskam, E. E (ed), *Measurement and Personality Assessment*. Amsterdam: Elsevier.

Bock, R. D., & Aitkin, M. 1981. Marginal maximum likelihood estimation of item parameters: Application of an EM algorithm. *Psychometrika*, **46**, 443–459.

Bock, R. D., & Lieberman, M. 1970. Fitting a response model for N dichotomously scored items. *Psychometrika*, **35**, 179–197.

Bock, R. D., & Mislevy, R. J. 1982. Adaptive EAP estimation of ability in a microcomputer environment. *Applied Psychological Measurement*, **6**, 431–444.

Bock, R. D., & Schilling, S. G. 1997. High-dimensional full-information item factor analysis. *Pages 164–176 of:* Berkane, M. (ed), *Latent Variable Modelling and Applications to Causality*. New York: Springer.

Bock, R. D., Gibbons, R. D., Schilling, S. G., Muraki, E., Wilson, D. T., & Wood, R. 1999. *TESTFACT: Test Scoring, Item Statistics, and Full Information Item Factor Analysis*. Chicago, IL: Scientific Software International.

Böckenholt, U. 1992. Thurstonian representation for partial ranking data. *British Journal of Mathematical and Statistical Psychology*, **45**, 31–39.

Böckenholt, U. 1993. Application of Thurstonian models to ranking data. *In:* Fligner, M. A., & Verducci, J. S. (eds), *Probability Models and Statistical Analysis for Ranking Data*. New York: Springer.

Böckenholt, U. 2001a. Hierarchical modeling of paired comparison data. *Psychological Methods*, **6**, 49–66.

Böckenholt, U. 2001b. Mixed-effects analyses of rank-ordered data. *Psychometrika*, **66**, 45–62.

Böhning, D. 1982. Convergence of Simar's algorithm for finding the maximum likelihood estimate of a compound Poisson process. *The Annals of Statistics*, **10**, 1006–1008.

Böhning, D. 2000. *Computer-Assisted Analysis of Mixtures and Applications. Meta-Analysis, Disease Mapping and Others.* London: Chapman & Hall.

Böhning, D., Ekkehart, D., Schlattmann, P., Mendonça, L., & Kircher, U. 1999. The zero-inflated Poisson model and the decayed, missing and filled teeth index in dental epidemiology. *Journal of the Royal Statistical Society, Series A*, **162**, 195–209.

Bohrnstedt, G. W. 1983. Measurement. *Pages 69–121 of:* Rossi, P. H., Wright, J. D., & Anderson, A. B. (eds), *Handbook of Survey Research.* New York: Academic.

Bollen, K. A. 1989. *Structural Equations with Latent Variables.* New York: Wiley.

Bollen, K. A. 2002. Latent variables in psychology and the social sciences. *Annual Review of Psychology*, **53**, 605–634.

Bollen, K. A., & Arminger, G. 1991. Observational residuals in factor analysis and structural equation models. *Pages 235–262 of:* Marsden, P. V. (ed), *Sociological Methodology 1991.* Oxford: Blackwell.

Bollen, K. A., & Long, J. S. (eds). 1993. *Testing Structural Equation Models.* Newbury Park, CA: Sage.

Booth, J. G., & Hobert, J. P. 1999. Maximizing generalized linear mixed model likelihoods with an automated Monte Carlo EM algorithm. *Journal of the Royal Statistical Society, Series B*, **62**, 265–285.

Booth, J. G., Hobert, J. P., & Jank, W. 2001. A survey of Monte Carlo algorithms for maximizing the likelihood of a two-stage hierarchical model. *Statistical Modelling*, **1**, 333–349.

Bottai, M. 2003. Confidence regions when the Fisher information is zero. *Biometrika*, **90**, 73–84.

Box, G. E. P., Hunter, W. G., & Hunter, J. S. 1978. *Statistics for Experimenters.* New York: Wiley.

Box, G. E. P., Jenkins, G. M., & Reinsel, G. C. 1994. *Time Series Analysis: Forecasting and Control.* Englewood Cliffs, NJ: Prentice-Hall.

Bozdogan, H. 1987. Model selection and Akaike's information criterion (AIC): The general theory and its analytical extensions. *Psychometrika*, **52**, 345–370.

Bradley, R. E., & Terry, M. E. 1952. Rank analysis of incomplete block designs I. The method of paired comparisons. *Biometrika*, **39**, 324–345.

Brady, H. E. 1989. Factor and ideal point analysis for interpersonally incomparable data. *Psychometrika*, **54**, 181–202.

Brady, H. E. 1990. Dimensional analysis of ranking data. *American Journal of Political Science*, **34**, 1017–1048.

Breckler, S. 1990. Applications of covariance structure modeling in psychology: Cause for concern? *Psychological Bulletin*, **107**, 260–273.

Breiman, L. 2001. Statistical modeling: The two cultures. *Statistical Science*, **16**, 199–215.

Brennan, R. L. 2001. *Generalizability Theory.* New York: Springer.

Breslow, N. E. 1974. Covariance analysis of censored survival data. *Biometrics*, **30**, 89–100.

Breslow, N. E. 2003. Whither PQL? *UW Biostatistics Working Paper Series*, 1–25. Downloadable from http://www.bepress.com/uwbiostat/paper192/.

Breslow, N. E., & Clayton, D. G. 1993. Approximate inference in generalized linear mixed models. *Journal of the American Statistical Association*, **88**, 9–25.

Breslow, N. E., & Day, N. 1980. *Statistical Methods in Cancer Research. Vol I – The Analysis of Case-Control Studies.* Lyon: IARC.

Breslow, N. E., & Day, N. 1987. *Statistical Methods in Cancer Research. Vol II – The Design and Analysis of Cohort Studies.* Lyon: IARC.

Breslow, N. E., & Lin, X. 1995. Bias correction in generalised linear mixed models with a single component of dispersion. *Biometrika*, **82**, 81–91.

Breslow, N. E., Leroux, B., & Platt, R. 1998. Approximate hierarchical modeling of discrete data in epidemiology. *Statistical Methods in Medical Research*, **4**, 49–62.

Brinch, C. 2001. *Non-parametric Identification of Mixed Hazards Models.* Ph.D. thesis, Department of Economics, University of Oslo.

Brody, R. A., & Page, B. I. 1972. Comment: The assessment of policy voting. *American Political Science Review*, **66**, 450–458.

Browne, M. W. 1984. Asymptotically distribution-free methods for the analysis of covariance structures. *British Journal of Mathematical and Statistical Psychology*, **37**, 62–83.

Browne, M. W. 2000. Cross-validation methods. *Journal of Mathematical Psychology*, **44**, 108–132.

Browne, M. W., & Arminger, G. 1995. Specification and estimation of mean- and covariance-structure models. *Pages 185–249 of:* Arminger, G., Clogg, C. C., & Sobel, M. E. (eds), *Handbook of Statistical Modelling for the Social and Behavioral Sciences*. New York: Plenum Press.

Browne, M. W., & Cudeck, R. 1989. Single sample cross-validation indices for covariance structures. *Multivariate Behavioral Research*, **24**, 445–455.

Browne, M. W., & Cudeck, R. 1993. Alternative ways of assessing model fit. *Pages 136–162 of:* Bollen, K. A., & Long, J. S. (eds), *Testing Structural Equation Models*. Newbury Park, CA: Sage.

Browne, W. J. 1998. *Applying MCMC Methods to Multi-Level Models*. Ph.D. thesis, Department of Statistics, University of Bath.

Browne, W. J. 2004. *MCMC estimation in MLwiN*. London: Instiute of Education. Downloadable from http://multilevel.ioe.ac.uk/download/manuals.html.

Browne, W. J., & Draper, D. 2004. A comparison of Bayesian and likelihood methods for fitting multilevel models. *Submitted*. Downloadable from http://multilevel.ioe.ac.uk/team/materials/wbrssa.pdf.

Bunch, D. S. 1991. Estimability in the multinomial probit model. *Transportation Research B*, **25**, 1–12.

Burnham, K. P., & Anderson, D. R. 2002. *Model Selection and Multimodel Inference. A Practical Information-Theoretic Approach (Second Edition)*. New York: Springer.

Buse, A. 1982. The Likelihood Ratio, Wald, and Lagrange Multiplier tests: An expository note. *The American Statistician*, **36**, 153–157.

Busemeyer, J. R., & Jones, L. E. 1983. Analysis of multiplicative combination rules when the causal variables are measured with error. *Psychological Bulletin*, **93**, 549–562.

Busing, F. M. T. A. 1993. *Distribution characteristics of variance estimates in two-level models; A Monte Carlo study. Preprint PRM 93-04*. Tech. rept. Department of Psychometrics and Research Methodology, Leiden University, Leiden, The Netherlands.

Busing, F. M. T. A., Meijer, E., & van der Leeden, R. 1994. *MLA. Software for multilevel analysis of data with two levels. Users guide for version 1.0b. Preprint PRM 94-01*. Tech. rept. Department of Psychometrics and Research Methodology, Leiden University, Leiden, The Netherlands.

Butler, J. S., & Moffitt, R. 1982. A computationally efficient quadrature procedure for the one-factor multinomial probit model. *Econometrica*, **50**, 761–764.

Butler, S. M., & Louis, T. A. 1992. Random effects models with nonparametric priors. *Statistics in Medicine*, **11**, 1981–2000.

Cameron, A. C., & Trivedi, P. K. 1998. *Regression Analysis of Count Data*. Cambridge: Cambridge University Press.

Campbell, A., Gurin, G., & Miller, W. E. 1954. *The Voter Decides*. Evanston IL: Row Peterson.

Campbell, A., Converse, P. E., & Rodgers, W. L. 1976. *The Quality of American Life*. New York, NY: Sage.

Cappellari, L., & Jenkins, S. P. 2003. Multivariate probit regression using simulated maximum likelihood. *The Stata Journal*, **3**, 278–294.

Carey, V. C., Zeger, S. L., & Diggle, P. J. 1993. Modelling multivariate binary data with alternating logistic regressions. *Biometrika*, **80**, 517–526.

Carlin, B. P., & Louis, T. A. 2000. *Bayes and Empirical Bayes Methods for Data Analysis (Second Edition)*. Boca Raton: Chapman & Hall/CRC.

Carlin, J. B. 1992. Meta-analysis for 2×2 tables: A Bayesian approach. *Statistics in Medicine*, **11**, 141–158.

Carpenter, J., Goldstein, H., & Rasbash, J. 1999. A non-parametric bootstrap for multilevel models. *Multilevel Modelling Newsletter*, **11**, 2–5.

Carroll, J. D., & Chang, J. J. 1972. *IDIOSCAL (Individual differences in orientation SCALing): Generalization of INDSCAL allowing IDIOsyncratic reference systems as well as analytic approximation to INDSCAL.* Paper presented at the meeting of the Psychometric Society.

Carroll, R. J., Gail, M. H., & Lubin, J. H. 1993. Case-control studies with errors in covariates. *Journal of the American Statistical Association*, **88**, 185–199.

Carroll, R. J., Ruppert, D., & Stefanski, L. A. 1995a. *Measurement Error in Nonlinear Models*. London: Chapman & Hall.

Carroll, R. J., Wang, S., & Wang, C. Y. 1995b. Prospective analysis of logistic case-control studies. *Journal of the American Statistical Association*, **90**, 157–169.

Carroll, R. J., Roeder, K., & Wasserman, L. 1999. Flexible parametric measurement error models. *Biometrics*, **86**, 44–54.

Casella, G., & George, E. I. 1992. Explaining the Gibbs sampler. *The American Statistician*, **46**, 167–174.

Cattell, R. B. 1949. The dimensions of culture patterns by factorization of national characters. *Journal of Abnormal and Social Psychology*, **44**, 443–469.

Cattell, R. B. 1966. The scree test for the number of factors. *Multivariate Behavioral Research*, **1**, 245–276.

Chalmers, I. 1993. The Cochrane Collaboration: Preparing, maintaining, and disseminating systematic reviews of the effects of health care. *Annals of the New York Academy of Science*, **703**, 156–165.

Chamberlain, G. 1980. Analysis of covariance with qualitative data. *Review of Economic Studies*, **47**, 225–238.

Chan, W., & Bentler, P. M. 1998. Covariance structure analysis of ordinal ipsative data. *Psychometrika*, **63**, 369–399.

Chao, A., Tsay, P. K., Lin, S.-H., Shau, W.-Y., & Chao, D.-U. 2001. Tutorial in biostatistics. The applications of capture-recapture models to epidemiological data. *Statistics in Medicine*, **20**, 3123–3157.

Chapman, R. G., & Staelin, R. 1982. Exploiting rank ordered choice set data within the stochastic utility model. *Journal of Marketing Research*, **14**, 288–301.

Chatterjee, S., & Hadi, A. S. 1988. *Sensitivity Analysis in Linear Regression*. London: Wiley.

Chechile, R. 1977. Likelihood and posterior identification: Implications for mathematical psychology. *British Journal of Mathematical and Statistical Psychology*, **30**, 177–184.

Chen, C.-F. 1981. The EM approach to the multiple indicators and multiple causes model via the estimation of the latent variable. *Journal of the American Statistical Association*, **76**, 704–708.

Chesher, A., & Irish, M. 1987. Residual analysis in the grouped and censored linear model. *Journal of Econometrics*, **34**, 33–61.

Chib, S., & Greenberg, E. 1995. Understanding the Metropolis-Hastings algorithm. *The American Statistician*, **49**, 327–335.

Chib, S., & Hamilton, B. H. 2002. Semiparametric Bayes analysis of longitudinal data treatment models. *Journal of Econometrics*, **110**, 67–89.

Christensen, E., Schlichting, P., Andersen, P. K., Fauerholdt, L., Schou, G., Pedersen, B. V., Juhl, E., Poulsen, H., & Thygstrup, N. 1986. Updating prognosis and therapeutic effect evaluation in cirrhosis using Cox's multiple regression model for time dependent variables. *Scandinavian Journal of Gastroenterology*, **21**, 163–174.

Christofferson, A. 1975. Factor analysis of dichotomized variables. *Psychometrika*, **40**, 5–32.

Claeskens, G., & Hjort, N. L. 2003. The focussed information criterion (with discussion). *Journal of the American Statistical Association*, **98**, 900–916.

Clarkson, D. B., & Zhan, Y. H. 2002. Using spherical-radial quadrature to fit generalized linear mixed effects models. *Journal of Computational and Graphical Statistics*, **11**, 639–659.

Clayton, D. G. 1978. A model for association in bivariate life tables and its application in epidemiological studies of familial tendency in chronic disease incidence. *Biometrika*, **65**, 141–151.

Clayton, D. G. 1988. The analysis of event history data: A review of progress and outstanding problems. *Statistics in Medicine*, **7**, 819–841.

Clayton, D. G. 1992. Models for the analysis of cohort and case-control studies with inaccurately measured exposures. *Pages 301–331 of:* Dwyer, J. H., Feinlieb, M., Lippert, P., & Hoffmeister, H. (eds), *Statistical Models for Longitudinal Studies on Health*. New York: Oxford University Press.

Clayton, D. G. 1996a. Contribution to the discussion of a paper by Y. Lee and J. A. Nelder. *Journal of the Royal Statistical Society, Series B*, **58**, 657–659.

Clayton, D. G. 1996b. Generalized linear mixed models. *In:* Gilks, W. R., Richardson, S., & Spiegelhalter, D. J. (eds), *Markov Chain Monte Carlo in Practice*. London: Chapman & Hall.

Clayton, D. G., & Cuzick, J. 1985. Multivariate generalizations of the proportional hazards model. *Journal of the Royal Statistical Society, Series A*, **148**, 82–108.

Clayton, D. G., & Hills, M. 1993. *Statistical Models in Epidemiology*. Oxford: Oxford University Press.

Clayton, D. G., & Kaldor, J. 1987. Empirical Bayes estimates of age-standardized relative risks for use in disease mapping. *Biometrics*, **43**, 671–681.

Clayton, D. G., & Rasbash, J. 1999. Estimation in large crossed random-effect models by data augmentation. *Journal of the Royal Statistical Society, Series A*, **162**, 425–436.

Cliff, N. 1983. Some cautions concerning the application of causal modeling methods. *Multivariate Behavioral Research*, **18**, 115–126.

Clogg, C. C. 1979. Some latent structure models for the analysis of Likert-type data. *Social Science Research*, **8**, 287–301.

Clogg, C. C. 1988. Latent class models for measuring. *Pages 173–205 of:* Langeheine, R., & Rost, J. (eds), *Latent Trait and Latent Class Models*. New York: Plenum Press.

Clogg, C. C. 1995. Latent class models. *Pages 311–359 of:* Arminger, G., Clogg, C. C., & Sobel, M. E. (eds), *Handbook of Statistical Modelling for the Social and Behavioral Sciences*. New York: Plenum Press.

Clogg, C. C., & Shihadeh, E. S. 1994. *Statistical Models for Ordinal Variables*. Thousand Oaks, CA: Sage.

Clogg, C. C., Rubin, D. B., Schenker, N., Schultz, B., & Weidman, L. 1991. Multiple imputation of industry and occupation codes in census public-use samples using Bayesian logistic-regression. *Journal of the American Statistical Association*, **86**, 68–78.

Cochran, W. G. 1950. The comparison of percentages in matched samples. *Biometrika*, **37**, 256–266.

Collett, D. 2002. *Modelling Binary Data (Second Edition)*. London: Chapman & Hall/CRC.

Collett, D. 2003. *Modelling Survival Data in Medical Research (Second Edition)*. Boca Raton, FL: Chapman & Hall/CRC.

Congdon, P. 2001. *Bayesian Statistical Modelling*. Chichester: Wiley.

Converse, P. E. 1972. Change in the American electorate. *In:* Campbell, A., & Converse, P. E. (eds), *The Human Meaning of Social Change*. New York, NY: Sage.

Cook, R. D., & Weisberg, S. 1982. *Residuals and Influence in Regression*. London: Chapman & Hall.

Cook, T. D., & Campbell, D. T. 1979. *Quasi-Experimentation*. Boston: Houghton-Mifflin.

Coombs, C. H. 1964. *A Theory of Data*. New York: Wiley.

Corbeil, R. R., & Searle, S. R. 1976. A comparison of variance component estimators. *Biometrics*, **32**, 779–791.

Cormack, R. M. 1992. Interval estimation for mark-recapture studies of closed populations. *Biometrics*, **48**, 567–576.

Costa, P. T., & McCrae, R. R. 1985. *The NEO Personality Inventory Manual*. Odessa, FL: Psychological Assessment Resources.

Costner, H. L., & Schoenberg, R. 1973. Diagnosing indicator ills in multiple indicator models. *Pages 167–199 of:* Goldberger, A. S., & Duncan, O. D. (eds), *Structural Equation Models in the Social Sciences.* New York: Seminar.

Coull, B. A., & Agresti, A. 1999. The use of mixed logit models to reflect heterogeneity in capture-recapture studies. *Biometrics,* **55,** 294–301.

Coull, B. A., & Agresti, A. 2000. Random effects modeling of multiple binomial responses using the multivariate binomial logit-normal distribution. *Biometrics,* **56,** 73–80.

Cox, D. R. 1961. Tests of separate families of hypotheses. *Pages 105–123 of:* Neyman, J. (ed), *Proceedings of the Fourth Berkeley Symposium on Mathematical Statistics and Probability.* Berkeley, CA: University of California Press.

Cox, D. R. 1962. Further results on tests of separate families of hypotheses. *Journal of the Royal Statistical Society, Series B,* **24,** 406–424.

Cox, D. R. 1972. Regression models and life tables. *Journal of the Royal Statistical Society, Series B,* **34,** 187–203.

Cox, D. R. 1990. Role of models in statistical analysis. *Statistical Science,* **5,** 169–174.

Cox, D. R., & Hinkley, D. V. 1974. *Theoretical Statistics.* London: Chapman & Hall.

Cox, D. R., & Oakes, D. 1984. *Analysis of Survival Data.* London: Chapman & Hall.

Cox, D. R., & Solomon, P. J. 2002. *Components of Variance.* Boca Raton, FL: Chapman & Hall /CRC.

Cox, D. R., & Wermuth, N. 1996. *Multivariate Dependencies.* London: Chapman & Hall.

Craig, S. C., & Maggiotto, M. A. 1982. Measuring political efficacy. *Political Methodology,* **8,** 85–109.

Crainiceanu, C. M., & Ruppert, D. 2004. Likelihood ratio tests in linear mixed models with one variance component. *Journal of the Royal Statistical Society, Series B,* **66,** 165–185.

Crocker, L., & Algina, J. 1986. *Introduction to Classical & Modern Test Theory.* Orlando: Holt, Rinehart and Winston.

Cronbach, L. J. 1971. Test validation. *Pages 443–507 of:* Thorndike, R. L. (ed), *Educational Measurement.* Washington: American Council on Education.

Cronbach, L. J., & Meehl, P. E. 1955. Construct validity in psychological tests. *Psychological Bulletin,* **52,** 281–302.

Cronbach, L. J., Gleser, G. C., Nanda, H., & Rajaratnam, N. 1972. *The Dependability of Behavioral Measurements: Theory of Generalizability for Scores and Profiles.* London: Wiley.

Croon, M. A. 1989. Latent class models for the analysis of rankings. *Pages 99–121 of:* De Soete, G., Feger, H., & Klauer, K. C. (eds), *New Developments in Psychological Choice Modeling.* Amsterdam: Elsevier.

Croon, M. A. 2002. Using predicted latent scores in general latent structure models. *Pages 195–224 of:* Marcoulides, G., & Moustaki, I. (eds), *Latent Variable and Latent Structure Models.* Mahwah, NJ: Erlbaum.

Crouch, E. A. C., & Spiegelman, D. 1990. The evaluation of integrals of the form $\int f(t)exp(-t^2)dt$: Application to logistic-normal models. *Journal of the American Statistical Association,* **85,** 464–469.

Crouchley, R., & Davies, R. B. 1999. A comparison of population average and random effects models for the analysis of longitudinal count data with base-line information. *Journal of the Royal Statistical Society, Series A,* **162,** 331–347.

Crowder, M. J., & Hand, D. J. 1990. *Analysis of Repeated Measures.* London: Chapman & Hall.

Cudeck, R., & Henly, S. 1991. Model selection in covariance structure analysis and the problem of sample size: A clarification. *Psychological Bulletin,* **109,** 512–519.

Daganzo, C. 1979. *Multinomial Probit: The Theory and its Application to Demand Forecasting.* New York: Academic.

Dale, J. R. 1986. Global cross-ratio models for bivariate, discrete, ordered responses. *Biometrics,* **42,** 909–917.

Danahy, D. T., Burwell, D. T., Aranov, W. S., & Prakash, R. 1976. Sustained hemodynamic and antianginal effect of high dose oral isosorbide dinitrate. *Circulation*, **55**, 381–387.

Daniels, M. J., & Gatsonis, C. 1997. Hierarchical polytomous regression models with applications to health services research. *Statistics in Medicine*, **16**, 2311–2325.

David, H. A. 1988. *The Method of Paired Comparisons*. Oxford: Oxford University Press.

Davidian, M., & Gallant, A. R. 1992. Smooth nonparametric maximum likelihood estimation for population pharmacokinetics, with application to quindine. *Journal of Pharmacokinetics and Biopharmaceutics*, **20**, 529–556.

Davidson, R., & MacKinnon, J. G. 1993. *Estimation and Inference in Econometrics*. Oxford: Oxford University Press.

Davies, R. B. 1993. Statistical modelling for survey analysis. *Journal of the Market Research Society*, **35**, 235–247.

Davies, R. B., & Pickles, A. 1987. A joint trip timing store-type choice model for grocery shopping, including inventory effects and nonparametric control for omitted variables. *Transportation Research A*, **21**, 345–361.

Davies, R. B., Elias, P., & Penn, R. 1992. The relationship between a husband's unemployment and his wife's participation in the labour force. *Oxford Bulletin of Economics and Statistics*, **54**, 145–171.

Davis, P. J., & Rabinowitz, P. 1984. *Methods of Numerical Integration*. Orlando: Academic.

Dayton, C. M., & MacReady, G. B. 1988. Concomitant variable latent class models. *Journal of the American Statistical Association*, **83**, 173–178.

De Boeck, P., & Wilson, M. 2004. *Explanatory Item Response Models: A Generalized Linear and Nonlinear Approach*. New York: Springer.

de Leeuw, J., & Verhelst, N. D. 1986. Maximum likelihood estimation in generalized Rasch models. *Journal of Educational Statistics*, **11**, 183–196.

Deely, J. J., & Lindley, D. V. 1981. Bayes empirical Bayes. *Journal of the American Statistical Association*, **76**, 833–841.

DeGroot, M. H. 1970. *Optimal Statistical Decisions*. New York: McGraw-Hill.

Dellaportas, P., & Smith, A. F. M. 1993. Bayesian inference for generalized linear and proportional hazards model via Gibbs sampling. *Journal of the Royal Statistical Society, Series C*, **42**, 443–460.

Demidenko, E. 2004. *Mixed Models. Theory and Applications*. New York: Wiley.

Dempster, A. P., Laird, N. M., & Rubin, D. B. 1977. Maximum likelihood from incomplete data via the EM algorithm. *Journal of the Royal Statistical Society, Series B*, **39**, 1–38.

Dempster, A. P., Rubin, D. B., & Tsutakawa, R. D. 1981. Estimation in covariance components models. *Journal of the American Statistical Association*, **76**, 341–353.

DerSimonian, R., & Laird, N. M. 1986. Meta-analysis in clinical trials. *Controlled Clinical Trials*, **7**, 1777–1788.

Dey, D. K., Gelfand, A. E., Swartz, T. B., & Vlachos, P. K. 1998. Simulation based model checking for hierarchical models. *Test*, **7**, 325–346.

Diggle, P. J. 1988. An approach to analysis of repeated measures. *Biometrics*, **44**, 959–971.

Diggle, P. J., & Kenward, M. G. 1994. Informative drop-out in longitudinal data analysis (with discussion). *Journal of the Royal Statistical Society, Series C*, **43**, 49–93.

Diggle, P. J., Heagerty, P. J., Liang, K.-Y., & Zeger, S. L. 2002. *Analysis of Longitudinal Data*. Oxford: Oxford University Press.

Dillon, W. R., Kumar, A., & Deborrero, M. S. 1993. Capturing individual-differences in paired comparisons - an extended BTL model incorporating descriptor variables. *Journal of Marketing Research*, **30**, 42–51.

DiPrete, T. A. 2002. Life course risks, mobility regimes and mobility consequences: A comparison of Sweden, Germany, and the United States. *American Journal of Sociology*, **108**, 267–309.

Dobbie, M. J., & Welsh, A. H. 2001. Modelling correlated zero-inflated count data. *Australian and New Zealand Journal of Statistics*, **43**, 431–444.

Dolan, C. V., & van der Maas, H. L. J. 1998. Fitting multivariate normal finite mixtures subject to structural equation modeling. *Psychometrika*, **63**, 227–253.

Du Toit, M. (ed). 2003. *IRT from SSI*. Lincolnwood, IL: Scientific Software International.

DuMouchel, W., Waternaux, C., & Kinney, D. 1996. Hierarchical Bayesian linear models for assessing the effect of extreme cold weather on schizophrenic births. *In:* Berry, D., & Stangl, D. (eds), *Bayesian Biostatistics*. New York: Marcel Dekker.

Duncan-Jones, P., Grayson, D. A., & Moran, P. A. P. 1986. The utility of latent trait models in psychiatric epidemiology. *Psychological Medicine*, **16**, 391–405.

Dunn, G. 1992. Design and analysis of reliability studies. *Statistical Methods in Medical Research*, **1**, 123–157.

Dunn, G. 2004. *Statistical Evaluation of Measurement Errors: Design and Analysis of Reliability Studies (Second Edition)*. London: Arnold.

Dunn, G., Everitt, B. S., & Pickles, A. 1993. *Modelling Covariances and Latent Variables using EQS*. London: Chapman & Hall.

Dupacŏvá, J., & Wold, H. 1982. On some identification problems in ML modeling of systems with indirect observation. *Pages 293–315 of:* Jöreskog, K. G., & Wold, H. (eds), *Systems under Indirect Observation - Causality ⋆ Structure ⋆ Prediction - Part II*. Amsterdam: North-Holland.

Durbin, J. 1951. Incomplete blocks in ranking experiments. *British Journal of Psychology*, **4**, 85–90.

Durbin, J., & Watson, G. S. 1950. Testing for serial correlation in least squares regression. *Biometrika*, **37**, 409–428.

Ecochard, R., & Clayton, D. G. 2001. Fitting complex random effect models with standard software using data augmentation: Application to a study of male and female fecundability. *Statistical Modelling*, **1**, 319–331.

Edwards, J. R., & Bagozzi, R. P. 2000. On the nature and direction of relationships between constructs and measures. *Psychological Methods*, **5**, 155–174.

Efron, B. 1977. The efficiency of Cox's likelihood function for censored data. *Journal of the American Statistical Association*, **72**, 557–565.

Efron, B., & Hinkley, D. V. 1978. Assessing the accuracy of the maximum likelihood estimator: Observed versus expected Fisher information. *Biometrika*, **65**, 475–487.

Efron, B., & Morris, C. 1973. Stein's estimation rule and its competitors - an empirical Bayes approach. *Journal of the American Statistical Association*, **68**, 117–130.

Efron, B., & Morris, C. 1975. Data analysis using Stein's estimator and its generalizations. *Journal of the American Statistical Association*, **70**, 311–319.

Efron, B., & Tibshirani, R. J. 1993. *An Introduction to the Bootstrap*. London: Chapman & Hall.

EGRET for Windows User Manual. 2000. *EGRET for Windows User Manual*. Cambridge, MA: Citel Software Corporation.

Einhorn, H. 1969. Alchemy in the behavioral sciences. *Public Opinion Quarterly*, **36**, 367–378.

Eklöf, M., & Weeks, M. 2003. Estimation of Discrete Choice Models using DCM for Ox: A Manual. *Technical Report*. Downloadable from www.econ.cam.ac.uk/faculty/weeks/DCM/DCMManual/DCMManual.pdf.

Elrod, T., & Keane, M. P. 1995. A factor analytic probit model for representing the market structure in panel data. *Journal of Marketing Research*, **32**, 1–16.

Elston, R. C. 1964. On estimating time-response curves. *Biometrics*, **20**, 643–647.

Embretson, S. E., & Reise, S. P. 2000. *Item Response Theory for Psychologists*. Mahwah, NJ: Erlbaum.

Engel, B., & Keen, A. 1994. A simple approach for the analysis of generalized linear mixed models. *Statistica Neerlandica*, **48**, 1–22.

Engle, R. F. 1984. Wald, likelihood-ratio, and Lagrange multiplier tests in econometrics. *Pages 776–826 of:* Griliches, Z., & Intriligator, M. D. (eds), *Handbook of Econometrics, Volume II*. Amsterdam: North-Holland.

Engle, R. F., Hendry, D. F., & Richard, J. F. 1983. Exogeneity. *Econometrica*, **51**, 277–304.

Etezadi-Amoli, J., & McDonald, R. P. 1983. A second generation nonlinear factor analysis. *Psychometrika*, **48**, 315–342.

Evans, M., & Swartz, T. B. 2000. *Approximating Integrals via Monte Carlo and Deterministic Methods*. Oxford: Oxford University Press.

Everitt, B. S. 1987. *Introduction to Optimization Methods and their Application in Statistics*. London: Chapman & Hall.

Everitt, B. S. 1988. A Monte Carlo investigation of the likelihood ratio test for number of classes in latent class analysis. *Multivariate Behsvioral Research*, **23**, 531–538.

Everitt, B. S., & Hand, D. J. 1981. *Finite Mixture Distributions*. London: Chapman & Hall.

Everitt, B. S., & Pickles, A. 1999. *Statistical Aspects of the Design and Analysis of Clinical Trials*. London: Imperial College Press.

Ezzet, F., & Whitehead, J. 1991. A random effects model for ordinal responses from a cross-over trial. *Statistics in Medicine*, **10**, 901–907.

Fahrmeir, L., & Tutz, G. 2001. *Multivariate Statistical Modelling Based on Generalized Linear Models (Second Edition)*. New York: Springer.

Falconer, D.S. 1981. *Introduction to Quantitative Genetics*. New York: Longman.

Fang, K.-T., & Wang, Y. 1994. *Number-Theoretic Methods in Statistics*. London: Chapman & Hall.

Farewell, V. T. 1979. Some results on the estimation of logistic models based on retrospective data. *Biometrika*, **66**, 27–32.

Farewell, V. T., & Prentice, R. L. 1980. The approximation of partial likelihood with emphasis on case-control studies. *Biometrika*, **67**, 273–278.

Faucett, C. L., & Thomas, D. C. 1996. Simultaneously modeling censored survival data and repeatedly measured covariates: A Gibbs sampling approach. *Statistics in Medicine*, **15**, 1663–1685.

Fayers, P. M., & Hand, D. J. 2002. Causal variables, indicator variables, and measurement scales. *Journal of the Royal Statistical Society, Series A*, **165**, 233–261.

Finney, D. J. 1971. *Probit Analysis*. Cambridge: Cambridge University Press.

Fischer, G. H. 1977. Linear logistic trait models: Theory and application. *Pages 203–225 of:* Spada, H., & Kempf, W. F. (eds), *Structural Models of Thinking and Learning*. Bern: Huber.

Fischer, G. H. 1995. Derivations of the Rasch model. *Pages 15–38 of:* Fischer, G. H., & Molenaar, I. W. (eds), *Rasch models. Foundations, Recent Developments, and Applications*. New York: Springer.

Fisher, F. M. 1966. *The Identification Problem in Econometrics*. New York: McGraw-Hill.

Fisher, R. A. 1922. On the mathematical foundations of theoretical statistics. *Philosophical Transactions of the Royal Society of London, Series A*, **222**, 309–368.

Fitzmaurice, G. M. 1995. A caveat concerning independence estimating equations with multivariate binary data. *Biometrics*, **51**, 309–317.

Fitzmaurice, G. M., Laird, N. M., & Rotnitzky, A. G. 1993. Regression models for discrete longitudinal responses. *Statistical Science*, **8**, 284–309.

Fitzmaurice, G. M., Laird, N. M., & Ware, J. H. 2004. *Applied Longitudinal Analysis*. New York: Wiley.

Fleiss, J. L. 1993. The statistical basis of meta-analysis. *Statistical Methods in Medical Research*, **2**, 121–145.

Fletcher, R. 1987. *Practical Methods of Optimization*. New York: Wiley.

Fokoué, E., & Titterington, D. M. 2003. Mixtures of factor analysers. Bayesian estimation and inference by stochastic simulation. *Machine Learning*, **50**, 73–94.

Follmann, D. A., & Lambert, D. 1989. Generalizing logistic regression by nonparametric mixing. *Journal of the American Statistical Association*, **84**, 295–300.

Formann, A. K. 1992. Linear logistic latent class analysis for polytomous data. *Journal of the American Statistical Association*, **87**, 476–486.

Formann, A. K., & Kohlmann, T. 1996. Latent class analysis in medical research. *Statistical Methods in Medical Research*, **5**, 179–211.

Fornell, C. 1983. Issues in the application of covariance structure analysis: A comment. *Journal of Consumer Research*, **9**, 443–450.

Fox, J. P. 2001. *Multilevel IRT: A Bayesian Perspective on Estimating Parameters and Testing Statistical Hypotheses*. Ph.D. thesis, Twente University, Department of measurement and data analysis.

Fox, J. P., & Glas, C. A. W. 2001. Bayesian estimation of a multilevel IRT model using Gibbs sampling. *Psychometrika*, **66**, 271–288.

Francis, B. J., Stott, D. N., & Davies, R. B. 1996. *SABRE: A guide for users, version 3.1*. Lancaster: Centre for Applied Statistics, Lancaster University. Downloadable from http://www.cas.lancs.ac.uk/software/sabre3.1/sabre.html.

Freedman, D. A. 1983. A note on screening regression equations. *The American Statistician*, **37**, 152–155.

Freedman, D. A. 1985. Statistics and the scientific method. *Pages 345–390 of:* Mason, M., & Fienberg, S. E. (eds), *Cohort Analysis in Social Research: Beyond the Identification Problem*. New York: Springer.

Freedman, D. A. 1986. As others see us: A case study in path analysis (with discussion). *Journal of Educational Statistics*, **12**, 101–128.

Freedman, D. A. 1992. Statistical methods and shoe leather. *Pages 291–313 of:* Marsden, P. V. (ed), *Sociological Methodology 1992*. Oxford: Blackwell.

Fruchter, B. 1954. *Introduction to Factor Analysis*. Princeton: Van Nostrand.

Fuller, W. A. 1987. *Measurement Error Models*. New York: Wiley.

Gabriel, K. R. 1962. Ante-dependence analysis of an ordered set of variables. *Annals of Mathematical Statistics*, **33**, 201–212.

Gabrielsen, A. 1978. Consistency and identifiability. *Journal of Econometrics*, **8**, 261–263.

Gallant, A. R., & Nychka, D. W. 1987. Semi-nonparametric maximum likelihood estimation. *Econometrica*, **55**, 363–390.

Gart, J. J., & Zweifel, J. R. 1967. On the bias of various estimators of the logit and its variance, with application to quantal bioassay. *Biometrika*, **54**, 181–187.

Geisser, S. 1975. The predictive sample reuse method with applications. *Journal of the American Statistical Association*, **70**, 320–328.

Gelman, A., & Rubin, D. B. 1996. Markov chain Monte Carlo methods in biostatistics. *Statistical Methods in Medical Research*, **5**, 339–355.

Gelman, A., Goegebeur, Y., Tuerlinckx, F., & Van Mechelen, I. 2000. Diagnostic checks for discrete data regression models using posterior predictive simulations. *Journal of the Royal Statistical Society, Series C*, **49**, 247–268.

Gelman, A., Carlin, J. B., Stern, H. S., & Rubin, D. B. 2003. *Bayesian Data Analysis (Second Edition)*. Boca Raton, FL: Chapman & Hall/CRC.

Geman, S., & Geman, D. 1984. Stochastic Relaxation, Gibbs Distributions, and the Bayesian Restoration of Images. *IEEE Transactions On Pattern Analysis and Machine Intelligence*, **6**, 721–741.

Geraci, V. J. 1976. Identification of simultaneous equation models with measurement error. *Journal of Econometrics*, **4**, 262–283.

Geweke, J. 1989. Bayesian inference in econometric models using Monte Carlo integration. *Econometrica*, **57**, 1317–1340.

Geweke, J., Keane, M. P., & Runkle, D. 1994. Alternative computational approaches to statistical inference in the multinomial probit model. *Review of Economics and Statistics*, **76**, 609–632.

Geys, H., Molenberghs, G., & Ryan, L. M. 2002. Pseudo-likelihood estimation. *Pages 90–114 of:* Aerts, M., Geys, H., Molenberghs, G., & Ryan, L. M. (eds), *Topics in Modelling Clustered Data*. Boca Raton, FL: Chapman & Hall/CRC.

Gibson, W. A. 1959. Three multivariate models: factor analysis, latent structure analysis and latent profile analysis. *Psychometrika*, **24**, 229–252.

Gilks, W. R. 1998. Markov chain Monte Carlo. *Pages 2415–2423 of: Encyclopedia of Biostatistics*. London: Wiley.

Gilks, W. R., & Wild, P. 1992. Adaptive rejection sampling for Gibbs sampling. *Journal of the Royal Statistical Society, Series C*, **41**, 337–348.

Gilks, W. R., Richardson, S., & Spiegelhalter, D. J. 1996. Introducing Markov chain Monte Carlo. *Pages 1–19 of:* Gilks, W. R., Richardson, S., & Spiegelhalter, D. J. (eds), *Markov Chain Monte Carlo in Practice*. Boca Raton: Chapman & Hall/CRC.

Glas, C. A. W. 1988. The derivation of some tests for the Rasch model from the multinomial distribution. *Psychometrika*, **53**, 525–546.

Glass, G. V. 1976. Primary, secondary and meta-analysis of research. *Educational Research*, **5**, 3–8.

Glass, G. V., & Maguire, T. 1966. Abuses of factor scores. *American Educational Research Journal*, **3**, 297–304.

Goldberger, A. S. 1962. Best linear unbiased prediction in the generalized linear regression model. *Journal of the American Statistical Association*, **57**, 369–375.

Goldberger, A. S. 1971. Econometrics and psychometrics: A survey of communalities. *Psychometrika*, **36**, 83–107.

Goldberger, A. S. 1972. Structural equation methods in the social sciences. *Econometrica*, **40**, 979–1001.

Goldberger, A. S. 1991. *A Course in Econometrics*. Cambridge, MA: Harvard University Press.

Goldstein, H. 1986. Multilevel mixed linear model analysis using iterative generalised least squares. *Biometrika*, **73**, 43–56.

Goldstein, H. 1987. Multilevel covariance component models. *Biometrika*, **74**, 430–431.

Goldstein, H. 1989. Restricted unbiased iterative generalised least squares estimation. *Biometrika*, **76**, 622–623.

Goldstein, H. 1991. Nonlinear multilevel models, with an application to discrete response data. *Biometrika*, **78**, 45–51.

Goldstein, H. 1994. Recontextualizing mental measurement. *Educational Measurement: Issues and Practice*, **13**, 16–19, 43.

Goldstein, H. 2003. *Multilevel Statistical Models (Third Edition)*. London: Arnold.

Goldstein, H., & Browne, W. J. 2002. Multilevel factor analysis modelling using Markov Chain Monte Carlo (MCMC) estimation. *Pages 225–243 of:* Marcoulides, G., & Moustaki, I. (eds), *Latent Variable and Latent Structure Models*. Mahwah, NJ: Erlbaum.

Goldstein, H., & McDonald, R. P. 1988. A general model for the analysis of multilevel data. *Psychometrika*, **53**, 455–467.

Goldstein, H., & Rasbash, J. 1996. Improved approximations for multilevel models with binary responses. *Journal of the Royal Statistical Society, Series A*, **159**, 505–513.

Goldstein, H., & Spiegelhalter, D. J. 1996. League tables and their limitations: Statistical issues in comparisons of institutional performance. *Journal of the Royal Statistical Society, Series A*, **159**, 385–409.

Goldstein, H., Browne, W. J., & Rasbash, J. 2002. Partitioning variation in multilevel models. *Understanding Statistics*, **1**, 223–232.

Goodman, L. A. 1974. Exploratory latent structure analysis using both identifiable and unidentifiable models. *Biometrika*, **61**, 215–231.

Goodman, L. A. 1979. Simple models for the analysis of association in cross-classifications having ordered categories. *Journal of the American Statistical Association*, **74**, 537–552.

Goodman, L. A. 1983. The analysis of dependence in cross-classifications having ordered categories, using log-linear models for frequencies and log-linear models for odds. *Biometrics*, **39**, 149–160.

Gorsuch, R. L. 1983. *Factor Analysis*. Hillsdale, NJ: Erlbaum.

Gould, W., Pitblado, J., & Sribney, W. 2003. *Maximum Likelihood Estimation with Stata*. College Station, TX: Stata Press.

Gourieroux, C., & Monfort, A. 1996. *Simulation-Based Econometric Methods*. Oxford: Oxford University Press.

Gourieroux, C., Monfort, A., Renault, E., & Trognon, A. 1987a. Generalised residuals. *Journal of Econometrics*, **34**, 5–32.

Gourieroux, C., Monfort, A., Renault, E., & Trognon, A. 1987b. Simulated residuals. *Journal of Econometrics*, **34**, 201–252.

Granger, C. W. J. 2002. Some comments on risk. *Journal of Applied Econometrics*, **17**, 447–456.

Green, P. E., & Srinivasan, V. 1990. Conjoint analysis in marketing: New developments with implications for research and practice. *Journal of Marketing*, **54**, 3–19.

Greene, V. L., & Carmines, E. G. 1980. Assessing the reliability of linear composites. *Pages 160–175 of:* Schuessler, K. F. (ed), *Sociological Methodology 1980*. San Francisco, CA: Jossey-Bass.

Greene, W. H. 2002a. *LIMDEP 8.0 Econometric Modeling Guide*. Plainview, NY: Econometric Software.

Greene, W. H. 2002b. *NLOGIT 3.0 Reference Guide*. Plainview, NY: Econometric Software.

Greene, W. H. 2003. *Econometric Analysis (Fifth Edition)*. Englewood Cliffs, NJ: Prentice Hall.

Greenland, S., & Brumback, B. 2002. An overview of relations among causal modelling methods. *International Journal of Epidemiology*, **31**, 1030–1037.

Greenwood, M., & Yule, G. U. 1920. An inquiry into the nature of frequency distributions of multiple happenings, with particular reference to the occurence of multiple attacks of disease or repeated accidents. *Journal of the Royal Statistical Society, Series A*, **83**, 255–279.

Griliches, Z. 1974. Errors in variables and other unobservables. *Econometrica*, **42**, 971–998.

Grilli, L., & Rampichini, C. 2003. Alternative specifications of bivariate multilevel probit ordinal response models. *Journal of Educational and Behavioral Statistics*, **28**, 31–44.

Gueorguieva, R., & Sanacora, G. 2003. A latent variable model for joint analysis of repeatedly measured ordinal and continuous outcomes. *Pages 171–176 of:* Verbeke, G., Molenberghs, G., Aerts, A., & Fieuws, S. (eds), *Proceedings of the 18th International Workshop on Statistical Modelling*. Leuven, Belgium: Katholieke Universiteit Leuven.

Gulliksen, H. 1950. *Theory of Mental Tests*. New York: Wiley.

Gustafson, P. 2004. *Measurement Error and Misclassification in Statistics and Epidemiology. Impacts and Bayesian Adjustments*. Boca Raton, FL: Chapman & Hall/CRC.

Guttmann, L. 1955. The determinacy of factor score matrices with implications for five other basic problems of common factor theory. *British Journal of Mathematical and Statistical Psychology*, **8**, 65–81.

Guttmann, L. 1977. What is not what in statistics. *Journal of the Royal Statistical Society, Series D*, **26**, 81–107.

Haaijer, M. E., Wedel, M., Vriens, M., & Wansbeek, T. J. 1998. Utility covariances and context effects in conjoint MNP models. *Marketing Science*, **17**, 236–252.

Haberman, S. J. 1977. Maximum likelihood estimates in exponential response models. *The Annals of Statistics*, **5**, 815–841.

Haberman, S. J. 1979. *Analysis of Qualitative Data: New Developments*. Vol. 2. New York: Academic.

Haberman, S. J. 1989. A stabilized Newton-Raphson algorithm for log-linear models for frequency tables derived by indirect observation. *Pages 193–211 of:* Clogg, C. C. (ed), *Sociological Methodology 1988*. Washington: American Sociological Association.

Hagenaars, J. A. 1988. Latent structure models with direct effects between indicators: Local dependence models. *Sociological Methods & Research*, **16**, 379–405.

Hagenaars, J. A. 1993. *Loglinear Models with Latent Variables*. Sage University Paper Series on Quantitative Applications in the Social Sciences. Newbury Park, CA: Sage.

Hajivassiliou, V. A., & Ruud, P. A. 1994. Classical estimation methods for LDV models using simulation. *Pages 2383–2441 of:* Engle, R. F., & McFadden, D. L. (eds), *Handbook of Econometrics*, vol. IV. New York: Elsevier.

Hajivassiliou, V. A., McFadden, D. L., & Ruud, P. A. 1996. Simulation of multivariate normal rectangle probabilities and their derivatives - Theoretical and computational results. *Journal of Econometrics*, **72**, 85–134.

Hall, D. B. 1997. Zero-inflated Poisson and binomial regression with random effects: A case study. *Biometrics*, **56**, 1030–1039.

Halperin, S. 1976. The incorrect measurement of components. *Educational and Psychological Measurement*, **36**, 347–353.

Hambleton, R. K., & Swaminathan, H. 1985. *Item Response Theory: Principles and Applications*. Boston: Kluwer.

Hambleton, R. K., Swaminathan, H., & Rogers, H. J. 1991. *Fundamentals of Item Response Theory*. Newbury Park, CA: Sage.

Hamerle, A., & Rönning, G. 1995. Panel analysis for qualitative variables. *Pages 401–451 of:* Arminger, G., Clogg, C. C., & Sobel, M. E. (eds), *Handbook of Statistical Modelling for the Social and Behavioral Sciences*. New York: Plenum Press.

Han, A., & Hausman, J. A. 1990. Flexible parametric estimation of duration and competing risk models. *Journal of Applied Econometrics*, **5**, 1–29.

Hand, D. J., & Crowder, M. J. 1996. *Practical Longitudinal Data Analysis*. London: Chapman & Hall.

Hardin, J., & Hilbe, J. 2002. *Generalized Estimating Equations*. Boca Raton, FL: Chapman & Hall/CRC.

Harman, H. H. 1976. *Modern Factor Analysis*. Chicago: University of Chicago Press.

Harper, D. 1972. Local dependence latent structure models. *Psychometrika*, **37**, 53–59.

Harrell, F. E. 2001. *Regression Modeling Strategies. With Application to Linear Models, Logistic Regression and Survival Analysis*. New York: Springer.

Harris, C. W. 1967. On factors and factor scores. *Psychometrika*, **32**, 363–379.

Hartzel, J., Agresti, A., & Caffo, B. 2001. Multinomial logit random effects models. *Statistical Modelling*, **1**, 81–102.

Harvey, A. 1976. Estimating regression models with multiplicative heteroscedasticity. *Econometrica*, **44**, 461–465.

Harville, D. A. 1977. Maximum likelihood approaches to variance components estimation and related problems. *Journal of the American Statistical Association*, **72**, 320–340.

Hauck, W. W., & Donner, A. 1977. Wald's test as applied to hypotheses in logit analysis. *Journal of the American Statistical Association*, **72**, 851–853.

Hauser, R. M., & Goldberger, A. S. 1971. The treatment of unobservable variables in path analysis. *Pages 81–117 of:* Costner, H. L. (ed), *Sociological Methodology 1971*. San Francisco, CA: Jossey-Bass.

Hausman, J. A. 1978. Specification tests in econometrics. *Econometrica*, **46**, 1251–1271.

Hausman, J. A., & Ruud, P. A. 1987. Specifying and testing econometric models for rank-ordered data. *Journal of Econometrics*, **34**, 83–103.

Hausman, J. A., & Wise, D. A. 1978. A conditional probit model for qualitative choice: Discrete decisions recognizing interdependence and heterogeneous preferences. *Econometrica*, **46**, 403–426.

Hausman, J. A., & Wise, D. A. 1979. Attrition bias in experimental and panel data: The Gary income maintenance experiment. *Econometrica*, **47**, 455–473.

Hausman, J. A., Hall, B. H., & Griliches, Z. 1984. Econometric models for count data with an application to the patents R and D relationship. *Econometrica*, **52**, 909–938.

Heagerty, P. J., & Zeger, S. L. 2000. Marginalized multilevel models and likelihood inference. *Statistical Science*, **15**, 1–26.

Heath, A., Curtice, J. K., Jowell, R., Evans, G., Fields, J., & Witherspoon, S. 1991. *Understanding Political Change: The British Voter 1964-1987*. Oxford: Pergamon Press.

Heath, A., Jowell, R., Curtice, J. K., Brand, J. A., & Mitchell, J. C. 1993. *British Election Panel Study, 1987-1992 [Computer file] SN: 2983*. Colchester, Essex: The Data Archive [Distributor].

Heath, A., Jowell, R., & Curtice, J. K. (eds). 1994. *Labour's Last Chance? The 1992 Election and Beyond*. Aldershot: Darthmouth Publishing Company.

Heckman, J. J. 1978. Dummy endogenous variables in a simultaneous equation system. *Econometrica*, **46**, 931–959.

Heckman, J. J. 1979. Sample selection bias as a specification error. *Econometrica*, **47**, 153–161.

Heckman, J. J. 1981a. Heterogeneity and state dependence. *Pages 91–139 of:* Rosen, S. (ed), *Studies in Labor Markets*. Chicago: Chicago University Press.

Heckman, J. J. 1981b. The incidental parameters problem and the problem of initial conditions in estimating a discrete stochastic process and some Monte Carlo evidence on their practical importance. *Pages 179–196 of:* Manski, C. F., & McFadden, D. L. (eds), *Structural Analysis of Discrete Data with Econometric Applications*. Cambridge: MIT Press.

Heckman, J. J. 1981c. Statistical models for discrete panel data. *Pages 114–178 of:* Manski, C. F., & McFadden, D. L. (eds), *Structural Analysis of Discrete Data with Econometric Applications*. Cambridge: MIT Press.

Heckman, J. J., & Singer, B. 1982. The identification problem in econometric models for duration data. *Pages 39–77 of:* Hildenbrand, W. (ed), *Advances in Econometrics, Proceedings of the Fourth World Congress of the Econometric Society*. Cambridge: Cambridge University Press.

Heckman, J. J., & Singer, B. 1984. A method of minimizing the impact of distributional assumptions in econometric models for duration data. *Econometrica*, **52**, 271–320.

Heckman, J. J., & Willis, R. J. 1977. A beta-logistic model for the analysis of sequential labor force participation by married women. *Journal of Political Economy*, **85**, 27–58.

Hedeker, D. 1999. MIXNO: A computer program for mixed-effects logistic regression. *Journal of Statistical Software*, 1–92.

Hedeker, D. 2003. A mixed-effects multinomial logistic regression model. *Statistics in Medicine*, **22**, 1433–1466.

Hedeker, D., & Gibbons, R. D. 1994. A random-effects ordinal regression model for multilevel analysis. *Biometrics*, **50**, 933–944.

Hedeker, D., & Gibbons, R. D. 1996a. MIXOR: A computer program for mixed-effects ordinal probit and logistic regression analysis. *Computer Methods and Programs in Biomedicine*, **49**, 157–76.

Hedeker, D., & Gibbons, R. D. 1996b. MIXREG: A computer program for mixed-effects regression analysis with autocorrelated errors. *Computer Methods and Programs in Biomedicine*, **49**, 229–252.

Hedeker, D., Siddiqui, O., & Hu, F. B. 2000. Random-effects regression analysis of correlated grouped-time survival data. *Statistical Methods in Medical Research*, **9**, 161–179.

Hedges, L. V., & Olkin, I. 1985. *Statistical Methods for Meta-analysis*. Orlando, FL: Academic Press.

Heinen, T. 1996. *Latent Class and Discrete Latent Trait Models: Similarities and Differences*. Thousand Oaks, CA: Sage.

Henderson, C. R. 1975. Best linear unbiased estimation and prediction under a selection model. *Biometrics*, **31**, 423–447.

Hershberger, S. L. 1994. The specification of equivalent models before the collection of data. *Pages 68–108 of:* von Eye, A., & Clogg, C. C. (eds), *Latent Variable Analysis: Applications to Developmental Research*. Thousand Oaks, CA: Sage.

Hilden-Minton, J. A. 1995. *Multilevel Diagnostics for Mixed and Hierarchical Linear Models*. Ph.D. thesis, Department of Mathematics, University of California, Los Angeles.

Hildesheim, A., Mann, V., Brinton, L. A., Szklo, M., Reeves, W. C., & Rawls, W. E. 1991. Herpes simplex virus type 2: A possible interaction with human papillomavirus types 16/18 in the development of invasive cervical cancer. *International Journal of Cancer*, **49**, 335–340.

Hill, P. W., & Goldstein, H. 1998. Multilevel modeling of educational data with cross-classification and missing identification of units. *Journal of Educational and Behavioral Statistics*, **23**, 117–128.

Hobcraft, J., & Murphy, M. 1987. Demographic event history analysis: a selective review. *In:* Crouchley, R. (ed), *Longitudinal Data Analysis*. Aldershot: Avebury.

Hobert, J. P. 2000. Hierarchical models: A current computational perspective. *Journal of the American Statistical Association*, **95**, 1312–1316.

Hodges, J. S., & Sargent, D. J. 2001. Counting degrees of freedom in hierarchical and other richly-parameterised models. *Biometrika*, **88**, 367–379.

Hogan, J. W., & Laird, N. M. 1997a. Mixture models for the joint distribution of repeated measures and event times. *Statistics in Medicine*, **16**, 239–257.

Hogan, J. W., & Laird, N. M. 1997b. Model-based approaches to analyzing incomplete repeated measures and failure time data. *Statistics in Medicine*, **16**, 259–271.

Hoijtink, H. 2001. Confirmatory latent class analysis: model selection using Bayes factors and (pseudo) likelihood ratio statistics. *Multivariate Behavioral Research*, **36**, 563–588.

Hoijtink, H., & Boomsma, A. 1995. On person parameter estimation in the dichotomous Rasch model. *Pages 53–68 of:* Fischer, G. H., & Molenaar, I. W. (eds), *Rasch Models: Foundations, Recent Developments, and Applications*. New York: Springer.

Holford, T. R. 1976. Life tables with concomitant variables. *Biometrics*, **32**, 587–597.

Holford, T. R. 1980. The analysis of rates and survivorship using log-linear models. *Biometrics*, **36**, 299–305.

Holford, T. R., White, C., & Kelsey, J. L. 1978. Multivariate analysis for matched case-control studies. *American Journal Epidemiology*, **107**, 245–256.

Holland, P. W. 1988. Causal inference, path analysis, and recursive structural equation models. *Pages 449–493 of:* Clogg, C. C. (ed), *Sociological Methodology 1988*. Washington: American Sociological Association.

Holmås, T. H. 2002. Keeping nurses at work: A duration analysis. *Health Economics*, **11**, 493–503.

Holzinger, K. L., & Harman, H. H. 1941. *Factor Analysis: A Synthesis of Factorial Methods*. Chicago: University of Chicago Press.

Horowitz, J. L., Sparmann, J. M., & Daganzo, C. 1982. An investigation of the accuracy of the Clark approximation for the multinomial probit model. *Transportation Science*, **16**, 382–401.

Hosmer, D. A., & Lemeshow, S. A. 1999. *Applied Survival Analysis: Regression Modeling of Time to Event Data*. New York: Wiley.

Hosmer, D. A., & Lemeshow, S. A. 2000. *Applied Logistic Regression (Second Edition)*. New York: Wiley.

Hotelling, H. 1957. The relations of the newer multivariate statistical methods to factor analysis. *British Journal of Statistical Psychology*, **10**, 69–79.

Hougaard, P. 1986a. A class of multivariate failure time distributions. *Biometrika*, **73**, 671–678.

Hougaard, P. 1986b. Survival models for heterogeneous populations derived from stable distributions. *Biometrika*, **73**, 387–396.

Hougaard, P. 2000. *Analysis of Multivariate Survival Data*. New York: Springer.

Hox, J. 2002. *Multilevel Analysis: Techniques and Applications*. Mahwah, NJ: Erlbaum.

Hsiao, C. 1983. Identification. *Pages 223–283 of:* Griliches, Z., & Intriligator, M. D. (eds), *Handbook of Econometrics, Volume I*. Amsterdam: North-Holland.

Hsiao, C. 2002. *Analysis of Panel Data (Second Edition)*. Cambridge, UK: Cambridge University Press.

Hu, P., Tsiatis, A. A., & Davidian, M. 1998. Estimating the parameters in the Cox model when the covariate variables are measured with error. *Biometrics*, **54**, 1407–1419.

Huber, P. J. 1967. The behavior of maximum likelihood estimates under non-standard conditions. *Pages 221–233 of: Proceedings of the Fifth Berkeley Symposium on Mathematical Statistics and Probability*, vol. I. Berkeley: University of California Press.

Huggins, R. M. 1989. On the statistical analysis of capture experiments. *Biometrika*, **76**, 133–140.

Hunt, M. 1997. *How Science Takes Stock: The Story of Meta-analysis*. New York, NY: Sage.

Imbens, G. W., & Rubin, D. B. 1997a. Bayesian inference for causal effects in randomized experiments with non-compliance. *The Annals of Statistics*, **25**, 305–327.

Imbens, G. W., & Rubin, D. B. 1997b. Estimating outcome distributions for compliers in instrumental variable models. *Review of Economic Studies*, **64**, 555–574.

Inglehart, R. 1977. *The Silent Revolution: Changing Values and Political Styles Among Western Publics*. Princeton, NJ: Princeton University Press.

James, L. R., Mulaik, S. A., & Brett, J. M. 1982. *Causal Analysis: Assumptions, Models, and Data*. Beverly Hills, CA: Sage.

James, W., & Stein, C. 1961. Estimation with quadratic loss. *Pages 361–379 of: Proceedings of the Fourth Berkeley Symposium on Mathematical Statistics and Probability*, vol. I. Berkeley: University of California Press.

Jedidi, K., Jagpal, H. S., & DeSarbo, W. S. 1997. Finite-mixture structural equation models for response-based segmentation and unobserved heterogeneity. *Marketing Science*, **16**, 39–59.

Jenkins, S. P. 1995. Easy estimation methods for discrete-time duration models. *Oxford Bulletin of Economics and Statistics*, **57**, 129–137.

Jewell, N. P. 1982. Mixtures of exponential distributions. *The Annals of Statistics*, **10**, 479–484.

Jo, B. 2002. Model misspecification sensitivity in estimating causal effects of interventions with noncompliance. *Statistics in Medicine*, **21**, 3161–3181.

Johansen, S. 1983. An extension of Cox's regression model. *International Statistical Review*, **51**, 258–262.

Johnson, N. L., Kotz, S., & Balakrishnan, N. 1994. *Distributions in Statistics: Continuous Univariate Distributions*. Vol. 1. New York: Wiley.

Johnson, R. A., & Wichern, D. W. 1983. *Applied Multivariate Statistical Analysis*. Englewood Cliffs: Prentice-Hall.

Johnson, V. E., & Albert, J. H. 1999. *Ordinal Data Modelling*. New York: Springer.

Joliffe, I. 1986. *Principal Components Analysis*. New York: Springer.

Jones, R. H. 1993. *Longitudinal Data with Serial Correlation: A State-Spece Approach*. London: Chapman and Hall.

Jöreskog, K. G. 1967. Some contributions to maximum likelihood factor analysis. *Psychometrika*, **32**, 443–482.

Jöreskog, K. G. 1969. A general approach to confirmatory maximum likelihood factor analysis. *Psychometrika*, **34**, 183–202.

Jöreskog, K. G. 1971a. Simultaneous factor analysis in several populations. *Psychometrika*, **36**, 409–426.

Jöreskog, K. G. 1971b. Statistical analysis of sets of congeneric tests. *Psychometrika*, **36**, 109–133.

Jöreskog, K. G. 1973. A general method for estimating a linear structural equation system. *Pages 85–112 of:* Goldberger, A. S., & Duncan, O. D. (eds), *Structural Equation Models in the Social Sciences*. New York: Seminar.

Jöreskog, K. G. 1974. Analysing psychological data by structural analysis of covariance matrices. *Pages 1–56 of:* Krantz, D. H., Atkinson, R. C., Luce, R. D., & Suppes, P. (eds), *Contemporary Developments in Mathematical Psychology*, vol. II. San Francisco, CA: Freeman.

Jöreskog, K. G. 1978. Structural analysis of covariance and correlation matrices. *Psychometrika*, **43**, 443–477.

Jöreskog, K. G. 1993. Testing structural equation models. *Pages 294–316 of:* Bollen, K. A., & Long, J. S. (eds), *Testing Structural Equation Models*. Newbury Park, CA: Sage.

Jöreskog, K. G. 1998. Interaction and nonlinear modeling. *Pages 239–250 of:* Schumaker, R. E., & Marcoulides, G. A. (eds), *Interaction and Nonlinear Effects in Structural Equation Modeling*. Mahwah, NJ: Erlbaum.

Jöreskog, K. G., & Sörbom, D. 1989. *LISREL 7: A Guide to the Program and Applications*. Chicago, IL: SPSS Publications.

Jöreskog, K. G., & Sörbom, D. 1990. Model search with TETRAD and LISREL. *Sociological Methods & Research*, **19**, 93–106.

Jöreskog, K. G., & Sörbom, D. 1994. *LISREL 8 and PRELIS 2: Comprehensive Analysis of Linear Relationships in Multivariate Data. LISREL 8 User's Guide.* Hillsdale, NJ: Erlbaum.

Jöreskog, K. G., Sörbom, D., Du Toit, S. H. C., & Du Toit, M. 2001. *LISREL 8: New Statistical Features.* Lincolnwood, IL: Scientific International.

Judge, G. G., Griffiths, W. E., Hill, R. C., Lütkepol, H., & Lee, T.-C. 1985. *The Theory and Practice of Econometrics.* New York: Wiley.

Kalbfleisch, J. D., & Prentice, R. L. 2002. *The Statistical Analysis of Failure Time Data (Second Edition).* New York: Wiley.

Kamakura, W. A., Kim, B. D., & Lee, J. 1996. Modeling preference and structural heterogeneity in consumer choice. *Marketing Science*, **15**, 152–172.

Kass, R. E., & Raftery, A. E. 1995. Bayes factors. *Journal of the American Statistical Association*, **90**, 773–795.

Kass, R. E., & Steffey, D. 1989. Approximate Bayesian inference in conditionally independent hierarchical models (parametric empirical Bayes models). *Journal of the American Statistical Association*, **84**, 717–726.

Kass, R. E., & Wasserman, L. 1995. A reference Bayesian test for nested hypotheses and its relationship to the Schwarz criterion. *Journal of the American Statistical Association*, **90**, 928–934.

Katz, E. 2001. Bias in conditional and unconditional fixed effects logit estimation. *Political Analysis*, **9**, 379–384.

Keane, M. P. 1992. A note on identification in the multinomial probit model. *Journal of Business and Economic Statistics*, **10**, 193–200.

Keane, M. P. 1994. A computationally practical simulation estimator for panel data. *Econometrica*, **62**, 95–116.

Keiding, N. 1992. Independent delayed entry (with discussion). *Pages 309–326 of:* Klein, J. P., & Goel, P. K. (eds), *Survival Analysis: State of the Art.* Dordrecht: Kluwer.

Kelly, P. J., & Lim, L. L-Y. 2000. Survival analysis for recurrent event data: An application to childhood infectious diseases. *Statistics in Medicine*, **19**, 13–33.

Kemp, I., Boyle, P., Smans, M., & Muir, C. 1985. *Atlas of Cancer in Scotland, 1975-1980. Incidence and Epidemiologic Perspective.* Scientific Publication 72. Lyon, France: International Agency for Research on Cancer.

Kempthorne, O. 1980. The term "design matrix". *The American Statistician*, **34**, 249.

Kendall, M., & Stuart, A. 1979. *The Advanced Theory of Statistics.* Vol. II. New York: Macmillan.

Kenkel, D. S., & Terza, J. V. 2001. The effect of physician advice on alcohol consumption: count regression with an endogenous treatment effect. *Journal of Applied Econometrics*, **16**, 165–184.

Kenny, D. A., & Judd, C. M. 1984. Estimating the non-linear and interactive effects of latent variables. *Psychological Bulletin*, **96**, 201–210.

Kestelman, H. 1952. The fundamental equation of factor analysis. *British Journal of Psychology, Statistical Section*, **5**, 1–6.

Kim, J.-O., & Mueller, C. W. 1978. *Factor Analysis. Statistical Methods and Practical Issues.* Sage University Paper Series on Quantitative Applications in the Social Sciences. Beverly Hills, CA: Sage.

Kim, J.-O., & Rabjohn, J. 1978. Binary items and index construction. *Pages 120–159 of:* Schussler, K. F. (ed), *Sociological Methodology 1978.* San Francisco, CA: Jossey-Bass.

King, G., Murray, C. J. L., Salomon, J. A., & Tandon, A. 2003. Enhancing the validity of cross-cultural comparability of survey research. *American Political Science Review*, **97**, 567–583.

Kirisci, L., & Hsu, T.-C. 2001. Robustness of item parameter estimation programs to assumptions of unidimensionality and normality. *Applied Psychological Measurement*, **25**, 146–162.

Klecka, W. R. 1980. *Discriminant Analysis.* Sage University Paper Series on Quantitative Applications in the Social Sciences. Beverly Hills, CA: Sage.

Klein, J. P., & Moeschberger, M. L. 2003. *Survival Analysis: Techniques for Censored and Truncated Data (Second Edition).* New York: Springer.

Knorr-Held, L., & Best, N. G. 2001. A shared component model for detecying joint and selective clustering of two diseases. *Journal of the Royal Statistical Society, Series A,* **164**, 73–85.

Knott, M., Albanese, M. T., & Galbraith, J. I. 1990. Scoring attitudes to abortion. *Journal of the Royal Statistical Society, Series D,* **40**, 217–223.

Koopmans, T. C., & Reiersøl, O. 1950. The identification of structural characteristics. *Annals of Mathematical Statistics,* **21**, 165–181.

Krane, W. R., & McDonald, R. P. 1978. Scale invariance and the factor analysis of correlation matrices. *British Journal of Mathematical and Statistical Psychology,* **31**, 218–228.

Kreft, I. G. G., & de Leeuw, J. 1994. The gender gap in earnings. *Sociological Methods & Research,* **22**, 319–341.

Kreft, I. G. G., & de Leeuw, J. 1998. *Introducing Multilevel Modeling.* London: Sage.

Kruskal, J. B., & Wish, M. 1978. *Multidimensional Scaling.* Sage University Paper Series on Quantitative Applications in the Social Sciences. Beverly Hills, CA: Sage.

Kuk, A. Y. C. 1995. Asymptotically unbiased estimation in generalized linear models with random effects. *Journal of the Royal Statistical Society, Series B,* **57**, 395–407.

Kuk, A. Y. C. 1999. Laplace importance sampling for generalized linear mixed models. *The Journal of Statistical Computation and Simulation,* **63**, 143–158.

Kuk, A. Y. C., & Cheng, Y. W. 1997. The Monte Carlo Newton-Raphson algorithm. *The Journal of Statistical Computation and Simulation,* **59**, 233–250.

Küsters, U. 1987. *Hierarchische Mittelwert- und Kovarianzstrukturmodelle mit Nicht-metrischen Endogenen Variablen.* Heidelberg: Physica.

Kvalem, I. L., & Træen, B. 2000. Self-efficacy, scripts of love and intention to use condoms among Norwegian adolescents. *Journal of Youth and Adolescents,* **29**, 337–353.

Lader, D., & Matheson, J. 1991. *Smoking among Secondary School Children in 1990.* London: HMSO.

Laird, N. M. 1978. Nonparametric maximum likelihood estimation of a mixing distribution. *Journal of the American Statistical Association,* **73**, 805–811.

Laird, N. M. 1982. Empirical Bayes estimates using the nonparametric maximum likelihood estimate for the prior. *The Journal of Statistical Computation and Simulation,* **15**, 211–220.

Laird, N. M. 1988. Missing data in longitudinal studies. *Statistics in Medicine,* **7**, 305–315.

Laird, N. M., & Olivier, D. 1981. Covariance analysis of censored survival data using log-linear analysis techniques. *Journal of the American Statistical Association,* **76**, 231–240.

Laird, N. M., & Ware, J. H. 1982. Random effects models for longitudinal data. *Biometrics,* **38**, 963–974.

Lambert, D. 1992. Zero-inflated Poisson-regression with an application to defects in manufacturing. *Technometrics,* **34**, 1–14.

Lancaster, T. 1990. *The Econometric Analysis of Transition Data.* Cambridge: Cambridge University Press.

Lancaster, T. 2000. The incidental parameter problem since 1948. *Journal of Econometrics,* **95**, 391–413.

Lange, N., & Ryan, L. M. 1989. Assessing normality in random effects models. *The Annals of Statistics,* **17**, 624–642.

Langford, I. H., & Lewis, T. 1998. Outliers in multilevel data. *Journal of the Royal Statistical Society, Series A,* **161**, 121–160.

Langford, I. H., Leyland, A. H., Rasbash, J., & Goldstein, H. 1999. Multilevel modeling of the geographical distributions of rare diseases. *Journal of the Royal Statistical Society, Series C,* **48**, 253–268.

Lawley, D. N. 1942. Further investigations in factor estimation. *Pages 176–185 of: Proceedings of the Royal Society of Edinburgh*, vol. 61.

Lawley, D. N., & Maxwell, A. E. 1971. *Factor Analysis as a Statistical Method.* London: Butterworths.

Lawson, A. B., Browne, W. J., & Vidal-Rodeiro, C. L. 2003. *Disease Mapping with WinBUGS and MLwiN.* New York: Wiley.

Lazarsfeld, P. F. 1950. The logical and mathematical foundation of latent structure analysis. *Pages 362–412 of:* Stouffer, S. A., Guttmann, L., Suchman, E. A., Lazarsfeld, P. F., Star, S. A., & Clausen, J. A. (eds), *Studies in Social Psychology in World War II*, vol. 4, Measurement and Prediction. Princeton, NJ: Princeton University Press.

Lazarsfeld, P. F. 1959. Latent structure analysis. *In:* Koch, S. (ed), *Psychology: A Study of A Science*, vol. III. New York: McGraw-Hill.

le Cessie, S., & van Houwelingen, J. C. 1994. Logistic regression for correlated binary data. *Journal of the Royal Statistical Society, Series C*, **43**, 95–108.

Leamer, E. E. 1978. *Specification Searches.* New York: Wiley.

Lee, E. W., Wei, L. J., & Amato, D. A. 1992. Cox-type regression analysis for large numbers of small groups of correlated failure time observations. *Pages 237–247 of:* Klein, J. P., & Goel, P. K. (eds), *Survival Analysis: State of the Art.* Dordrecht: Kluwer.

Lee, L.-F. 1982. Health and wage: A simultanous equation model with multiple discrete indicators. *International Economic Review*, **23**, 199–221.

Lee, S.-Y., & Shi, J.-Q. 2001. Maximum likelihood estimation of two-level latent variable models with mixed continuous and polytomous data. *Biometrics*, **57**, 787–794.

Lee, Y., & Nelder, J. A. 1996. Hierarchical generalized linear models (with discussion). *Journal of the Royal Statistical Society, Series B*, **58**, 619–678.

Lee, Y., & Nelder, J. A. 2001. Hierarchical generalised linear models: A synthesis of generalised linear models, random-effect models and structured dispersions. *Biometrika*, **88**, 987–1006.

Lehmann, E. L. 1990. Model specification: The views of Fisher and Neyman, and later developments. *Statistical Science*, **5**, 160–168.

Lenk, P. J., & DeSarbo, W. S. 2000. Bayesian inference for finite mixtures of generalized linear models with random effects. *Psychometrika*, **65**, 93–119.

Leppik, I. E., Dreifuss, F. E., Porter, R., Bowman, T., Santilli, N., Jacobs, M., Crosby, C., Cloyd, J., Stackman, J., Graves, N., Sutula, T., Welty, T., Vickery, J., Brundage, R., Gates, J., Gumnit, R. J., & Guttierrez, A. 1987. A controlled-study of progabide in partial seizures - methodology and results. *Neurology*, **37**, 963–968.

Lerman, S., & Manski, C. F. 1981. On the use of simulated frequencies to approximate choice probabilities. *In:* Manski, C. F., & McFadden, D. L. (eds), *Structural Analysis of Discrete Data with Econometric Applications.* Cambridge, MA: MIT Press.

Lesaffre, E., & Spiessens, B. 2001. On the effect of the number of quadrature points in a logistic random-effects model: An example. *Journal of the Royal Statistical Society, Series C*, **50**, 325–335.

Lesaffre, E., & Verbeke, G. 1998. Local influence in linear mixed models. *Biometrics*, **54**, 570–582.

Lewis, T., & Langford, I. H. 2001. Outliers, robustness and detection of discrepant data. *Pages 75–90 of:* Leyland, A. H., & Goldstein, H. (eds), *Multilevel Modelling of Health Statistics.* Chichester, UK: Wiley.

Leyland, A. H. 2001. Spatial analysis. *Pages 127–140 of:* Leyland, A. H., & Goldstein, H. (eds), *Multilevel Modelling of Health Statistics.* Chichester: Wiley.

Liang, K.-Y., & Zeger, S. L. 1986. Longitudinal data analysis using generalized linear models. *Biometrika*, **73**, 13–22.

Liang, K.-Y., Zeger, S. L., & Qaqish, B. 1992. Multivariate regression analysis for categorical data. *Journal of the Royal Statistical Society, Series B*, **54**, 3–40.

Likert, R. 1932. A technique for the measurement of attitudes. *Archives of Psychology*, **140**, 5–53.

Lillard, L. A. 1993. Simultaneous-equations for hazards - marriage duration and fertility timing. *Journal of Econometrics*, **56**, 189–217.

Lillard, L. A., & Panis, C. W. A. 2000. *aML User's Guide and Reference Manual.* Los Angeles, CA: EconWare.

Lin, X., & Breslow, N. E. 1996. Bias correction in generalized linear mixed models with multiple components of dispersion. *Journal of the American Statistical Association*, **91**, 1007–1016.

Linda, N. Y., Lee, S.-Y., & Poon, W.-Y. 1993. Covariance structure analysis with three level data. *Computational Statistics & Data Analysis*, **15**, 159–178.

Lindley, D. V. 1969. Discussion of Copas (1969): Compound decisions and empirical Bayes. *Journal of the Royal Statistical Society, Series B*, **31**, 419–421.

Lindley, D. V. 1971. The estimation of many parameters. *Pages 435–455 of:* Godambe, V. P., & Sprott, D. A. (eds), *Foundations of Statistical Inference.* Toronto: Holt, Rinehart and Winston.

Lindley, D. V., & Smith, A. F. M. 1972. Bayes estimates for the linear model (with discussion). *Journal of the Royal Statistical Society, Series B*, **34**, 1–41.

Lindsay, B. G. 1983. The Geometry of Mixture Likelihoods: A General Theory. *The Annals of Statistics*, **11**, 86–94.

Lindsay, B. G. 1995. *Mixture Models: Theory, Geometry and Applications.* NSF-CBMS Regional Conference Series in Probability and Statistics, vol. 5. Hayward, CA: Institute of Mathematical Statistics.

Lindsay, B. G., Clogg, C. C., & Grego, J. 1991. Semiparametric estimation in the Rasch model and related exponential response models, including a simple latent class model for item analysis. *Journal of the American Statistical Association*, **86**, 96–107.

Lindsey, J. K. 1996. *Parametric Statistical Inference.* Oxford, UK: Oxford University Press.

Lindsey, J. K. 1999. *Models for Repeated Measurements (Second Edition).* Oxford, UK: Oxford University Press.

Lindsey, J. K., & Lambert, P. 1998. On the appropriateness of marginal models for repeated measurements in clinical trials. *Statistics in Medicine*, **17**, 447–469.

Lipsitz, S., Laird, N. M., & Harrington, D. 1991. Generalized estimating equations for correlated binary data: Using odds ratios as a measure of association. *Biometrika*, **78**, 153–160.

Listhaug, O. 1989. *Citizens, Parties and Norwegian Electoral Politics 1957-1985.* Trondheim: Tapir.

Little, R. J. A. 1995. Modeling the drop-out mechanism in repeated measures studies. *Journal of the American Statistical Association*, **90**, 1112–1121.

Little, R. J. A., & Rubin, D. B. 1983. Models for nonresponse in sample surveys. *The American Statistician*, **37**, 218–220.

Little, R. J. A., & Rubin, D. B. 2002. *Statistical Analysis with Missing Data (Second Edition).* New York: Wiley.

Little, R. J. A., & Yau, L. H. Y. 1998. Statistical techniques for analyzing data from prevention trials. *Psychological Methods*, **3**, 147–159.

Liu, C., Rubin, D. B., & Wu, Y. 1998. Parameter expansion to accelerate EM: The PX-EM algorithm. *Biometrika*, **85**, 755–770.

Liu, Q., & Pierce, D. A. 1994. A note on Gauss-Hermite quadrature. *Biometrika*, **81**, 624–629.

Lohmöller, J. 1989. *Latent Variable Path Modeling with Partial Least Squares.* Heidelberg: Physica-Verlag.

Long, J. S. 1997. *Regression Models for Categorical and Limited Dependent Variables.* Thousand Oaks, CA: Sage.

Long, J. S., & Trivedi, P. K. 1993. Some specification tests for the linear regression model. *In:* Bollen, K. A., & Long, J. S. (eds), *Testing Structural Equation Models.* Newbury Park, CA: Sage.

Longford, N. T. 1987. A fast scoring algorithm for maximum likelihood estimation in unbalanced mixed models with nested random effects. *Biometrika*, **74**, 817–827.

Longford, N. T. 1993. *Random Coefficient Models*. Oxford: Oxford University Press.

Longford, N. T. 1994. Logistic regression with random coefficients. *Computational Statistics & Data Analysis*, **17**, 1–15.

Longford, N. T. 2001. Simulation-based diagnostics in random-coefficient models. *Journal of the Royal Statistical Society, Series A*, **164**, 259–273.

Longford, N. T., & Muthén, B. O. 1992. Factor analysis for clustered observations. *Psychometrika*, **57**, 581–597.

Lord, F. M. 1953. The relation of test score to the trait underlying the test. *Educational and Psychological Measurement*, **13**, 517–549.

Lord, F. M. 1980. *Applications of Item Response Theory to Practical Testing Problems*. Hillsdale, NJ: Erlbaum.

Lord, F. M., & Novick, M. R. 1968. *Statistical Theories of Mental Test Scores*. Reading: Addison-Wesley.

Louis, T. A. 1982. Finding the observed information matrix when using the EM algorithm. *Journal of the Royal Statistical Society, Series B*, **44**, 226–233.

Louis, T. A. 1984. Bayes and empirical Bayes estimates of a population of parameter values. *Journal of the American Statistical Association*, **79**, 393–398.

Luce, R. D. 1959. *Individual Choice Behavior*. New York: Wiley.

Luce, R. D., & Suppes, P. 1965. Preference, utility and subjective probability. *In:* Luce, R., Bush, R., & Galanter, E. (eds), *Handbook of Mathematical Psychology III*. New York: Wiley.

Luijben, T. C. W. 1991. Equivalent models in covariance structure analysis. *Psychometrika*, **56**, 653–665.

MacCallum, R. C., Wegener, D. T., Uchino, B. N., & Fabrigar, L. R. 1993. The problem of equivalent models in applications of covariance structure analysis. *Psychological Bulletin*, **114**, 185–199.

MacDonald, I. L., & Zucchini, W. 1997. *Hidden Markov and Other Models for Discrete-valued Time Series*. London: Chapman & Hall/CRC.

MacReady, G. B., & Dayton, C. M. 1992. The application of latent class models in adaptive testing. *Psychometrika*, **57**, 71–88.

Maddala, G. S. 1971. The use of variance components models in pooling cross-section and time series data. *Econometrica*, **39**, 341–358.

Maddala, G. S. 1977. *Econometrics*. New York: McGraw-Hill.

Maddala, G. S. 1983. *Limited Dependent and Qualitative Variables in Econometrics*. Cambridge: Cambridge University Press.

Magder, S. M., & Zeger, S. L. 1996. A smooth nonparametric estimate of a mixing distribution using mixtures of Gaussians. *Journal of the American Statistical Association*, **11**, 86–94.

Magidson, J., & Vermunt, J. K. 2002. Nontechnical introduction to latent class models. *Statistical Innovations White Paper 1*.

Manski, C. F. 1981. Structural models for discrete data: The analysis of discrete choice. *Pages 58–109 of:* Leinhardt, S. (ed), *Sociological Methodology 1981*. San Francisco, CA: Jossey-Bass.

Maple 9 Learning Guide. 2003. *Maple 9 Learning Guide*. Waterloo, ON: Maplesoft.

Marden, J. I. 1995. *Analyzing and Modeling Rank Data*. London: Chapman & Hall.

Maritz, J. S., & Lwin, T. 1989. *Empirical Bayes Methods*. London: Chapman & Hall.

Marsh, H. W., Balla, J. W., & McDonald, R. P. 1988. Goodness-of-fit indices in confirmatory factor analysis: Effects of sample size. *Psychological Bulletin*, **103**, 391–411.

Marshall, E. C., & Spiegelhalter, D. J. 2003. Simulation-based tests for divergent behaviour in hierarchical models. Submitted for publication.

Martin, J. K., & McDonald, R. P. 1975. Bayesian estimation in unrestricted factor analysis: A treatment for Heywood cases. *Psychometrika*, **40**, 505–517.

Marx, K. 1970. A note on classes. *Pages 5–6 of:* E. Laumann et al. (ed), *The Logic of Social Hierarchies*. Chicago, IL: Markham.

Mason, W. M., House, J. S., & Martin, S. S. 1985. On the dimensions of political alienation in America. *Pages 111–151 of:* Tuma, N. B. (ed), *Sociological Methodology 1985.* San Francisco, CA: Jossey-Bass.

Masters, G. N. 1982. A Rasch model for partial credit scoring. *Psychometrika,* **47,** 149–174.

Masters, G. N. 1985. A comparison of latent-trait and latent-class analyses of Likert-type data. *Psychometrika,* **50,** 69–82.

Maxwell, A. E. 1971. Estimating true scores and their reliabilities in the case of composite psychological tests. *British Journal of Mathematical and Statistical Psychology,* **24,** 195–204.

McCullagh, P. 1980. Regression models for ordinal data (with discussion). *Journal of the Royal Statistical Society, Series B,* **42,** 109–142.

McCullagh, P. 1983. Quasi-likelihood functions. *The Annals of Statistics,* **11,** 59–67.

McCullagh, P., & Nelder, J. A. 1989. *Generalized Linear Models (Second Edition).* London: Chapman & Hall.

McCulloch, C. E. 1994. Maximum likelihood variance components estimation for binary data. *Journal of the American Statistical Association,* **89,** 330–335.

McCulloch, C. E. 1997. Maximum likelihood algorithms for generalized linear mixed models. *Journal of the American Statistical Association,* **92,** 162–170.

McCulloch, C. E., & Searle, S. R. 2001. *Generalized, Linear and Mixed Models.* New York: Wiley.

McCulloch, C. E., Lin, H., Slate, E. H., & Turnbull, B. W. 2002. Discovering subpopulation structure with latent class mixed models. *Statistics in Medicine,* **21,** 417–429.

McCutcheon, A. L. 1987. *Latent Class Analysis.* Sage University Paper Series on Quantitative Applications in the Social Sciences. Beverly Hills, CA: Sage.

McDonald, R. P. 1967. *Nonlinear Factor Analysis.* Bowling Green, OH: Monograph 15, Psychometric Society.

McDonald, R. P. 1979. The simultanous estimation of factor loadings and scores. *British Journal of Mathematical and Statistical Psychology,* **32,** 212–228.

McDonald, R. P. 1981. The dimensionality of tests and items. *British Journal of Mathematical and Statistical Psychology,* **34,** 100–117.

McDonald, R. P. 1982. A note on the investigation of local and global identifiability. *Psychometrika,* **47,** 101–103.

McDonald, R. P. 1985. *Factor Analysis and Related Methods.* Hillsdale, NJ: Erlbaum.

McDonald, R. P. 1989. An index of goodness-of-fit based on noncentrality. *Journal of Classification,* **6,** 97–103.

McDonald, R. P., & Goldstein, H. 1989. Balanced and unbalanced designs for linear structural relations in two-level data. *British Journal of Mathematical and Statistical Psychology,* **42,** 215–232.

McDonald, R. P., & Krane, W. R. 1977. A note on local identifiability and degrees of freedom in the asymptotic maximum likelihood test. *British Journal of Mathematical and Statistical Psychology,* **30,** 198–203.

McFadden, D. L. 1973. Conditional logit analysis of qualitative choice behavior. *Pages 105–142 of:* Zarembka, P. (ed), *Frontiers in Econometrics.* New York: Academic Press.

McFadden, D. L. 1989. A method of simulated moments for estimation of discrete response models without numerical integration. *Econometrica,* **57,** 995–1026.

McFadden, D. L., & Train, K. E. 2000. Mixed MNL models for discrete choice. *Journal of Applied Econometrics,* **15,** 447–470.

McGilchrist, C. A. 1994. Estimation in generalized mixed models. *Journal of the Royal Statistical Society, Series B,* **56,** 61–69.

McGilchrist, C. A., & Aisbett, C. W. 1991. Regression with frailty in survival analysis. *Biometrics,* **47,** 461–466.

McGraw, K. O., & Wong, S. P. 1996. Forming inferences about some intraclass correlation coefficients. *Psychological Methods,* **1,** 30–46.

McKelvey, R., & Zavoina, W. 1975. A statistical model for the analysis of ordinal dependent variables. *Journal of Mathematical Sociology,* **4,** 103–120.

McKendrick, A. G. 1926. Applications of mathematics to medical problems. *Proceedings of the Edinburgh Mathematical Society*, **44**, 98–130.

McLachlan, G., & Basford, K. E. 1988. *Mixture Models. Inference and Applications to Clustering*. New York, NY: Dekker.

McLachlan, G., & Krishnan, T. 1997. *The EM Algorithm and Extensions*. Chichester: Wiley.

McLachlan, G., & Peel, D. A. 2000. *Finite Mixture Models*. New York, NY: Wiley.

Meijer, E., van der Leeden, R., & Busing, F. M. T. A. 1995. Implementing the bootstrap for multilevel models. *Multilevel Modelling Newsletter*, **7**, 7–11.

Mendonça, L., & Böhning, D. 1994. Die Auswirkung von Gesundheitsunterricht und Mundspühlung mit Na-Flourid auf die Prävention von Zahnkaries: eine Kohortenstudie mit urbanen Kindern in Brasilien. *In: 39th Conf. Deutsche Gesellschaft für Medizinische Informatik, Biometrie und Epidemiologie*.

Meng, X.-L., & Rubin, D. B. 1991. Using EM to obtain asymptotic variance-covariance matrices: The SEM Algorithm. *Journal of the American Statistical Association*, **86**, 899–909.

Meng, X-L., & Rubin, D. B. 1992. Performing likelihood ratio tests with multiply-imputed data sets. *Biometrika*, **79**, 103–111.

Meng, X-L., & Rubin, D. B. 1993. Maximum likelihood estimation via the ECM algorithm: A general framework. *Biometrika*, **80**, 267–278.

Meng, X-L., & Schilling, S. G. 1996. Fitting full-information item factor models and an empirical investigation of bridge sampling. *Journal of the American Statistical Association*, **91**, 1254–1267.

Meredith, W. 1964. Notes on factorial invariance. *Psychometrika*, **29**, 177–185.

Meredith, W. 1993. Measurement invariance, factor analysis and factor invariance. *Psychometrika*, **58**, 525–544.

Meredith, W., & Tisak, J. 1990. Latent curve analysis. *Psychometrika*, **55**, 107–122.

Messick, S. 1981. Constructs and their vicissitudes in educational and psychological measurement. *Psychological Bulletin*, **89**, 575–588.

Miller, A. J. 1984. Selection of subsets of regression variables (with discussion). *Journal of the Royal Statistical Society, Series B*, **147**, 389–425.

Miller, J. J. 1977. Asymptotic properties of maximum likelihood estimates in the mixed model of the analysis of variance. *The Annals of Statistics*, **5**, 746–762.

Miller, W. E., Miller, A. H., & Schneider, E. J. 1980. *American National Election Studies Data Source-Book: 1952-1978*. Cambridge: Harvard University Press.

Mislevy, R. J. 1985. Estimation of latent group effects. *Journal of the American Statistical Association*, **80**, 993–997.

Molenaar, P. C. W., & von Eye, A. 1994. On the arbitrary nature of latent variables. *Pages 226–242 of:* von Eye, A., & Clogg, C. C. (eds), *Latent Variable Analysis: Applications to Developmental Research*. Thousand Oaks, CA: Sage.

Molenberghs, G. 2002. Generalized estimating equations. *Pages 47–75 of:* Aerts, M., Geys, H., Molenberghs, G., & Ryan, L. M. (eds), *Topics in Modelling Clustered Data*. Boca Raton, FL: Chapman & Hall/CRC.

Mollié, A. 1996. Bayesian mapping of disease. *Pages 359–379 of:* Gilks, W. R., Richardson, S., & Spiegelhalter, D. J. (eds), *Markov Chain Monte Carlo in Practice*. London: Chapman & Hall.

Moran, P. A. P. 1971. Maximum likelihood estimation in non-standard conditions. *Proceedings of the Cambridge Philosophical Society*, **70**, 441–450.

Morris, C. 1983. Parametric empirical Bayes inference, theory and applications. *Journal of the American Statistical Association*, **78**, 47–65.

Morris, J. N., Marr, J. W., & Clayton, D. G. 1977. Diet and heart: Postscript. *British Medical Journal*, **2**, 1307–1314.

Moustaki, I. 1996. A latent trait and a latent class model for mixed observed variables. *British Journal of Mathematical and Statistical Psychology*, **49**, 313–334.

Moustaki, I. 2000. A latent variable model for ordinal variables. *Applied Psychological Measurement*, **24**, 211–223.

Moustaki, I., & Knott, M. 2000. Generalized latent trait models. *Psychometrika*, **65**, 391–411.

Mukherjee, B. N. 1973. Analysis of covariance structures and exploratory factor analysis. *British Journal of Mathematical and Statistical Psychology*, **26**, 125–154.

Mulaik, S. A. 1972. *The Foundations of Factor Analysis*. New York: McGraw-Hill.

Mulaik, S. A. 1988a. A brief history of the philosophical foundations of exploratory factor analysis. *Multivariate Behavioral Research*, **23**, 267–305.

Mulaik, S. A. 1988b. Confirmatory factor analysis. *Pages 259–288 of:* Nesselroade, J. R., & Cattell, R. B. (eds), *Handbook of Multivariate Experimental Psychology*. New York: Plenum.

Mulaik, S. A., James, L. R., Alstine, J. Van, Bennett, N., Lind, S., & Stillwell, C. D. 1989. An evaluation of goodness of fit indices for structural equation models. *Psychological Bulletin*, **105**, 430–445.

Müller, P., & Roeder, K. 1997. A Bayesian semiparametric model for case-control studies with errors in variables. *Biometrika*, **84**, 523–537.

Muraki, E. 1990. Fitting a polytomous item response model to Likert-type data. *Applied Psychological Measurement*, **14**, 59–71.

Muraki, E., & Bock, R. D. 1993. *PARSCALE*. Chicago, IL: Scientific Software.

Muraki, E., & Engelhard Jr., G. 1985. Full-information item factor analysis: Applications of EAP scores. *Applied Psychological Measurement*, **9**, 417–430.

Muthén, B. O. 1977. *Statistical Methodology for Structural Equation Models involving Latent Variables with Dichotomous Indicators*. Ph.D. thesis, University of Uppsala.

Muthén, B. O. 1978. Contributions to factor analysis of dichotomous variables. *Psychometrika*, **43**, 551–60.

Muthén, B. O. 1981. Factor analysis of dichotomous variables. *In:* Jackson, D. F., & Borgatta, E. F. (eds), *Factor Analysis and Measurement in Sociological Research: A Multidimensional Perspective*. London: Sage.

Muthén, B. O. 1982. Some categorical response models with continuous latent variables. *Pages 65–79 of:* Jöreskog, K. S., & Wold, H. (eds), *Systems under Indirect Observation - Causality ★ Structure ★ Prediction - Part II*. Amsterdam: North-Holland.

Muthén, B. O. 1983. Latent variable structural modelling with categorical data. *Journal of Econometrics*, **22**, 43–65.

Muthén, B. O. 1984. A general structural equation model with dichotomous, ordered categorical and continuous latent indicators. *Psychometrika*, **49**, 115–132.

Muthén, B. O. 1985. A method for studying the homogeneity of test items with respect to other relevant variables. *Journal of Educational Statistics*, **10**, 121–132.

Muthén, B. O. 1988a. *LISCOMP: Analysis of Linear Structural Equations with a Comprehensive Measurement Model*. Mooresville, IN: Scientific Software.

Muthén, B. O. 1988b. Some uses of structural equation modelling in validity studies: Extending IRT to external variables. *Pages 213–238 of:* Wainer, H., & Braun, H. (eds), *Test Validity*. Hillsdale, NJ: Erlbaum.

Muthén, B. O. 1989a. Dichotomous factor analysis of symptom data. *Sociological Methods & Research*, **18**, 19–65.

Muthén, B. O. 1989b. Multiple-group structural modelling with non-normal continuous variables. *British Journal of Mathematical and Statistical Psychology*, **42**, 55–62.

Muthén, B. O. 1989c. Tobit factor analysis. *British Journal of Mathematical and Statistical Psychology*, **42**, 241–250.

Muthén, B. O. 1989d. Using item-specific instructional information in achievement modeling. *Psychometrika*, **54**, 385–396.

Muthén, B. O. 1993. Goodness of fit with categorical and other nonnormal variables. *Pages 205–234 of:* Bollen, K. A., & Long, J. S. (eds), *Testing Structural Equation Models*. Newbury Park, CA: Sage.

Muthén, B. O. 1994. Multilevel covariance structure analysis. *Sociological Methods & Research*, **22**, 376–398.

Muthén, B. O. 2001. Second-generation structural equation modeling with a combination of categorical and continuous latent variables: New opportunities for latent class/latent growth modeling. *Pages 291–322 of:* Collins, L. M., & Sayer, A. (eds), *New Methods for the Analysis of Change*. Washington, DC: APA.

Muthén, B. O. 2002. Beyond SEM: General latent variable modeling. *Behaviormetrika*, **29**, 81–117.

Muthén, B. O., & Kaplan, D. 1992. A comparison of some methodologies for the factor-analysis of nonnormal likert variables - a note on the size of the model. *British Journal of Mathematical and Statistical Psychology*, **45**, 19–30.

Muthén, B. O., & Satorra, A. 1995. Complex sample data in structural equation modeling. *Pages 267–316 of:* Marsden, P. (ed), *Sociological Methodology 1995*. Cambridge, MA: Blackwell.

Muthén, B. O., & Satorra, A. 1996. Technical aspects of Muthén's LISCOMP approach to estimation of latent variable relations with a comprehensive measurement model. *Psychometrika*, **60**, 489–503.

Muthén, B. O., & Shedden, K. 1999. Finite mixture modeling with mixture outcomes using the EM algorithm. *Biometrics*, **55**, 463–469.

Muthén, B. O., Du Toit, S. H. C., & Spisic, D. 1997. *Robust inference using weighted least squares and quadratic estimating equations in latent variable modeling with categorical and continuous outcomes*. Unpublished paper.

Muthén, L. K., & Muthén, B. O. 1998. *Mplus User's Guide*. Los Angeles, CA: Muthén & Muthén.

Muthén, L. K., & Muthén, B. O. 2003. *Mplus Version 2.13. Addendum to the Mplus User's Guide*. Los Angeles, CA: Muthén & Muthén. Downloadable from http://www.statmodel.com/support/download.

Myles, J., & Clayton, D. G. 2001. *The GLMMGibbs package: Generalised linear mixed models by Gibbs sampling*. Downloadable from http://lib.stat.cmu.edu/R/CRAN/doc/packages/GLMMGibbs.pdf.

Nagin, D. S., & Land, K. C. 1993. Age, criminal careers, and population heterogeneity: Specification and estimation of a nonparametric mixed Poisson model. *Criminology*, **31**, 327–362.

Natarajan, R., & Kass, R. E. 2000. Reference Bayesian methods for generalized linear mixed models. *Journal of the American Statistical Association*, **95**, 227–237.

Naylor, J. C., & Smith, A. F. M. 1982. Applications of a method for the efficient computation of posterior distributions. *Journal of the Royal Statistical Society, Series C*, **31**, 214–225.

Naylor, J. C., & Smith, A. F. M. 1988. Econometric illustrations of novel numerical integration strategies for Bayesian inference. *Journal of Econometrics*, **38**, 103–125.

Neale, M. C., & Cardon, L. R. 1992. *Methodology for Genetic Studies of Twins and Families*. London: Kluwer.

Neale, M. C., Boker, S. M., Xie, G., & Maes, H. H. 2002. *Mx: Statistical Modeling (Sixth Edition)*. Richmond, VA: Virginia Commonwealth University, Department of Psychiatry. Downloadable from http://www.vipbg.vcu.edu/mxgui/.

Nelder, J. A., & Wedderburn, R. W. M. 1972. Generalised linear models. *Journal of the Royal Statistical Society, Series A*, **135**, 370–384.

Neuhaus, J. M. 1992. Statistical methods for longitudinal and clustered designs with binary responses. *Statistical Methods in Medical Research*, **1**, 249–273.

Neuhaus, J. M., & Jewell, N. P. 1990. The effect of retrospective sampling on binary regression models for clustered data. *Biometrics*, **46**, 977–990.

Neuhaus, J. M., Hauck, W. W., & Kalbfleisch, J. D. 1992. The effects of mixture distribution misspecification when fitting mixed-effects logistic models. *Biometrika*, **79**, 755–762.

Neyman, J. 1939. On a new class of 'contagious' distributions, applicable in entomology and bacteriology. *Annals of Mathematical Statistics*, **10**, 35–57.

Neyman, J., & Scott, E. L. 1948. Consistent estimates based on partially consistent observations. *Econometrica*, **16**, 1–32.

Niemi, R. G., Craig, S. C., & Mattei, F. 1991. Measuring internal political efficacy in the 1988 National Election Study. *American Political Science Review*, **85**, 1407–1413.

Normand, S.-L. T. 1999. Tutorial in biostatistics. Meta-analysis: Formulating, evaluating, combining, and reporting. *Statistics in Medicine*, **18**, 321–359.

Novick, M. R., & Lewis, C. 1967. Coefficient alpha and the reliability of composite measurements. *Psychometrika*, **32**, 1–13.

Novick, M. R., Lewis, C., & Jackson, P. H. 1973. The estimation of proportions in M groups. *Psychometrika*, **38**, 19–46.

Nunnally, J. C., & Durham, R. L. 1975. Validity, reliability, and special problems of measurement in evaluation research. *In:* Struening, E. L., & Guttentag, M. (eds), *Handbook of Evaluation Research*, vol. 1. London: Sage.

Oakes, D. 1999. Direct calculation of the information matrix via the EM algorithm. *Journal of the Royal Statistical Society, Series B*, **61**, 479–482.

Oakes, M. 1993. The logic and role of meta-analysis in clinical research. *Statistical Methods in Medical Research*, **2**, 147–160.

O'Hagan, A. 1976. On posterior joint and marginal modes. *Biometrika*, **63**, 329–333.

Olsson, U. 1979. Maximum likelihood estimation of the polychoric correlation coefficient. *Psychometrika*, **44**, 443–460.

Olsson, U., Dragsow, F., & Dorans, N. J. 1982. The polyserial correlation coefficient. *Psychometrika*, **47**, 337–347.

Orchard, T., & Woodbury, M. A. 1972. Missing information principle: Theory and applications. *Pages 697–715 of: Proceedings of the 6th Berkeley Symposium on Mathematical Statistics and Probability*. Berkeley, CA: University of California Press.

Otis, D. L., Burnham, K. P., White, G. C., & Anderson, D. R. 1978. Statistical inference from capture data on closed animal populations. *Wildlife Monographs*, **62**, 1–135.

Ouwens, M. J. N. M., Tan, F. E. S., & Berger, M. P. R. 2001. Local influence to detect influential data structures for generalized linear mixed models. *Biometrics*, **57**, 1166–1172.

Pan, J.-X., & Fang, K.-T. 2002. *Growth Curve Models and Statistical Diagnostics*. New York: Springer.

Pan, J. X., & Thompson, R. 2003. Gauss-Hermite quadrature approximation for estimation in generalised linear mixed models. *Computational Statistics*, **18**, 57–78.

Parke, W. P. 1986. Pseudo maximum likelihood estimation: The asymptotic distribution. *The Annals of Statistics*, **14**, 355–357.

Patterson, B. H., Dayton, C. M., & Graubard, B. I. 2002. Latent class analysis of complex sample survey data: Application to dietary data. *Journal of the American Statistical Association*, **97**, 721–741.

Patterson, H. D., & Thompson, R. 1971. Recovery of inter-block information when block sizes are unequal. *Biometrika*, **58**, 545–554.

Pawitan, Y. 2001. *In All Likelihood: Statistical Modelling and Inference Using Likelihood*. Oxford: Oxford University Press.

Payne, R. W. (ed). 2002. *The Guide to GenStat Release 6.1 – Part 2: Statistics*. Hemel Hempstead: VSN International.

Pearl, J. 2000. *Causality. Models, Reasoning, and Inference*. Cambridge: Cambridge University Press.

Pearson, K. 1901. Mathematical contributions to the theory of evolution VII. On the inheritance of characters not capable of exact quantitative measurement. *Philosophical Transactions of the Royal Society A*, **195**, 79–150.

Pearson, K., & Heron, D. 1913. On theories of association. *Biometrika*, **9**, 159–315.

Pendergast, J. F., Gange, S. J., Newton, M. A., Lindstrom, M. J., Palta, M., & Fisher, M. R. 1996. A survey of methods for analyzing clustered binary response data. *International Statistical Review*, **64**, 89–118.

Peto, R. 1972. Contribution to the discussion of a paper by D. R. Cox. *Journal of the Royal Statistical Society, Series B*, **34**, 205–207.

Pfeffermann, D. 2002. Small area estimation – new developments and directions. *International Statistical Review*, **70**, 125–143.

Pickles, A. 1998. Generalized estimating equations. *Pages 1626–1636 of: Encyclopedia of Biostatistics*. London: Wiley.

Pickles, A., & Crouchley, R. 1994. Generalizations and applications of frailty models for survival and event data. *Statistical Methods in Medical Research*, **3**, 263–278.

Pickles, A., & Crouchley, R. 1995. A comparison of frailty models for multivariate survival data. *Statistics in Medicine*, **14**, 1447–1461.

Pickles, A., Pickering, K., & Taylor, C. 1996. Reconciling recalled dates of developmental milestones, events and transitions: a mixed generalized linear model with random mean and variance functions. *Journal of the Royal Statistical Society, Series A*, **159**, 225–234.

Pickles, A., Pickering, K., Taylor, C., Sutton, C., & Yang, S. 2001. Multilevel risk models for retrospective age-of-onset data: School children's first cigarette. *Journal of Adolescent Research*, **16**, 188–204.

Pinheiro, J. C., & Bates, D. M. 1995. Approximations to the log-likelihood function in the nonlinear mixed-effects model. *Journal of Computational Graphics and Statistics*, **4**, 12–35.

Pinheiro, J. C., & Bates, D. M. 2000. *Mixed-Effects Models in S and S-PLUS*. New York, NY: Springer.

Pinheiro, J. C., Liu, C., & Wu, Y. 2001. Efficient algorithms for robust estimation in linear mixed-effects models using the multivariate *t*-distribution. *Journal of Computational Graphics and Statistics*, **10**, 249–276.

Plackett, R. L. 1975. The analysis of permutations. *Journal of the Royal Statistical Society, Series C*, **24**, 193–202.

Poon, W.-Y., & Lee, S.-Y. 1992. Maximum likelihood and generalized least squares analyses of two-level structural equation models. *Statistics and Probability Letters*, **14**, 25–30.

Pregibon, D. 1981. Logistic regression diagnostics. *The Annals of Statistics*, **9**, 705–724.

Prentice, R. L. 1988. Correlated binary regression with covariates specific to each binary observation. *Biometrics*, **44**, 1033–1048.

Prentice, R. L. 1989. Surrogate endpoints in clinical trials: Definition and operational criteria. *Statistics in Medicine*, **8**, 431–440.

Prentice, R. L., Williams, N. J., & Peterson, A. V. 1981. On the regression analysis of multivariate failure times. *Biometrika*, **68**, 373–379.

Prochaska, J. O., & DiClemente, C. C. 1983. Stages and processes of self-change of smoking: Toward an integrative model of change. *Journal of Consulting and Clinical Psychology*, **51**, 390–395.

Qu, Y., Tan, M., & Kutner, M. H. 1996. Random effects models in latent class analysis for evaluating accuracy of diagnostic tests. *Biometrics*, **52**, 797–810.

Quandt, R. E. 1972. A new approach to estimating switching regressions. *Journal of the American Statistical Association*, **67**, 306–310.

Rabe-Hesketh, S., & Everitt, B. S. 2003. *Handbook of Statistical Analyses using Stata (Third Edition)*. Boca Raton, FL: Chapman & Hall/CRC.

Rabe-Hesketh, S., & Pickles, A. 1999. Generalised linear latent and mixed models. *Pages 332–339 of:* Friedl, H., Berghold, A., & Kauermann, G. (eds), *14th International Workshop on Statistical Modeling*.

Rabe-Hesketh, S., & Skrondal, A. 2001. Parameterization of multivariate random effects models for categorical data. *Biometrics*, **57**, 1256–1264.

Rabe-Hesketh, S., Pickles, A., & Taylor, C. 2000. sg129: Generalized linear latent and mixed models. *Stata Technical Bulletin*, **53**, 47–57.

Rabe-Hesketh, S., Pickles, A., & Skrondal, A. 2001a. GLLAMM: A general class of multilevel models and a Stata program. *Multilevel Modelling Newsletter*, **13**, 17–23.

Rabe-Hesketh, S., Pickles, A., & Skrondal, A. 2001b. *GLLAMM Manual.* Tech. rept. 2001/01. Department of Biostatistics and Computing, Institute of Psychiatry, King's College, University of London. Downloadable from http://www.gllamm.org/docum.html.

Rabe-Hesketh, S., Touloupulou, T., & Murray, R. M. 2001c. Multilevel modeling of cognitive function in schizophrenics and their first degree relatives. *Multivariate Behavioral Research*, **36**, 279–298.

Rabe-Hesketh, S., Yang, S., & Pickles, A. 2001d. Multilevel models for censored and latent responses. *Statistical Methods in Medical Research*, **10**, 409–427.

Rabe-Hesketh, S., Skrondal, A., & Pickles, A. 2002. Reliable estimation of generalized linear mixed models using adaptive quadrature. *The Stata Journal*, **2**, 1–21.

Rabe-Hesketh, S., Pickles, A., & Skrondal, A. 2003a. Correcting for covariate measurement error in logistic regression using nonparametric maximum likelihood estimation. *Statistical Modelling*, **3**, 215–232.

Rabe-Hesketh, S., Skrondal, A., & Pickles, A. 2003b. Maximum likelihood estimation of generalized linear models with covariate measurement error. *The Stata Journal*, **3**, 386–411.

Rabe-Hesketh, S., Skrondal, A., & Pickles, A. 2004a. Generalized multilevel structural equation modeling. *Psychometrika*, **69**, 167–190.

Rabe-Hesketh, S., Skrondal, A., & Pickles, A. 2004b. *GLLAMM Manual.* Tech. rept. 160. U.C. Berkeley Division of Biostatistics Working Paper Series. Downloadable from http://www.gllamm.org/docum.html.

Rabe-Hesketh, S., Skrondal, A., & Pickles, A. 2005a. Maximum likelihood estimation of limited and discrete dependent variable models with nested random effects. *Journal of Econometrics*, in press.

Rabe-Hesketh, S., Pickles, A., & Skrondal, A. 2005b. *Multilevel and Structural Equation Modeling of Continuous, Categorical and Event Data.* College Station, TX: Stata Press.

Raftery, A. E. 1993. Bayesian model selection in structural equation models. *Pages 163–180 of:* Bollen, K. A., & Long, J. S. (eds), *Testing Structural Equation Models.* Newbury Park, CA: Sage.

Raftery, A. E. 1995. Bayesian model selection in social research. *Pages 111–164 of:* Raftery, A. E. (ed), *Sociological Methodology 1995.* Oxford: Blackwell.

Rao, C. R. 1975. Simultaneous estimation of parameters in different linear models and applications to biometric problems. *Biometrics*, **31**, 545–554.

Rasbash, J., & Browne, W. J. 2001. Modelling non-hierarchical structures. *Pages 91–106 of:* Leyland, A. H., & Goldstein, H. (eds), *Multilevel Modelling of Health Statistics.* Chichester, UK: Wiley.

Rasbash, J., Steele, F., Browne, W., & Prosser, B. 2004. *A User's Guide to MLwiN Version 2.0.* London: Institute of Education. Downloadable from http://multilevel.ioe.ac.uk/download/manuals.html.

Rasch, G. 1960. *Probabilistic Models for Some Intelligence and Attainment Tests.* Copenhagen: Danmarks Pædagogiske Institut.

Rasch, G. 1961. On general laws and the meaning of measurement in psychology. *Pages 321–333 of: Proceedings of the IV Berkeley Symposium on Mathematical Statistics and Probability.* Berkeley, CA: University of California Press.

Rasch, G. 1967. An informal report on the theory of objectivity in comparisons. *Pages 1–19 of:* Van der Kamp, L. J. T., & Vlek, C. A. J. (eds), *Psychological Measurement Theory.* Leiden: University of Leiden.

Raudenbush, S. W., & Bryk, A. S. 2002. *Hierarchical Linear Models.* Thousand Oaks, CA: Sage.

Raudenbush, S. W., & Sampson, R. 1999a. Assessing direct and indirect effects in multilevel designs with latent variables. *Sociological Methods & Research*, **28**, 123–153.

Raudenbush, S. W., & Sampson, R. 1999b. Ecometrics: Toward a science of assessing ecological settings, with application to the systematic social observation of neighborhoods. *Pages 1–41 of:* Marsden, P. V. (ed), *Sociological Methodology 1999.* Oxford: Blackwell.

Raudenbush, S. W., & Yang, M.-L. 1998. Numerical integration via high-order, multivariate LaPlace approximation with application to multilevel models. *Multilevel Modelling Newsletter*, **10**, 11–14.

Raudenbush, S. W., Yang, M.-L., & Yosef, M. 2000. Maximum likelihood for generalized linear models with nested random effects via high-order, multivariate Laplace approximation. *Journal of Computational and Graphical Statistics*, **9**, 141–157.

Raudenbush, S. W., Bryk, A. S., Cheong, Y. F., Fai, Y., & Congdon, R. 2004. *HLM 6: Hierarchical Linear and Nonlinear Modeling*. Lincolnwood, IL: Scientific Software International.

Raykov, T., & Penev, S. 1999. On structural equation model equivalence. *Multivariate Behavioral Research*, **34**, 199–244.

Reboussin, B. A., & Liang, K.-Y. 1998. An estimating equations approach for the LISCOMP model. *Psychometrika*, **63**, 165–182.

Reiser, M. 1996. Analysis of residuals for the multinomial item response model. *Psychometrika*, **61**, 509–528.

Reiser, M., & Lin, Y. 1999. A goodness-of-fit test for the latent class model when expected frequencies are small. *Pages 81–111 of:* Sobel, M. E., & Becker, M. P. (eds), *Sociological Methodology 1999*. Cambridge: Blackwell.

Revelt, D., & Train, K. E. 1998. Mixed logit with repeated choices: Household's choices of appliance efficiency level. *Review of Economics and Statistics*, **80**, 647–657.

Richardson, S., & Gilks, W. R. 1993. A Bayesian approach to measurement error problems in epidemiology using conditional independence models. *American Journal of Epidemiology*, **138**, 430–442.

Richardson, S., Leblond, L., Jaussent, I., & Green, P. 2002. Mixture models in measurement error problems, with reference to epidemiological studies. *Journal of the Royal Statistical Society, Series A*, **165**, 549–556.

Rijmen, F., Tuerlinckx, F., De Boeck, P., & Kuppens, P. 2003. A nonlinear mixed model framework for item response theory. *Psychological Methods*, **8**, 185–205.

Rindskopf, D. 1992. A general approach to categorical data analysis with missing data, using generalized linear models with composite links. *Psychometrika*, **57**, 29–42.

Rindskopf, D., & Rindskopf, W. 1986. The value of latent class analysis in medical diagnosis. *Statistics in Medicine*, **5**, 21–27.

Ritz, J., & Spiegelman, D. 2004. A note about the equivalence of conditional and marginal regression models. *Statistical Methods in Medical Research*, **13**, 309–323.

Rivers, D. 1988. Heterogeneity in models of electoral choice. *American Journal of Political Science*, **32**, 737–757.

Roberts, J. S., Donoghue, J. R., & Laughlin, J. E. 2000. A general item response theory model for unfolding unidimensional polytomous responses. *Applied Psychological Measurement*, **24**, 3–32.

Robins, J. M., Rotnitzky, A. G., & Zhao, L. P. 1994. Estimation of regression coefficients when some regressors are not always observed. *Journal of the American Statistical Association*, **89**, 826–866.

Robins, P. K., & West, R. W. 1977. Measurement errors in the estimation of home value. *Journal of the American Statistical Association*, **72**, 290–294.

Robinson, G. K. 1991. That BLUP is a good thing: The estimation of random effects. *Statistical Science*, **6**, 15–51.

Robinson, P. M. 1974. Identification, estimation, and large sample theory for regressions containing unobservable variables. *International Economic Review*, **15**, 680–692.

Robinson, W. S. 1950. Ecological correlations and the behavior of individuals. *American Sociological Review*, **15**, 351–357.

Rodriguez, G., & Goldman, N. 1995. An assessment of estimation procedures for multilevel models with binary responses. *Journal of the Royal Statistical Society, Series A*, **158**, 73–89.

Rodriguez, G., & Goldman, N. 2001. Improved estimation procedures for multilevel models with binary response: A case study. *Journal of the Royal Statistical Society, Series A*, **164**, 339–355.

Roeder, K., Carroll, R. J., & Lindsay, B. G. 1996. A semiparametric mixture approach to case-control studies with errors in covariables. *Journal of the American Statistical Association*, **91**, 722–732.

Rosett, R. N., & Nelson, F. D. 1975. Estimation of a two-limit probit regression model. *Econometrica*, **43**, 141–146.

Rosner, B., Spiegelman, D., & Willett, W. C. 1990. Correction of logistic regression relative risk estimates and confidence intervals for measurement error: The case of multiple covariates measured with error. *American Journal of Epidemiology*, **132**, 734–745.

Rothenberg, T. 1971. Identification in parametric models. *Econometrica*, **39**, 577–591.

Rotnitzky, A. G., & Wypij, D. 1994. A note on the bias of estimators with missing data. *Biometrics*, **50**, 1163–1170.

Royston, P., & Altman, D. G. 1994. Regression using fractional polynomials of continuous covariates - parsimonious parametric modeling. *Journal of the Royal Statistical Society, Series C*, **43**, 429–467.

Rubin, D. B. 1976. Inference and missing data. *Biometrika*, **63**, 581–592.

Rubin, D. B. 1983. Some applications of Bayesian statistics to educational data. *Journal of the Royal Statistical Society, Series D*, **32**, 155–167.

Rubin, D. B. 1984. Bayesianly justifiable and relevant frequency calculations for the applied statistician. *The Annals of Statistics*, **12**, 1151–1172.

Rubin, D. B. 1987. *Multiple Imputation for Nonresponse in Surveys*. New York: Wiley.

Rubin, D. B. 1991. EM and beyond. *Psychometrika*, **56**, 241–254.

Rubin, D. B., & Thayer, D. T. 1982. EM algorithms for ML factor analysis. *Psychometrika*, **47**, 69–76.

Rubin, D. B., Stern, H. S., & Vehovar, V. 1995. Handling "Don't Know" survey responses: The case of the Slovenian plebiscite. *Journal of the American Statistical Association*, **90**, 822–828.

Rummel, R. J. 1967. *Applied Factor Analysis*. Evanston: Northwestern University Press.

Sackett, D. L., Haynes, R. B., Guyatt, G. H., & Tugwell, P. 1991. *Clinical Epidemiology*. Massachusetts: Little Brown & Company.

Samejima, F. 1969. *Estimation of Latent Trait Ability Using A Response Pattern of Graded Scores*. Bowling Green, OH: Psychometric Monograph 17, Psychometric Society.

Samejima, F. 1972. *A General Model for Free-Response Data*. Bowling Green, OH: Psychometric Monograph 18, Psychometric Society.

Sammel, M. D., Ryan, L. M., & Legler, J. M. 1997. Latent variable models for mixed discrete and continuous outcomes. *Journal of the Royal Statistical Society, Series B*, **59**, 667–678.

Sanathanan, L. 1972. Estimating the size of a multinomial population. *The Annals of Mathematical Statistics*, **43**, 142–152.

SAS/Stat User's Guide, version 8. 2000. *SAS/Stat User's Guide, version 8*. Cary, NC: SAS Publishing.

Satorra, A. 1990. Robustness issues in structural equation modeling: A review of recent developments. *Quality and Quantity*, **24**, 367–386.

Satorra, A. 1992. Asymptotic robust inferences in the analysis of mean and covariance structures. *Pages 249–278 of:* Marsden, P. V. (ed), *Sociological Methodology 1992*. Washington, DC: American Sociological Association.

Satorra, A., & Bentler, P. M. 1994. Corrections to test statistics and standard errors in covariance structure analysis. *Pages 285–305 of:* von Eye, A., & Clogg, C. C. (eds), *Latent Variable Analysis: Applications to Developmental Research*. Newbury Park, CA: Sage.

Satten, G. A., & Kupper, L. L. 1993. Inferences about exposure-disease association using probability-of-exposure information. *Journal of the American Statistical Association*, **88**, 200–208.

Schafer, D. W. 2001. Semiparametric maximum likelihood for measurement error regression. *Biometrics*, **57**, 53–61.

Schafer, J. L. 1997. *Analysis of Incomplete Multivariate Data*. London: Chapman & Hall.

Schall, R. 1991. Estimation in generalized linear-models with random effects. *Biometrika*, **78**, 719–727.

Schlesselman, J. J. 1982. *Case-Control Studies: Design, Conduct, Analysis*. Oxford: Oxford University Press.

Schluchter, M. D. 1988. Analysis of incomplete multivariate data using linear models with structured covariance matrices. *Statistics in Medicine*, **7**, 317–324.

Schoenberg, R. 1985. Latent variables in the analysis of limited dependent variables. *Pages 213–242 of:* Tuma, N. B. (ed), *Sociological Methodology 1985*. San Francisco, CA: Jossey-Bass.

Schoenberg, R. 1996. *Constrained Maximum Likelihood*. Maple Valley, WA: Aptech Systems and The University of Washington. Downloadable from http://faculty.washington.edu/rons/.

Schoenberg, R., & Richtand, C. 1984. An application of the EM method to the maximum likelihood estimation of multiple indicator and factor analysis models. *Sociological Methods & Research*, **13**, 127–150.

Schumaker, R. E., & Marcoulides, G. A. (eds). 1998. *Interaction and Nonlinear Effects in Structural Equation Modeling*. Mahwah, NJ: Erlbaum.

Schwarz, G. 1978. Estimating the dimension of a model. *The Annals of Statistics*, **6**, 461–464.

Searle, S. R., Casella, G., & McCulloch, C. E. 1992. *Variance Components*. New York: Wiley.

Self, S. G., & Liang, K.-Y. 1987. Asymptotic properties of maximum likelihood estimators and likelihood ratio tests under non-standard conditions. *Journal of the American Statistical Association*, **82**, 605–610.

Seong, T.-J. 1990. Sensitivity of marginal maximum likelihood estimation of item and ability parameters to the characteristics of the prior ability distributions. *Applied Psychological Measurement*, **14**, 299–311.

Serfling, R. J. 1980. *Approximation Theorems of Mathematical Statistics*. New York: Wiley.

Shavelson, R. J., & Webb, N. M. 1991. *Generalizability Theory: A Primer*. Newbury Park, CA: Sage.

Shaw, J. E. H. 1988. A quasirandom approach to integration in Bayesian statistics. *The Annals of Statistics*, **16**, 895–914.

Shepard, R. N. 1974. Representation of structure in similarity data: Problems and prospects. *Psychometrika*, **39**, 373–421.

Shrout, P. E., & Fleiss, J. L. 1979. Intraclass correlations: Uses in assessing rater reliability. *Psychological Bulletin*, **86**, 420–428.

Sijtsma, K., & Molenaar, I. W. 2002. *Introduction to Nonparametric Item Response Theory*. Thousand Oaks, CA: Sage.

Silagy, C. 2003. Nicotine replacement therapy for smoking cessation (Cochrane review). *In: The Cochrane Library, Issue 4*. Chichester: Wiley.

Silva, F. 1993. *Psychometric Foundations and Behavioral Assessment*. Newbury Park, CA: Sage.

Simar, L. 1976. Maximum likelihood estimation of a compound Poisson process. *The Annals of Statistics*, **4**, 1200–1209.

Singer, J. D., & Willett, J. B. 1993. It's about time: Using discrete-time survival analysis to study duration and the timing of events. *Journal of Educational Statistics*, **18**, 155–195.

Singer, J. D., & Willett, J. B. 2003. *Applied Longitudinal Data Analysis: Modeling Change and Event Occurrence*. New York: Oxford University Press.

Skaug, H. J. 2002. Automatic differentiation to facilitate maximum likelihood estimation in nonlinear random effects models. *Journal of Computational & Graphical Statistics*, **11**, 458–470.

Skellam, J. G. 1948. A probability distribution derived from the binomial distribution by regarding the probability of a success as variable between the sets of trials. *Journal of the Royal Statistical Society, Series B*, **10**, 257–261.

Skrondal, A. 1996. *Latent Trait, Multilevel and Repeated Measurement Modelling with Incomplete Data of Mixed Measurement Levels.* Oslo: Section of Medical Statistics, University of Oslo.

Skrondal, A. 2000. Design and analysis of Monte Carlo experiments: Attacking the conventional wisdom. *Multivariate Behavioral Research,* **35**, 137–167.

Skrondal, A. 2003. Interaction as departure from additivity in case-control studies: A cautionary note. *American Journal of Epidemiology,* **158**, 251–258.

Skrondal, A., & Laake, P. 1999. Factorial invariance in confirmatory factor analysis. *Technical Report.*

Skrondal, A., & Laake, P. 2001. Regression among factor scores. *Psychometrika,* **66**, 563–576.

Skrondal, A., & Rabe-Hesketh, S. 2003a. Generalized linear mixed models for nominal data. *American Statistical Association, Proceedings of the Biometrics Section,* 3931–3936.

Skrondal, A., & Rabe-Hesketh, S. 2003b. Multilevel logistic regression for polytomous data and rankings. *Psychometrika,* **68**, 267–287.

Skrondal, A., & Rabe-Hesketh, S. 2003c. Some applications of generalized linear latent and mixed models in epidemiology: Repeated measures, measurement error and multilevel modelling. *Norwegian Journal of Epidemiology,* **13**, 265–278.

Skrondal, A., & Rabe-Hesketh, S. 2004. Generalised linear latent and mixed models with composite links and exploded likelihoods. *Pages 27–39 of:* A.Biggeri, E.Dreassi, C.Lagazio, & M.Marchi (eds), *19th International Workshop on Statistical Modeling.* Florence, Italy: Firenze University Press.

Skrondal, A., Eskild, A., & Thorvaldsen, J. 2000. Changes in condom use after HIV diagnosis. *Scandinavian Journal of Public Health,* **28**, 71–76.

Skrondal, A., Rabe-Hesketh, S., & Pickles, A. 2002. Informative dropout and measurement error in cluster randomised trials. *Page 61 of: Proceedings of the XXIth International Biometric Conference (Abstracts).*

Smith, A. F. M. 1973. A general Bayesian linear model. *Journal of the Royal Statistical Society, Series B,* **35**, 61–75.

Smith, T. W. 1988. *Rotation Design of the GSS.* Tech. rept. 52. GSS Methodological Report, Chicago.

Snijders, T. A. B., & Berkhof, J. 2004. Diagnostic checks for multilevel models. *Page Chapter 4 of:* de Leeuw, J, & Kreft, I. (eds), *Handbook of Quantitative Multilevel Analysis.* Thousand Oaks, CA: Sage.

Snijders, T. A. B., & Bosker, R. J. 1999. *Multilevel Analysis.* London: Sage.

Sobel, M. E. 1995. Causal inference in the social and behavioral sciences. *Pages 1–38 of:* Arminger, G., Clogg, C. C., & Sobel, M. E. (eds), *Handbook of Statistical Modelling for the Social and Behavioral Sciences.* New York: Plenum Press.

Sobel, M. E., & Bohrnstedt, G. W. 1985. Use of null models in evaluating the fit of covariance structure models. *Pages 152–178 of:* Tuma, N. B. (ed), *Sociological Methodology 1985.* San Francisco, CA: Jossey-Bass.

Social and Community Planning Research. 1987. *British Social Attitudes Panel Survey, 1983-1986 [Computer file] SN: 2197.* Colchester, Essex: The Data Archive [Distributor].

Sommer, A., Katz, J., & Tarwotjo, I. 1983. Increased mortality in children with slight vitamin A deficiency. *American Journal of Clinical Nutrition,* **40**, 1090–1095.

Spearman, C. 1904. General intelligence, objectively determined and measured. *American Journal of Psychology,* **15**, 201–293.

Spearman, C. 1927. *The Abilities of Man: Their Nature and Measurement.* London: Macmillan.

Spiegelhalter, D. J., Freedman, L. S., & Parmar, M. K. B. 1994. Bayesian approaches to randomized trials (with discussion). *Journal of the Royal Statistical Society, Series A,* **157**, 357–416.

Spiegelhalter, D. J., Thomas, A., Best, N. G., & Gilks, W. R. 1996a. *BUGS 0.5 Bayesian Analysis using Gibbs Sampling. Manual (version ii).* Cambridge: MRC-Biostatistics Unit. Downloadable from http://www.mrc-bsu.cam.ac.uk/bugs/documentation/contents.shtml.

Spiegelhalter, D. J., Thomas, A., Best, N. G., & Gilks, W. R. 1996b. *BUGS 0.5 Examples, Volume 1.* Cambridge: MRC-Biostatistics Unit.

Spiegelhalter, D. J., Thomas, A., Best, N. G., & Gilks, W. R. 1996c. *BUGS 0.5 Examples, Volume 2.* Cambridge: MRC-Biostatistics Unit.

Spiegelhalter, D. J., Best, N. G., Carlin, B. P., & van der Linde, A. 2002. Bayesian measures of model complexity and fit (with discussion). *Journal of the Royal Statistical Society, Series B*, **64**, 583–639.

SPSS Inc. 2001. *SPSS 11.0 Advanced Models.* Prentice Hall: Englewood Cliffs, NJ.

Stata Cross-Sectional Time-Series. 2003. *Cross-Sectional Time-Series Reference Manual.* College Station, TX: Stata Press.

StataCorp. 2003. *Stata Statistical Software: Release 8.0.* College Station, TX.

Steiger, J. H. 1990. Structural model evaluation and modification: An interval estimation approach. *Multivariate Behavioral Research*, **25**, 173–180.

Stelzl, I. 1986. Changing causal relationships without changing the fit: Some rules for generating equivalent path models. *Multivariate Behavioral Research*, **21**, 309–331.

Sterlin, T. D. 1959. Publication decisions and their possible effects on inferences drawn from tests of significance – or vice-versa. *Journal of the American Statistical Association*, **54**, 30–34.

Stern, S. 1992. A method for smoothing simulated moments of discrete probabilities in multinomial probit models. *Econometrica*, **60**, 943–952.

Stern, S. 1997. Simulation-based estimation. *Journal of Economic Literature*, **35**, 2006–2039.

Stevens, S. S. 1951. Mathematics, measurement and psychophysics. *In:* Stevens, S. S. (ed), *Handbook of Experimental Psychology.* New York: Wiley.

Stewart, M. B. 1983. On the least squares estimation when the dependent variable is grouped. *Review of Economic Studies*, **50**, 737–753.

Stiratelli, R., Laird, N. M., & Ware, J. H. 1984. Random effects models for serial observations with binary responses. *Biometrics*, **40**, 961–971.

Stone, M. 1974. Cross-validatory choice and assessment of statistical predictions (with discussion). *Journal of the Royal Statistical Society, Series B*, **36**, 111–147.

Stram, D. O., & Lee, J. W. 1994. Variance components testing in the longitudinal mixed effects model. *Biometrics*, **50**, 1171–1177.

Stram, D. O., & Lee, J. W. 1995. Correction to 'Variance components testing in the longitudinal mixed effects model'. *Biometrics*, **51**, 1196.

Streiner, D. L., & Norman, G. R. 1995. *Health Measurement Scales: A Practical Guide to Their Development and Use.* Oxford: Oxford University Press.

Strenio, J. L. F., Weisberg, H. I., & Bryk, A. S. 1983. Empirical Bayes estimation of individual growth curve parameters and their relations to covariates. *Biometrics*, **39**, 71–86.

Stroud, A. H. 1971. *Approximate Calculation of Multiple Integrals.* Englewood Cliffs, NJ: Prentice-Hall.

Stroud, A. H., & Secrest, D. 1966. *Gaussian Quadrature Formulas.* Englewood Cliffs, NJ: Prentice-Hall.

Sudman, S., & Bradburn, N. M. 1973. Effects of time and memory factor on response in surveys. *Journal of the American Statistical Association*, **63**, 805–815.

Sutton, A. J., Song, F., Gilbody, S. M., & Abrams, K. R. 2000. Modelling publication bias in meta-analysis: A review. *Statistical Methods in Medical Research*, **9**, 421–445.

Swaminathan, H., & Gifford, J. A. 1982. Bayesian estimation in the Rasch model. *Journal of Educational Statistics*, **7**, 175–192.

Swaminathan, H., & Gifford, J. A. 1985. Bayesian estimation in the two-parameter logistic model. *Psychometrika*, **50**, 349–364.

Swaminathan, H., & Gifford, J. A. 1986. Bayesian estimation in the three-parameter logistic model. *Psychometrika*, **51**, 589–601.

Takane, Y. 1987. Analysis of covariance structures and probabilistic binary data. *Communication & Cognition*, **20**, 45–62.

Takane, Y., & de Leeuw, J. 1987. On the relationship between item response theory and factor analysis of discretized variables. *Psychometrika*, **52**, 393–408.

Tanaka, J. S. 1993. Multifaceted conceptions of fit in structural equation models. *Pages 10–39 of:* Bollen, K. A., & Long, J. S. (eds), *Testing Structural Equation Models*. Newbury Park, CA: Sage.

Tanaka, J. S., Panter, A. T., Winborne, W. C., & Huba, G. J. 1990. Theory testing in personality and personality psychology with structural equation models. A primer in twenty questions. *Pages 217–242 of:* Hendrick, C., & Clark, M. S. (eds), *Review of Personality and Social Psychology*. Newbury Park, CA: Sage.

Tanner, M. A. 1996. *Tools for Statistical Inference (Third Edition)*. New York: Springer.

Tanner, M. A., & Wong, W. H. 1987. The calculation of posterior distributions by data augmentation. *Journal of the American Statistical Association*, **82**, 528–540.

Ten Berge, J. M. F. 1983. On Green best linear composites with a specified structure, and oblique estimates of factor scores. *Psychometrika*, **48**, 371–375.

Ten Berge, J. M. F., Krijnen, W. P., Wansbeek, T. J., & Shapiro, A. 1999. Some new results on correlation-preserving factor scores prediction methods. *Linear Algebra And Its Applications*, **289**, 311–318.

Ten Have, T. R., & Localio, A. R. 1999. Empirical Bayes estimation of random effects parameters in mixed effects logistic regression models. *Biometrics*, **55**, 1022–1029.

Terza, J. V. 1985. Ordinal probit: A generalization. *Communications in Statistics, Theory and Methods*, **14**, 1–11.

Terza, J. V. 1998. Estimating count data models with endogenous switching: Sample selection and endogenous treatment effects. *Journal of Econometrics*, **84**, 129–154.

Thall, P. F., & Vail, S. C. 1990. Some covariance models for longitudinal count data with overdispersion. *Biometrics*, **46**, 657–671.

Therneau, T. M., & Grambsch, P. M. 2000. *Modeling Survival Data*. New York: Springer.

Thissen, D. 1982. Marginal maximum likelihood estimation for the one-parameter logistic model. *Psychometrika*, **47**, 175–186.

Thissen, D., & Steinberg, L. 1986. A taxonomy of item response models. *Psychometrika*, **51**, 567–577.

Thissen, D., & Steinberg, L. 1988. Data analysis using item response theory. *Psychological Bulletin*, **104**, 385–395.

Thisted, R. A. 1987. *Elements of Statistical Computing*. London: Chapman and Hall.

Thompson, B. 1984. *Canonical Correlation Analysis: Uses and Interpretation*. Sage University Paper Series on Quantitative Applications in the Social Sciences. Beverly Hills, CA: Sage.

Thompson, R., & Baker, R. J. 1981. Composite link functions in generalized linear models. *Journal of the Royal Statistical Society, Series C*, **30**, 125–131.

Thompson, S. G., & Pocock, S. J. 1991. Can meta-analysis be trusted? *The Lancet*, **338**, 1127–1130.

Thompson, S. G., Turner, R. M., & Warn, D. E. 2001. Multilevel models for meta-analysis, and their application to absolute risk differences. *Statistical Methods in Medical Research*, **10**, 375–392.

Thompson, W. A. 1977. On the treatment of grouped observations in life studies. *Biometrics*, **33**, 463–470.

Thomson, G. H. 1938. *The Factorial Analysis of Human Ability*. London: University of London Press.

Thoresen, M., & Laake, P. 2000. A simulation study of measurement error correction methods in logistic regression. *Biometrics*, **56**, 868–872.

Thurstone, L. L. 1935. *The Vectors of Mind*. Chicago: University of Chicago Press.

Thurstone, L. L. 1947. *Multiple-Factor Analysis: A Development And Expansion of the Vectors of Mind.* Chicago: University of Chicago Press.

Tierney, L., & Kadane, J. B. 1986. Accurate approximations for posterior moments and marginal densities. *Journal of the American Statistical Association,* **81**, 82–86.

Titterington, D. M., Smith, A. F. M., & Makov, U. E. 1985. *Statistical Analysis of Finite Mixture Distributions.* Chichester: Wiley.

Tobin, J. 1958. Estimation of relationships for limited dependent variables. *Econometrica,* **26**, 24–36.

Torgerson, W. S. 1958. *Theory and Methods of Scaling.* New York: Wiley.

Træen, B. 2003. Effect of an intervention to prevent unwanted pregnancy in adolescents. A randomised, prospective study from Nordland County, Norway, 1999-2001. *Journal of Community & Applied Social Psychology,* **13**, 224-239.

Train, K. E. 1986. *Qualitative Choice Analysis.* Cambridge, MA: MIT Press.

Train, K. E. 2002. *Mixed Logits with Bounded Distributions.* Berkeley, CA: Department of Economics, University of California Berkeley. Downloadable from http://elsa.berkeley.edu/~train/software.html.

Train, K. E. 2003. *Discrete Choice Methods with Simulation.* Cambridge: Cambridge University Press.

Train, K. E., Revelt, D., & Ruud, P. A. 1999. *Mixed Logit Estimation Routine for Panel Data.* Berkeley, CA: Department of Economics, University of California Berkeley. Downloadable from http://elsa.berkeley.edu/~train/software.html.

Tucker, L. R. 1972. Relations between multidimensional scaling and three-mode factor analysis. *Psychometrika,* **37**, 3–27.

Turner, R. M., Omar, R. Z., Yang, M., Goldstein, H., & Thompson, S. G. 2000. A multilevel framework for meta-analysis of clinical trials with binary outcomes. *Statistics in Medicine,* **19**, 3417–3432.

U. S. Department of Defense. 1982. *Profile of American Youth.* Washington DC: Office of the Assistant Secretary of Defense for Manpower, Reserve Affairs, and Logistics.

Uebersax, J. S. 1993. Statistical modeling of expert ratings on medical treatment appropriateness. *Journal of the American Statistical Association,* **88**, 421–427.

Uebersax, J. S., & Grove, W. M. 1993. A latent trait finite mixture model for the analysis of rating agreement. *Biometrics,* **49**, 823–835.

Vaida, F., & Blanchard, S. 2004. Conditional Akaike information for mixed effects models. *Submitted for publication.*

Vaida, F., & Xu, R. 2000. Proportional hazards model with random effects. *Statistics in Medicine,* **19**, 3309–3324.

van de Pol, F., & Langeheine, R. 1990. Mixed Markov latent class models. *Pages 213–247 of:* Clogg, C. C. (ed), *Sociological Methodology 1990.* Oxford: Blackwell.

van den Berg, G. J. 2001. Duration models: specification, identification, and multiple durations. *In:* Heckman, J. J., & Leamer, E. (eds), *Handbook of Econometrics,* vol. V. Amsterdam: North-Holland.

van der Leeden, R., & Busing, F. M. T. A. 1994. *First iteration versus IGLS/RIGLS estimates in two-level models: A Monte Carlo study with ML3. Preprint PRM 94-03.* Tech. rept. Psychometrics and Research Methodology, Leiden, The Netherlands.

van der Linden, W. J., & Hambleton, R. K. 1997. *Handbook of Modern Item Response Theory.* New York, NY: Springer.

van Houwelingen, J. C., Arends, L. R., & Stijnen, T. 2002. Tutorial in biostatistics. Advanced methods in meta-analysis: multivariate approach and meta-regression. *Statistics in Medicine,* **21**, 589–624.

Vassend, O., & Skrondal, A. 1995. Factor-analytic studies of the NEO personality-inventory and the five-factor model - the problem of high structural complexity and conceptual indeterminacy. *Personality and Individual Differences,* **19**, 135–147.

Vassend, O., & Skrondal, A. 1997. Validation of the NEO personality inventory and the five-factor model. Can findings from exploratory and confirmatory factor analysis be reconciled? *European Journal of Personality,* **11**, 147–166.

Vassend, O., & Skrondal, A. 1999. The problem of structural indeterminacy in multidimensional symptom report instruments. The case of SCL-90-R. *Behaviour Research and Therapy*, **37**, 685–701.

Vassend, O., & Skrondal, A. 2004. The five-factor model of personality. Language, factor structure, and the problem of generalization. *Submitted for publication*.

Vaupel, J. W., Manton, K. G., & Stallard, E. 1979. The impact of heterogeneity in individual frailty on the dynamics of mortality. *Demography*, **16**, 439–454.

Venables, W. N., & Ripley, B. D. 2002. *Modern Applied Statistics with S (Fourth Edition)*. New York, NY: Springer.

Verbeke, G., & Lesaffre, E. 1996. A linear mixed-effects model with heterogeneity in the random-effects population. *Journal of the American Statistical Association*, **91**, 217–221.

Verbeke, G., & Molenberghs, G. 1997. *Linear Mixed Models in Practice : A SAS-Oriented Approach*. New York, NY: Springer.

Verbeke, G., & Molenberghs, G. 2000. *Linear Mixed Models for Longitudinal Data*. New York, NY: Springer.

Verbeke, G., & Molenberghs, G. 2003. The use of score tests for inference on variance components. *Biometrics*, **59**, 254–262.

Verbyla, A. P. 1993. Modeling variance heterogeneity: residual maximum likelihood and diagnostics. *Journal of the Royal Statistical Society, Series B*, **55**, 493–508.

Vermunt, J. K. 1997. *Log-Linear Models for Event Histories*. Thousand Oaks, CA: Sage.

Vermunt, J. K. 2001. The use of restricted latent class models for defining and testing nonparametric item response theory models. *Applied Psychological Measurement*, **25**, 283–294.

Vermunt, J. K. 2003. Multilevel latent class models. *Pages 213–239 of:* Stolzenberg, R. M. (ed), *Sociological Methodology 2003*, vol. 33. Oxford: Blackwell.

Vermunt, J. K. 2004. An EM algorithm for the estimation of parametric and nonparametric hierarchical nonlinear models. *Statistica Neerlandica*, **58**, 220–233.

Vermunt, J. K., & Magidson, J. 2000. *Latent Gold User's Guide*. Belmont, MA: Statistical Innovations.

Vermunt, J. K., & Magidson, J. 2003a. *Addendum to the Latent GOLD Users Guide: Upgrade Manual for Version 3.01*. Belmont, MA: Statistical Innovations.

Vermunt, J. K., & Magidson, J. 2003b. Latent class models for classification. *Computational Statistics & Data Analysis*, **41**, 531–537.

Vinokur, A. D., Price, R. H., & Schul, Y. 1995. Impact of JOBS intervention on unemployed workers varying in risk for depression. *American Journal of Community Psychology*, **19**, 543–562.

Vonesh, E. F., & Chinchilli, V. M. 1997. *Linear and Nonlinear Models for the Analysis of Repeated Measurements*. New York: Marcel Dekker.

Waclawiw, M. A., & Liang, K.-Y. 1994. Empirical Bayes estimation and inference for the random effects model with binary response. *Statistics in Medicine*, **13**, 541–551.

Wainer, H. 1976. Estimating coefficients in linear models: It don't make no nevermind. *Psychological Bulletin*, **83**, 213–217.

Wainer, H., & Kiely, G. L. 1987. Item clusters and computerized adaptive testing: A case for testlets. *Journal of Educational Measurement*, **24**, 185–201.

Wainer, H., & Thissen, D. 1982. Some standard errors in item response theory. *Psychometrika*, **47**, 397–412.

Wald, A. 1950. Note on the identification of economic relations. *In:* Koopmans, T. C. (ed), *Statistical Inference in Dynamic Economic Models*. New York: Cowles Commission Monograph 10, Wiley.

Wang, M. W., & Stanley, J. C. 1970. Differential weighting: A review of methods and empirical studies. *Review of Educational Research*, **40**, 663–705.

Wansbeek, T. J., & Meijer, E. (eds). 2002. *Measurement Error and Latent Variables in Econometrics*. Amsterdam: North Holland.

Warm, T. A. 1989. Weighted likelihood estimation of ability in item response models. *Psychometrika*, **54**, 427–450.

Wasserman, L. 2000. Bayesian model selection and model averaging. *Journal of Mathematical Psychology*, **44**, 92–107.

Waternaux, C., Laird, N. M., & Ware, J. H. 1989. Methods for analysis of longitudinal data: bloodlead concentrations and cognitive development. *Journal of the American Statistical Association*, **84**, 33–41.

Wedderburn, R. W. M. 1974. Quasi-likelihood functions, generalized linear models, and the Gauss-Newton method. *Biometrika*, **61**, 439–447.

Wedel, M. 2002a. Concomitant variables in finite mixture models. *Statistica Neerlandica*, **56**, 362–375.

Wedel, M. 2002b. *GLIMMIX. A Program for Estimation of Latent Class Mixture and Mixture Regression Models. User's Guide Manual.* Groningen: ProGamma.

Wedel, M., & Kamakura, W. A. 2000. *Market Segmentation: Conceptual and Methodological Foundations (Second Edition).* Dordrecht: Kluwer.

Wedel, M., & Kamakura, W. A. 2001. Factor analysis with (mixed) observed and latent variables. *Psychometrika*, **66**, 515–530.

Wedel, M., Böckenholt, U., & Kamakura, W. A. 2003. Factor models for multivariate count data. *Journal of Multivariate Analysis*, **87**, 356–369.

Wei, G. C. G., & Tanner, M. A. 1990. A Monte Carlo implementation of the EM algorithm and the poor man's augmentation algorithms. *Journal of the American Statistical Association*, **86**, 669–704.

Wei, L. J., Lin, D. Y., & Weissfeld, L. 1989. Regression analysis of multivariate incomplete failure time data by modeling marginal distributions. *Journal of the American Statistical Association*, **84**, 1065–1073.

Werts, C. E., Linn, R. L., & Jöreskog, K. G. 1974. Intraclass reliability estimates: Testing structural assumptions. *Educational and Psychological Measurement*, **34**, 25–33.

Wheaton, B., Muthén, B. O., Alwin, D. F., & Summers, G. 1977. Assessing reliability and stability in panel models. *In:* Heise, D. R. (ed), *Sociological Methodology 1977.* San Francisco, CA: Jossey-Bass.

White, H. 1982. Maximum likelihood estimation of misspecified models. *Econometrica*, **50**, 1–25.

Whitehead, A. 2002. *Meta-analysis of Controlled Clinical Trials.* Chichester: Wiley.

Whitehead, J. 1980. Fitting Cox's regression model to survival data using GLIM. *Journal of the Royal Statistical Society, Series C*, **29**, 269–275.

Wieken-Mayser, M. 1979. *Political Action: An Eight Nation Study 1973-1976.* Köln: Zentralarchiv für Empirische Sozialforschung.

Wiggins, R. D., Ashworth, K., O'Muircheartaigh, C. A., & Galbraith, J. I. 1990. Multilevel analysis of attitudes to abortion. *Journal of the Royal Statistical Society, Series D*, **40**, 225–234.

Wiley, D. E. 1973. The identification problem for structural equation models with unmeasured variables. *Pages 69–83 of:* Goldberger, A. S., & Duncan, O. D. (eds), *Structural Equation Models in the Social Sciences.* New York: Academic Press.

Wilks, S. S. 1938. Weighting systems for linear functions of correlated variables when there is no dependent variable. *Psychometrika*, **3**, 23–40.

Williams, D. A. 1975. The analysis of binary responses from toxological experiments involving reproduction and teratogenecity. *Journal of the Royal Statistical Society, Series C*, **31**, 949–952.

Williams, R. L. 2000. A note on robust variance estimation for cluster-correlated data. *Biometrics*, **56**, 645–646.

Wilson, M., & Adams, R. J. 1995. Rasch models for item bundles. *Psychometrika*, **60**, 181–198.

Winkelmann, R. 2003. *Econometric Analysis of Count Data.* New York: Springer.

Wolfe, R., & Firth, D. 2002. Modelling subjective use of an ordinal response scale in a many period crossover experiment. *Journal of the Royal Statistical Society, Series C*, **51**, 245–255.

Wolfinger, R. D. 1999. *Fitting Non-Linear Mixed Models with the New NLMIXED procedure*. Tech. rept. SAS Institute, Cary, NC.

Wolfinger, R. D., & O'Connell, M. 1993. Generalized linear mixed models: A pseudo-likelihood approach. *Journal of Statistical Computing and Simulation*, **48**, 233–243.

Wolfram, S. 2003. *The Mathematica Book 5 (Fifth Edition)*. Champaign, IL: Wolfram Media.

Wooldridge, J. M. 2002. *Econometric Analysis of Cross Section and Panel Data*. Cambridge, MA: The MIT Press.

Woolf, B. 1955. On estimating the relationship between blood groups and disease. *Annals of Human Genetics*, **19**, 251–253.

Wright, B. D. 1977. Solving measurement problems with the Rasch model. *Journal of Educational Measurement*, **14**, 97–116.

Wu, M. C., & Carroll, R. J. 1988. Estimation and comparison of change in the presence of informative right censoring by modeling the censoring process. *Biometrics*, **44**, 175–188.

Wulfsohn, M. S., & Tsiatis, A. A. 1997. A joint model for survival and longitudinal data measured with error. *Biometrics*, **53**, 330–339.

Xiang, L., Tse, S.-K., & Lee, A. H. 2002. Influence diagnostics for generalized linear mixed models: Applications to clustered data. *Computational Statistics & Data Analysis*, **40**, 759–774.

Xu, J., & Zeger, S. L. 2001. Joint analysis of longitudinal data comprising repeated measures and times to events. *Journal of the Royal Statistical Society, Series C*, **50**, 375–387.

Yang, M. 2001. Multinomial regression. *Pages 107–123 of:* Leyland, A. H., & Goldstein, H. (eds), *Multilevel Modelling of Health Statistics*. Chichester: Wiley.

Yang, S., Pickles, A., & Taylor, C. 1999. Multilevel Latent Variable Model for Analysing Two-phase Survey data. *Pages 402–408 of: Proceedings of the 14th International Workshop on Statistical Modeling*.

Yau, K. K. W., & Kuk, A. Y. C. 2002. Robust estimation in generalized linear mixed models. *Journal of the Royal Statistical Society, Series B*, **54**, 101–117.

Yellott, J. 1977. The relationship between Luce's choice axiom, Thurstone's theory of comparative judgement, and the double exponential distribution. *Journal of Mathematical Psychology*, **15**, 109–144.

Yule, G. U. 1912. On the methods of measuring association between two attributes. *Journal of the Royal Statistical Society*, **75**, 579–652.

Yung, Y.-F. 1997. Finite mixtures in confirmatory factor-analysis models. *Psychometrika*, **62**, 297–330.

Zeger, S. L., & Karim, M. R. 1991. Generalized linear models with random effects: A Gibbs sampling approach. *Journal of the American Statistical Association*, **86**, 79–86.

Zeger, S. L., & Liang, K.-Y. 1986. Longitudinal data analysis for discrete and continuous outcomes. *Biometrics*, **42**, 121–130.

Zeger, S. L., Liang, K.-Y., & Albert, P. S. 1988. Models for longitudinal data: A generalized estimating equation approach. *Biometrics*, **44**, 1049–1060.

Zellner, A. 1970. Estimation of regression relationships containing unobservable variables. *International Economic Review*, **11**, 441–454.

Zellner, A. 1971. *An Introduction to Bayesian Inference in Econometrics*. New York: Wiley.

Zhao, L. P., & Prentice, R. L. 1990. Correlated binary regression using a quadratic exponential model. *Biometrics*, **77**, 642–648.

Zimowski, M. F., Muraki, E., Mislevy, R. J., & Bock, R. D. 1996. *BILOG-MG 3: Multiple-group IRT analysis and test maintenance for binary items*. Chicago, IL: Scientific Software International.

Zucchini, W. 2000. An introduction to model selection. *Journal of Mathematical Psychology*, **44**, 41–61.

Author index

Index